Following the 1952 reorganization of the Portuguese Air Force from the army and naval air arms, Portugal now had an entity dedicated solely to aviation that would bring it into line with its new NATO commitment. As it proceeded to develop a competence in modern multi-engine and jet fighter aircraft for its NATO role and train a professional corps of pilots, it was suddenly confronted in 1961 with fighting insurgencies in all three of its African possessions. This development forced it to acquire an entirely new and separate air force, the African air force, to address this emerging danger. This is the story of just how Portuguese leadership anticipated and dealt with this threat, and how it assembled an air force from scratch to meet it. The aircraft available at the time were largely cast-offs from the larger, richer, and more sophisticated air forces of its NATO partners and not designed for counterinsurgency. Yet Portugal adapted them to the task and effectively crafted the appropriate strategies and tactics for their successful employment. The book explores the vicissitudes of procurement, an exercise fraught with anti-colonial political undercurrents, the imaginative modification and adaptation of the aircraft to fight in the African theaters, and the development of tactics, techniques, and procedures for their effective employment against an elusive, clever, and dangerous enemy. Advances in weaponry, such as the helicopter gunship, were the outgrowth of combat needs. The acquired logistic competences assured that the needed fuel types and lubricants, spare parts, and qualified maintenance personnel were available in even the most remote African landing sites. The advanced flying skills, such as visual reconnaissance and air-ground coordinated fire support, were honed and perfected. All of these aspects and more are explored and hold lessons in the application of airpower in any insurgency today.

I0894074

John P. Cann is a Research Fellow and retired Professor of National Security Studies at Marine Corps University, a former member of the research staff at the Institute for Defense Analyses, and Scholar-in-Residence at the University of Virginia. He earned his doctorate in War Studies at King's College London in 1996, published *Counterinsurgency in Africa* in 1997, *Memories of Portugal's African Wars, 1961–1975* (ed.) in 1998, *The Brown Waters of Africa* in 2008, and numerous articles on small wars over the years. He is a retired naval captain and flight officer specializing in open ocean reconnaissance aviation and served in a variety of aviation assignments, including command. He has been awarded the Portuguese Navy Cross Medal and the Medal of Dom Afonso Henriques for his writings on conflict in Lusophone Africa.

FLIGHT PLAN AFRICA

Wolverhampton Military Studies

www.helion.co.uk/wolverhamptonmilitarystudies

Editorial board

Submissions

The publishers would be pleased to receive submissions for this series. Please contact us via email (info@helion.co.uk), or in writing to Helion & Company Limited, 26 Willow Road, Solihull, West Midlands, B91 1UE.

Titles

No.1 *Stemming the Tide. Officers and Leadership in the British Expeditionary Force 1914* Edited by Spencer Jones (ISBN 978-1-909384-45-3)

No.2 *'Theirs Not To Reason Why'. Horsing the British Army 1875-1925* Graham Winton (ISBN 978-1-909384-48-4)

No.3 *A Military Transformed? Adaptation and Innovation in the British Military, 1792-1945* Edited by Michael LoCicero, Ross Mahoney and Stuart Mitchell (ISBN 978-1-909384-46-0)

No.4 *Get Tough Stay Tough. Shaping the Canadian Corps, 1914-1918* Kenneth Radley (ISBN 978-1-909982-86-4)

No.5 *A Moonlight Massacre: The Night Operation on the Passchendaele Ridge, 2 December 1917. The Forgotten Last Act of the Third Battle of Ypres* Michael LoCicero (ISBN 978-1-909982-92-5)

No.6 *Shellshocked Prophets. Former Anglican Army Chaplains in Interwar Britain* Linda Parker (ISBN 978-1-909982-25-3)

No.7 *Flight Plan Africa: Portuguese Airpower in Counterinsurgency, 1961-1974* John P. Cann (ISBN 978-1-909982-06-2)

Flight Plan Africa

Portuguese Airpower in Counterinsurgency, 1961–1974

Wolverhampton Military Studies No.7

John P. Cann

Helion & Company Limited

Helion & Company Limited
26 Willow Road
Solihull
West Midlands
B91 1UE
England
Tel. 0121 705 3393
Fax 0121 711 4075
Email: info@helion.co.uk
Website: www.helion.co.uk
Twitter: @helionbooks
Visit our blog http://blog.helion.co.uk/

Published by Helion & Company 2015

Designed and typeset by Bookcraft Ltd, Stroud, Gloucestershire
Cover designed by Euan Carter, Leicester (www.euancarter.com)
Printed by Lightning Source, Milton Keynes, Buckinghamshire

Front cover – top: a Sud-Aviation Puma helicopter collecting troops following an operation in the Chana da Cameia in the east of Angola (Source: Forca Aerea Portuguesa); bottom – a North American T-6G Harvard on patrol in the north of Angola (Source: Forca Aerea Portuguesa). Rear cover - A Sud-Aviation Alouette III refueling at a tactical base in the Téte District of Mozambique (Source: Forca Aerea Portuguesa).

ISBN 978 1 909982 06 2

British Library Cataloguing-in-Publication Data.
A catalogue record for this book is available from the British Library.

For details of other military history titles published by Helion & Company Limited contact the above address, or visit our website: http://www.helion.co.uk.

We always welcome receiving book proposals from prospective authors.

For my dear friend
General Jorge Manuel Brochado de Miranda

Contents

List of Photographs

List of Maps

Preface

During the research for my first book, *CounterInsurgency in Africa, The Portuguese Way of War, 1961–1974*, the door opened and revealed to me the unusual importance of airpower in the Portuguese campaigns to retain their African possessions. The air dimension, normally a support for ground forces, played more than this traditional limited role. It was indeed a key part of the equation because of its flexibility, imaginative application, and aviator courage in responding to the varied insurgent threats and ground force requirements throughout the three theaters of Angola, Guiné, and Mozambique.

Normally counterinsurgency is not conducive to the centerpiece employment of airpower, as it is a labor-intensive enterprise that requires vast numbers of light infantry who are busy bringing security to the population and winning the "hearts and minds" competition. Yet in this case, the Portuguese Air Force was unwilling to cede the focus of the debate entirely to ground forces.

In each theater Portugal was faced with a very different enemy in his aggressiveness, imagination, competence, and size. Airpower was able to adjust and adapt to these changing battlefield mosaics and remain for Portugal more than the usual complement to ground forces. Normally insurgents operate within populations, and Portuguese Africa was no exception, so any attempt to destroy them from the air was fraught with potential repercussions. This meant that targeting had to be effective in preventing the insurgents from reaching the vulnerable populations, and here we see throughout the book examples of just how this was done. Between the "piracy" operations in the east of Angola and the heli-borne tactical trackers and envelopment groups in the north, the insurgents were never able to reach the targeted populations and proselytize them successfully in Angola. In Mozambique it was the armed reconnaissance missions and patient intelligence gathering that enabled targeted strikes on insurgent columns and restricted them to the border districts of the theater. In the cramped battleground of Guiné, the insurgents were very aggressive, and the conflict bordered on the conventional in its violence. Yet time and again the aviators reduced insurgent forces insistent on exposing themselves. Airpower was the bane of their existence in all the theaters, and they constantly sought solutions to this persistent and dangerous threat. The industry of the insurgents and their Soviet, Chinese, and Cuban advisors in addressing the airpower scourge led to the deployment of the

Strela, a Soviet-designed, heat-seeking, air-ground, hand-held missile. Despite its initial success and surprise, the Portuguese quickly developed effective countermeasures and reasserted the constant pressure of airpower.

The remarkable results from the air effort came from a force with one of the smallest presences in Portuguese Africa, and yet its leverage and impact belied its modest numbers. Out of perhaps 158,000 men under arms in the *ultramar* at the height of the fighting in 1973, the Air Force numbered about 6,000 men and women, and this figure included about 1,900 paratroops and their para-nurses. The navy was the smallest of the three at about 3,000, while the army was the largest at about 149,000. Out of the 4,100-man dedicated aviation force, there were never more than 300 pilots at any one time, and these bore the burden of all aviation missions. Their enormous impact was disproportionate to their small numbers.

The book explores just how Portugal went about developing an air strategy and executing and adjusting it to achieve a new approach to counterinsurgency that could employ airpower more effectively and at less proportional cost than comparable counterinsurgencies. This air strategy was based on a network of airfields that offered varying levels of support across the vast reaches of Portuguese Africa and a fleet of aircraft with unimproved field capability to extend the airpower reach. The airfields were based on a spoke pattern around a key installation that could sustain the more primitive fields that allowed wide-ranging support of ground forces, independent air operations, and wounded air evacuation. The aircraft were proven, simple, and inexpensive. The infrastructure and fleet were designed both to support the population and destroy the insurgent, and in both instances they worked well.

Unfortunately, long wars are frustrating to airmen because of the highly complex and technical nature of aviation and the expertise needed to manage even routine operations. It took the initial years to develop an effective air force and adapt it to fighting insurgents successfully in the difficult terrain. The Air Force in this case began with a sound foundation grounded in NATO and adapted remarkably well. It developed its own unique approach, and the book describes how it managed this success. The achievement of the Portuguese Air Force in counterinsurgency establishes it as a relevant model for those who are disinclined to cede the debate to ground forces on the importance and criticality of airpower in counterinsurgency.

The Wolverhampton Military Studies Series
Series Editor's Preface

As series editor, it is my great pleasure to introduce the *Wolverhampton Military Studies Series* to you. Our intention is that in this series of books you will find military history that is new and innovative, and academically rigorous with a strong basis in fact and in analytical research, but also is the kind of military history that is for all readers, whatever their particular interests, or their level of interest in the subject. To paraphrase an old aphorism: a military history book is not less important just because it is popular, and it is not more scholarly just because it is dull. With every one of our publications we want to bring you the kind of military history that you will want to read simply because it is a good and well-written book, as well as bringing new light, new perspectives, and new factual evidence to its subject.

In devising the *Wolverhampton Military Studies Series*, we gave much thought to the series title: this is a *military* series. We take the view that history is everything except the things that have not happened yet, and even then a good book about the military aspects of the future would find its way into this series. We are not bound to any particular time period or cut-off date. Writing military history often divides quite sharply into eras, from the modern through the early modern to the mediaeval and ancient; and into regions or continents, with a division between western military history and the military history of other countries and cultures being particularly marked. Inevitably, we have had to start somewhere, and the first books of the series deal with British military topics and events of the twentieth century and later nineteenth century. But this series is open to any book that challenges received and accepted ideas about any aspect of military history, and does so in a way that encourages its readers to enjoy the discovery.

In the same way, this series is not limited to being about wars, or about grand strategy, or wider defence matters, or the sociology of armed forces as institutions, or civilian society and culture at war. None of these are specifically excluded, and in some cases they play an important part in the books that comprise our series. But there are already many books in existence, some of them of the highest scholarly standards, which cater to these particular approaches. The main theme of the *Wolverhampton Military Studies Series* is the military aspects of wars, the preparation for wars or their prevention, and their aftermath. This includes some books whose main theme is the technical details of how armed forces have worked, some books on wars and battles, and some books that re-examine the evidence about the existing stories, to show in a different light what everyone thought they already knew and understood.

As series editor, together with my fellow editorial board members, and our publisher Duncan Rogers of Helion, I have found that we have known immediately and almost by instinct the kind of books that fit within this series. They are very much the kind of well-written and challenging books that my students at the University of Wolverhampton would want to read. They are books which enhance knowledge, and offer new perspectives. Also, they are books for anyone with an interest in military history and events, from expert scholars to occasional readers. One of the great benefits of the study of military history is that it includes a large and often committed section of the wider population, who want to read the best military history that they can find; our aim for this series is to provide it.

Stephen Badsey
University of Wolverhampton

Acknowledgements

This book would not have been possible without the unstinting support of my friend General Jorge Manuel Brochado de Miranda, former Chief of Staff of the Portuguese Air Force and former Chief of the Historical Archive of the Air Force, for which I am most grateful. He made virtually every resource at his disposal available to me and for that I am most appreciative. General Tómas George Conceição Silva, likewise a dear friend, took an active interest in the project and shared many of his personal experiences in Africa in addition to a personal tour of the Air Force Museum at Sintra. I am indebted to the very professional staffs of the Air Force Historical Archive and *Revista Mais Alto* who gave their time selflessly to helping with the research and particularly to Lieutenant General António Mimoso e Carvalho, Chief of the Historical Archives. I am likewise indebted to Dr. José Manuel Correia and Lieutenant Colonel (pilot aviator) João José Brandão Ferreira, both good friends who generously spent many hours explaining the details of air operations in Africa.

Additionally, I am indebted to Major General José Krus Abecasis, who shared many insights on operations in Guiné in the early years; to General José Armando Vizela Cardoso, who patiently explained air reconnaissance in Mozambique; to General José Augusto Barrigas Queiroga, who shared his personal archives and experiences as a helicopter pilot in Angola and Mozambique; to Lieutenant General António de Jesus Bispo, who over lunch shared his insights on the overall air campaign in Africa; to Major António Lobato, who some years ago talked with me on his accident and rescue from Conakry; to the aviation scholar José Matos for his help with the story on alternative F-86 sources and the potential Mirage replacement; and to Drs. Mark Jacobsen and Wray R. Johnson, both former colleagues and professors at Marine Corps University, who shared their views on airpower in counterinsurgency over many years. Finally, I owe a particular debt to my wife Anne, who lived patiently with the domestic chaos of this work.

Abbreviations

AB	*Aeródromo Base* or Aerodrome Base
ABAKO	*Alliance des Ba-Kongo* or Alliance of the Bakongo
ACB	Air Club of Beira
AgrAr	*Agrupamento Aéreo* or Air Group
AI	*Aeródromo de Instrução* or Training Aerodrome
AIL	*Áreas de Intervenção Livre* or Free Fire Zones
AL	*Aeródromo Logístico* or Logistical Aerodrome
ALAT	*Aviation Légère de l'Armée de Terre* or Army Light Aviation
AM	*Aeródromo de Manobra* or Maneuver Aerodrome
ANC	African National Congress
AOA	Angle of Attack
AR	*Aeródromo de Recurso* or Alternate Aerodrome
AT	*Aeródromo de Trânsito* or Transit Aerodrome
ATIP	*Ataque Independento Preparado* or Planned Attack
ATIR	*Ataque Independente em Reconhecimento* or Armed Reconnaissance
AU	African Union
BA	*Base Aérea* or Air Base
BCP	*Batalhão de Caçadores Pára-quedistas* or Paratroop Battalion
BE	*Batalhão Eventual* or Contingent Battalion
CAS	Close Air Support
CAT	Civil Air Transport
CCFAA	Commander-in-Chief Armed Forces of Angola
CCIVM	*Orgão de Controlo e Informação de Voo Militar* or Military Flight Information and Control Authority
CECI	*Comando Especial de Contra Infiltração* or Special Command for Counter-Infiltration
CFB	*Caminho de Ferro de Benguela* or Benguela Railroad
CIA	Central Intelligence Agency
CCAA	*Centro Conjunto de Apoio Aéreo* or Joint Air Support Centre
CCE	Company of *Caçadores Especiais* or Special Hunters
CCMDS	Company of Commandos

CMIG-Zero	*Centre Militaire d'Instruction Guerrilla–Zero* or Military Center for Instruction in Guerrilla Warfare, Number Zero
COFI	*Comando Operacional das Forçes de Intervenção* or Operational Command of the Intervention Forces
COMSECAR	*Comando do Sector Aéreo Leste* or Eastern Air Sector Command
CONAKAT	*Confédération des Associations Tribales du Katanga* or Confederation of the Tribal Associations of Katanga
Cotonang	*Companhia Geral dos Algodões de Angola* or Cotton Company of Angola
DC	*Destacamento de Cooperação* or Detachment of Cooperation
DGS	*Direcção-Geral Segurança* or General Security Directorate
Diamang	*Companhia de Diamantes de Angola* or Diamond Company of Angola
DIH	*Détachements d'Intervention Héliportés* or Heliborne Intervention Detachments
DISC	Directorate of the Service of Management and Accounting (*Direcção do Serviço de Intendência e Contabilidade*)
DLOC	*Destruição Localizada* or Localized Destruction
DRIL	*Directório Revolucionário Ibérico de Libertação*, or the Iberian Revolutionary Directorate of Freedom
DShK	Degtyarev-Shpagin Large-Calibre (12.7mm)
DTA	*Divisão dos Transportes Aéros* or Air Transport Division
EALA	*Escadrille d'Aviation Légère d'Appui* or Light Support Aviation Squadron
EEC	European Economic Community
EPLA	*Exército Popular de Libertação de Angola* or Popular Army for the Liberation of Angola
ETAP	École des Troupes Aeroportées
EVA	*Esquadrilha dos Voluntários do Ar* or Flight of Air Volunteers
FAC	Forward Air Controller
FAP	*Força Aérea Portuguesa* or Portuguese Air Force
FARP	*Forças Armadas Revolucionárias de Povo* or Revolutionary Armed Forces of the People
FAV	*Formações Aéreas de Voluntários* or Volunteer Air Formations
FNLA	*Frente Nacional de Libertção de Angola* or National Front for the Liberation of Angola
FOD	Foreign Object Damage
FRELIMO	*Frente de Libertação de Moçambique* or Front for the Liberation of Mozambique
GALA	*Groupe d'Aviation Légère d'Appui* or Light Support Aviation Group
GATac	*Groupements Aériens Tactiques* or Tactical Air Command
GHAN-1	*Groupe d'Hélicoptères de l'Aéronautique Navale NO 1* or Naval Helicopter Group No 1
GO 1001	Operational Group 1001

GP	General Purpose Bomb
GRAE	*Governo da República de Angola no Exílio* or Government of the Republic of Angola in Exile
HE	High Explosive
HF	High Frequency Radio
IAEM	*Instituto de Altos Estudos Militares* or Institute for Higher Military Studies
IFR	Instrument Flight Rules
IP	Initial Point
JASC	Joint Air Support Centre
LDM	*Launcha de Desembarque Média* or Medium Landing Craft
LZ	Landing Zone
MAAG	U.S. Military Advisory Assistance Group
MANU	Please see UNAM
MAP	U.S. Military Assistance Program
MDAP	Mutual Defense Assistance Program
MEDEVAC	Medical Evacuation (Aviation)
MNC	*Mouvement National Congolais* Congolese National Movement
MPLA	*Movimento Popular de Libertação de Angola* or Popular Movement for the Liberation of Angola
NATO	North Atlantic Treaty Organization
OAS	Organization of African States
OAU	Organization of African Unity
OPC	Ovambo People's Congress
ORMIS	*Ordem de Missão* or Mission Order
PAC	Pan African Congress
PAIGC	*Partido Africano da Independência da Guiné e Cabo Verde* or African Party for the Independence of Guiné and Cape Verde
PAC	Pan African Congress
PCA	*Posto de Comando Avançado* or Forward Command Post
PIDE	*Polícia Internacional de Defesa do Estado* or International Police for Defense of the State
PLAN	People's Liberation Army of Namibia
PoAF	Portuguese Air Force
PSA	*Partido de Solidariedade Africana, Parti Solidaire Africain* or African Solidarity Party
RA	*Região Aérea* or Air Region
RAF	Royal Air Force
RCAR	*Reconhecimento Armado* or Armed Reconnaissance
RDF	Radio Direction Finder (Radio Compass)
RECA	*Reconhecimento Armado* or Armed Reconnaissance
RFOT	*Reconhecimento Fotográfico* or Photographic Reconnaissance
RhAF	Rhodesian Air Force

RLI	Rhodesian Light Infantry
RPG	Rocket-Propelled Grenade
RVIS	*Reconhecimento Visual* or Aerial Visual Reconnaissance
SAA	South African Airways
SAAF	South African Air Force
SADF	South African Defence Force
SAM	Surface-to-Air Missile
SAP	South African Police
SAR	Search and Rescue
SAS	Special Air Service
SCCI	*Serviço de Centralização e Coordenação de Informações* or Service of Centralization and Coordination of Intelligence
SFERMA	Société Française d'Entretien et de Réparation de Matériel Aéronautique
SIGG	*Serviço de Informações do Governo-Geral* or Intelligence Service of the Governor-General
SIGINT	Signals Intelligence
SIPA	Société Industrielle Pour l'Aéronautique
STOL	Short Takeoff and Landing
SWAPO	South West Africa People's Organization
TAP	*Transportes Aéreos Portugueses* or Portuguese Air Transport
TAM	*Transporte Aéreo Militar* or Military Air Transport
TLCA	*Transporte Ligeiro de Carga Aérea* or Air Cargo Light Transport
UAT	Union Aéromartime du Transport
UDI	Rhodesian Unilateral Declaration of Independence
UHF	Ultra High Frequency Radio
UNAM	*União Nacional Africana Moçambique* or Mozambican African National Union
UNAMI	*União Nacional do Moçambique Independente*, or National Union of Independent Mozambique
UDENAMO	*União Democrática Nacional de Moçambique* or the National Democratic Union of Mozambique
UN	United Nations
UNITA	*União Nacional para a Independência Total de Angola* or the National Union for the Total Independence of Angola
UPA	*Uniao das Populações de Angola* or Union of Angolan Peoples
UTCI	*Unidade Táctica de Contra Infiltração* or Tactical Counter-Infiltration Unit
VHF	Very High Frequency Radio
ZA	*Zona Aérea* or Air Zone
ZACVG	*Zona Aérea de Cabo Verde e Guiné* or Air Zone of Cape Verde and Guiné
ZANU	Zimbabwe African National Union

ZAPU	Zimbabwe African People's Union
ZIL	*Zona de Intervenção Leste* or Eastern Intervention Zone
ZIN	*Zona de Intervenção Norte* or Northern Intervention Zone
ZL	*Zona de Largada* or Landing Zone
ZML	*Zona Militar Leste* or Eastern Military Zone
ZPU	*зенитная пулемётная установка* or Antiaircraft Machinegun Mount

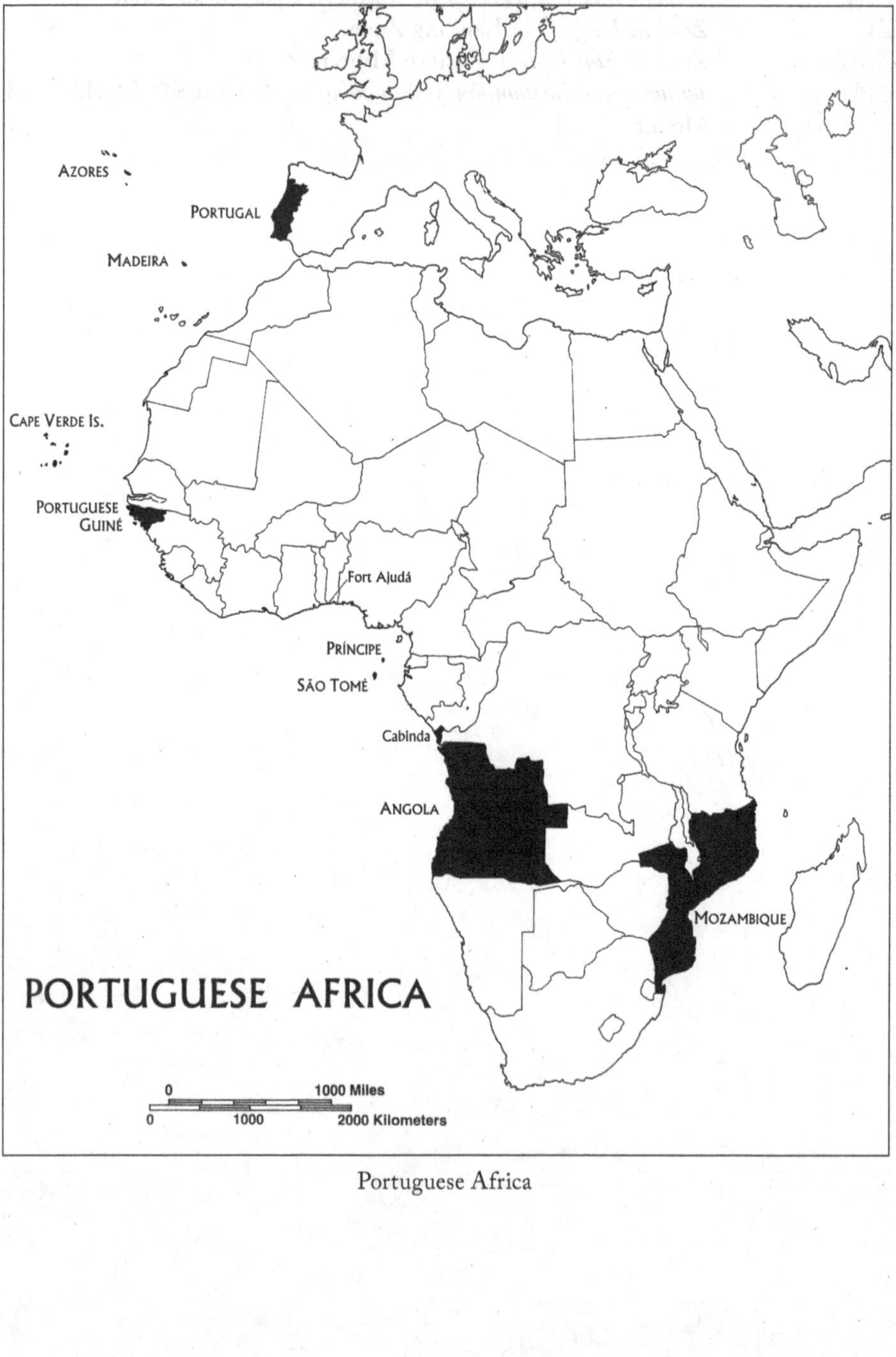

Portuguese Africa

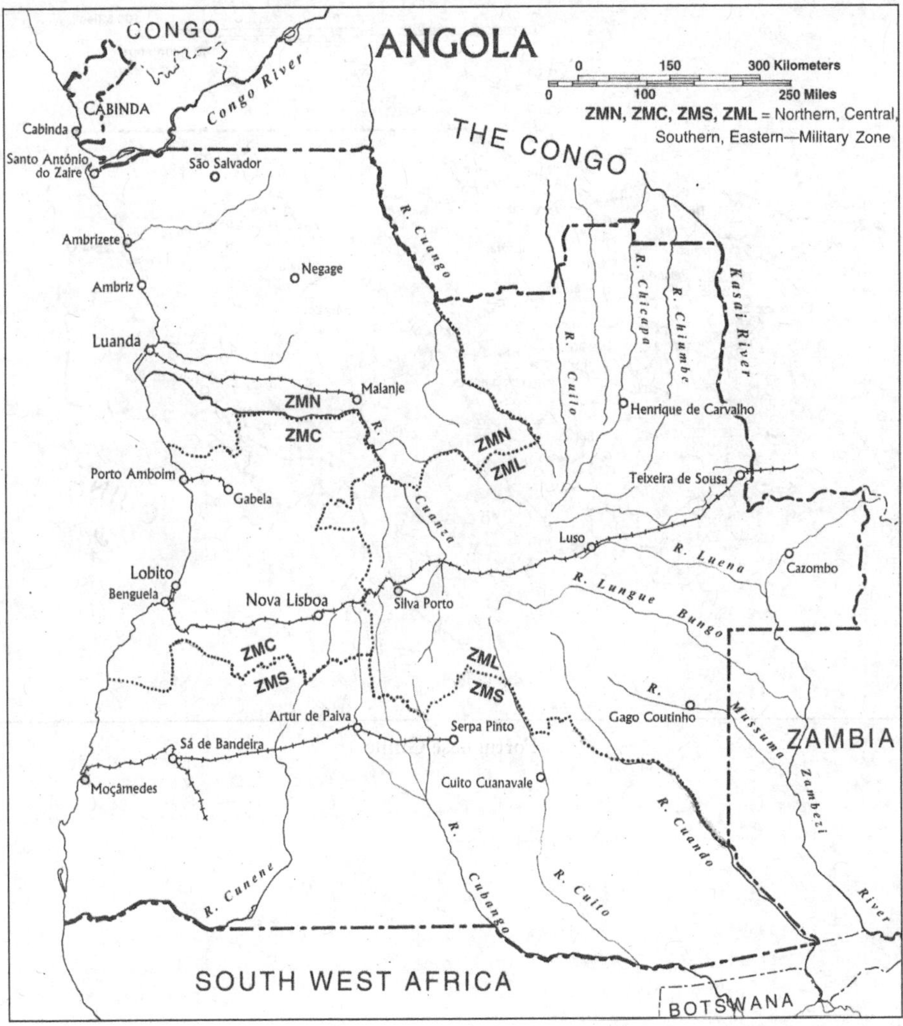
ANGOLA
0 150 300 Kilometers
0 100 250 Miles
ZMN, ZMC, ZMS, ZML = Northern, Central, Southern, Eastern—Military Zone
CONGO
CABINDA
Cabinda
Congo River
Santo António do Zaire
São Salvador
THE CONGO
Ambrizete
Negage
Ambriz
Luanda
R. Cuango
R. Culo
R. Chicapa
R. Chiumbe
Kasai River
Malanje
Henrique de Carvalho
ZMN
ZMC
ZML
Porto Amboim
Gabela
Teixeira de Sousa
R. Cuanza
Luso
R. Luena
Cazombo
Lobito
Benguela
Nova Lisboa
Silva Porto
R. Lungue Bungo
ZMS
R. Mussuma
Gago Coutinho
Artur de Paiva
Serpa Pinto
ZAMBIA
Sá de Bandeira
Moçâmedes
Culto Cuanavale
Zambezi River
R. Cuando
R. Cubango
R. Culto
R. Cunene
SOUTH WEST AFRICA
BOTSWANA

Angola

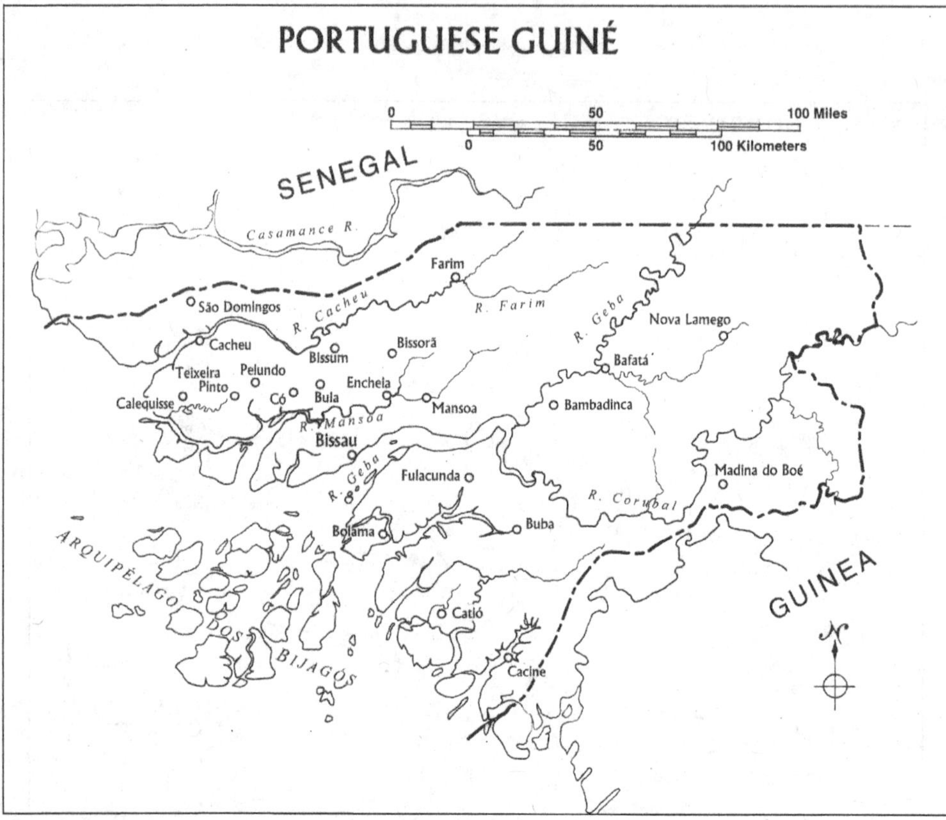

Portuguese Guiné

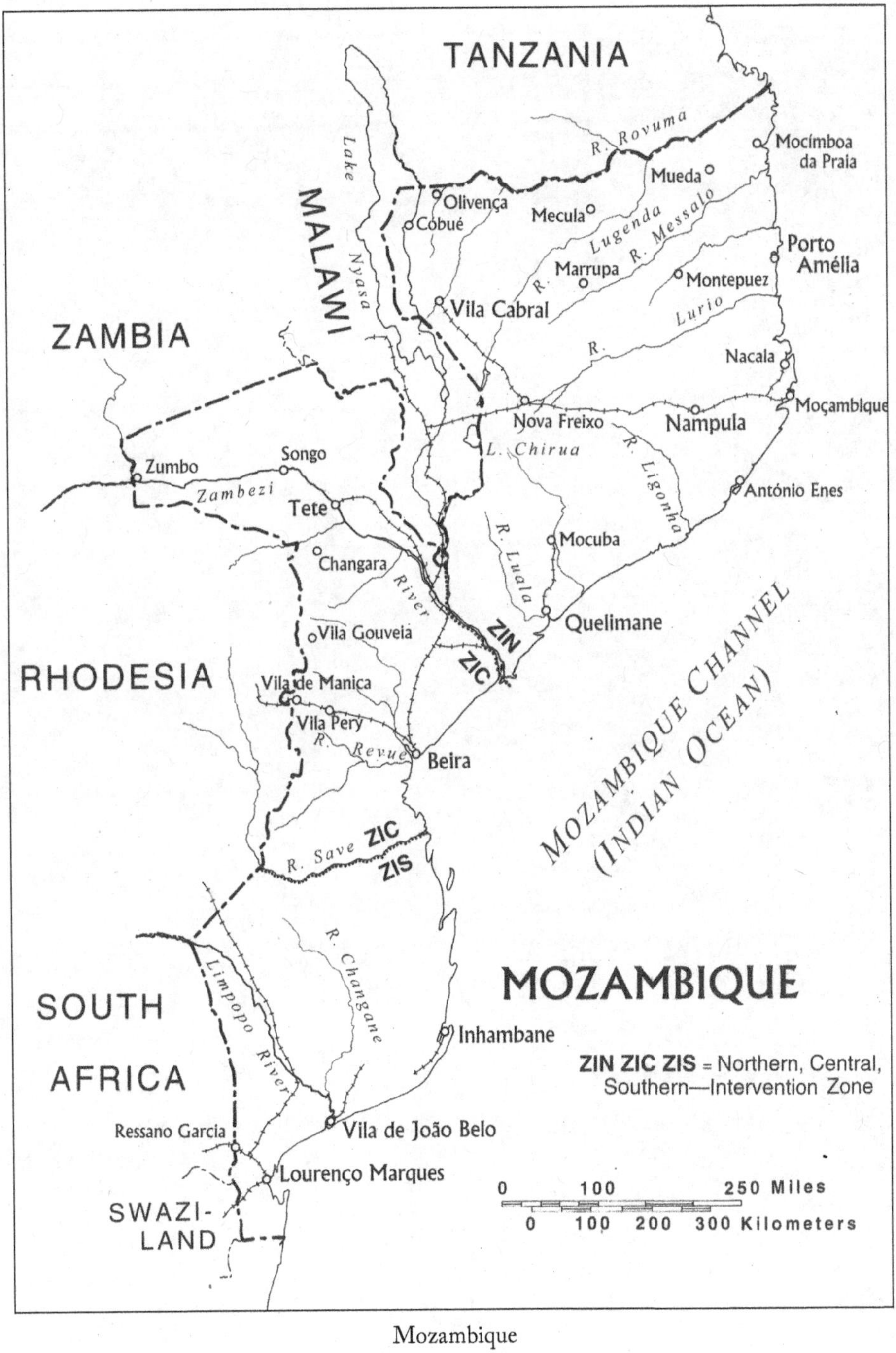
TANZANIA
ZAMBIA
MALAWI
RHODESIA
SOUTH AFRICA
SWAZI-LAND
MOZAMBIQUE
MOZAMBIQUE CHANNEL (INDIAN OCEAN)
R. Rovuma
Mocímboa da Praia
Mueda
Lake Nyasa
Olivença
Cóbué
Mecula
R. Lugenda
R. Messalo
Porto Amélia
Marrupa
Montepuez
Vila Cabral
R. Lurio
Nacala
Moçambique
Nova Freixo
Nampula
L. Chirua
R. Ligonha
Zumbo
Songo
Zambezi
Tete
António Enes
Mocuba
R. Luala
Changara
River
Quelimane
Vila Gouveia
ZIN
ZIC
Vila de Manica
Vila Pery
R. Revue
Beira
R. Save
ZIC
ZIS
R. Changane
Limpopo River
Inhambane
ZIN ZIC ZIS = Northern, Central, Southern—Intervention Zone
Ressano Garcia
Vila de João Belo
Lourenço Marques
0 100 250 Miles
0 100 200 300 Kilometers

Mozambique

1

Portugal Besieged

Since the introduction of aviation in Portugal in 1910, its aviators have shown great energy, imagination, and courage and undertook some of the most daring exploratory flights of the pioneering age. Indeed, from these early days they have had an illustrious history in the spirit of the caravels. Such luminaries as Gago Coutinho, Sacadura Cabral, Brito Pais, Saramento de Beires, and Manuel Gouveia became aviation household names for their trans-South Atlantic and trans-Asiatic flights between 1919 and the early 1930s. These and other feats earned international acknowledgment, and the Clifford Harmon Trophy, a prize created to celebrate aeronautical accomplishments worldwide, was awarded four times in five years to Portuguese pilots. These feats were accomplished by army and naval aviators; however, when Portugal joined NATO in 1949 as one of its founding members, the national vision of military aviation changed from one fragmented between these two services to a unified concept. This thinking led to the establishment of the Portuguese Air Force in 1952 as an entity dedicated to aviation that would bring the needed flexibility, economy, and control to mobilize and concentrate aviation resources where they were most needed.

Because of the nature of World War II, air forces in general in the immediate post-war years were oriented toward the classic interstate conventional conflict. There was little attention, if any, paid to intrastate unconventional conflict, the one that would threaten the *ultramar* or overseas territories of the Portuguese Empire. This would be a conflict that required air support to the primary ground force, the police, and government agencies that were busy protecting the population from insurgents and supporting people in their day-to-day lives, and not to a conventional army in a symmetrical conflict. The Portuguese Air Force under able leadership anticipated such a conflict and began in its formative years to lay the groundwork in Africa to defend the *ultramar*. In the following pages we shall explore just how this occurred and how in the subsequent conflict the Portuguese Air Force learned by doing. It identified the operating problems and found appropriate solutions to overcoming the vast distances and forbidding flying conditions in Africa to bring the critically needed support to both its military and civil colleagues.

In performing this vital function, it was constrained by the Portuguese colonial position in a shifting geopolitical climate that viewed empires as out of step with the new political awareness sweeping through the undeveloped world. These "winds of change" spawned new enemies and battles that would be fought not only in the swamps and savannas of Africa but also in the diplomatic forums of the United Nations and the Organization of African States.

"Winds of change" and the Cold War

In the decade of the 1950s, the "winds of change" were blowing through Africa. The largest colonial power, Britain, was freeing its colonies at a regular pace in step with its reluctant promise made to the United States during World War II. Following the independence of British India in August 1947, its empire wound steadily down, and by the mid-1960s all of its African possessions were independent countries for all practical purposes. France, despite its reduced circumstances, remained adamant in retaining its colonies and fought two losing conflicts, Indochina between 1946 and 1954 and Algeria between 1954 and 1963. During the Algerian conflict, an exhausted France began to grant independence to the remainder of its colonies at a rapid pace. With the 1959 débâcle in the Belgian Congo and its independence the following year, Portugal then possessed the last intact colonial empire.

Following World War II, the colonial troops of both Britain and France who had helped their masters defeat the two imperial regimes of the Axis Powers returned to their role as lesser citizens in the colonial empires of their mother countries. There was a conflict in their minds between the ideals for which they had fought and their continuing underprivileged "native" status. They had difficulty in reconciling this discrepancy, and this led to the notion that perhaps their colonial home should be an independent country with the political freedoms that they were currently being denied. The answer to this search for equal status in the face of intransigent regimes was to form resistance movements. As the British, French, and Belgian colonies gained independence, the indigenous movements became political parties in the bequeathed nominal democracies. While these were generally short-lived and rapidly became oligarchies or dictatorships with fewer economic or political freedoms than before, they were, nevertheless, now independent states and comprised a large part of Africa. Their national and foreign policies reflected their heritage of resistance to colonialism. Portugal thus found itself surrounded in Africa by ideologically hostile countries that would readily serve as insurgent sanctuaries and provide other practical support for resistance movements within its colonies. Angola initially faced such regimes in the Congo to the north and later Zambia to the east. Guiné was sandwiched between and indeed besieged by the former French Guinée to its south and a compliant Senegal to its north. Mozambique had Tanzania to its north and an ambivalent Malawi to the northwest. All of these neighbors were either sources or avenues of armed incursions.

In the context of the Cold War, the Soviet Union, the Peoples Republic of China, and their various satellites were ready to provide important support to insurgent movements hostile to Portugal and other colonial powers in the direct form of military training, advice, and weapons and indirectly in the moral dimension. Their encouragement to Portugal's colonial opponents developed across a broad spectrum from armed conflict in the field to support in the diplomatic arenas of the United Nations (UN), the Organization of African States (OAS), and numerous other international fora. This hostile environment was clearly developing in line with the solidarity that the newly independent former colonies, now largely despotic regimes, felt with each other and the uncritical tyrannical communist bloc. Portugal, one of the early members of NATO, was staunchly anticommunist, and so there was now an additional dimension to the philosophical divergence with its African neighbors.

This development reflected the competing forces of the Cold War, which lasted some forty-four years from the end of World War II in 1945 to the fall of the Berlin Wall in 1989 and pitted the largely former Western Allies against the Soviet Union, communist China and their satellite states. The initial of several phases ran until 1960 and was characterized by the creation of the Western and Eastern Blocs in a simple and starkly contrasting international bipolar system, with various states aligning themselves largely through alliances with either the principal "free world" or "communist" powers. The Western Bloc led by the United States followed a long-term policy of containment towards its Eastern counterpart and managed to extend and consolidate its Western system in this context. This consolidation was made through such vehicles as the Marshall Plan, the Atlantic Alliance, the Rio Pact, the Japanese-American Alliance, and the creation of the Federal Republic of Germany. In a very short time the United States had marshaled the forces of the developed world against its communist bloc foes in support of this containment policy.[1] The Soviet Union in response to this increased encirclement initiated two crises during the period in an attempt to increase its hegemony, the Berlin Blockade of 1948–1949 and the Korean War of 1950–1952. In the first instance the United States organized the Berlin Airlift as a response to a Soviet rail and road blockade of the city. This response avoided an obvious Allied ground confrontation, one that had the potential to be a messy troop encounter without solving the problem of sustaining an Allied presence. The Soviets lifted their blockade once the success of the airlift became obvious and demonstrated Allied determination to remain.

In the second instance the Soviet Union tried the less direct method of using a proxy force to gain communist hegemony on the Korean peninsula. The Allies had demobilized their forces substantially following the war and were clearly vulnerable in peripheral areas, such as Korea. The Soviet Union and China encouraged and supported North Korea in its desire to unify the Korean peninsula under a communist regime and in a surprise attack on the South to achieve this aim. The Allies again

1 António José Telo, *História da Marinha Portuguesa: Doutrinas e Organização, 1924–1974* [History of the Portuguese Navy: Doctrine and Organization, 1924–1974] (Lisbon: Academia de Marinha, 1999), 515.

scrambled to counter this aggression and under a UN resolution fielded a force that eventually restored South Korea under its legitimate government. The naïve post-war notion in many quarters that communists were "simple peace-loving people" was forever dispelled here.[2] The effect of this realization was a serious rearmament in the West, an unfortunate and debilitatingly expensive development for the Soviets.

The second "rearmament" phase ran from 1960 to 1973 and corresponded to the period of the Portuguese wars in Africa, which were directly affected by U.S. and Soviet superpower confrontation. Following an established pattern, the Soviet Union initiated two crises during this period, the Berlin Crisis of 1961 and the Cuban Missile Crisis of 1962. These crises followed on the heels and in the latter instance were a direct result of the Pay of Pigs fiasco. This disaster had its origins in a failed plan that was developed during the last months of the Eisenhower and the first weeks of the Kennedy administration by the Central Intelligence Agency (CIA) to invade Cuba with about 1,000 Cuban expatriates and overthrow its dictator Fidel Castro, a continuing thorn to the United States. It was based on the hopeful premise that the unhappy Cuban population would rally around the U.S.-trained and -armed invaders and help defeat Castro and his 200,000-man militia. When the CIA put the force ashore in April 1961, it was smothered by the militia.[3] Later that summer Berlin was thrown into crisis with the building of the Berlin Wall by its communist East to isolate it as part of Soviet premier Nikita Khrushchev's continuing efforts to intimidate the Allies into abandoning the city. Things were not proceeding well for the new Kennedy administration.

Khrushchev had twice in 1960 pledged to defend Cuba from U.S. aggression, and after the Bay of Pigs threat, took this commitment to heart. Instead of simply installing defensive hardware there, he decided to use Cuba as an opportunity to bring much of the United States within reliable nuclear range of Soviet attack by deploying both 1,000-mile medium-range and 2,200-mile intermediate-range missiles and 700-mile range Ilyushin-28 (IL-28) Beagle jet bombers. While strategically it would hardly matter whether the United States was struck by an attack from Cuba or an intercontinental ballistic missile from the Soviet Union proper, the now off-balance Kennedy administration could accept neither the domestic nor the international political consequences of Soviet-Cuban fortification. Kennedy consequently instituted a naval blockade of Cuba to prevent any further Soviet build-up after publicly revealing the existence of the missiles and generating diplomatic support for their removal. He directly threatened the island with a U.S. invasion, an option that the Soviets could not counter successfully so far from home. Khrushchev had thus gambled and lost.

2 George F. Kennan, "Long Telegram" from Moscow, February 22, 1946, in *Foreign Relations of the United States, 1946* (Washington, D.C.: U.S. Government Printing Office, 1969), vol. VI, 666–709.

3 Russell F. Weigley, *The American Way of War* (New York: Macmillan, 1973; Indiana University Press, 1977), 450.

The Soviet troops and matériel were removed, and two years later Khrushchev paid the price with his fall from power. This military humiliation and the compensation that the Soviet armed forces extracted for it marked the beginning of the accelerated advance in Soviet naval power over the next decade. It was to develop into a force formidable enough to challenge the free hand that the U.S. and Allied navies had previously enjoyed. By the 1970s Soviet fleets were a significant presence in every ocean of the world. For Portugal this increased presence meant that communist influence could be extended to remote peripheral areas and that outside support could now be delivered more readily to subversive, anti-Western resistance movements and their insurgent elements.

A second development stemming from the rise of Soviet power was a diffusion of the Atlantic Alliance. In this post-war era the United States had been presiding over world affairs in a way not previously available to any nation. It possessed only a relatively small portion of the global population and natural resources but had proceeded to dominate military and commercial power with a vast margin of superiority over both its friends and foes. With the relative economic recovery of Europe after the war and the birth of the European Economic Community (EEC), many European nations had begun to chafe at the U.S. direction and now sought greater autonomy. U.S. leaders took for granted the uniformity of interests among members of the Alliance and appeared to forget in their position of dominance the long history of European dynamism, political philosophy, and style of diplomacy. This heritage was bound to reassert itself, and no one felt this need more than Charles de Gaulle, the President of France. In an attempt to give new stature to his country, one that was suffering from a deep sense of failure and vulnerability after its devastation in both world wars and defeat in Indochina and Algeria, he took France on its own course. As he was unable to bridge the psychological gulf between his country and the United States, he first ordered the removal of all nuclear weapons from French soil, then withdrew the French fleet from the integrated NATO command, and finally in 1966 separated France from the NATO military structure altogether. In a turnabout France began to seek stronger collaboration with the Federal Republic of Germany. De Gaulle wanted France to be perceived by Germany as a more reliable ally than the United States and over time to displace U.S. with French leadership. The deviation from U.S. influence by these two countries was to give Portugal two critical allies in the prosecution of its wars in Africa.

This crack in the bipolar solidarity was not restricted to the West. China developed a far deeper breach with its Soviet neighbor, so much so in fact that forty Soviet divisions faced the Chinese army across the 4,000-mile Sino-Soviet border from 1969 until the Soviet Union in its economic distress unilaterally abandoned the arms race in 1988. The potentially violent and confrontational nature of this breach played into Allied hands and was fully exploited in the U.S. pursuit of its containment policy. Immediately the United States sought to open contacts with China and in February 1972 signed the Shanghai Communiqué formalizing relations with this formerly hostile state.

There were other divergences and crosscurrents in the bipolar concept, and none were more apparent than the behavior of the nonaligned nations. In 1955 a group of Asian, African and European nonaligned and communist countries met in Indonesia at the Bandung Conference and passed a resolution condemning colonialism "in all its manifestations" and declaring that "the subjection of peoples to alien domination and exploitation constitutes a denial of fundamental human rights, is contrary to the Charter of the United Nations, and is an impediment to the promotion of world peace and cooperation."[4] This declaration signaled a new offensive against the Western powers and was greatly encouraged by the retreat of Britain and France from Suez in 1956. This retreat not only destroyed the Great Power status of these two and prestaged the wholesale abandonment of their colonies but also left the United States alone outside of Europe to man the ramparts of the Cold War.

This anticolonial momentum was reflected in a further targeting of Portugal, as it increasingly assumed the role of sole colonial power. The Organization of African Unity (OAU) was formed in 1963 with its genesis in the Bandung Conference and held its first meeting in Addis Ababa. It approved a resolution there authorizing armed violence to overthrow existing colonial regimes. In the following year it specifically advocated the expulsion of Portugal from Africa by force, and in 1968 at its Algerian meeting endorsed concrete military support for all liberation movements.[5] This Third World rhetoric gained added legitimacy through its treatment in the United Nations.

Portugal had joined the United Nations in 1955 following a number of vetoes by the Soviet Union, and as if these Soviet rejections were not enough, was immediately made to feel isolated by the international community over its colonies. There had been noisy pressure from a contingent of UN members for Portugal to grant its colonies self-rule. This agitation had been fed by the newly independent states joining the United Nations and by the writings of a number of authors exposing the human abuses in Lisbon policy toward its colonies, most notably the Galvão report. Henrique Galvão, an army captain, was chief inspector of the Colonial Administration and had authored a report in 1947 describing labor conditions in Angola and cautioning against their continuance.[6] In a rather obtuse decision, considering the developing international climate, António de Oliveira Salazar, the Portuguese prime minister, had the report banned and in 1952 arrested Galvão on charges of treason. With this reaction the incident gained international notoriety and fueled the opponents' case. Additionally Portugal refused to submit the periodic technical reports on its colonies

4 David M. Abshire and Michael A. Samuels, *Portuguese Africa, A Handbook* (New York: Praeger Publishers, 1969), 375.

5 Sérgio Augusto Margarido Lima Bacelar, *A Guerra em África 1961–1974: Estratégias Adoptadas pelas Forças Armadas* [The War in Africa 1961–1974: Strategies Adopted by the Armed Forces] (Porto: Liga dos Amigos do Museu Militar do Porto and Universidade Portucalenese Infante D. Henrique, 2000), 41.

6 Captain Henrique Galvão, *Report on Native Problems in the Portuguese Colonies* (Lisbon: Ministry of the Colonies, 1947).

that were requested by the United Nations for non-selfgoverning territories. While other colonial powers were unhappy about this requirement, they complied. Portugal managed to attract additional unwanted attention in its refusal. So Portugal, in fighting the highly vocal anticolonial lobby, seemed to play into its hands with these ill-conceived decisions.

Likewise in 1955, a group of UN members advocating colonial independence managed to pass a resolution with the help of the Soviet Union condemning colonialism as a violation of human rights and the UN Charter. In response, Portugal claimed that it had no colonies and that all of its overseas provinces were part of a single state with one constitution. It likewise claimed that the United Nations had neither jurisdiction nor competence in this matter, as it was an internal affair. The issue festered in debate for four years, and finally in December 1960 the UN General Assembly, pushed by this group and the Soviet Union, voted against Portugal. While Portugal was to an extent a victim of its own intransigence, it saw itself as targeted and refused to accept the resolution. Its NATO allies rallied to prevent a catastrophe; however, this support began to fray by 1961.[7] To maintain its position successfully as a colonial power, Portugal would have to become more flexible and imaginative in both its international and domestic approaches to the problem.

By 1961, the Portuguese *ultramar* had become an important and integral part of the country's overall economic health. Its economy by that time had shifted from a partial autarky under orthodox economic practices to a fledgling but rapidly growing industrialized one. The shift away from an agriculturally based economy in both the *metrópole,* or continental Portugal, and the *ultramar* meant that there was a decreasing dependence on peasant labor and its attendant policies. As the *metrópole* developed in this direction, so the first moves were made to foster complementary development in the colonies. Mining, oil exploration and refining, textiles, and cashew processing were in place by 1961, and other basic industries were in the planning stages. These activities reflected a break with the past and a new Salazar policy fueled by colonial promise. Education received renewed and expanded attention, as literate workers with skills were in increasing demand. The paranoia regarding foreign investment had evaporated, and French, German, U.S., Rhodesian, and South African participation in the economy was welcomed. The gathering momentum of the colonial economies continued to accelerate well past 1961 and became a welcome support for the political element in the counterinsurgency campaigns. The colonies were thus developing into substantial economic engines in their own right, and not only were their citizens beginning to benefit individually, but Portugal itself was also reaping substantial rewards from this growing prosperity. The historic potential of the colonies was being realized, a fact that reinforced their long-time importance to Portugal and Salazar's commitment of the nation to their defense.

7 Joaquim Moreira da Silva Cunha, *O Ultramar, a Nação e o "25 de Abril"* [The Overseas Provinces, the Nation and the "25th of April"] (Coimbra: Atlântida Editora, 1977), 13–14.

Colonial Resistance

At the time when Portugal's colonial commitment was being strengthened, local resistance within its African population was increasing. During this period the democratized European powers in Africa were freeing their colonial possessions in step with the post-World War II trend. This development put increasing pressure on Salazar to move in line with the Western European forms of government and to allow the Portuguese colonies to do so as well. Revolts and the war enabled Salazar to exploit a tide of Portuguese nationalist fervor in preserving the *status quo* and his personal regime, the *Estado Novo* or New State. Consequently, the nationalist resistance and its challenge to his colonial vision had the effect of reinforcing the Portuguese commitment rather than the opposite. While the economy was deemed important, Salazar's personal position of authority was overriding.[8] Further, his hatred and mistrust of communism played an important role. He was mindful of the impotence of the Western powers to contain the economically bankrupt but politically ascendant communism.[9] In a speech to army and naval officers on 6 July 1936, Salazar described communism as "systems of ideals which are literally systems of crime" and was so convinced of the threat of this ideology that he believed "Western civilization is at stake."[10] His worst fears were realized when Daniel Semenovich Solod, the "brilliant organizer and expert in the tactics of infiltration and subversion," was assigned to Guinée in 1960.[11] Ambassador Solod had established an impressive reputation for increasing Soviet influence in the Middle East and North Africa, and now began to work on the Portuguese colonies and to nurture their long-standing dissident undercurrent of nationalism.

The nationalist movements and their military wings of insurgents that challenged Portuguese ownership of its colonies had their origins in the 1930s. The emergence of modern-day black opposition to Portuguese rule began with the repressive practices of Salazar's *Estado Novo* toward any form of dissent, particularly political. This attitude extended from the *metrópole* to the colonies. Resistance began slowly, as there was a practical barrier to any such opposition in the ethnic and social fragmentation of the overseas non-white community. Without strong leadership there would be

8 W. Gervase Clarence-Smith, *The Third Portuguese Empire 1825–1975: A Study in Economic Imperialism* (Manchester: Manchester University Press, 1985), 193. Internal and external pressures to democratize Portugal and put it in step with Western Europe had increased since World War II, and Salazar's rule had become tenuous in the late 1950s.

9 Hugh Kay, *Salazar and Modern Portugal* (London: Eyre and Spottswoode, 1970), 69, 133. The author argues that even in England communism had become "the rallying cry of the revolutionary instincts of our age" and had done so "by power of words, by sheer bluff, perhaps by the voluptuousness of contrast." It was largely the success of communist "evangelism" that made Salazar so suspicious of it.

10 Salazar quoted in Ibid., 133.

11 Ibid., 238.

no nationalist movement able to gain the necessary momentum in reconciling these divergent viewpoints and crystallizing resistance to the Salazar regime. Local African grievances were long-standing and had come to the fore during the early twentieth century with the influx of white settlers and abusive labor practices. This indigenous resentment was publicly evident in 1932 when an independent Mozambican newspaper, *O Brado Africano* (The African Cry) slipped through Salazar's censorship and published a scathing editorial titled "Enough." Thereafter this feeling was never far below the surface, and the apparent calm was illusory.

Following World War II, nationalist sentiments grew among the *mestiços* (mixed-race peoples) and *assimilados* (mostly *mestiços* who were legally assimilated into Portuguese culture). However, these groups were largely urban and thus did not represent the greater population. As they were located in cities, they were in a hostile environment for two reasons: the majority of their opponents, the white population, lived in cities, and the national police or PIDE (*Polícia Internacional de Defesa do Estado*, or International Police for Defense of the State) operated most effectively there. Consequently, they were either short-lived or dormant.[12] By 1956 the young Marxists of the Angolan Communist Party contributed to the formation of the MPLA (*Movimento Popular de Libertação de Angola*, or Popular Movement for the Liberation of Angola). The MPLA developed roots among the urban and largely radical intellectuals of Luanda, among its slum dwellers, and to a lesser extent, eastward from the capital among the Mbundu, the second largest ethnolinguistic group in Angola, and the Chokwe people. These urban roots were composed largely of *mestiços*, who controlled the party. The movement had little in common with the rural peasants of the east and south of Angola and made little effort to gain their true devotion. In December 1956 the initial MPLA manifesto was openly published in a direct frontal assault on the government. Predictably the PIDE reacted adversely, and a number of the MPLA leaders were forced to flee into exile. From 1957 onward PIDE action was so successful "that the nationalists were not able to maintain more than the most rudimentary organization inside the colonies and could not communicate with those cells that did exist."[13] The parties were forced to conduct their affairs from neighboring states and were deeply influenced by their foreign connections.

The presidential election in May 1958 gave all of Portugal some opportunity to express its dissatisfaction with the *status quo*. Elections under the Salazar regime as a rule were perfunctory, colorless, cosmetic affairs with foreseeable results. In 1958, however, General Humberto Delgado's high-profile and emotionally charged challenge to Salazar's candidate, Admiral Américo Tomás, excited all of Portugal. This

12 John A. Marcum, *The Angolan Revolution, Vol. I, The Anatomy of an Explosion (1950–1962)* (Cambridge, Mass.: MIT Press, 1969), 347–351. Marcum lists some fifty-nine groups affecting Angola alone beginning in the 1940s and either merging with one another or vanishing by 1962.

13 Malyn Newitt, *Portugal in Africa: The Last Hundred Years* (London: C. Hurst & Co., 1981), 190.

taste of partial suffrage awakened dissatisfaction within the *mestiços* and *assimilados*, and a number of small parties were formed in Angola, only to be shattered through arrests in March, May, and July of 1959. As the PIDE systematically wrecked the MPLA organization, it became progressively weaker and isolated from its leadership that was now abroad. In this deteriorating position it supported an uprising in February 1961 that stood no chance of a lasting success. It was doomed to be transient, for it occurred in Luanda, center of Portuguese police and military strength, and the MPLA had no constituency or bases elsewhere among the rural population.

The MPLA in exile established itself initially in Léopoldville and aligned itself not only with other independent African nations and their socialist philosophy but also with the communist bloc, including the Italian and French communist parties. The leadership was consequently familiar with the communist theory in wars of national liberation and organized itself accordingly. The MPLA found that it was in competition with the other prominent Angolan nationalist group at the time, the UPA (*Uniao das Populações de Angola*, or Union of Angolan Peoples), for acceptance as the leading representative of the Angolan people. In 1962 the MPLA formed its military wing, EPLA (*Exército Popular de Libertação de Angola*, or Popular Army for the Liberation of Angola), to project its influence into Angola. This nascent force numbered between 250 and 300 young men who had undergone military training in Ghana and Morocco. The EPLA sought to expand the conflict across the northern border of Angola with this force and penetrate the entire country, publicizing the MPLA manifesto. Recruiting proved to be difficult because of ethnic rivalries, and military action was thwarted by the competing UPA. The UPA through its influence with the Congo leadership forced the MPLA to leave Léopoldville in 1963 and reestablish itself in Brazzaville, from which it was difficult to conduct a campaign across an unenthusiastic third country and into a now distant Angola. As a result northern Angola proved to be barren for the MPLA, and it was not until 1966, with the opening of its second front from Zambia, that some success would come. The most consequential development from the Portuguese perspective was the capture in July 1963 of various 35mm films that described the MPLA military doctrine of revolutionary warfare. It paralleled the Maoist creed by reiterating that the movement was a people's war and that the struggle would be protracted.[14] The first priority would be indoctrination and organization of the masses, and next the establishment of rural

14 Willem S. van der Waals, *Portugal's War in Angola 1961–1974* (Rivonia: Ashanti Publishing, 1993), 103. Neto is quoted: "If the enemy presently possesses more forces than we do, and this is so, then it is correct that we should prepare for a protracted war. The misconception that we should be able to execute a war of rapid decisions should be removed once and for all." See also Alpoim Calvão, "Guerra Revolucionária e Guerra Subversiva" [Revolutionary War and Subversive War], *Anais do Clube Militar Naval* (October-December 1967): 713-720.

bases and resistance areas.[15] This pseudo-nationalistic doctrine was flawed in that the MPLA had little in common with the larger population of Angola. Nevertheless, it would serve the MPLA until 1974, and as we shall see in future chapters, the Portuguese correctly anticipated the weakness of this approach.

The UPA was formed by Barros Nekaka in the mid-1950s from a number of small groups with conflicting goals. In 1958, he passed leadership to his nephew Holden Roberto. UPA strength rested in the rural populations of the Bakongo ethnolinguistic region of Angola. These people straddled the border area that reached into the Belgian Congo, Angola, Cabinda and the French Congo, the footprint of the ancient Kongo kingdom. Roberto unequivocally held the view that not just a "Bakongo kingdom" or some other entity but all of Angola must be freed. An ardent anticolonialist, Roberto had been born in Angola but had lived his adult life in the Belgian Congo. He had been educated in the Baptist Church missionaries and employed in the Belgian colonial economy as an accountant between 1941 and 1949. Northern Angola was an area that had become more politically aware in the 1950s through white settlement, Baptist missionary influence, and an easy access to the developing political activities of the Belgian Congo. Roberto thus felt a close kinship with the peoples immediately across the border. The UPA was able to develop a following there because of the relatively open frontier, and this loyal cadre became the basis for the uprising in March 1961. Portuguese presence in this area took the form of *chefes do posto* (heads of posts) and administrators, as opposed to PIDE, and these officials were so sparse that it was physically impossible for them to maintain anything but the most casual control over their districts.[16]

While Roberto was relatively well educated, he was a member of the Bakongo ethnolinguistic group, was not a *mestiço*, and consequently did not share the more European cultural perspective of the MPLA. He was also tribally oriented in contrast to the non-tribal declarations of the MPLA. As a result, the personality and leadership philosophy of the UPA contrasted clearly with the MPLA and its sophisticated *mestiço* leadership, which was left-wing, intellectual, and acculturalatively Portuguese. Funding and support also glaringly contrasted, the MPLA actually being linked with the Eastern bloc. The UPA received financial support from the American Committee on Africa and from various African governments, preponderantly that of

15 Região Militar de Angola, *Supintrep No. 19: Guerra Revolucionária* [Supplemental Intelligence Report No 19: Revolutionary War], July 1963, Luanda, quoted in van der Waals, 103.

16 Douglas L. Wheeler and René Pélissier, *Angola* (London: Pall Mall Press, 1971), 167. The authors cite as an example the Congo district in 1960. For its 37,000 square miles it had fourteen *concelhos* (basic urban or semi-urban administrative unit) or *circunscrições* (basic rural administrative division) and thirty-seven posts, for an average of 725 square miles per administrative division. This presence would hardly be effective in controlling a frontier, as the posts would be dozens of miles apart. Large numbers of people could and did cross undetected.

Léopoldville.[17] Accordingly they were never able to resolve their differences and join forces effectively.

When the Belgian Congo became independent on 30 June 1960, its new government began to give Roberto practical assistance, including permission to establish a radio station and a training camp within its borders. This sanctuary was an important facet of UPA operations in its early years. Roberto had witnessed the long series of Congolese crises that had begun with the violent political rioting on 4 January 1959 and had led to the accelerated Belgian push toward Congo self-government and independence in eighteen months. By December 1960, he believed that just as the Belgians had quickly grown weary of armed conflict, so would the Portuguese when it was initiated. He consequently used his Congo sanctuary and the porous common border to set the stage for an end to relative colonial tranquility for Portugal.

The UPA formed its military wing, the ELNA (*Exército de Libertação Nacional de Angola*, or Army of National Liberation of Angola), in June 1961 after the March attacks did not achieve a Portuguese withdrawal. Roberto was its commander-in-chief and its other two leaders were Portuguese Army deserters, Marcos Xavier Kassanga, its chief of staff in Léopoldville, and João Batista, its operational commander in Angola with headquarters near Bembe. This leadership was ineffective. The "fiery-tongued" Roberto was so autocratic that he would accept little more than arms and money.[18] Indeed, he appeared hard, introverted, and even sinister from behind his apparently ever-present sunglasses, which he seemed to wear even in dark rooms and on cloudy days.[19] Without training, the ELNA "set a demoralizing example of politico-military incompetence and indiscipline."[20] The South African Defence Force vice-consul in Luanda, Brigadier Willem S. van der Waals, noted that the ELNA "involved itself in military activities in the narrowest sense...but avoided contact with the Portuguese security forces as far as possible."[21] Van der Waals further argues that in the traditional exercise of the informational lever of national power, Portuguese propaganda and social work among the refugees returning to Angola after the March 1961 attacks persuaded most of these displaced people to move into controlled settlements or

17 Hélio Felgas, "Angola e a Evolução Política dos Territórios Vizinhos" [Angola and the Political Evolution of the Neighboring Territories], *Revista Militar* (December 1965): 706.

18 Shola Adenekan, "Holden Roberto," *The New Black Magazine*, 22 October 2007, http://www.thenewblackmagazine.com/view.aspx?index=1042 (accessed 23 March 2009).

19 Ibid.

20 van der Waals, 96.

21 Ibid., 97. The author argues that Portuguese propaganda and social work among the refugees in Angola persuaded most of these displaced people to move into controlled settlements. This development deprived ELNA of popular support. ELNA had concentrated on military action in a human desert and on preventing MPLA infiltration. It had neglected to indoctrinate, organize, and win recruits among refugees returning to Angola and thus missed an opportunity to undermine Portuguese authority. Consequently no ELNA internal political infrastructure was established in Angola. Portugal gained the upper hand and maintained superior momentum until 1974.

aldeamentos. This development deprived the ELNA of popular support, and its military action consequently occurred in a human desert. It was thus unable to proselytize the population, as in a classic insurgency, and as an alternative focused on preventing competitive MPLA infiltration. It had also neglected to indoctrinate, organize, and win recruits among refugees in the Congo returning to Angola and had thus missed an important opportunity to undermine Portuguese authority. There was accordingly no ELNA internal political infrastructure in Angola. Portugal gained the upper hand and dominated the human terrain until 1974 in a classic example of successful informational warfare with civil support and population proselytizing.

ELNA training was so poor that despite its expansion to about 6,200 troops, their deportment at such camps as Kinkuzu in the Congo was cause for alarm.[22] Andreas Shipango, South West Africa Peoples Organization representative in Léopoldville, made an appraisal during a 1963 visit: "With representatives from a number of other liberation movements, I visited Holden Roberto's training camps near the Angolan border with a view to sending our young men there. But the atmosphere in Roberto's training camps was very bad, and I could not recommend such a course."[23] Roberto had only a weak military program unsupported by political indoctrination, and indeed there was no talk of winning the population to the UPA point of view, which was simply that Angola should be an independent country with Roberto as head of state. The approach was arrogant and naïve and, as it proved, totally ineffective.

This lack of direction caused great rifts in the UPA leadership. Despite the UPA reorganization in March 1962 at the behest of Mobutu Sese Seko, president of the Republic of the Congo, to include additional groups, to rename itself FNLA (*Frente Nacional de Libertção de Angola*, or National Front for the Liberation of Angola), and to establish a government in exile named GRAE (*Governo da República de Angola no Exílio*, or Government of the Republic of Angola in Exile), little of substance was accomplished. A frustrated Jonas Savimbi, Roberto's "foreign minister" and an Ovimbundu, formally broke with the UPA/FNLA in July 1964, labeling Roberto a "corrupt racist," and eventually formed the third nationalist movement in Angola, UNITA (*União Nacional para a Independência Total de Angola*, or the National Union for the Total Independence of Angola). Savimbi had publically announced his break at the 1964 Organization of African States (OAU, now the African Union or AU) meeting in Cairo, Egypt, and alleged that Roberto had set up a "commercial empire in the Congo" and that FNLA administrators were "wage earners and profiteers who enriched themselves on the money of New York financial circles and other international organizations."[24] Within two years, Savimbi had built his meagre 12-man force into a sizeable army, gaining popularity and support as the only leader to work within the country alongside his men in battle against the Portuguese. "Leaders must fight

22 Neil Bruce, "Portugal's African Wars," *Conflict Studies*, ⊠ 34 (March 1973): 22.
23 Sue Armstrong, *In Search of Freedom* (Gibraltar: Ashanti Publishing, 1989), 71.
24 Adenekan.

alongside the people and not stay abroad, sending 'second-class' fighters to face the Portuguese," Savimbi proclaimed as he denounced Roberto.[25] While Roberto was renowned for his aloofness, the bearded Savimbi mixed often and easily with ordinary people as well as his military. The next year Alexandre Taty, "minister of armaments," after challenging Roberto in an unsuccessful coup, defected to the Portuguese in Cabinda with a substantial number of his followers.

Indeed, Roberto seemed more interested in his personal power than in a war of national liberation, and according to U.S. intelligence sources, was "subservient" to Mobutu, who protected him from internal challenges to his leadership.[26] As the Portuguese weekly *Expresso* observed in 1974, "The FNLA is Holden Roberto and Holden Roberto belongs to Mobutu, to whom he is connected by an umbilical cord."[27] Mobutu was playing both ends against the middle in loudly proclaiming his support for the FNLA while discretely cultivating good relations with Portugal, for like Zambia, the Congo depended on the Benguela Railway (*Caminho de Ferro de Benguela* or CFB), which carried more than half of its foreign trade. There were also dissident elements in Angola that, if unleashed in cross-border operations, could make considerable trouble for him. Consequently he kept a tight control over the FNLA activities both within the Congo and without. He provided just enough political and material support to give it international credibility and to provide the Congo with a stake in Angola should the Portuguese eventually leave.[28] As for Roberto, after 1966 he started a property business in Léopoldville and was often seen driving a shiny black Mercedes-Benz about the city, just as Savimbi had predicted. He clearly became less interested in running a nationalist movement and, according to U.S. intelligence, had not set foot inside Angola since 1956.[29] As Henry Kissinger observed at the time, "The strength of the FNLA continues to suffer from Holden's refusal to move from Zaïre (the Congo) to Angola to take direct control of FNLA activities."[30] Indeed, Roberto spoke French and English far better than Portuguese.[31] CIA Luanda station chief at the time, Robert Hultslander, wrote later that, "This organization was led by corrupt, unprincipled men who represented the very worst of radical black African racism.... It was a squalid spectacle: a corrupt leader dancing to the tune of a foreign master."[32] As for his troops, they have been described as "underfed, ragged, and villainous,"

25 Ibid.
26 Piero Gleijeses, *Conflicting Missions: Havana, Washington, and Africa, 1959–1976* (Chapel Hill: University of North Carolina Press, 2002), 239.
27 *Expresso* (Lisbon), 17 September 1974, 18.
28 Gleijeses, 238.
29 Alex Vines, "Holden Roberto," *The Independent* (London), 8 August 2007, http://www.independent.co.uk/news/obituaries/holden-roberto-460658.hmtl (accessed 23 March 2009).
30 Ibid. Mobutu renamed the Congo in 1971 as Zaïre.
31 Adenekan.
32 Gleijeses, 238.

and were hardly a credible army.[33] The Portuguese, it seems, had little to fear from Roberto and his organization after the March 1961 attacks.

John Marcum described the situation as it existed in 1963: "Whether by the inaction or heavy hand of shortsighted leadership, one opportunity after another was lost, one potential source of support after another was alienated."[34] The political crosscurrents within the UPA/FNLA, the lack of training for ELNA cadres, and major competition from MPLA and UNITA activities reduced the UPA/FNLA to a spent force within two years of initiating the conflict. Roberto followed no sophisticated guerrilla creed other than the initiation of violence in the hope that the Portuguese would become weary with it and capitulate. The approach was amateurish and ineffective alongside that of PAIGC (*Partido Africano da Independência da Guiné e Cabo Verde*, or African Party for the Independence of Guiné and Cape Verde) and the work of its founder, Amílcar Cabral.

Aside from Angola, there were nationalist movements associated with Guiné and Mozambique that prior to the events of 1961 were hoping to negotiate concessions with the Portuguese on self-determination. In Guiné efforts by local nationalists to organize began in the early 1950s. The PAIGC was founded in September 1956 by local *assimilados* and educated Cape Verdeans. Its initial political organization prompted an aggrieved dockworkers' strike on 3 August 1959, which ended in a violent disaster when it was broken with excessive military force. Fifty workers were killed, and the incident became known as the "Pidjiguiti dock massacre." PAIGC leadership quickly realized that peaceful protest would not achieve its objective of self-rule and independence. Accordingly, it shifted its strategy to one of clandestinely organizing the rural population for an insurgency.[35] The PAIGC had learned hard lessons in 1959 well ahead of the MPLA and UPA/FNLA experiences of 1961, and had shifted its approach. It was not prepared to begin guerrilla war in Guiné until January 1963, when all of the elements for success were in place, including firm sanctuaries in adjacent countries.

The driving force behind the PAIGC was Amílcar Cabral, who was born in Guiné of Cape Verdean parents. Cabral was an agronomist by profession, having been educated in Lisbon, served the Portuguese administration in Guiné (1952–1955), and worked for various agricultural institutions in the *metrópole* (1955–1959) with research trips to Angola. His political awareness came at an early age and matured during his academic time in Lisbon. While influenced by Marxist-Leninist ideas of the time, Cabral was primarily a nationalist and developed his own variant of both the PAIGC political message and its associated military dimension. In his own words: "It is good [for all nationalist movements] to remember…that regardless of how similar are their

33 Ibid, 295.
34 John A. Marcum, *The Angolan Revolution, Vol. II, Exile Politics and Guerrilla Warfare (1962–1976)* (Cambridge, Mass.: MIT Press, 1978), 113.
35 Patrick Chabal, *Amílcar Cabral* (Cambridge: Cambridge University Press, 1983), 56–57.

struggles and their enemies to one another, national liberation and social revolution cannot be exported. They are…the products of local and national forces. While somewhat influenced by external factors, they are largely determined and tempered by the particular culture of a country's people and its unique local characteristics."[36] It was in this context that Cabral began to prepare the political landscape for guerrilla warfare.

Following his experience in the Pidjiguiti dock demonstration, Cabral realized that the Portuguese would not negotiate and that an armed struggle was the only way to achieve PAIGC ends. Cabral had received no known military training and had little interest in such affairs prior to 1959. It is possible that he had some such exposure during his visit to China in 1960, and certainly Chinese influence was seen in the training of his guerrilla army. It was known that elements of PAIGC also underwent courses in guerrilla warfare and subversion in Algeria, Russia, and Czechoslovakia.[37] Notwithstanding this lack of military experience, the mantle of undisputed commander and tactician fitted him well, and his imagination and flexibility were evident in the conduct of his campaign.

Cabral became quite attuned to the requirement for population indoctrination and keenly aware of the need to bridge the gap between the urban intellectual and the traditional Guinean. His two-year preparation of the political battlefield was classic in its effort to draw the population together in a common ideology that would transcend tribal and ethnic divisions. His investigation into local grievances was the most thorough of any of the nationalist movements. Cabral faced a difficult task in convincing the population that they were being oppressed. The land, for instance, already belonged to the peasants and was generally village property. Guiné had no concentration of foreign settlers who were seemingly exploiting the population. In Cabral's own words: "We were not able to mobilize the people by telling them: 'The land to him that works it.' Because here land is not lacking. … We were never able to mobilize the people on a basis of the struggle against colonialism. This yielded nothing. To speak of the struggle against imperialism yielded nothing between us. … This proved the necessity of having each peasant find his own formula to mobilize for the fight."[38] He thus sought to couch his revolutionary message in terms that would address the daily concerns of the rural population: "Remember always that the people do not fight for ideas, for things that only exist in the heads of individuals. The people fight and they accept the necessary sacrifices. But they do it in order to gain material advantages, to live in peace and to improve their lives, to experience progress, and to

36 Amílcar Cabral, *Guiné-Bissau–Nação Africana Forjada na Luta* [Guiné-Bissau–African Nation Forged in Struggle] (Lisbon: Publicações Nova Auora, 1974), 39.
37 Hélio Felgas, *Os Movimentos Terroristas* [The Terrorist Movements] (Lisbon: Privately printed, 1966), 57.
38 Amílcar Cabral, *Textos Políticos* [Political Texts] (Porto: Edições Afrontamento, 1974), 19–20.

be able to guarantee a future to their children."[39] This approach was far more fruitful than comparable activities in Angola.

Cabral met and worked closely with MPLA leaders and established his exile headquarters in Conakry, the capital of the ex-French Republic of Guinée. From here he conducted his confrontational campaign against the Portuguese. The other prominent nationalist movement in Guiné was FLING (*Frente de Luta pela Independência da Guiné*, or Front for the Struggle for the Independence of Guiné). Led by Benjamin Bull, Jonas Fernandes, and Henry Labery, it was an amalgamation of a number of smaller, more moderate movements and was given sanctuary in Senegal by its leader Léopold Sédar Senghor to operate on the northern border of Guiné. Senghor, a moderate himself, was at odds with Cabral and the PAIGC. While he did not want this movement using Senegal as a sanctuary to launch attacks into Portuguese territory, there was little that he could do about it, given the difficult physical and political terrain of the isolated Casamance region between the Gambia River and the border and his lack of a force to prevent a PAIGC presence. Again as in Angola, the philosophies of the PAIGC and FLING leaders were so disparate that there was little common ground for agreement and cooperation, although Cabral attempted to compromise and remained in contact with Senghor.

Following the unsuccessful employment of autonomous guerrilla groups in the first year of the conflict, Cabral held the Cassacá Congress in February 1964 to reorganize the war effort and establish a national army in the FARP (*Forças Armadas Revolucionárias de Povo*, or Revolutionary Armed Forces of the People). Because the Portuguese had shown that they were not going to negotiate, Cabral's only option was to win in the field. He thoughtfully assembled and implemented the proper elements of guerrilla warfare, particularly that of political indoctrination, to achieve PAIGC ends.

In Mozambique there were a number of exile, very small, nationalist organizations prior to 1961. The Portuguese government made every effort to dampen the spirit of nationalism in its formative stages; however, there were at any one time perhaps half a million Mozambicans, or about 10 percent of the population, working in neighboring countries. These expatriates were exposed to new political ideas and in the period 1958–1960 began to organize themselves into associations with the goals of social contact, self-help, and ultimately national politics. The first true nationalist organization was UDENAMO (*União Democrática Nacional de Moçambique*, or the National Democratic Union of Mozambique) and was established in Southern Rhodesia in October 1960, moving to Dar-es-Salaam in February 1961. Two additional groups of note appeared at the time, MANU or UNAM (*União Nacional Africana Moçambique*, or Mozambican African National Union), depending on the English or Portuguese conformation, founded among Makonde emigrants in Kenya

39 Amílcar Cabral, *Palavras de Ordem Gerais* [Speeches on Overall Methods] (Bissau: PAIGC/Secretariado Geral, 1976), 34.

and Tanzania, and UNAMI (*União Nacional do Moçambique Independente*, or National Union of Independent Mozambique), established in Malawi. Both MANU and UDENAMO set aside their differences and attended the 1961 Casablanca conference of African states, where the nationalist movements in the Portuguese colonies consolidated their front to become a coalition. Subsequently, in September 1962, elements of MANU, UNAMI, and UDENAMO were united in FRELIMO (*Frente de Libertação de Moçambique*, or Front for the Liberation of Mozambique) at the urging of Julius Nyerere, the Tanganyikan leader, making it the strongest and most important movement.

Eduardo Mondlane assumed its leadership. He, too, absorbed the lessons of 1961 and was not prepared to launch a guerrilla war until some three years later in September 1964, after his small army was trained. Mondlane, in coming late to the nationalist movements against Portugal, was very much influenced by the trend in Angola and Guiné. His organization initially developed similarly to the PAIGC and experienced the same sort of problems in subordinating military operations to political leadership. His doctrine paralleled that of the MPLA and particularly of the PAIGC with its emphasis on political indoctrination, and it was along these lines that he sought to conduct his military campaign. Mozambicans belonging to MANU had already tried peaceful demonstrations with the same consequences that had occurred in Angola and Guiné. At Mueda on 16 June 1960, it attempted to petition the local Portuguese administrator, who responded by using troops against unarmed peasants. Reputedly seventeen Africans were killed in this demonstration.[40] This "massacre" at Mueda politicized the Makonde and affected the development of FRELIMO and its campaign. FRELIMO now felt that armed struggle was the only answer, as Portugal would not grant self-determination and would destroy those who demonstrated for political freedom.[41]

Political expression was forbidden both in the *ultramar* and in the *metrópole* except in the narrow context of the Portuguese staged elections every four years. The airing of grievances or liberal political views attracted a heavy hand from the authorities. In the case of the 1959 strike by the Pidjiguiti dockworkers, the PAIGC had hoped that it would lead to negotiations with the authorities and a redress of grievances. When this traditional procedure failed, Cabral had no choice but to seek an alternative in an insurgency with the goal of opening negotiations or ultimately of gaining control of the country. The same can be said for the MPLA and its earlier flight from Angola in 1956 and 1957.

40 Imigrantes Somos Todas, "Moçambique: A Guerra de Liberação (1964–1974)" [Mozambique: The War of Liberation (164–1974)]: http://imigrantes.no.sapo.pt/page2mocGLibert.html, accessed 4 February 2010.

41 Eduardo Mondlane, *The Struggle for Mozambique* (London: Zed Press, 1969), 125.

Peasant populations are not normally a revolutionary force, and such was the case in Portuguese Africa.[42] They are conservative by nature and find security and comfort in the routine of their lives and the socioeconomic institutions that govern them. Change is resisted and outsiders are viewed with suspicion. The guerrillas of the various nationalist movements in Portuguese Africa represented change that the populations were not prepared to accept readily. Thus, despite the justified grievances of Portuguese black African citizens, overall they appeared to be loyal to Portugal and to suspect the activities of the nationalists. This apparent support reinforced the commitment of Portugal to its colonies and their people. As Portugal entered 1961, its internal confidence in its position was as strong as ever.

42 Gerald J. Bender, "The Limits of Counterinsurgency: An African Case," *Comparative Politics*, Vol. 4, No 3 (April 1972): 357.

2

Over the Brink

Beginning in January 1961, there occurred a series of events that had the cumulative effect not only of reinforcing Portugal's internal confidence but of hardening it into an irreversible position that would clearly lead to war. This series in order began with the unrest of the Angolan cotton growers in Cassange in the controversial "cotton war," the humiliating hijacking of the Portuguese ship *Santa Maria* through a reappearance of the activist Galvão, the affront of the attacks in the north of Angola, and an attempted coup against Salazar in an effort to moderate the government position toward the demands of the nationalist movements. This failure at moderation alongside the message from the United Nations served to silence voices of reason. The final blow was the débâcle of Goa, which made any further loss unthinkable. Each of these and its role in moving Portugal to war is examined below.

Baixa do Cassange

On 4 January 1961 in Baixa do Cassange, in the north of Angola, the blacks who worked in the cotton fields began a strike and launched what became known as "Maria's War" after one of its instigators, António Mariano. Protest in the old kingdom of Kasanje had its origins in the resistance to colonial rule that prolonged pacification until the autumn of 1911, when the population capitulated, and opposition "finally fell like a ripe pear," according to Réne Pélissier.[1] Nevertheless, from that point on resentfulness was never far below the surface, and as General Fernando Pinto de Resende, commander of the 2nd Air Region in Angola, observed regarding the obtuseness of the government to such feelings, "the unawareness was such that no consideration was given to the dominating and war-like characteristics of the Maholo people, inhabitants of Baixa do Cassange, who during the Portuguese pacification of the last century showed themselves as difficult conquests and extremely dangerous.

1 Réne Pélissier, *Les Guerres Grises: Resistance et Revoltes en Angola (1845–1941)* (Orgeval: Pélissier, 1977), 298.

We made them cotton farmers, clearly by force, and now we are bombing them from the sky."[2]

Compulsory cotton growing was introduced into Angola in March 1947, when the Companhia Geral dos Algodões de Angola, a Luso-Belgian company founded in 1926 and commonly known as Cotonang, was granted a concession in the vast cotton zone east and west of Malange covering some 80,000 square kilometers.[3] Another, more recent firm, Lagos & Irmão, was also a concessionaire in the area, which extends over a wide plain along the Cuango River and is known to the Portuguese as Baixa do Cassange. Resende calls these the "two great 'trusts' of the cotton feudal lords," *os senhores feudais*.[4] The profitable cotton economy in the area was based on compulsive cultivation that was organized on the basis of an impressment system and yielded a harvest of about 5,000 tons per year.[5] African villagers, some 150,000 people in 35,000 families, were hauled from their homes and forced to grow cotton on designated patches of land.[6] There were no wages for this work, and when the crop was harvested at the end of the season, the Africans were forced to sell it to Cotonang and Lagos & Irmão at fixed, below-market prices, a value between five and six times less than the world price.[7] The crop risk was assumed involuntarily by the impressed workers, who simply starved if there were crop failure and barely survived if it were a success. The concession companies thus took no commodity risk and purchased the crop from the growers with a certain built-in and generous profit margin, a near riskless arbitrage. As with other similar enterprises, such as Companhia de Diamantes de Angola (Diamang), the diamond mining concession, the laborers earned a below-subsistence income, and the shareholders received an extraordinary return on their investment. Many of those profiting could not even identify Angola on a map.[8] As Resende writes, "The culture of cotton is an inflammatory exploitation of the indigenous people and, moreover, a major antagonism to them, for this type of obligatory work (400$00 escudos per year in the worst areas…for the farmer, when he has the misfortune to lose his complete crop—cases that can be verified—he then receives

2 Fernando Pinto de Resende (Brigadeiro), Commander, 2nd Air Region, to João Faustino de Albuquerque de Freitas (General), Chief of Staff Portuguese Air Force, 8 February 1961, personal correspondence regarding the "cotton war," personal papers archive, Vol. II, No Prov. 323, Arquivo Histórico de Força Aérea Portuguesa, Alfragide.

3 Francisco Camacho, "Baixa do Cassange: O Massacre que Veio do Ceu" [Baixa do Cassange: The Massacre that Came from the Sky], *Grande Reportagem* 137 (August 2002): 64.

4 Resende to Albuquerque de Freitas, 8 February 1961, personal correspondence. A trust in this sense is a combination of firms holding a monopoly.

5 Edgar Percira da Costa Cardoso, *Presença da Força Aérea em Angola* [Presence of the Air Force in Angola] (Lisbon: Secretaria de Estado de Aeronáutica, 1963), 79.

6 Camacho, 65.

7 Ibid.

8 Basil Davidson, *In the Eye of the Storm: Angola's People* (New York: Doubleday, 1972), 128.

zero for a year of work). This is but one example of the many villainies that the white tribe perpetrates on the black tribe."[9]

Uniformed Africans who were recruited from the army were used as enforcers to ensure that the conscripted laborers worked their plots and grew only cotton.[10] They were not allowed to grow food for their families on these plots, and when the land was exhausted, they were forced to move further and further from their homes to tend fresh plots. This distance could range up to fifty miles.[11] This form of serfdom created a virulent current of unrest, as indeed the people lived in conditions of absolute moral and material misery.[12] As Resende again writes, "The Governor [General Venâncio Augusto Deslandes] is the No 1 enemy of this system, which he has not received consent to end yet. In Mozambique it is still worse!… The chief agitators understand all of this well and give priority to proselytizing the populations in these areas. Lunda must, moreover, also be one of the targeted objectives, just as the diamonds."[13]

Local protest dated from 1927, and over the years there developed a number of fledgling protest movements. Prominent among these was the syncretic or composite religion known as Quimbanguismo, which was founded by Simão Quimbango (Simon Kimbangi) in 1921 in the Belgian Congo and took root in Baixa do Cassange in the late 1950s. According to John Marcum, it was an outlet for the repressed emotions of a subject people and a special blend of Christian and animist practice.[14] Its preachings were lifted principally from passages of the Bible teaching protest and revolt, and it was considered subversive and banned by the government.[15]

When Mariano left Malange in 1959, crossed the border into the Congo, and took a job as chauffeur to a local chief, Kizamba dos Maholos, he often drove to

9 Fernando Pinto de Resende (Brigadeiro), Commander, 2nd Air Region, to João Faustino de Albuquerque de Freitas (General), Chief of Staff Portuguese Air Force, 2 February 1961, personal correspondence regarding the "cotton war," personal papers archive, Vol. II, No Prov. 323, Arquivo Histórico de Força Aérea Portuguesa, Alfragide.

10 John A. Marcum, *The Angolan Revolution, Vol. II, Exile Politics and Guerrilla Warfare (1962–1976)* (Cambridge, Mass.: MIT Press, 1978), 47.

11 Ibid.

12 António Lopes Pires Nunes, *Angola 1961: Da Baixa do Cassange a Nambuangongo* [Angola 1961: From Baixa do Cassange to Nambuangongo] (Lisbon: Prefácio, 2005), 59.

13 Resende to Albuquerque de Freitas, 2 February 1961, personal correspondence.

14 António Luís Ferronha, "Sincretismos afro-cristãos" [Syncretisms Afro-Christian], *Grande Reportegem* (August 2002): 72. See also Marcum, 47–48; and Estado-Maior do Exército, *Resenha Histórico-Militar das Campanhas de África, Vol. VI, Aspectos da Actividade Operacional, Tomo I – Angola, Livro 1* [Historical-Military Report on the African Campaigns, Vol. VI, Aspects of Operational Activity, Tome I – Angola, Book 1] (Lisbon: Estado-Maior do Exército, 1998), 103–107. Quimbanguismo is sometimes known as Kasonzola.

15 Estado-Maior do Exército, *Resenha Histórico-Militar das Campanhas de África, Vol. II, Dispositivo das Nossas Forças Angola* [Historical-Military Report on the African Campaigns, Vol. II, Disposition of Our Angolan Forces] (Lisbon: Estado-Maior do Exército, 1989), 30.

Léopoldville, where he came into contact with the UPA. He returned to Malange in December 1960, and with his personal charisma, a bit of Quimbanguismo, and the nationalist doctrine from the UPA, emerged as the leader of a new "prophet-protest" movement known simply as "Maria." It was a sort of nationalist evangelism with the trappings of incense and the magic of "Maria water," which, when sprinkled on Africans in a "miraculous baptism," made them believe that they were impervious to Portuguese bullets.[16]

Alongside Mariano, there was also at play the activities of the PSA (*Partido de Solidariedade Africana, Parti Solidaire Africain,* or African Solidarity Party). This was a political movement founded in 1959 that advocated the independence of the Belgian Congo. It espoused a communist doctrine and was hostile to the West and to the presence of Europeans in Africa. It advocated violence and threats and through these means gained ascendancy over the other Congolese parties. When the Congo was granted its independence in June 1960, PSA agitation was directed across the Angolan border into the districts of Malange and Lunda, and particularly into the locales of Camaxilo, Lubalo, Milando, and Tembo Aluma adjacent to the Cuango River frontier. It was learned by the Portuguese authorities from captured documents and the interrogation of prisoners much later after the crisis had passed that the unrest had been planned through Maholo PSA agents, who were infiltrated into the frontier area and aided by their fellow Malholos on the Portuguese side.[17] The PSA advocated

16 Camacho, 67. This notion was not unique and was a characteristic of Congolese rebel mobilization. In July 1963, Pierre Mulele arrived in the Kwilu area of the Congo. He was a key revolutionary figure who had traveled widely in Eastern Europe before reaching China, where he received sustained training in guerrilla warfare. Upon arriving in Kwilu, Mulele recruited a solid phalanx of followers among members of his own ethnic group, the Mbunda, as well as among the Pende, both of whom had long been the target of government repression. The rebellion began in January 1964, when Mulelist insurgents attacked government outposts, mission stations, and company installations. Mulele's men, who called themselves *la Jeunesse*, were fired by a strange mixture of leftist dogma and African magic, which they used time and again to put the superstitious Congolese National Army to flight. With shouts of *Mulele mai* (Water of Mulele), they threw themselves into battle, convinced that they were impervious to bullets. The rebellion was subdued in December 1965. Similarly, in a concurrent eastern rebellion led by Gaston Soumialot, in January 1964, he was able to recruit thousands of dedicated supporters in eastern Kivu with backing from the Burundi. On 15 May, the town of Uvira fell to the rebels, and, shortly thereafter, Fizi. From then on, the rebels, who were now known as *Simbas* (lions in Swahili), made increasing use of magic to claim immunity to bullets. Panic-stricken, two heavily armed Congolese National Army battalions were routed by spear-toting *Simbas,* who believed that they had been rendered invincible by anti-bullet concoctions.

17 Luís Artur Carvalho Teixeira de Morais, correspondence with the author, 11 October 2010, Paço d'Arcos.

resistance to Portuguese rule through a refusal to work or pay taxes, among other things. Hence Mariano's proselytizing fell on prepared ground.[18]

There were two wedge issues that the PSA agents had used to generate animosity between the local population in Baixa do Cassange and the Portuguese government. The first was the division of the Maholo people, for as in many parts of Africa, the Maholos were divided by an international frontier, in this case by the border between Angola and the Belgian Congo. When the Congo became independent in 1960, some of the Maholos were "independent," and others were "dependent" of Portugal. This difference in status was promoted as a grievance by the PSA. The second was the longstanding notion that the ancient Congo Kingdom should be reestablished. This Maholo "dream" would involve annexing Baixa do Cassange and parts of the Malange District into this new "independent" entity, essentially a subterfuge for annexation by the new independent Republic of the Congo.[19]

The early months of Congo independence were chaotic, and Portuguese authorities believed that rogue elements of the PSA were pursuing agitation in Baixa do Cassange, as it was unlikely that the fragile government of President Joseph Kasavubu would officially court overt hostility with its neighbors, even if he disagreed with their politics. Agents of the PSA, apparently acting independently, saw an opportunity for anti-colonial gain and the potential to annex a portion of Angola for their newly-independent country by taking advantage of the historical unrest in Baixa do Cassange. The source of the grievances lay in the obligatory cotton cultivation, the scandalous abuse of the labor laws, and their exploitation by the concession companies. Promising these people a tempting and easy life, much as PSA agents had done in the Belgian Congo, they now agitated for revolt and expulsion of the European population and Portuguese authorities to achieve PSA aims.[20] As a consequence, the local people destroyed their plantations and killed their cattle, as they would no longer be needed. After "independence" a person would only have to scratch himself to have all that he wanted. Likewise, PSA agents in the region, known colloquially as *pessis*, encouraged the people to destroy all of their official Portuguese identity documents, to disobey any order given by the Portuguese, and to obey only Kasavubu. It is unlikely that Kasavubu had any knowledge of developments in Baixa do Cassange or that his name was being invoked by the *pessis*.[21]

Overt unrest began in the late months of 1960, and as the January planting time arrived, many of the workers burned their seed, heaped their farm tools along roadsides, and joined a near-religious crusade for their vision of independence. This protest against their Cotonang, Lagos & Irmão, and government oppressors was manifested in the destruction of barges at river crossings on the Cambo, Lui, and Cuango Rivers

18 Pires Nunes, 63–64.
19 Teixeira de Morais correspondence.
20 Ibid.
21 Ibid.

and bridges over the Luando and Lui Rivers; the barricading of various roads; the slaughter of livestock; breaking and entering of shops and Catholic missions; and the harassment of Europeans.[22] The protest originated around the village of Tempo Aluma next to the *posto* or government outpost at Mangando and spread from this remote border area to the heart of the Malange district. As the protest gained momentum, Cotonang and Lagos & Irmão pressured the government in Luanda to send troops with the objective of pacifying the region and opening its major roads to traffic.[23] Because the rainy season that runs from October to March had softened the roads to mud in many areas, this last request was academic. Accordingly, it was decided to send some platoons of the 4th Company of *Caçadores Especiais* (CCE) to augment the 3rd CCE at Malange, and these were to be supported by elements of the Portuguese Air Force (*Força Aérea Portuguesa* or FAP), which would provide air support and reconnaissance to the ground force.

On 11 January 1961, the 3rd CCE under the command of Captain Manuel Augusto Teixeira Teles Grilio sent a patrol from its headquarters at Malange towards Milando, where it arrived at dawn on the 12th. As it moved through Milando and towards the settlement of Ganga-Mexita, some 5 kilometers from Milando, the patrol encountered about 200 natives armed with *catanas* and *canhangulos*.[24] Once the peaceful intentions of the force were made known to the mutineers, three of them spoke with the officer leading the patrol and explained their situation. On the following morning when the patrol returned with a representative from Cotonang, they found the settlement abandoned. On the 22nd, identical rebellions occurred in Tembo Aluma, and then in Cunda-Ria-Baza, Quela, Marimba, Iongo, and Xá-Muteba, as shown on the nearby map. The revolt now included both the Bângala and Maholo peoples, who were heavily influenced by the PSA.[25] Although calm appeared to reign in Milando, a section of riflemen was installed there, and the 3rd CCE began extensive patrols in the border area extending eastward into Lunda as far as Camaxilo to determine more clearly the extent of the rebellion.

The situation in Baixa do Cassange worsened day by day. On 1 February more than a thousand Africans concentrated in the area of Cunda-Ria-Baza, and on the next day the Europeans fled to Malange. The 3rd CCE was unable to prevent expansion of the

22 Marcum, 125.

23 José Manuel Correia, "A Guerra do Algodão" [The Cotton War], *Mais Alto* (January–February 2006), 43. See also Fernando Pinto de Resende (Brigadeiro), Commander, 2nd Air Region, to João Faustino de Albuquerque de Freitas (General), Chief of Staff Portuguese Air Force, 17 February 1961, personal correspondence regarding the "cotton war," personal papers archive, Vol. II, No Prov. 323, Arquivo Histórico de Força Aérea Portuguesa, Alfragide.

24 *Catanas* are bush knives used for clearing dense vegetation, and *canhangulos* are home-made muzzle-loading hunting guns, at times a length of water pipe mounted on a rough stock and loaded with metal scrap.

25 Pires Nunes, 68.

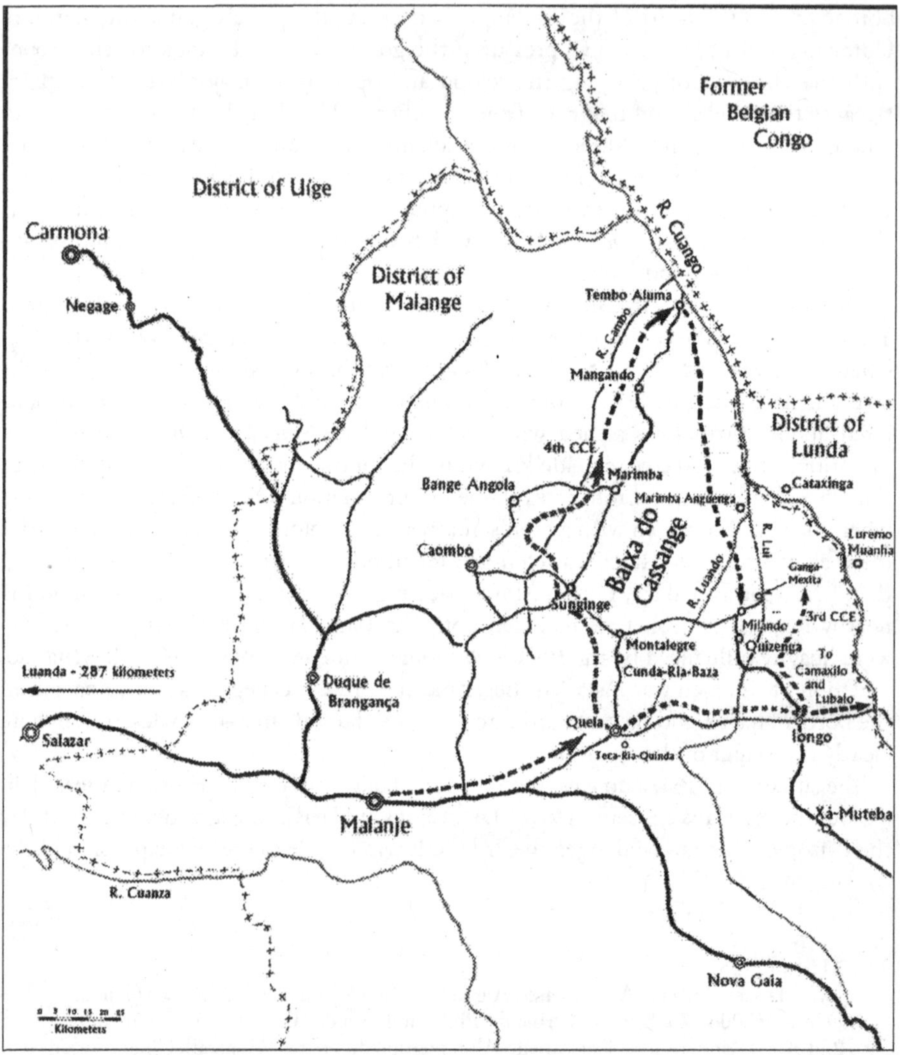

Map 2.1 Advance of 3rd and 4th Companies, Caçadores Especiais, Baixa do Cassange. (Drawn by the author)

revolt or bring peace to the area, and was clearly overwhelmed by the magnitude of the disturbance. Consequently, a larger force was assembled in the form of a Contingent Battalion (*Batalhão Eventual* or BE) with its headquarters in Malange and comprising the 3rd, 4th, and 5th CCEs.[26]

The battalion commander, Major Camilo Augusto Mitanda Rebocho Vaz, arrived in Malange on 3rd February; the 4th CCE commanded by Captain Luís Artur Carvalho Teixeira de Morais, on the 4th; and the 5th CCE, on the 18th. Rebocho Vaz's plan to restore public order was known as Operation *Cassange* and was aimed at suppressing the disturbance in Cunda-Ria-Baza just beyond Quela, disarming the groups there, rounding up the ringleaders, and discrediting them in the eyes of the population. This was the first military operation in several decades for Portuguese forces, so all was quite new to the troops.

At the briefing held on the night of the arrival of the 4th CCE in Malange, the battalion commander identified the air support as being Austers from Malange for reconnaissance and light emergency non-parachute supply drops, and PV-2 Harpoons flying from Luanda for armed support and more robust emergency non-parachute supply drops.[27] The Harpoons, originally open-ocean reconnaissance aircraft, had been adapted as "bombers" with 12.7mm machine guns. The shocking part of the brief occurred when the major informed the assembly that neither type of aircraft would be able to contact the ground forces, as the radios were incompatible. Hence communication between air and ground elements, such as it was to be, would be made through messages written on white cloth banners, by coded signals, or as in World War I, by hand-written paper messages tied to rocks and dropped into the midst of the ground elements. Captain Teixeira de Morais recalls that he could hardly believe his ears, but in fact this was the procedure used when the first Harpoon appeared. Several days afterward the Harpoons were able to correct the problem; however, throughout the operation, the Austers continued to use the World War I system.[28]

As the operation began, and the 3rd CCE passed through Quela and advanced towards the Cuango River on 9 February, it encountered a group of 300 to 400 armed natives who had cut the road and various bridges. It was first necessary to repair the road damage and then the bridges, as the line of communications back to Malange needed to be preserved, and, moreover, no advance could be made without passable

26 Ibid. 69.

27 These were Auster D.5/160 aircraft manufactured by the British firm Auster Aircraft Co., later the Beagle Aircraft Co. Twenty-eight were manufactured in Great Britain, and eleven of these were delivered to Portugal in July 1960. These were shipped to Angola, and four of them participated in the action around Baixa do Cassange. Beginning in 1962, Beagle-Auster shipped 147 Austers in kit form to be assembled at Oficinas Gerais de Material Aeronautico (OGMA) in Alverca, and these were completed and delivered between the years 1962 and 1966. See Mário Canongia Lopes, *Os Aviões da Cruiz de Cristo* [The Aircraft of the Cross of Christ] (Lisbon: Dinalivro, 2001), 333–337.

28 Teixeira de Morais correspondence.

roads. Always the reason for the hostility and destruction was "Maria." When the crowd refused to disperse at the sound of the shots being fired in the air and severely threatened the small element of troops, aim was adjusted in self-defense, resulting in the deaths of 8 mutineers and 42 wounded. The 11th and 12th were spent repairing the damage and then continuing the advance along the route to Cataxinga and Luremo Muanha through Iongo. By 23 February the operation for the 3rd CCE was completed, and the troops remained in the field until mid-March, providing support to the now-vulnerable population. After the insane act of destroying their cattle, the people began to starve following the pacification, and this situation forced the Portuguese government to supply them with many tons of food. This humanitarian operation, which would have been complicated under normal conditions, was made even more difficult by the weather and the destroyed infrastructure.

During the advance on Quela by the 4th CCE, the troops were constantly taunted by the mob with insults and disruptive demonstrations in the path of the vehicles. The situation required rigorous discipline and steady nerves, as the vehicles frequently became bogged in mud, and the immobilized force was vulnerable. This was the risk of conducting operations in the rainy season. As with the 3rd CCE, the natives destroyed the bridges and pontoon crossings over the numerous rivers, thereby isolating the force. Resupply became a serious problem and was resolved through air drops. The 4th CCE continued to encounter large concentrations of hostile natives who understandably had little respect for government authority and less for that of Cotonang and Lagos & Irmão. The acts of violence now extended to the destruction of Cotonang and Lagos & Irmão facilities, commercial enterprises, and government administrative posts.[29]

The troops initially made firing demonstrations to discourage the strikers, but such displays whether from ground troops or from aircraft only served to provoke the crowd, many of whose members still believed they were immune to bullets. On 6 February in Teca-Ria-Quinda, on the road just beyond Quela, the 4th CCE was confronted with a "frontier" protected by a formidable guard armed with *canhangulos*. Portuguese troops would only be allowed to pass with the "authorization of Kasavubu." Immediately following this confrontation, the 4th CCE intercepted a vast crowd of several thousand men armed with *canhangulos*, cutlasses, and other primitive weapons and headed to Quela to expel or kill the European population and Portuguese authorities. Following that, its plans were to visit the mayhem on the town of Malange. Unfortunately, it was necessary to use force in stopping the mob, and casualties were taken on both sides.[30] In the exchange, the company suffered one killed and four wounded, while the protesters suffered 71 dead and 41 wounded.[31] The natives were

29 Pires Nunes, 70–74.
30 Teixeira de Morais correspondence.
31 Pires Nunes, 70–74.

disarmed and convinced to return to their villages, where they were to remain, and this they did.

Also on the 6th for the first time, the FAP intervened with a PV-2 Harpoon in bombing the settlements opposing passage of the CCE on the road north from Quela towards Montalegre. These aircraft came from Air Base No 9 (*Base Aérea No 9* or BA 9) in Luanda. Lieutenant Carlos Alves flew the first Harpoon fire-support mission, and this and subsequent missions along the 64-kilometer Quela-Cunda-Ria-Baza route supported the slow advance of the ground forces. On 9 February Alves, in attacking a hostile concentration in the vicinity of Marimba in a PV-2, released a 100-pound general purpose (GP) bomb at 500 feet of altitude, apparently with a no-delay contact fuse, for he returned to base with the left wing and engine perforated with shrapnel in many places.[32] As Resende noted, the PV-2 crews were not trained for this air support mission, so if they saw "a large concentration of rebels identified as such, they may have the luck of a hit."[33] Indeed, the PV-2 crews, with little or no experience outside of the anti-submarine warfare and open-ocean reconnaissance missions for which they were trained, adapted well to land reconnaissance and low altitude aerial bombardment. There were some teething difficulties. Frequently the bombs buried themselves in mud when dropped and often failed to explode, to the frustration of the pilots.

On the other hand, the view from the ground represents a contrasting perspective. The Harpoon sortie on the 6th represented the first opportunity that the 4th CCE had to talk with the Harpoons by radio, and it occurred as the force was approaching a large armed group with an apparently "peaceful aspect," according to the Harpoon pilot's observation. This aerial appraisal missed the mark widely, and it annoyed Captain Teixeira de Morais that the Harpoons refused to disperse the crowd with a nearby shooting demonstration in light of his request and observation of the hostile crowd. Unfortunately, there were predictable problems with this "peacefully armed" group, and as we have seen, blood was shed that could have been avoided.[34]

Captain Teixeira de Morais reported the incident to his superiors and requested that the Harpoon crews be briefed about events on the ground, as there seemed to be an ocean of ignorance and a considerable misunderstanding on this matter. He specifically requested that air support loiter away from the crowds, as aircraft flying over the various gatherings of natives aggravated the situation with their "pacific" posture. In fact, up to this point the aircraft had made only reconnaissance flights, sometimes flying over large concentrations of people and reporting them as being "peaceful," for they waved to the aircraft and threw bits of earth at them as a sign of friendship when the Harpoons swooped low over the crowds. What the pilots did not know was that the *pessis* had convinced the people that the planes were from Kasavubu and were coming to help them. After this situation became known to the Harpoon crews,

32 Cardoso, 80–81.
33 Resende to Freitas, 17 February 1961, personal correspondence.
34 Teixeira de Morais correspondence.

Lockheed PV-2 Harpoon at Luanda. (Source: Força Aérea Portuguesa)

they responded quite effectively in delivering nearby demonstrations of gunfire, bomb drops, and low-level passes to disperse the groups successfully. It should be noted that never was direct fire on a crowd requested or undertaken.[35]

From Quela the 4th CCE had turned north and rapidly moved through Montalegre, Sunginge, Caombo, Marimba, Mangando, and ultimately arrived at Tambo Aluma on the Congo border on 17 February. On reaching the Cuango River and the Congo frontier, the 4th CCE found the *sanzala* or settlement of Bumba, King of the Mahalos, to have been abandoned in haste, as the cooking fires were still burning, and the cooking pots and pans were still on the fires. The troops could see the last of the villagers on the far side of the river emerging from the water. King Bumba was the only chief in Baixa do Cassange to have managed to escape to the Congo.[36] On completing its missions, the 4th CCE received permission to proceed south along the Lui River to help the population in the area of Milando. Originally this task had been assigned to the 3rd CCE; however, it was unable to cross the Lui River because of the destroyed bridges and ferries and was thus isolated to the east of the river, which was swollen with the rains.[37]

35 Ibid.
36 Ibid.
37 Ibid.

As the forces advanced across the land dominated by the "cottontocracy," they were at a loss to understand the revolts and the profound grief that they witnessed. Troops of the 4th CCE tried to calm the thousands of natives by promising reforms to their chiefs and by putting an end to the extortions of certain venal *chefes do posto*, or outpost chiefs.[38] Equally the troops attempted to impress the protesters with traditional shows of force, such as low-flying aircraft and shots fired overhead, but the return of the authorities and concession company agents immediately behind the force made any promises or displays seem hollow.[39] The view of the rebels was that there had been little to lose in pressing forward, particularly as all believed that any bullets fired at them would turn into water. Further, the promises of the military leaders were thought to have meant little, as all knew that the significant decisions would be made in Luanda or Lisbon.[40]

The Harpoons from Luanda had supported the advance of the ground forces, without the need to fire a shot, all the way to Tembo Aluma in the heart of the insurrection. As the crisis reached its height toward the middle of February, there had been a reduction in the number of PV-2 sorties from Luanda, and this had necessitated Austers filling the gap. Four Auster light reconnaissance aircraft had been operating from Carmona since 13 December and patrolling along the roads and border of the Congo District well to the north and west of Baixa do Cassange. Two of these were reassigned in support of the 3rd CCE and on 25 January began to fly reconnaissance missions from Malange over the troubled zones with army officers occupying the copilot's seat. These small, light aircraft dropped provisions and messages to the advancing ground force and served as its rather tentative backup communications link to the headquarters in Malange. In these missions the Austers flew in front of the columns and advised the troops of potential ambushes. The missions were long and exhausting in that they required nearly four hours of intense concentration.[41] This cycle was repeated more than twice a day, and indeed, the pilots flew until they were completely spent and there was no more daylight. Their only pause was to grab a quick drink and a sandwich during refueling stops between missions.[42] As the confrontations between the troops and the mutineers became more frequent and intense, the Austers assumed a more aggressive role and began using army officers as "bombardiers" to disperse hostile crowds. The officer would fly with a box of grenades between his feet and, when required, drop one into the center of a threatening mob from the open window of the aircraft as it swooped low. On three days in February, the 18th, the 23rd, and the 24th, grenades were dropped, causing losses and panic in the bands of rebels.

38 José Correia, FAP historian, correspondence with the author, 23 February 2010. Many of the *chefes de posto* were mulato (*mestiço*) Cape Verdians (*cabo-verdianos*) who had been imported. Being almost white, they were cruel to the black workers, who hated them.

39 Pélissier, 417.

40 Ibid.

41 Ibid., 43.

42 José Correia, FAP historian, correspondence with the author, 22 February 2010.

Again from the ground perspective, this "bombing" had not been requested and proved counterproductive, as it caused many unwanted casualties. On the morning of the 24th, for example, the "inoffensive" Austers swooped low over the large armed crowd that for many days had surrounded the European and African populations of Milando, which had taken refuge in the *posto*. The grenades that they dropped had had the arming pins removed and were then inserted into a glass jar to retain the safety lever in place. When the glass jar hit the ground and broke apart, the safety lever was released and the grenade exploded. This makeshift delay fusing allowed the relatively slow aircraft to escape the blast, but not the people in the crowd. The wounds suffered by them were quite grave and required extensive medical care, which was administered by army medical personnel. Captain Teixeira de Morais noted that never did the army leave a wounded person without treatment in the field or fail to evacuate the more seriously wounded to the hospital in Malange, and these actions greatly helped progress in quelling the rebellion and in improving government relations with the population.[43]

Although rebel activity ceased shortly afterwards, the FAP continued to conduct reconnaissance throughout the zone until operations were concluded.[44] The Austers at Malange rotated periodically with the other two at Carmona, and the selfless pilots of the four Austers, Lieutenant Colonel Augusto Soares de Moura, Lieutenants Corte Real Negrão and Elio Mesquita, and Sergeant Pedro Carvalhão, flew dozens of missions in marginal conditions without incident until the conclusion of air operations on 2 March.[45]

Indicative of Auster operations is the experience of Negrão in reconnoitering the primary road that runs east some 210 kilometers between Malange and Xá-Muteba. His timely and bold action at Xá-Muteba resulted in the calming of an excited crowd and the consequent saving of many lives. Originally his mission was to deliver the mail by air drop, making an initial low pass over the village to alert the people and then in the second pass making the drop directly into the small village square and the hands of the villagers. On this day no villagers appeared on Negrão's initial pass, and as he looked down, he noticed a large red stain of fresh blood in the square. Glancing astern following his pass, he saw a group of natives fleeing into the grass "like a band of antelopes."[46] In the road there was a prisoner held by two armed men, his arms tied. This raised the Auster crew's curiosity, so on the protective wrapping of the mail package Negrão wrote: "We are going to land."[47]

43 Teixeira de Morais correspondence.
44 Correia, 44.
45 Ibid.; José Correia, FAP historian, correspondence with the author, 23 February 2010; "AeródromoBase 3 – Negage, Súmula Histórica" [Airdrome Base 3 – Negage, Historical Summary], http://aerodino.no.sapo.pt/negage1.html (accessed 24 February 2010).
46 Cardoso, 85.
47 Ibid.

Portuguese Air Force Auster at Negage. (Source: Força Aérea Portuguesa)

There was in Xá-Muteba a long strip of terrain that was described as an "airfield" by the administration, but it was primitive. From the air it appeared clear and smooth and seemed to be suitable for landing and disembarking. The cleared field had been usable before the rains; however, the land was now a veritable sea. The Auster after a brief touchdown came to a stop in a mass of wet vegetation concluding in a bog. It became immediately obvious that there was an insurrection afoot in Xá-Muteba, as the men quickly surrounded the Auster in a rude and rebellious display. Negrão told them that there was a military force advancing on the village that would arrive in the morning, if they could not be persuaded to disband. The crowd still assumed an aggressive attitude, gripped their *catanas,* and awaited a signal. Apparently that day Negrão's words and manner were persuasive, for there would be only a single victim. The crowd eventually dispersed, and calm reigned. The casualty was buried, and only the blood remained. The rains delayed Negrão's departure, but eventually the sun came out, and with the help and "advice" of the villagers, the Auster was dragged to the edge of the field for a successful takeoff.[48]

During Operation *Cassange* the Harpoons flew 28 missions from BA9 in Luanda, and the Austers flew countless more from Malange. In reflecting on the operation, Resende observed how much more effective the missions of the four fragile Austers

48 Ibid.

were in ending the revolt compared to those flown by the worn out Harpoons.[49] While any casualties are unfortunate, in the case of the "cotton war" they were relatively light, two dead and four wounded among the army troops. For the mutineers there were 243 deaths.[50]

The population throughout the entire Baixo do Cassange was now vulnerable and needed food badly, as the partial destruction of its crops, the death of its cattle and

49 Correia, 44.

50 Camacho, 73; and Estado-Maior do Exército, *Resenha Histórico-Militar das Campanhas de África, Vol. VI, Aspectos da Actividade Operacional, Tomo I – Angola, Livro 1* [Historical-Military Report on the African Campaigns, Vol. VI, Aspects of Operational Activity, Tome I – Angola, Book 1] (Lisbon: Estado-Maior do Exército, 1998), 150. There are many overblown casualty figures, principally the 10,000 native dead offered by the UPA after it initially asserted 7,000. As the UPA was not involved, the credibility of its estimates is suspect. Others include 5,000 by José Ervedosa, of whom we will hear more in a moment, and "hundreds, perhaps thousands dead" by Pélissier, who vividly describes mass graves dug by bulldozers, of which there were none operating in Baixa do Cassange in the rainy season. There are also accusations that the FAP used napalm against the population, incinerating villages with women and children trapped in their huts. The origin of this falsehood is the February 1970 article by José Ervedosa in the French journal *Africasia*, page 31. It alleges that 17 villages were destroyed and that 5,000 men, women, and children were burned alive. This article also asserts, for instance, that on 8 February 1961 Captain Teixeira de Morais requested a Harpoon flown by Ervedosa bomb the *sanzala* Cunda-Ria-Base. On that day it rained torrentially, and the Harpoons did not fly, so Ervedosa's account is suspect in many ways. This inaccurate story was picked up René Pélissier in his book *La Colonie du Minotaure: Nationalismes et Revoltes en Angola (1926–1961)* (Orgeval:Pélissier, 1978), page 418, and repeated by Pélissier and Douglas L. Wheeler in their collaborative book *Angola* (London: Pall Mall Press, 1971), page 174. More recently there was an exposé of the "atrocity" in an article by the journalist Francisco Camacho, "Baixa do Cassange: O Massacre que Veio do Ceu" [Baixa do Cassange: The Massacre that Came from the Sky], *Grande Reportagem* 137 (August 2002): pages 63–77. The charge is without merit and a complete fabrication. Napalm was used in Angola; however, it was not present in the theater until after the uprising in Baixa do Cassange and came in late April, 1961, after the 15 March attacks. Further, a review of the log books of the small number of pilots involved in the "cotton war" indicates that no napalm ordnance loads were carried on any missions in Baxia de Cassange in February 1961. There were no NAP (napalm) entries but plenty of MET (*metralhamento* or machinegun) and BOP (*bombardeamento em picada* or low-angle attack bombing). Lastly, the source of the fabrication, José Ervedosa, was a former naval aviator who had come to the FAP at its inception; his behavior was inappropriate and questionable, particularly for an officer, and he was consequently subjected to regular disciplinary measures. His difficulties are ably recorded by his squadron commanding officer, António Silva Cardoso, in his book *Angola: Anatomia de uma tragédia* [Angola: Anatomy of a Tragedy] (Lisbon: Oficina do Livro, 2000), pages 38, 47, 191, 206, 208, and 209. Ervedosa exercised poor judgment, was involved in a number of dodgy dealings, and ultimately was relieved of his duties and dismissed from the squadron by Silva Cardoso on the occasion when he returned from Lisbon to find Ervedosa playing cards with the mechanics and his duties unperformed. Ervedosa subsequently deserted to Algeria and generated the fabrication from his new anti-colonial home.

dogs, and the seizure of its hunting arms had created a crisis. The misery was enormous, and the army would work diligently to feed them and alleviate their distress. While Operation *Cassange* would conclude on a humanitarian note, the situation would also have a decisive psychological effect on the people who had been reduced to this situation by the agitators, the concession company agents, and the government authorities. In fact, as the people became aware of the hollow promises of the *pessis* and understood that the army was attentively listening to their complaints and attempting to resolve them, they began to ask that the Maholos, the "cause of all of their suffering," be punished, and that the troops remain in the region to protect them from further attacks.[51] This was done.

The corrective legislation that lifted the native obligation to cultivate cotton and dealt a blow to the concession companies came four months too late, but the injustice was corrected.[52] Hence Resende's letters to Lisbon may have finally affected thinking at the highest levels of government. Certainly the experience portended events to come, both in further colonial unrest and in a radical, humanistic, and positive change in Portuguese native policy. Indicative of the spontaneous and localized nature of the uprising, no African nationalist movement claimed credit or publicized it at the time.

In this case, airpower was simply one of the resources available to Portugal in tackling the uprising. In such a situation, all the elements of national power, the entire political, economic and military mix, was used to achieve success. Although the Portuguese Air Force made a significant contribution in Baixa do Cassange, it was but a single component in the artful use of national power. Deslandes and Resende ultimately brought together the other, non-military elements in combination with the military to achieve a more lasting solution.

Santa Maria Incident

In January 1961, the Portuguese ship *Santa Maria* was seized by a group of twenty-four men belonging to DRIL (*Directório Revolucionário Ibérico de Libertação*, or the Iberian Revolutionary Directorate of Freedom) and led by the familiar activist Galvão. Galvão had been an exile in Venezuela after escaping from prison in 1959 and traveling through Argentina and Brazil. The *Santa Maria* was a liner carrying about 612 passengers with a crew of 350 and operating in the Caribbean. Galvão, his group, and their intentions had come to the renewed attention of Portuguese authorities as early as October 1960. Although their plans for a demonstration did not materialize then, there was significant support for their activities from the Portuguese expatriate opposition community and particularly from Delgado now residing in Brazil. On 20 January some of the DRIL operatives boarded the *Santa Maria* in the Venezuelan port of La Guaria, and a day later Galvão and the remainder clandestinely entered the

51 Teixeira de Morais correspondence.
52 Decree Law No 43 639 of 2 May 1961.

ship in Curaçao. Once at sea, the ship was easily seized on the evening of 22 January, resulting in one death and two wounded crewmembers in what was styled by DRIL as Operation *Dulcineia*. Galvão declared the ship to be the "first free piece of Portuguese territory" and rechristened it the "*Santa Liberdade*."[53]

The rather fanciful goal of the seizure announced by Galvão to Lieutenant Daniel Krauss, the commander of the U.S. Navy P2V-7 Neptune open ocean reconnaissance aircraft that located the ship on 25 January, was to take the vessel to the West Coast of Africa and there to conduct a surprise attack on the island of Fernando Pó in Spanish Guinea, just north of the Portuguese possessions of São Tomé and Príncipe.[54] Once taken, this island would then serve as a base for an attack on Luanda supported by undefined local rebel forces. Next Galvão would conduct a campaign to liberate part of Portuguese territory, install an opposition government in it, and from there mount a war against the Salazar regime. After Angola, he would free Mozambique and then the *metrópole*, presumably establishing his own regime.[55] The contradiction is immediately apparent in seizing a major vessel on the high seas in a noisy and widely publicized act of piracy and then an ocean away attempting to conduct a surprise attack in which a clandestine approach was vital to any success. This ridiculous plan was actually a ruse to gain time, for Galvão intended to take the hijacked ship to Brazil, where on 31 January the newly-elected president, Jânio Quadros, would take office. Quadros was a personal friend of Galvão and Delgado and would support their political adventure, while the current president, Juscelino Kubitsheck, would not.

While most of the search forces would anticipate intercepting the *Santa Maria* in the vicinity of the Cape Verde Archipelago on its way to Angola, in fact Galvão was attempting to time his arrival in Brazil to occur on or after 31 January, when new leadership would be installed. At that point Galvão would have realized his ultimate goal of showing the world in a highly public way that there was an active opposition to the Salazar regime. He did indeed accomplish this end.

As information on the *Santa Maria* was received in Lisbon, the Portuguese government created an emergency cabinet and decided on 23 January to enlist the help of the U.S. and British governments and accordingly approached both with initially favorable reactions. The United States saw the incident as one of piracy and launched a flotilla of ships and aircraft from various points around the Atlantic. This force was to locate and track the errant vessel with the intent of boarding it and returning it to the nearest North American port with the use of force, if necessary. The objective was to protect the passengers and ship and to return the vessel to its rightful owners. The

53 Pedro Manuel Ferreira, "*Santa Maria*, A Força Aérea na Caça ao Paquete 'Rebelde'" [*Santa Maria*, The Air Force in the Hunt for the "Rebel" Steamship], *Mais Alto* (November–December 2004): 12–22.

54 Ibid.

55 Luís Nuno Rodrigues, *Salazar–Kennedy: A Crise de uma Aliança* [Salazar–Kennedy: The Crisis of an Alliance] (Lisbon: Editorial Notícias, 2002), 35–36.

British and Spanish navies likewise responded to aid in fixing the *Santa Maria*, but the French and Dutch declined.

On 27 January, Portugal deployed a four-plane detachment of P2V-5 Neptunes from Squadron 61 to the airfield at Sal in the Cape Verde Archipelago in support of what had by then been named Operation *Galvanex*. The Neptunes had been recently acquired from the Royal Dutch Navy and carried the powerful 1.5 megawatt APS-20 surface search radar.[56] While their newly trained crews held various stages of proficiency in the aircraft, it had greater open ocean search capability that the aging PV-2 Harpoon of World War II vintage, the only other immediately available alternative.[57] The first mission was flown on 27 January. Two days later on 29 January, U.S. Navy Patrol Squadron Ten (VP-10) with eleven of its P2V-5s arrived in Sal accompanied by four C-54 Skymaster cargo aircraft for logistical support. The squadron commander put his force at the disposal of the Portuguese authorities.[58]

The new Kennedy administration, having come to office on 17 January, developed second thoughts about its initial reaction. On 24 January, Galvão had made a proclamation carried on a number of national news broadcasts that DRIL was pro-democracy and anti-totalitarianism and opposed to all forms of arbitrary government. Simultaneously in Brazil, Delgado chimed in to assert his support for DRIL and to say that he had personally ordered Galvão to seize the ship.[59] This position appealed to the liberal elements of Kennedy's political party and caused his administration to do some homework on Delgado and Galvão. After three days the Kennedy administration was not certain that a case of piracy could be made. Meanwhile, the U.S. Navy was pursuing the *Santa Maria* with the authorization to intercept and board it to protect U.S. citizens and to look after the humanitarian interests of the other passengers and crewmembers. In response to the political change of heart, the U.S. Navy suddenly received orders simply to follow the ship and report its movements. The Royal Navy too relaxed its stance.

Kennedy now faced an element in his party that accused him of making a pact with the "dictator" Salazar. Anxious to please his constituency and to resolve the problem of "political piracy," Kennedy gave the navy new instructions to negotiate the return of the ship and release of the passengers and crew.[60] From 25 January substantial communications developed between the ship and the naval reconnaissance aircraft in which Galvão was persuaded to free the U.S. passengers before continuing his supposed

56 António Silva Cardoso, *Angola: Anatomia de uma Tragédia* [Angola: Anatomy of a Tragedy] (Lisbon: Oficina de Livro, 2000), 75–80.

57 Ibid., 63–80.

58 Ferreira, 18

59 Rodrigues, 39.

60 António José Malheiro García, "A Operação *Santa Maria*" [Operation *Santa Maria*], in the unpublished collection *Participação da Armada na Defesa das Províncias Ultramarinas* [Participation of the Navy in the Defense of the Overseas Provinces], TMs [photocopy], 1972, 1.

voyage to Africa. Eventually an agreement was reached in which a U.S. emissary would board the ship at sea and negotiate a plan to discharge the passengers at the closest Brazilian port. Following these negotiations on 27 January, Galvão agreed to take the *Santa Maria* to Recife to disembark the passengers. After feverish diplomatic work to gain assurances over current treaty obligations, the *Santa Maria* entered Recife. Even so, there remained some confusion about the conditions of departure, and Galvão continued to detain the hostages on board. This standoff was brief and ended when Admiral Dias Fernandes, the naval district commander at Recife, and 60 of his marines acted to resolve it. Galvão and his DRIL compatriots spent another day alone on the ship pondering their situation before asking for political asylum in Brazil. Again with great fanfare, they disembarked into a crowd of journalists, reporters, and fellow oppositionists led by Delgado. Several days later the *Santa Maria* was returned to the Portuguese government.

This episode served to highlight the domestic political struggle against the Salazar regime. Galvão had succeeded in his main objective of bringing his crusade to international prominence and at the same time creating a political divide within the highest levels of the Portuguese government over his affront to the regime and insult to its leader. As we shall see, this divide within the Salazar government and the evolving position of the Kennedy administration on Portuguese colonial policy will both resurface.

Angola Uprisings of 1961

The Angola uprisings and their immediate events served as a warning of things to come and prompted Portugal to think more clearly about defending its colonies. Sporadic agitation and unrest had occurred throughout 1960 at the time when Portugal began to realize the rich economic potential of Angola and Mozambique. Yet European troop strength in Angola numbered under 1,000 in early 1958, and was reinforced only to about 3,000 by mid-1960.[61] Overall strength was 8,000, of which at least 5,000 were African troops.[62] These forces, while scattered throughout Angola, were confined to garrisons in the larger towns and accustomed merely to administering subjective rule.[63] This modest order of battle was neither properly organized, nor trained and led, nor adequately manned to face the uprisings in early 1961.

61 Douglas L. Wheeler, "The Portuguese Army in Angola," *Modern African Studies*, 7, no. 3 (October 1969): 430.

62 Douglas L. Wheeler, "African Elements in Portugal's Armies in Africa (1961–1974)," *Armed Forces and Society*, 2, no. 2 (February 1976): 237.

63 Estado-Maior do Exército, *Resenha Histórico-Militar das Campanhas de África, Vol. II, Dispositivo das Nossas Forças Angola* [Historical-Military Report on the African Campaigns, Vol. II, Disposition of Our Angolan Forces] (Lisbon: Estado-Maior do Exército, 1989), 63–65.

The opening challenge was instigated by the MPLA in Luanda on 4 February 1961 by a truly aggrieved group armed only with clubs and knives and driven by a frustration with their treatment. It entered the capital and attacked a number of police installations, a prison, and the radio station in an attempt to have 52 political prisoners released. Seven policemen and 40 of the group were killed. During the funeral for the policemen, shots were fired on the mourners, and whites attacked blacks in a display that incensed Portugal. The prison was suicidally attacked again on 10 February. Government forces overreacted in quelling the disturbance because of the strong emotional feelings that had accumulated in Luanda. Several hundred Africans were killed indiscriminately, their bodies being left to rot in the streets as a warning sign to aspiring revolutionaries. The events gained international attention and put Portugal firmly on the UN agenda, as the news reporters during the *Santa Maria* incident had raced to Luanda to be in position to cover its arrival there with Galvão. Instead, they ably covered the February disturbance. Race relations remained

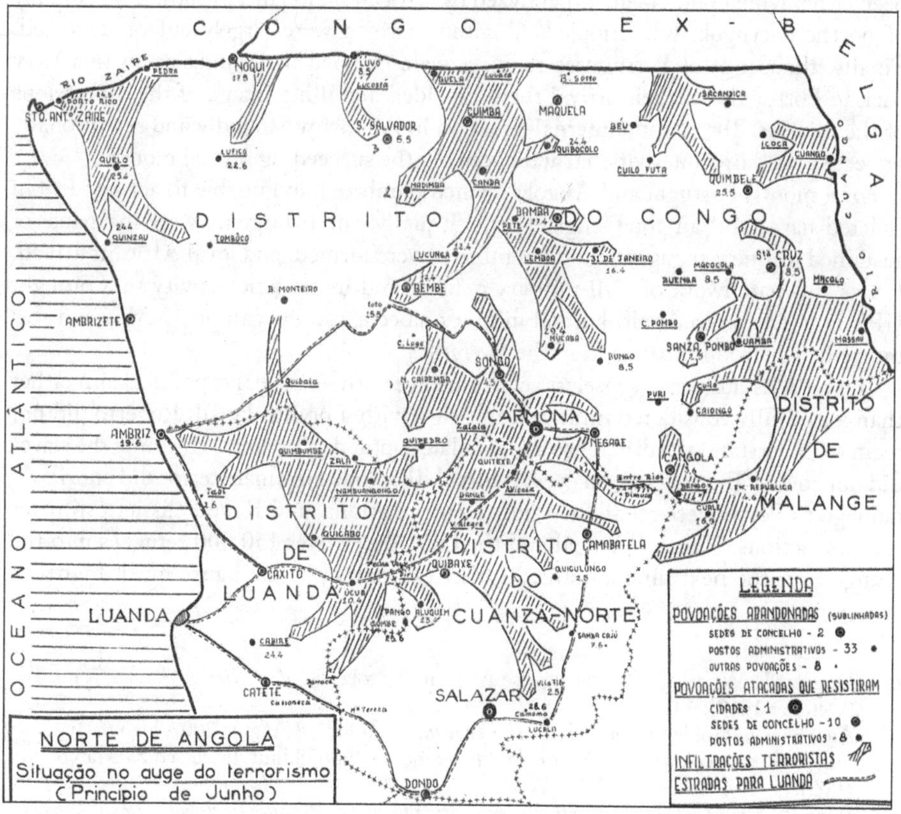

Map 2.2 Attacks in the North of Angola at their Height, Beginning of June 1961. (Source: Hélio Felgas, *Guerra em Angola* [War in Angola] (Lisbon: Livraria Clássica Editoria, 1961), 24.)

polarized in Luanda and would take years to rebuild. The MPLA instigated the mob action with little planning or consideration of the consequences. Picking the capital with its troop and national police concentration was an unfortunate choice, and the MPLA was quickly destroyed inside Angola.

These events were a prelude as well as a warning of further trouble. On 15 March 1961 shortly after the MPLA episode, the UPA seized on the confusion and launched a multipronged attack in northern Angola with a flood of 4,000 to 5,000 armed men across a 300-kilometer strip. This mob laid waste to whatever was in its path. All of Portugal was shocked at the horror, and Roberto's belief that Portugal would capitulate at the first sign of violence, as Belgium had done, proved false.

This savage foray occurred in an area demarcated by the Congolese frontier, the Cuango River, the Malange-Luanda railroad, and the Atlantic Ocean, as depicted in the nearby map. The attackers pushed nearly to Luanda. Military leaders faced a situation in which over 100 administrative posts and towns in the three districts of northern Angola from the Congo border to within 30 miles of Luanda, the capital, had been either wiped out, taken, or paralyzed by African nationalist groups. The economy of northern Angola was crippled. Communications were largely cut or damaged. Finally, thousands of Portuguese refugees were camped in Luanda or on their way back to Portugal and their arrival there provided a chilling image of the dimensions of the disaster. The chaotic internal situation in Angola was rapidly and sensationally projected to a large international audience over the succeeding several months.[64]

For a month Portugal and Angola seemed paralyzed and unable to act aside from isolated naval and air force engagement. Equally, the insurgents were incapable of sustained military engagement. Civil militias were formed, and loyal Africans armed. It was this patchwork of civil-military defense and its frenetic activity that brought UPA momentum to a halt. Formal military reoccupation began on 13 May and was intensified as troops arrived from the *metrópole*.

The UPA assault was a senseless act of violence with only a naïve political aim rather than a carefully considered military campaign with a political goal. Roberto did not seem to understand the difference and was later quoted as saying, "This time the slaves did not cower. They massacred everything."[65] This indiscriminate terror did enormous damage to Portuguese credibility and to race relations, and René Pélissier estimates that the actions of both Portugal and the UPA drove over 150,000 refugees into the Congo over the next nine months.[66] Other estimates put the figure much higher at

64 Douglas L. Wheeler, "The Portuguese Army in Angola," *Modern African Studies*, 7, no. 3 (October 1969): 431.

65 Alex Vines, "Holden Roberto," *The Independent* (London), 8 August 2007, http://www.independent.co.uk/news/obituaries/holden-roberto-460658.hmtl (accessed 23 March 2009).

66 René Pélissier, *La Colonie du Minotaure, Nationalismes et Révoltes en Angola (1926–1961)* [The Colony of the Minotaur, Nationalist Movements and Revolts in Angola (1926–1961)] (Orgeval: Editions Pélissier, 1978), 658.

250,000, and Alex Vines even estimates the figure to be as high as 400,000.[67] On 7 October, General Deslandes, the governor-general, announced that reoccupation was complete and that mop-up policing would begin. This series of events deeply shocked all of Portugal and hardened its colonial commitment and the restoration of order in Angola. To the Portuguese it was unthinkable that such lawlessness should be tolerated, and a strong, uncompromising reaction to the nationalist behavior was widely supported. These events with their horror also diverted domestic attention from Salazar's increasing political vulnerability.

The "Abrilada" of 1961

The fissure that had been opened within the Salazar government by the *Santa Maria* incident was plainly evident when on 17 February the U.S. Ambassador to Portugal, Charles Burke Elbrick, lunched with General Júlio Carlos Alves Dias Botelho Moniz, the Portuguese Minister of National Defense. The ambassador recorded that the general was very frank in his criticism of the Salazar regime and the impracticable and undesirable maintenance of the *status quo* in Portuguese Africa. He indicated that action must be taken to accelerate social reforms both in the *ultramar* and the *metrópole* and to "pry loose the iron grip of the few controlling families" over Portuguese affairs.[68] Moniz here claimed that he had discussed this position with the senior members of his staff, all of whom were in agreement, apparently including General João Faustino de Albuquerque de Freitas, Chief of Staff of the Air Force, and that at least 90 percent of the Portuguese armed forces supported this view. With "startling candor" he described Salazar's increasing inability to exercise true leadership and attributed this shortcoming to his advancing years.[69] Elbrick's final observation on the meeting was that Moniz was a man to watch.

Salazar's vulnerability had been revealed in the Delgado challenge of 1958 and its strong message that it was time for Portugal to move toward democracy. The colonies were seen by many as a liability in that their ownership under the current arrangement represented a major obstacle to Portuguese membership in the EEC and in trading with Third World nations, a direction that held great promise.[70] The strength of this view was manifested in the attempted coup of 13 April 1961, which was probably the

67 Ibid.

68 Charles Burke Elbrick, U.S. Ambassador to Portugal, Lisbon, Telegram 508 to U.S. Secretary of State, Washington, 18 February 1961.

69 Ibid.

70 W. Gervase Clarence-Smith, *The Third Portuguese Empire 1825–1975: A Study in Economic Imperialism* (Manchester: Manchester University Press, 1985), 193. The author argues that by 1968 the Portuguese economy was turning decisively from the colonies toward Europe. Portugal in the course of the 1960s and early 1970s became a booming and aggressive "newly industrialized country" and was described as a "Taiwan of southern Europe."

closest that Salazar came to being removed. The 1960 débâcle in the Belgian Congo prompted heated debate in the Portuguese Supreme Council for National Defense (*Conselho Superior de Defesa Nacional*) about the security of its overseas territories, as there were clear limits to the ability of Portugal to defend the *ultramar*. Colonel Kaúlza de Arriaga was notably vociferous in saying that the *ultramar* forces should be augmented, particularly in Angola.[71] He was opposed by Moniz and by fellow Colonels Afonso Magalhães de Almeida Fernandes and Francisco da Costa Gomes, who advocated a dose of the "winds of change" for the colonial situation.[72] Moniz consequently undertook to write a long, formal, and well-meaning letter to Salazar describing the concerns that he and many of his politico-military colleagues on the Council shared about a military solution to the colonial problem. They felt that success was very uncertain and that proceeding down such a road would lead Portugal to an untenable position in which it would find itself under attack with its forces dispersed on four continents.[73] He described the military course of action as suicidal because of a lack of resources and the ability neither to win on the battlefield nor achieve a favorable political solution to colonial ownership.[74] The pursuit of such a course would generate considerable ill-will, particularly in the United Kingdom, the United States, and now Brazil, where both Delgado and Galvão had been given political asylum. From here these two continued to produce a stream of invective that drew international attention to the corruption of the Salazar regime and its exploitation of the indigenous African population that was subject to enormous abuse and forced to work for "wages of misery."[75] A change of direction, it was argued, was needed to correct the moral injustices in the *ultramar* and at home and to regain international prestige, something the current line of diplomacy seemed incapable of doing. The political shackles needed to be removed so that Portugal could expand socially and economically with the Western world.

The uprisings themselves acted to bring the differing opinions into focus. After considerable debate, virtually the entire defense staff decided that a motion of no confidence in Salazar should be made at the approaching meeting of the Council on 8

71 António de Spínola, *País sem Rumo* [Country without Direction] (Lisbon: Editorial SCIRE, 1978), 83.

72 Douglas Porch, *The Portuguese Armed Forces and the Revolution* (Stanford: The Hoover Institution Press, 1977), 38.

73 José Krus Abecasis, *Bordo de Ataque, Memórias de uma Caderneta de Voo e um Contributo para a História* [Attack Heading, Memories from an Aviation Log Book and a Contribution to History] (Coimbra: Coimbra Editora, 1985), 245–248.

74 Sérgio Augusto Margarido Lima Bacelar, *A Guerra em África 1961–1974: Estratégias Adoptadas pelas Forças Armadas* [The War in Africa 1961–1974: Strategies Adopted by the Armed Forces] (Porto: Liga dos Amigos do Museu Militar do Porto and Universidade Portucalenese Infante D. Henrique, 2000), 39.

75 João José Brandão Ferreira, *A Evolução do Conceito Estratégico Ultramarino Português* [Evolution of the Portuguese *Ultramar* Strategic Concept] (Lisbon: Hugin Editores, 2002), 377.

April. Salazar, tipped to the plans by Arriaga, did not attend the meeting. As an alternative the plotters had asked President Tomás to dismiss Salazar. Tomás indicated that he would not dismiss the "greatest statesman of the century after Churchill." By then Salazar had identified the opposition, and on 13 April the relevant participants were reassigned or retired. The inevitable reshuffle ensued in which Salazar assumed the defense portfolio, and no change in policy was brooked. Indeed, three days after the reshuffle, Freitas writes to his friend Resende that refusing to countenance a change in policy and failing to consider the sound advice of his advisors is leading the country down a dangerous road. This is a very grave error.[76] Although Arriaga, as the Subsecretary of State for Air, displayed unrestrained public praise for Freitas, his alignment with Moniz and the clandestine activities of his friend placed Freitas in an unfortunate position. This conspiracy had grown, and its net extended to Angola, where the military was appalled at the "miserable exploitation of the blacks by the colonial capitalists," and for these capitalists the armed forces would be forced to shed its blood.[77]

Salazar had managed over the years of his rule to control and manipulate the military through a strategy of co-optation and "divide and rule."[78] Military pay was poor, and to advance in pay and promotion, ambitious officers were removed from the immediate military environment by posting them to lucrative and prestigious special positions. Normally these were at high levels of government in both the *metrópole* and the *ultramar*, and it was these postings, promotions, and pay that Salazar oversaw. Allegiances were ambivalent. This undermining of traditional military bonds enabled Salazar to create mistrust and fear within the armed forces to his advantage. Until the 1974 revolution when the coup participants had developed broad and determined backing, the Salazar regime would be safe from a military coup, and despite its enormous reservations and latent moral indignation, the armed forces would be forced to honor his commitment to the colonies and to fight his war in Africa.[79] With this event it became evident that the purpose of colonial policy was now in large part to preserve the Salazar regime. Advocates of any alternative colonial options had been neutralized, and commitment to war reinforced through destruction of the opposition. These events accented the increasing international isolation of Portugal.

76 João Faustino de Albuquerque de Freitas (General), Chief of Staff, Portuguese Air Force, to Fernando Pinto de Resende (Brigadeiro), Commander, 2nd Air Region, 16 April 1961, personal correspondence regarding the "Abrilada," personal papers archive, Vol. 2, No Prov. 323, Arquivo Histórico de Força Aérea Portuguesa, Alfragide.

77 Krus Abecasis, 246.

78 Douglas L. Wheeler, "The Military and the Portuguese Dictatorship, 1926–1974: 'The Honor of the Army,'" in *Contemporary Portugal*, ed. Lawrence S. Graham and Harry M. Makler (Austin: University of Texas Press, 1979), 199.

79 Porch, 26.

UN Reversal

Following the uprisings and the *Santa Maria* incident, the UN Security Council convened in May and June to discuss among other agenda items the events of Angola. On 19 May the UN Commission on Information on Autonomous Territories put forth a resolution that denounced Portugal for not cooperating and attempted to force Portuguese cooperation in reporting on its colonies in the future. The United States supported this action. In the following month the Security Council approved a new resolution condemning the massacres and the excessive force used against the Angolan population, saying that it "compromised the maintenance peace and international security."[80] The United States under the new Kennedy administration reversed support in this forum and sided with the Soviet Union in condemning Portuguese African policy.[81] After the change of heart in the *Santa Maria* incident, the Kennedy administration in the spring of 1961 was torn between the "Africanists" and the "Europeanists" in crafting its policy towards Portugal. The more liberal Africanists prevailed in advocating that the UN in the form of a conciliation commission initiate immediate negotiations between Portugal and the nationalist movements with the view of obtaining a cease-fire and instituting reforms leading to auto-determination for the colonies. This event was an enormous blow to Salazar, who criticized the United States for voting with the Soviet Union in the face of historical opposition to Soviet diplomatic and military activity. This also called into question the NATO relationship. The U.S. position was tempered later in the summer when it was necessary to renegotiate its lease on the strategically important base in the Azores. Nevertheless, the damage between two old friends had been done. Portugal found itself a semi-pariah state, politically isolated for the moment along with its colonial neighbors South Africa and Southern Rhodesia, and was forced to fight the ensuing war hobbled by this isolation. It felt beleaguered and betrayed by its long-term ally, the United States, and viewed its position as neither understood nor appreciated. As Resende again writes, "We must, in this fight here, play without the help of the Queen [of the chess board] of the West [the United States]. I do not know that they are not against us. As with the *Santa Maria*, it seemed that the Americans only supported freeing the passengers."[82] This ostracism served to harden the Portuguese commitment to its colonies, particularly as Portugal believed that it was acting properly and that it had nothing of which to be ashamed. It proceeded accordingly to defend its sovereign territory and interests. It also opened the way for India to take the Portuguese sovereign possessions on the Malabar Coast of the subcontinent.

80 Rodrigues, 92.
81 Ibid., 92–95.
82 Fernando Pinto de Resende (Brigadeiro), Commander, 2nd Air Region, to João Faustino de Albuquerque de Freitas (General), Chief of Staff Portuguese Air Force, 2 February 1961, personal correspondence regarding the reinforcing of Angola, personal papers archive, Vol. II, No Prov. 323, Arquivo Histórico de Força Aérea Portuguesa, Alfragide.

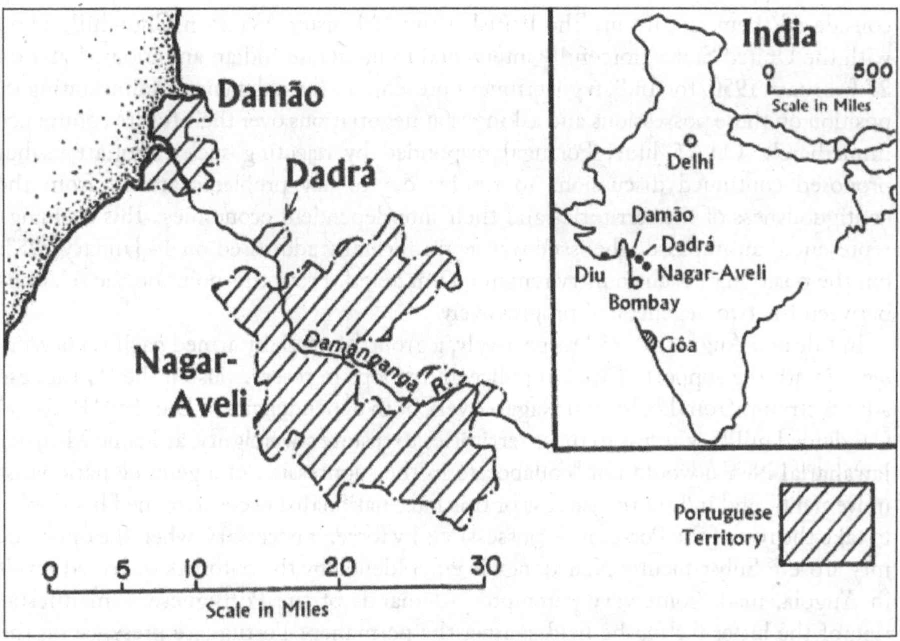

Portuguese India (Modified by the author from "A Free Access to Colonial Enclaves" by Pradyumna P. Karan, *Annals of the Association of American Geographers* 50, No 2 (June 1960), 189.

The Collapse of Goa

This beleaguered colonial position was reinforced by the actual loss of the Indian colonies in what was an act of war that Portugal was powerless to prevent. The events surrounding Goa in 1961 again hardened the Portuguese position by providing Salazar with a nationally distressing event to coalesce the population for the coming African conflict. While Portuguese Indian territory was relatively small at 4,194 square kilometers, its seizure by the Indian government was a blatant act of aggression against another state.

Portugal had three trading enclaves there: Goa, Damão, and Diu.[83] The fledgling Indian government had threatened to take these properties in the late 1940s, as it

83 The actual Portuguese possessions within the peninsula are Goa, Damão and Diu. Goa is a tract of picturesque and fertile country on the West Coast about 250 miles south of Bombay, measuring 63 miles in length by 40 miles in breadth. It comprises a nucleus of "old conquests," Goa, Bardez, and Salcete, an outer belt of "new conquests," and the Island of Angediva. The population borders on half a million, the majority of whom are native Catholics whose ancestors were converted centuries ago. Damão, 100 miles north of Bombay, a fortified Portuguese town with a small outlying district in the interior, has

considered them an affront. The British Prime Minister, Winston Churchill, along with the United States, forcefully intervened to moderate Indian ambitions. Later on 27 February 1950, the Indian government presented a formal memorandum stating its position on these possessions and asking that negotiations over their future commence immediately. On 15 June, Portugal responded by rejecting such negotiations but proposed continued discussions to resolve day-to-day problems arising from the contiguousness of the territories and their interdependent economies. This exchange represented an impasse. The issue was again formally addressed on 14 January 1953, but the position of each country remained unaltered. From this point on the relations between the two degenerated progressively.

In July and August of 1954 respectively, a group of partially armed pacifists or *satyagrahi* with the support of Indian police and troops in reserve ousted the Portuguese administrators from Dadrá and Nagar-Aveli, both dependencies of Damão.[84] Portugal was denied military access to these territories to restore sovereignty, as Prime Minister Jawaharlal Nehru would not "collaborate in the suppression of a genuine nationalist movement," and indeed the success of this faux-nationalist event hardened his resolve to take the rest of the Portuguese possessions by force, if necessary, when the opportunity arose.[85] Subsequently Nehru, newly emboldened by the outbreak of armed revolt in Angola, made some very peremptory demands of the Portuguese, a manifestation of the bitter feeling he held against the permanent Portuguese presence on the Malabar Coast.[86] When the Portuguese position was again made clear, India massed 30,000 troops, supported by tanks, aircraft, and warships, on the borders of the three territories. Governor-General Manuel António Vassalo e Silva could defend his territory with only about 3,000 ill-equipped troops without air cover or air defense, 900 Goan police, and an old frigate. On 11 December, Salazar attempted to invoke the Anglo-Portuguese Treaty without success. Great Britain had since 1954 refused to help, saying that the 600-year-old alliance had clear limitations, particularly since a

an area of 82 square miles with a total population of over 50,000. Diu is a small fortified island at the southern point of the Kathiawar coast, measuring about 7 miles by 2, with a population of something over 12,000.

84 In 1783 Nagar-Aveli was ceded to the Portuguese in compensation for the sinking of a Portuguese vessel by the Maratha navy. In 1785 the Portuguese purchased Dadrá. Both were administered by the governor of Damão until the 1954 ousting of the Portuguese administrators. These two territories functioned independently as the "Free Dadra and Nagar Haveli Administration" until they were merged with the Indian Union in August 1961.

85 Pedro Manuel Ferreira, "Ecos do Ataque ao Império: Invasão da Índia Portuguesa, O Confronto Diplomático" [Reflections on the Attack on the Empire: Invasion of Portuguese India, the Diplomatic Confrontation], *Mais Alto* 340 (November–December 2002): 37.

86 Richard Robinson, *Contemporary Portugal* (London: George Allen & Unwin, 1979), 103; and Krus Abecasis, 259. Nehru announced that India was "not prepared to tolerate the presence of the Portuguese in Goa, even if the Goans want them to be there."

member of the Commonwealth was involved. Britain no longer needed Portuguese ports for the transit to its colonies, and Portugal no longer needed the protection of the dwindling British navy. New alliance structures, such as NATO, had clearly replaced the Anglo-Portuguese Treaty in both intent and practice.

Salazar gave instructions that resistance must last at least eight days to mobilize international support. Should that not be forthcoming, then "total sacrifice" must be made to save Portuguese honor. Indian forces invaded in the morning of 17 December in a "blitzkrieg" that frustrated Salazar's intentions, and Vassalo e Silva capitulated in the pointless struggle on 19 December, far short of eight days and total sacrifice.[87] Vassalo e Silva and the other officers involved were dismissed from the army in 1963 in what appeared to be an effort to shift blame for the loss of Portuguese India to the military and to set an example for officers in similar future situations, such as the approaching colonial wars. The military at large was resentful at the unjust punishment and at making soldiers the scapegoats for civilian mistakes. The armed forces carried into the African campaigns this "ominous message that the government was prepared to manipulate and sacrifice them in hopeless missions and to court-martial virtually all survivors."[88] General António de Spínola, who was to be the Commander-in-Chief of Portuguese forces in Guiné (1968–1972), the Deputy Chief of the General Staff (1972–1974), and the first President of the new government following the revolution in 1974, described in his controversial book *Portugal and the Future* the anxiety permeating the entire Portuguese military on the eve of the African wars:

> India is a clear example of what we fear. Never was its inevitable loss believed possible. And yet when the tragedy happened, the Nation's attention was immediately focused on the narrow aspect of military conduct. The Armed Forces were accused of not having defended India heroically; when, in reality, no matter how effective its defense, India would have fallen in only a matter of days.[89]

The punishment for the leaders of the Indian garrison carried the broad message that there would be no turning back from the government decision to defend the colonies. Salazar's commitment to preserving his regime was unswerving in its purpose, and the events in Goa served to push Portugal further down the road to war.

87 Actually, not all Portuguese felt that the loss of Goa was a catastrophe. Father John Paul, an Anglican missionary in Mozambique, writes, "One of my more enlightened friends in Vila Cabral said, however, that he thought it quite a good thing Portugal had been forced to abandon Goa, since the millions of escudos that had been poured into the place to keep it going for the sake of Portugal's prestige could now be diverted to the more worthy cause of developing Portugal and her overseas territories in Africa." See John Paul, *Mozambique, Memoirs of a Revolution* (Harmondsworth: Penguin Books, 1975), 87.

88 Porch, 36.

89 António de Spínola, *Portugal e o Futuro* [Portugal and the Future] (Lisbon: Editoria Arcádia, 1974), 235.

Portugal's commitment to the defense of the *ultramar* had its origins in a search for the renewed greatness of an earlier era as espoused by the Salazar government vision of empire. The modern-day African colonies had largely been an irregular economic and political burden until the eve of the wars, and until after World War II, had held only a promise of any substantial economic benefit. Political opposition to Salazar was tolerated neither at home nor in the *ultramar*. The long-standing abuses of Portuguese African populations thus created widespread dissatisfaction with no outlet. Between the intransigent Salazar and an aggrieved African population an explosion was inevitable. When it happened in 1961, the events in Angola along with the *Santa Maria* incident, the "*abrilada*," the isolation in the United Nations, and the seizing of Goa pushed Salazar to solidify the commitment of the Portuguese people to defend the colonies and hence preserve his regime. This national commitment was a reflection of his own personal commitment and his propensity to brook no opposition, particularly from seemingly upstart nationalist movements and elements of his military. So strong was this feeling that it defied any voice of reason and foreclosed any retreat or compromise over African affairs. The armed forces and treasure of Portugal were thus pledged in full as the ultimate manifestation of this promise to make the colonial system work in Salazar's concept of empire.[90]

90 Manuel Limpo Serra, "Imperativo da Continuação de Portugal em África" [The Imperative of the Continuation of Portugal in Africa], *Anais do Clube Militar Naval* (July–September 1961): 455.

3

The New Air Strategy

When the troubles in Angola began during the opening months of 1961, the Portuguese Air Force had anticipated such a development almost a decade earlier. In 1952 the Military Aeronautics (*Aeronáutica Militar*) branch of the Army and the Naval Aviation (*Aviação Naval*) branch of the Navy were consolidated into a single, dedicated service under the Subsecretary of State for Aeronautics, a position equivalent to that of the Ministers of the Army and the Navy. Consolidation moved slowly at first, as the position of subsecretary went unfilled for the next three years. In the interim, the Chief of Staff of the Air Force performed these duties on a day-to-day basis, and when required, the Minister of Defense aided in the role. This makeshift arrangement was merely caretaker guidance, and the FAP needed an energetic visionary to fashion it into a modern force capable of defending the vast reaches of a far flung Empire.

Lieutenant Colonel Kaúlza de Arriaga was chief of cabinet to Prime Minister Salazar at the time, and hence the two knew each other well. From this relationship Salazar's confidence in Arriaga grew, and ultimately he was appointed to fill the vacant position of subsecretary. Salazar had made an excellent choice, for Arriaga was not only a strategic thinker of the first order but also a man who understood and would easily adapt to the workings of a highly technical air service. The relationship turned out to be symbiotic, as Arriaga was to prove a solver of problems for which no book solutions existed and would unerringly guide the FAP into modernity in the crucible of war.

The first task was to fashion a tailored strategy for the use of the FAP in supporting Portuguese national strategy in light of its circumscribed resources and in the fulfillment of its duty in fighting a counterinsurgency in defense of the *ultramar*. The FAP actually crafted no written doctrine, such as the famous "blue books" of the army that contained its *Army in Subversive War* or the more specific naval document *Marine Manual*.[1] There were, however, just as in all of the services, efforts to codify critical

1 Estado-Maior do Exército, *O ExércitoGuerra Subversiva* [The Army in Subversive War] (Lisbon: Estado-Maior do Exército, 1966); and Escola de Fuzileiros, *Manual do Fuzileiro* [Marine Manual] (Lisbon: Ministério da Marinha, 1962).

and specific experiences at the tactical level as informal local instructions, checklists, and guidance on actions to be taken or avoided, a sort of accumulation of "best practices." Direction at all levels of the FAP was shaped by several forces, primarily the shifting world order following World War II and its effect on the Portuguese defense establishment. The French conflicts in Indochina and Algeria likewise had a specific influence in its preparation for Africa, and here we will examine these important forces and conclude with an imputed air strategy.

The New World Order

With the creation of the FAP as an independent service, there arose within its leadership great questions as to how it would fulfill its new and expanded mission, a mission well beyond the relatively straightforward and discrete tasks of the earlier Naval Aviation and Military Aeronautics service branches. This anxiety was, in all fairness, caused by a shifting national policy from one of looking toward Europe and the United States within the Atlantic Alliance membership to one of renewed concern for the *ultramar.* These colonial domains, which were designated as overseas provinces under the constitutional reform of 1951, were considered an integral part of Portugal, and as such, sovereign national territory. The FAP was charged, together with its sister services, with the defense of all of Portuguese national territory, which according to the concept now defined by the New State, included the colonies. This expansion of responsibilities was enormous, indeed daunting, as Angola was about fourteen times the size of continental Portugal, and Mozambique about nine times that of its mother country.

These vast new responsibilities would bring Portugal into divergence with the stated purpose of NATO, which was to confront the Soviet Union and its immediate threat to Western Europe. Because of this orientation, the NATO area of operations was limited to the south by the Tropic of Cancer, a parallel of latitude well short of the Southern Hemisphere, where the vulnerable colonies needed defending. This meant that Portugal would have to find ways both to defend the *ultramar* and to meet its NATO commitment, something that would be difficult to do with the finite resources at its disposal. Moreover, Portugal could not count on wholehearted support of its NATO allies in countering any threat to its colonies. This situation meant that a careful and complete study of the colonial security problem was required, as the military strategy for defense of the *ultramar* and the needed wherewithal to implement it had not been put in place. These voids affected all three services, and hence preparation for the coming conflict became a preoccupation of their leadership.[2]

2 João José Brandaõ Ferreira, *En Nome da Pátria: Portugal, o Ultramar e a Guerra Justa* [In the Name of the Fatherland: Portugal, the *Ultramar* and the Just War] (Alfragide: Publicações Dom Quixote, 2009), 133.

National policy is a statement of national interests, and in its pursuit it will create and direct war, if it indeed has to be initiated to realize national goals. National strategy is that activity that strives directly to attain the objectives of policy and is the method of applying and coordinating all of the elements of national power, military, diplomatic, economic, and ideological, to form a cohesive effort in realizing policy. The military component of national strategy, the one that concerns us here, is the applied or threatened use of military force to impose policy.[3] Aviation policy and strategy are a further component of military strategy. With these concepts and their relationship established, it is important to put Portuguese national policy and strategy in context as a prelude to examining the resultant aviation subset.

The national policy of Portugal and the rationale for the deployment and use of the Portuguese armed forces in Africa was always to attain the political objective of maintaining the integrity of the Empire. This policy reflected the Portuguese politico-strategic philosophy of holding its widely dispersed territories as an insoluble traditional empire in a geopolitical environment generally hostile to such a concept.[4] The Empire was officially seen as a heritage and a promise of a richer future. This notion was captured in an editorial by Dr. Marcello Caetano in *O Mundo Português* [Portuguese World] that appeared in 1935: "Africa is for us a moral justification and a *raison d'être* as a power. Without it we would be a small nation; with it, we are a great country."[5] From this perspective any fragmentation would not only betray the past but would reduce the Empire to a modest European country. This seeming catastrophe was to be avoided at all costs.

Salazar was particularly sensitive to the vulnerability of Portuguese Africa to the nationalist movements that had gained ascendancy following World War II, and particularly to the local resistance within its African population. During this period, the democratized European powers in Africa were freeing their colonial possessions in step with the post-war trend. There was thus increasing pressure for Portugal to move in line with the Western European governments and to allow its colonies to become independent. The sensitivity to this pressure was likewise reinforced by a very deep and ingrained anticommunist Cold War stance by Portugal. Salazar's hatred and mistrust of communism played an important role, as he was very much aware of the impotence of the West to contain the "economically bankrupt but politically ascendant" communism.[6] Salazar was of the common view following World War II that with the German bulwark dismantled, an unchecked Soviet Union would engulf

3 U.S. Marine Corps, *Campaigning* (Washington: U.S. Marine Corps, 1990), 3.

4 Estado-Maior do Exército (EME), *Subsídios para o Estudo da Doutrina Aplicada nas Campanhas de África (1961–1974)* [Aid to the Study of Doctrine Applied in the African Campaigns (1961–1974)] (Lisbon: Estado-Maior do Exército, 1990), 30.

5 Marcello Caetano, "Editorial," *O Mundo Português* [Portuguese World], 2 (1935): 218.

6 Hugh Kay, *Salazar and Modern Portugal* (London: Eyre and Spottswoode, 1970), 69 and 133. The author argues that even in England communism had become "the rallying cry of the revolutionary instincts of our age" and had done so "by power of words, by sheer

Europe. Few, he believed, understood the acquisitive and imperialistic character of Soviet leadership despite the fact that Stalin in the aftermath of World War II had annexed over 600,000 square kilometers of Eastern Europe and had brought an additional 87 million non-Russians under his rule. The place of Portugal was alongside the democratic nations that had won the war, and the collective security arrangement of NATO was the only logical way to contain the threat. This national policy had led Portugal and its armed forces into a NATO orientation at the expense of colonial policing.

When the Soviet Union found itself checked in Europe, it sought other avenues of expansion to realize its national policy of creating a world communist order directed from the Kremlin. The Soviets regarded Africa as a fruitful ground for subversion despite the Congo setback, which we will examine shortly, and Portuguese Africa was one of their valued targets.[7] Characteristic of this African policy was the disturbing development from the 6 December 1960 meeting in Moscow of the "Congress of 81" communist countries. At first, little attention was paid to the congress by the West, as its last meeting in 1957 had attracted only thirteen countries. However, as the minutes of this widely publicized recent gathering disclosed, Portugal was specifically targeted for subversive activity.[8] It was asserted that the way to topple the Salazar government, which had refused even to establish diplomatic ties with the Soviet Union, and to replace it with one friendly to the communist sphere was through the vulnerability of its colonies. According to Soviet logic, if Portugal could be separated from its colonies in a debilitating war as well as politically undermined at home, then a weakened Salazar would topple. Nikita Khrushchev seemed to confirm the first of this two-pronged strategy in his famously disturbing speech on "wars of national liberation" delivered at this noisy and threatening forum. This direct challenge was a clear danger to Portuguese Africa.

In addressing this threat, the Portuguese national strategy shifted from one of supporting a policy of benignly maintaining the Empire with all of its promise to one of supporting a mobilization of national power in its defense against nationalist movements advocating colonial liberation. The new strategy had three primary components:

1. Military action would maintain order in the colonies and expel or defeat in combat the military wings of the nationalist movements.
2. Diplomatic efforts would support military action by seeking to nullify the activities of Portuguese adversaries, strengthen ties with allies, and forge new alliances, and while not accepting the legitimacy of the nationalist movements, seek

bluff, perhaps by voluptuousness of contrast." It was largely the success of communist "evangelism" that made Salazar so suspicious of it.

7 Ibid., 238; and Edgar O'Balance, *Tracks of the Bear* (Novato, Calif.: Presidio Press, 1982), 103–112.

8 Jean Valluy, "Au Subjet du Manifeste des 81," *Revue Militaire Générale* (October 1961): 282.

to open a dialogue with them toward a peaceful solution that did not compromise Portuguese sovereignty.

3. Socioeconomic development within the colonies would improve the lot of the colonial population and make it feel ideologically a part of greater Portugal, thereby neutralizing arguments of the nationalist movements.[9]

The FAP and Insurgency

The FAP was initially as ignorant of subversive war or insurgency as were its sister services, for since World War II the focus had been on a conventional conflict in Europe and not one against an array of insurgent movements in Africa. Indeed, Portuguese military officers of the time might be known as the "NATO generation" in that their careers had been largely spent solidifying the Portuguese position in the Atlantic Alliance. NATO had provided direction with its Eurocentric strategy, new infrastructure in an economically recovering continent, and modern equipment that was compatible and complementary to that of other NATO partners. Europe had been the logical direction for Portugal with its expanding economy, continental location, and the palpable Soviet threat developing at sea, on land and in the air to the East. Yet, the number and severity of insurgencies had in the post-war period increased to the point that this form of conflict had become the most prevalent in the world. Hence, despite the European Soviet threat, it seemed imperative that Portugal prepare for the reality of an insurgency, one that would most likely require the mobilization of large numbers of troops and other national resources.[10]

One of the dilemmas facing this NATO generation and particularly its FAP component was that conventional war logic does not apply to insurgencies. First, insurgency has as its center of gravity the population on the ground and is focused on human interaction. The primary activities of counterinsurgency are the targeting of groups that threaten a population, and the protection and control of that population. This very high level of human interaction means that the primary effort in counterinsurgency lies with the ground forces, police, and government agencies. In all insurgencies of the last sixty years, the primary role of airpower has been to act in support of forces on the ground.[11] While airpower has traditionally played a key function as a force enhancer in counterinsurgency, by its very nature it is unlikely to be featured

9 Joaquim Moreira da Silva Cunha, *O Ultramar, a Nação e o "25 de Abril"* [The Overseas Provinces, the Nation and the 25th of April] (Coimbra: Atlântida Editora, 1977), 31–32.

10 Kaúlza de Arriaga, *Guerra e Política* [War and Policy] (Lisbon: Edições Refrendo, 1987), 119–120.

11 James S. Corum, "Airpower Roles and Missions in Counterinsurgency" in *No Clear Flight Plan*, eds. James Fergusson and William March (Winnipeg: Centre for Defence and Security Studies, University of Manitoba, 2008), 205.

as the star but rather as a sustainment force tailored to the conflict, its venue, and the resources of the government.

Prior to the conflict and during its initial years, there were two illustrative counterinsurgencies fought by the French, Indochina and Algeria, as that country attempted to preserve its own empire. In the first, there was a lengthy search for the proper aircraft and an enormous amount of experimentation with these craft to develop an air strategy and supporting doctrine for their most effective employment in support of ground forces. In the second, the search for the proper mix of aircraft continued with the addition of the helicopter, and strategy and doctrine were further refined. The FAP was influenced substantially by the French experiences, and its adaptation of these to the Portuguese landscape was an important feature of airpower projection in Lusophone Africa.

In reality little or almost nothing was known in Portugal about the employment of airpower in this type of conflict, according to those familiar with doctrinal issues at the time.[12] Slowly, thinking evolved on how aviation should be used in a counterinsurgency, although it was only a sketch, and nothing was yet officially written.[13] Locally, small manuals appeared in various regions on how aviation was used in Malaya, Algeria, and Kenya and how it might be applied in the *ultramar.* These were more theoretical than practical, as each insurgency is different, and adaptations must be made.[14] Nevertheless, there were writings and conferences on the topic reflecting official thinking, and these mined the rich vein of French experience in addition to other post World War II insurgencies. First of all, it was acknowledged, as the French had discovered, that the principal mission of the air force was the support of ground forces and all that this entails.[15] The refinement of this notion in the FAP context was advanced by Lieutenant Colonel Amadeu José Ferreira in a March 1960 article in which he acknowledges that despite the new type of war, close air support of ground forces remains the primary role of the air force. In his words, the air force is to serve as a "multiplicator of land forces."[16]

The following month the French air ace of World War II fame and a participant in the Algerian war, Lieutenant Colonel Pierre Clostermann, was the featured speaker at

12 Luís Alves de Fraga, *A Força Aérea na Guerra em África: Angola, Guiné e Moçambique (1961–1974)* [The Air Force in the War in Africa: Angola, Guiné and Mozambique (1961–1974)] (Lisbon: Prefácio, 2004), 43.

13 Ibid., 44.

14 Ibid.

15 Philippe Gras, *L'Armée de l'air en Indochine (1945–1954): L'impossible mission* (Paris: L'Harmattan, 2001), 237; Charles Christienne and Patrice Buffotot, "L'Adaption de l'Armée de l'Air Française a un Type de Guerre Nouveau: La Guerre d'Indochine (1946–1954)" in *Regards sur L'Aviation Militaire Française en Indochine, 1940–1954* (Château de Vincennes: Service Historique de l'Armée de l'Air, 1999), 112.

16 Amadeu José Ferreira, "Contribuição da Força Aérea para uma doutrina e uma solução military" [Contribution of the Air Force towards a doctrine and a military solution], *Revista do Ar* 270 (March 1960): 344–348.

the Institute for Higher Military Studies (*Instituto de Altos Estudos Militares* or IAEM) in Pedrouços, a suburb west of Lisbon. Before a large gathering of FAP officers he delivered a paper "Aviation in Subversive War," which reflected the French experience in Algeria.[17] Clostermann had strong opinions on airpower in counterinsurgency: "The air force is one part of a team of forces; not only is it key to winning in subversive war, but without it the war will be lost. Its contribution is, moreover, enormous."[18] Clostermann was full of specifics from the French experience and shared them with his Portuguese hosts. He detailed what he believed were the key air functions: transport, fire support, command and control, observation, and helicopter operations. His vision also included elaborate infrastructure centered around key logistics and maintenance bases, each supporting a network of less-capable, operational-specific bases of varying capacities to extend the air umbrella to remote corners of a theater of operations. While his shared qualitative notions were valuable, his specific requirements for aircraft numbers and mix were much less so, as each insurgency is different in so many respects that such calculations cannot be universal. Nevertheless, Clostermann's shared experiences helped either to reinforce or to refine Portuguese decisions already underway. His paper was widely read by Portuguese aviators, and its reprints reached the farthermost corners of the Empire.[19]

Prior to Clostermann's visit, the FAP had begun to answer questions about its air role in counterinsurgency and had reached several conclusions on projecting airpower in Africa. The first was that there would have to be an aviation infrastructure in each of the three threatened theaters that was designed to support the ground forces in the context of the unique operating environment. Each of the territories had its own characteristics, including unequal development in their economies and in social and human factors, and these needed to be considered in addressing their defense. Portugal would build from virtually nothing, as there were no airfields of any note, no navigation aids, no aeronautical charts, no radios, inadequate aircraft in numbers and type, and shortfalls of almost every other kind. In its search for ideas on how best to accomplish the complete establishment of airpower in the *ultramar*, Portugal looked primarily to its long-term ally France to benefit from its experience in Indochina and Algeria. The lessons gleaned from these two French counterinsurgencies give rise to a new understanding of the projection of airpower in a remote and hostile environment and the purpose, design, and use of aircraft in counterinsurgency, as no other country contributed more to this development than France. Indeed, the airmobile helicopter operations that the United States employed in South Vietnam were a direct outgrowth of French helicopter employment in Algeria. Hence, a look at the French experience

17 Pierre Clostermann, "A Aviação na Guerra Subversiva" [Aviation in Subversive War], paper presented at the Instituto de Altos Estudos Militares, Lisbon, 29 April 1960.

18 Pierre Clostermann, "A Aviação na Guerra Subversiva" [Aviation in Subversive War], *Boletim do Comando Militar de Moçambique*, 5 (May 1961): 59.

19 Ibid., 55–75.

provides useful guidance in constructing a new air strategy, selecting aircraft, modifying airframes, training aircrew, developing tactics, and employing the new force.

For France in both Indochina and Algeria, the campaigns were ones of poverty; that is, wars conducted with very limited resources. It was this characteristic and the imaginative management of two air wars under severe political and resource constraints that provide thoughtful and applicable lessons for today. France struggled to fund and supply its war effort in Indochina half a world away, and Algeria became not only an enormous drain in public treasure, manpower, and matériel, but also a highly divisive political problem. In both cases, airpower held the promise of delivering victory against all odds and of solving the political and resource issues. In Indochina, acquisition of the needed aircraft, operating knowledge, and experience came too slowly to realize this promise. In Algeria, the French embraced the lessons from Indochina and were quite successful and innovative in the use of airpower; however, that conflict was a classic case of flawed strategy in that no amount of tactical success, however well supported by ample airpower, could gain the strategic objective of retaining Algeria as a part of France against the will of a majority of Algerians.

The French View

Militarily, counterinsurgency requires, among other things, nearly endless numbers of light infantry to protect the population and infrastructure of the threatened country, for the insurgents can strike when and where they choose. The government, on the other hand, needs to defend everything, something that is almost impossible to do. To address the manpower shortage, the government generally establishes small outposts within or near villages or other population settlements and organizes the local population into a self-defense militia to hold off an insurgent attack until help can arrive. The insurgents will seek to attack and overwhelm these outposts with superior numbers, and without outside help the outposts would succumb in time. To avoid this eventuality, the government maintains a quick-reaction force at a central and tactically propitious location to counter the unpredictability of an insurgent strike on its scattered outposts. The effectiveness of such a force clearly depends on its ability to move promptly to the fight and destroy the attackers. Thus troop mobility becomes a vital consideration. Ground forces travelling to aid these outposts are subject to the Clausewitzian dynamic of friction, as their road-bound route can be made impassable through enemy ambush, mining, destruction of bridges or other infrastructure, weather, or a combination of these. Consequently, the ability of ground forces to reinforce their beleaguered mates by land can be catastrophically delayed or foreclosed. At least in theory, aviation allows the quick-reaction force to overcome these frictional impediments and speed to the aid of a besieged outpost, achieving localized superiority in numbers and firepower against the attacking insurgents. Aviation also permits, at least in theory, overwhelming destruction to be rained on the attackers and their lines of communication.

The French knew that rapid mobility over land could not be attained in the terrain of Indochina and Algeria, as ground transport in these jungle, desert, and mountain environments was limited to primitive, rudimentary, and dangerous roads. These roads, or tracks in many cases, were not designed for speed or heavy vehicles. While seapower helped with mobility somewhat, as both theaters had coastlines, it was limited in its ability to address the insurgent threat. Airpower was thus seen as the crucial counterinsurgency element and force multiplier.

In the minds of the French planners, aircraft would replace the relatively slow, vulnerable, and ambush-prone road convoy as a means of power projection. The ability to react rapidly to insurgent attacks, to reinforce outposts quickly, and to provide a protective umbrella of firepower would be vastly improved with aviation, and the likelihood of success in countering an attack would increase proportionately with its use. When the insurgents sought to disperse and evade destruction, these airmobile forces would also be able to overtake and destroy the fleeing enemy. Because there would apparently be no aerial opposition, aircraft would be unhindered in supporting ground forces with enormous firepower in the form of various types of bombs, rockets, and automatic weapons. Indeed, aviators would be able to do most of the killing from above and could track, harass, and eliminate forces attempting to escape. This new process would also require fewer troops, as land route security would no longer be required. This would in turn reduce the political and fiscal costs of such conflict by decreasing the need for traditionally large and expensive numbers of troops, minimizing casualties, and eliminating large-scale and politically divisive conscription. Indeed, as James Eliot Cross observed:

> Aviation, and the development of air power, has certainly affected the course of unconventional war more than any other technological development of the past generation. In general, air operations have strengthened the hand of the government in handling rebellions, and in some areas it has provided a decisive superiority. It became clear a long time ago that insurgents and guerrillas could not long survive in regions where they were subject to attack from the air and where the terrain provided little or no shelter from such attack.[20]

Despite this optimistic strategy, the French experiences would show that projecting airpower in a primitive and harsh environment is fraught with obstacles, and the enemy is far from defenseless against a celestial threat.

20 James Eliot Cross, *Conflict in the Shadows: The Nature and Politics of Guerrilla War* (New York: Doubleday & Company, 1963), 75.

French Indochina

When this new concept was introduced in Indochina in 1945, it proved difficult to execute, as operating conditions were very severe, resources limited, and the enemy proved extraordinarily resourceful. The primordial geographic and climatic environment of Indochina proved an enormous impediment to military operations, and particularly aviation. The monsoons, typhoons, heavy cloud cover, metrological instability, torrential rains, electrical storms, high temperatures and humidity, all made flying difficult and aircraft maintenance a nightmare, and indeed, the enemy Viêt-Minh adroitly planned operations to take advantage of bad flying weather.[21] These extreme climatic conditions generated metrological phenomena unknown in Europe, and for which the majority of French airmen were unprepared. Not only did this physical environment adversely affect electronic and mechanical equipment, it also caused significant health problems and discomfort for the air and ground crews. The enervating heat and saturating humidity coupled with poor hygiene and diet provided ideal conditions for all sorts of diseases and infections, particularly as wounds either did not heal in such circumstances or required a lengthy period to do so.[22]

Maintenance was nightmarish, for many of the aircraft used were no longer in production, and thus spare parts were not generally available. Also, many were not designed for a tropical environment, and as a result, their useful lives were severely curtailed. Mechanics could work on the aircraft only in the early morning or the evening, because under the tropical sun all metal became burning hot.

The airfields in Indochina in 1945 were primitive with the exception of Tan Son Nhut outside of Saïgon, and the support infrastructure for modern aviation needed to be built largely from scratch to project airpower.[23] Consequently, the refurbishment of airbases became a priority, as this supported a logistic network that successfully employed transport aircraft to maintain forces engaged in the field. This "umbilical cord" of the French expeditionary corps became an attractive target of the Viêt-Minh, who sought constantly to cut it. In the strategy of fortified outposts, the local airfield became a symbol of survival, and its rupture synonymous with extinction. Among the numerous examples of this principle are the battles of Na San in 1952 and later the U.S. experience at Khé Sanh in 1968, where they survived with logistical air support, and Điên Biên Phû in 1954, where the landing strip was lost, and defeat became inevitable.

21 Christienne and Buffotot, 109–112; and Patrick Facon, "L'armée de l'air et la guerre d'Indochine" in *Regards sur L'Aviation Militaire Française en Indochine, 1940–1954* (Château de Vincennes: Service Historique de l'Armée de l'Air, 1999), 105.

22 Gras, 170–173.

23 Patrick Facon, "Reconstitution de l'armée de l'air en Indochine, 1945–1946" in *Regards sur L'Aviation Militaire Française en Indochine, 1940–1954* (Château de Vincennes: Service Historique de l'Armée de l'Air, 1999), 129.

Toucan. (Source: Archives de France)

On top of this, there were too few aircraft to provide the level of support the ground forces required, whether outposts or maneuver units in the field, and those available suffered from obsolescence and wear.[24] Until the emergency at Điên Biên Phû in March 1954, there were never more than 275 aircraft in all of Indochina.[25] Last, the aircraft were designed for a different kind of war and not easily adapted to the counterinsurgency environment.[26] Their generally limited range, small ordinance loads, and fragility in the heat and humidity of Indochina severely circumscribed airpower projection. Indicative of this limitation was the attempt to provide air support for the ground forces at Điên Biên Phû. The typical fighter-bomber had barely enough fuel for the 300-mile round trip from its base outside of Haïphong, and this constraint severely limited its bomb load and loiter time to deliver its ordnance under a withering Viêt-Minh antiaircraft defense, one akin to the German *flak* experienced in World War II.[27] In this vein, the enemy developed a defensive competence that made life miserable for low-flying aircraft and exploded the naïve notion there would be no aerial opposition.

When France reoccupied Indochina in 1945, strategically it was a distraction to *l'Armée de l'air*, the French air force. Initially, only a small group of transport and liaison aircraft was envisioned, and twelve Douglas C-47 Skytrain or Dakota transports were designated to support the mission.[28] This was aligned with French planning that anticipated a benign return to Indochina. Reflective of this complacent

24 James S. Corum and Wray R. Johnson, *Airpower in Small Wars, Fighting Insurgents and Terrorists* (Lawrence: University Press of Kansas, 2003), 145–148.
25 Ibid. 160. At the moment of the cease fire there were 388 aircraft: 180 fighters and bombers, 100 transport, and 108 liaison. This represented about 20% of *l'Armée de l'air* inventory.
26 Christienne and Buffotot , 110–117.
27 Ibid., 115–116; Corum and Johnson, 157–160.
28 Facon, "L'armée de l'air et la guerre d'Indochine," 87. The Douglas C-47 Skytrain was also called the Dakota by the British Commonwealth forces after the acronym DACoTA (Douglas Aircraft Company Transport Aircraft). Some 10,692 of the C-47s were produced during World War II, and additional numbers were produced in the Soviet Union and Japan. The civilian version was the venerable DC-3.

attitude, a mere six Dakotas were sent in August 1945, when *l'Armée de l'air* established a presence in Saïgon.[29] From the Indochina perspective, the situation was more complicated. The new air command there proceeded to supplement these transports with two makeshift squadrons of abandoned Japanese aircraft to meet the need for fighter and ground-support capability, and did so until they could be replaced with Spitfires left by the departing British. By January 1946, after the first hostilities with the Viêt-Minh, it became apparent that the primary mission of *l'Armée de l'air* in the Far East would be entirely reflective of ground force requirements.[30] The command was consequently expanded to comprise three air groups: one of C-47 transports and two of fighter-bombers. Soon thereafter it was reinforced with a fourth: a transport group of French-built Junkers Ju-52s.[31]

By the fall of 1947, there arose a need to coordinate and control air operations in support of ground offensives and to allocate the available resources more effectively. This was attempted through the establishment of two tactical commands, one that operated in the north and the other in the south. Experience showed that this arrangement required refinement, and in March 1950 all theater air forces were organized into three Tactical Air Commands (*Groupements Aériens Tactiques* or GATac) to facilitate their principal mission of supporting ground forces.[32] Each of these was defined and organized according to natural geographical boundaries and threats, and was given an inventory of aircraft best suited to supporting ground operations in its area. The staffs of each GATac became experts in the local operating conditions, enemy activity, and the supported land forces.[33] GATac/South covered all of Cochinchina, Cambodia, and southern Annam. GATac/Center covered northern Annam and southern Laos, and GATac/North had jurisdiction over Tonkin and northern Laos. There was an additional GATac/Laos established in 1953. The officer commanding each GATac exercised full tactical control over all aircraft operating in his area. The single exception was the Army Light Aviation (*Aviation Légère de l'Armée de Terre* or ALAT), which oversaw the small artillery spotter and observation planes and assigned them in direct support under the local control of specific army units.[34] While the GATac structure was organizationally sound, *l'Armée de l'air* still struggled, throwing available aircraft into the breach on a routine basis. Each GATac executed four types of missions: 1. Transport, 2. Reconnaissance and surveillance, 3. Direct support (close air support or CAS), and 4. Indirect support (strategic bombing).[35] The liaison mission was handled by the relevant ground command, with the support of ALAT.

29 Christienne and Buffotot , 113.
30 Bernard B. Fall, *Street Without Joy* (New York: Schocken Books, 1972), 261.
31 Facon, "L'Armée de l'air et la guerre d'Indochine," 88.
32 Gras, 237.
33 Ibid., 242.
34 Fall, 262. The ALAT inventory consisted of the Piper L4 Grasshopper, the Morane-Saulner 500 Criquet, and the Cessna L19 *Bird Dog*.
35 Gras, 243–244.

Coordination between air and ground forces was considered very good. *L'Armée de l'air* officers were assigned to most combat units as forward air controllers (FACs), and each year several training courses were conducted for ground force officers to apprise them of air support capabilities and limitations.[36] Indochina, however, was clearly considered a sideshow, as evidenced by the fact that from the authorized air force personnel strength of 65,000 men, only 4,812 or 7.4 percent were destined for the Far East.[37] Eventually this number would rise to 10,657, still only about 7.5 percent of the total force a decade later and still well short of the number thought to be needed by those in Indochina.[38]

Following the violent Viêt-Minh offensive that General Vö Nguyên Giáp launched on the northern province of Tonkin in March 1949, *l'Armée de l'air* was reinforced with a fighter group, a transport group, and a reconnaissance squadron. This force level represented the logistical limit of support in light of the resources available to *l'Armée de l'air* and the distance from France.[39]

The United States, initially a reluctant player, soon became heavily involved in Indochina, supporting the French in the Cold War context. It offered overt aid in the form of large quantities of arms and aircraft, such as the Grumman F8F Bearcats and Douglas B-26 Invaders, and small numbers of aircraft maintenance technicians. Also in 1952, the United States underwrote a fleet of fifty C-47s, which came from U.S. Air Force stocks in Japan and France. These replaced the Ju-52s and gave *l'Armée de l'air* the capability of dropping 3,200 paratroops in a single operation.[40] Later, the Eisenhower Administration offered covert support by providing not only Fairchild C-119 Flying Boxcar transports but also the pilots to fly them in support of French forces at Điên Biên Phû. It further provided strategic airlift to bolster the limited French capability. This help came in the form of U.S. Air Force Douglas C-124 Globemaster II transports making the long-haul shuttle between France and Indochina to move troops and supplies.

An important part of air reinforcement after 1950 was the air arm of the French navy (*Aéronautique Navale*), which usually maintained two carriers off the coast of Indochina, one being the *Belleau Wood* on loan from the United States.[41] Both of these had a normal complement of twelve Grumman F6F-5 Hellcat or F8F Bearcat fighter-bombers and nine Curtiss SB2C-5 Helldiver dive bombers. Aside from the carrier force there was a detachment of twelve Grumman JRF-5 Goose amphibians based ashore that supported maritime reconnaissance, coastal surveillance, and

36 Fall, 262.
37 Facon, "L'Armée de l'air et la guerre d'Indochine," 88–89.
38 Ibid., 103–104.
39 Ibid., 96.
40 Ibid., 103.
41 The USS *Belleau Wood* (CVL-24) was transferred to the French in September 1953 and renamed *Bois Belleau* (R-97). The ship was returned to the United States in September 1960, stricken from the fleet, and sold for scrap.

Invader over the Red River Delta. (Source: Archives de France)

medical evacuation.[42] Because of the limited capacity of the carriers, most naval aircraft flew from land bases, particularly after the United States began a serious reinforcement of the French naval air arm in late 1951. The land bases were also nearer the fighting, so the aircraft could be on station more quickly and stay longer.

As often happens in a war, one uses what is available, and throughout the conflict France did just this. In Indochina the French initially employed small numbers of abandoned World War II Japanese aircraft, such as the Nakajima Ki-43 Hayabusa (Peregrine Falcon), Allied code name Oscar, a single-engine land-based tactical fighter, and the Aichi E13A1a, Allied code name Jake, a reconnaissance seaplane.[43] These provided the first attack capability until alternatives could be found. Transport capacity was supplemented by the German Junkers Ju-52, and liaison by the Fieseler Fi-156 Storch (Stork) aircraft. Both had been manufactured in France during the German occupation of World War II as the Amiot AAC-1 Toucan (Toucan) and the Morane-Saulnier MS-500 Criquet (Cricket) respectively. The Junkers/Amiot, a small tri-motor transport, was used for troop and cargo transport, for paratroop drops, for medical evacuation (MEDEVAC), and as a makeshift bomber adapted with wing rails to carry 50-kilogram bombs.[44] Prior to the installation of wing rails, "bombardiers" simply lobbed bombs and napalm canisters from its cargo door.[45] It was not a

42 Corum and Johnson, 152–153.

43 Thirteen of the Nakajima Ki-43 Hayabusa (Peregrine Falcon) fighters and four of the Aichi E13A seaplanes were used.

44 Alexander Zervoudakis, "L'Emploi de l'Armee de l'Air en Indochine, 1951–1952" in *Regards sur L'Aviation Militaire Française en Indochine, 1940–1954* (Château de Vincennes: Service Historique de l'Armée de l'Air, 1999), 168. The Ju-52 had a gross weight of 20,270 pounds and carried 18 troops, or 12 litter patients, or about 7,000 pounds of cargo.

45 Fall, 261.

particularly satisfactory craft, as for example, its BMW 132 engines needed a major overhaul every 450 flight hours, compared to the 1,500-hour maintenance cycle of the Pratt and Whitney R-1830 Wasp engines on the C-47.[46]

The Criquet was a small, light liaison aircraft noted for its remarkable short take-off and landing (STOL) capabilities and was used for reconnaissance, communication, forward air control, and road convoy escort. While the Criquet was an ideal aircraft, French air reconnaissance units were singularly ineffective, as in jungle areas there is little or no chance of seeing much of what is happening on the ground through the blanket of vegetation.[47] To compound this difficulty, the Viêt-Minh became masters of camouflage, painstakingly hiding troops and supplies. They moved by night, or if by day, they regularly changed camouflage to blend with the varying foliage. Aerial reconnaissance was thus at an extreme disadvantage.[48] The Portuguese would learn from this difficulty and develop their own more successful and effective reconnaissance techniques based on a new generation of light observation aircraft.

Despite these drawbacks, the role of the Criquets became indispensable, as the ground forces faced an increasingly substantial and capable Viêt-Minh after 1949 and came to rely heavily on CAS. Equipped with new radios to coordinate with both tactical aircraft and ground troops, the Criquets proved effective in bringing important and often critical firepower to hard-pressed units. The tropical environment was harsh on their wooden and canvass components, and hence, they were supplemented by an array of similar liaison aircraft as the conflict progressed, such as the Nord 1001 Pingouin (Razorbill), the Aérocentre NC-702 Martinet (Swift), and the Dassault MD-315 Flamant (Flamingo).

On 31 January 1946, when Royal Air Force 273 Squadron was decommissioned at Tan Son Nhut, it transferred eight Vickers Supermarine Mk. VIII Spitfires to the French.[49] These were shortly replaced by large numbers of Mk IX Spitfires, which were transferred from Europe, raising the total to sixty.[50] Unfortunately these were not designed for tropical operations and not well adapted to local combat conditions in Indochina with their limited range of 450 miles and ordnance load of two 250-pound bombs. With few suitable airfields to extend their coverage, they lacked the range to reach the widely dispersed ground forces. Their armament was also less than ideal for ground-support. By 1950 their wood and canvas components were literally rotting from the aircraft in midflight, and they were removed from service.[51]

46 Corum and Johnson, 148.
47 Cross, 76.
48 Corum and Johnson, 150.
49 Tan Son Nhut Airport was built by the French colonial government in the 1920s as an unpaved airfield in the village of Tan Son Nhut to serve the nearby city of Saïgon. During World War II, the Imperial Japanese Army used Tan Son Nhut as a transport base.
50 Fall, 261.
51 René J. Francillon, *Vietnam: The War in the Air* (New York: Arch Cape Press, 1987), 16.

In view of the limited range and armament of the Spitfire, *l'Armée de l'air* introduced at the beginning of 1947 about 180 De Havilland DH98 Mosquito FB Mk VI fighter-bombers to supplement the Spitfires. The Mosquito, variously known as the "Wooden Wonder" and the "Balsa Bomber" because of its bonded-plywood construction, had a robust undercarriage and a large disposable load. Unfortunately, as the British and the Australians had already discovered during World War II operations in the South Pacific, its wooden structure suffered severely from the high humidity, heat and invasive tropical insects. This, coupled with the age of its airframe, led to serious maintenance difficulties and hence availability was generally low. Further, all had to be based at Tan Son Nhut, as they required long runways. After only six months of operation they were returned to France in July 1947.[52]

The French situation in Indochina was not seriously threatened until September 1949, when the People's Republic of China was formed. The Viêt-Minh now had a sanctuary for rest and training and a source of supplies and arms. It also meant that France would begin to face large concentrations of enemy troops dominating vast areas of Tonkin and driving its forces into the Tonkin Gulf littoral. In light of this development, the United States now saw France as a staunch anticommunist ally and began to help in significant ways. Among the first was the release of equipment held in Europe under the North Atlantic Alliance, and fifty Bell P-63C Kingcobras were hurriedly dispatched to Indochina. These proved generally effective, and were well suited to the climate and the prevailing type of operations. While their range at 320 miles was considerably less than that of the Spitfires at 450 miles, they were highly resistant to the ever increasing volume of ground fire being experienced. The *Kings*, as their French crews called them, required well prepared airfields and carried a relatively light ordnance load of three 500-pound bombs. While the Kingcobra was an improvement in many ways, it was not the answer.

Next, the United States helped *l'Armée de l'air* to equip its fighter-bomber units with Chance Vought F4U-4 Corsairs, Grumman F6F-5 Hellcats, and F8F Bearcats. These carrier-based aircraft had STOL characteristics that were an important feature in Indochina, for now air support could be based at less elaborate airfields nearer the battlefront. Further, the Corsair had a 450-mile combat radius and carried 4,000 pounds of external stores; the Hellcat had a 945-mile combat radius and carried 1,000 pounds of bombs or rockets; and the Bearcat had a combat radius of 1,105 miles and likewise carried 1,000 pounds of ordnance.[53] This sort of capability was ideal for counterinsurgency; however, by the start of the conflict in the *ultramar*, these aircraft were worn out and not available to Portugal. Nevertheless, this experience would hold valuable lessons for FAP operations.

These advanced capabilities made an enormous difference in the CAS that *l'Armée de l'air* could deliver for the ground forces; however, more effective airpower was yet

52 Ibid, 17.
53 Ibid., 19 and 233.

Corsair. (Source: Archives de France)

to come. The French had long wanted a medium bomber, and with the U.S. change of heart, the first four of 113 Douglas B-26 Invaders on loan were delivered to Tan Son Nhut in November 1950. These aircraft began operating in February 1951 from Tourane, later known as Da Nang, in Annam, and they were formidable with their 8,000-pound ordnance load of internal and external bombs, rockets, and napalm canisters. Their forward-firing armament of up to fourteen .50-caliber machineguns and 1,480-mile combat range made them undoubtedly the most effective combat aircraft operated by *l'Armée de l'air* in Indochina.[54] Some 300 U.S. Air Force mechanics were quietly assigned to *l'Armée de l'air* to maintain the B-26 force and supposedly served only in secure areas where they could not be captured or noticed by news reporters.[55] The B-26s were returned to the United States when the French departed Indochina. The FAP would attempt unsuccessfully to acquire this potent weapon in numbers.

In November 1950, French naval aviation also obtained a highly capable aircraft with the initial delivery of four Consolidated PB4Y-2S Privateers. These replaced the Consolidated PBY-5A Catalina amphibians, which had been doing this bombing and reconnaissance duty since October 1945. This addition was so important that these aircraft with their 12,000-pound bomb load and twelve .50-caliber machineguns participated in all major operations from their appearance through the end of the war. Eventually there were 22 Privateers stationed at Tan Son Nhut, with a small but growing detachment at Cat Bi outside of Haïphong. These aircraft made extensive use of the 500-pound antipersonnel cluster bomb that, with its 11,200 finned missiles, proved effective against massed Viêt-Minh troops. Unfortunately, even with this lethality there were never enough Privateers or cluster bombs to destroy the Viêt-Minh. At the end of the war, six aircraft were returned to the United States and

54 Ibid., 20–21.
55 Ibid., 37

Privateer at Cat Bi. (Source: Archives de France)

twelve were flown to North Africa. Four were lost in the war.[56] Heavy bombers, such as the Privateer, would not be needed in Algeria or the *ultramar*, as there would be no massed enemy troops of any significance in either theater. Virtually no other insurgent leader could muster the numbers amassed by Giáp, and thus none could afford to waste their small bands of followers using massed formations with their attendant vulnerability to heavy casualties from airpower.

The British were the first to use the helicopter in counterinsurgency, with the formation in Malaya in April 1950 of the Far East Casualty Air Evacuation Flight with three Westland S-51 Dragonfly helicopters. While this force eventually grew to thirty-one, it was never enough.[57] The French were only a month behind. *L'Armée de l'air* had been prevented by budgetary constraints from acquiring helicopters, and thus the first two Hiller H-23A Ravens to appear in Indochina in May 1950 were owned by the *Service de Santé d'Indochine* (Indochina Health Service). Crewed by military pilots, they were extraordinarily useful in evacuating wounded personnel. Following

56 Ibid., 30.

57 John P. Cann, *Counterinsurgency in Africa: The Portuguese Way of War, 1961–1974* (Westview, Conn.: Greenwood Publishing, 1997), 129.

this success, *l'Armée d'air* acquired four Westland S-51 Dragonflies and proceeded to accumulate an inventory of forty Westland and Hiller rotorcraft.[58] This number was only adequate for rescue operations, but the French would use helicopters for combat operations in Algeria with a vengeance, amassing a force of over 250 by 1959 and expanding this to 600 by 1962.[59] The medical evacuation pilots and aircrews logged 4,189 flight hours, evacuated 3,987 wounded, and flew 1,984 missions to save the lives of their countrymen and would show the way for similar operations in Algeria and the *ultramar*.[60] Most notably were the courageous actions of Capitaine Valérie André, a young woman who was not only a pilot but also a neurosurgeon and qualified parachutist. Piloting her Hiller Raven, she rescued 67 men from behind Việt-Minh lines.[61]

Another less well known nurse, paratrooper, and pilot was Paule Dupont d'Isigny, a product of the French Resistance and a Nazi prison camp. She logged more than 4,200 flight hours and thirty combat missions, and was awarded the *Croix de Guerre* with two palms in Indochina, the *Croix de Guerre* with one palm in World War II, and Legion of Honor for military valor.[62] Few of those rescued would ever forget the soft and comforting touch of the female helicopter medics.

As 1952 drew to a close, the Toucans had reached the end of their useful lives, and France began to struggle with a patchwork of air transports until May 1953, when the United States lent six state-of-the-art Fairchild C-119C Flying Boxcars to the French. It followed shortly with another twelve flown by Civil Air Transport (CAT) crews. The CAT was originally the Central Intelligence Agency (CIA) air force and operated in support of Nationalist China. When the Nationalist cause collapsed in 1949, the CAT was renamed Air America. It generally employed seconded U.S. Air Force personnel who were "sheep dipped"—stripped of the most obvious signs of their continuing service connection. The aircraft were likewise sanitized in that they were rendered anonymous and, to the extent possible, untraceable to give the United States plausible deniability of any involvement in the activities of the aircrews and aircraft.[63]

Fifteen additional C-119s were flown from Clark Air Force Base in the Philippines by U.S. crews in support of the French, bringing the total number of reflagged Boxcars to thirty-three. With its maximum payload of 30,000 pounds, the C-119 added a welcome boost to French theater airlift capability. Alongside the obsolete C-47s with their single, relatively narrow cargo door, the Boxcars were a quantum leap forward. During the siege of Điện Biên Phủ, for instance, the French operated them with their

58 Francillon, 24–25
59 Tom Cooper, "Algerian War, 1954–1962" in *Western and Northern Africa Database*, Air Combat Information Group, http://www.acig.org/artman/publish/article_354.shtml (accessed 24 May 2010).
60 *L'Indochine, Les GAOA, Le GFHATI*, http://doumelico.perso.sfr.fr/indochine.htm (accessed 10 June 2010).
61 Fall, 136; and Valérie André, *Ici ventilateur* (Paris: Calmann-Lévy, 1954).
62 Fall, 137.
63 Ibid., 107–112.

Boxcar at Cat Bi with rear doors removed. (Source: Archives de France)

rear cargo doors removed, and this modification enabled the Boxcars to dispatch their entire 6-ton load in a single pass over a drop zone. Conversely, the C-47s required twelve passes to dispatch their more modest load of 5,000 pounds.[64] All Boxcars were returned to the U.S. Air Force at the end of hostilities.

The airmen who flew the Boxcars played a crucial and unsung support role, often at considerable personal risk. In most instances, the C-119s flew in a benign environment; however, during the siege of Điên Biên Phû, when the stranglehold on the base was such that it could only be supplied through difficult and hazardous airdrops, CAT pilots were unquestionably flying in harm's way. Their approach to and cargo release over the drop zone at the besieged garrison was subject to withering antiaircraft fire. These CAT flights, known as Operations *Squaw I* and *Squaw II*, used a dozen of the reflagged C-119s and 37 pilots to accomplish an astonishing 682 missions between 13 March and 6 May 1954.[65] Indicative of the enormous risk in these missions was the one flown by James McGovern, known as "Earthquake McGoon," and Wallace Burford on the evening of 6 May. It was part of a flight of six C-119s that departed

64 Ibid., 263.

65 Central Intelligence Agency, "A Look Back…Earthquake McGoon's Final Flight, "Featured Story Archive, https://www.cia.gov/news-information/featured -story-archive/earthquake-mcgoons-final-flight.html (accessed 22 September 2010).

Cat Bi on that afternoon and carried a load of ammunition. The first aircraft in the flight dropped its cargo and was away safely, and McGovern and Buford followed. During the approach, their aircraft was hit in the port engine by a 37mm antiaircraft round and immediately afterwards sustained considerable damage to the horizontal stabilizer. After releasing the cargo, McGovern and Buford followed the lead aircraft for 40 minutes toward a remote landing strip 75 miles to the southwest in Laos, all the while struggling to maintain flight on a single engine in a crippled machine. Just a few hundred yards short of the landing strip, the wingtip of the C-119 clipped a tree, and the aircraft cartwheeled, broke in half, and burned, killing McGovern and Buford.[66] In February 2005, McGovern and Buford posthumously along with the surviving CAT pilots were awarded the Legion of Honor for military valor on behalf of France for their actions at Điên Biên Phû.

In June 1954, the Nord 2501 Noratlas was introduced into Indochina with its 18,647-pound payload capacity. Unfortunately, this capability came too late to affect the outcome, for the Geneva accords were signed a month later in July 1954. The Noratlas, with a design similar to that of the Boxcar, would reappear as an ideal intra-theater counterinsurgency medium-lift transport in Algeria and later in the *ultramar*.

Overall, *l'Armée de l'air* fulfilled its mission in Indochina as well as could have been expected. What it lacked in aircraft it more than made up in the skill and dedication of its pilots. Even so, it could never bring adequate firepower the distances required to interdict the Viêt-Minh lines of communication coming from China and Laos. Nor could it overcome the nationalist ideology and the enormous "butcher's bill" Giáp was willing to pay with the lives of his Viêt-Minh troops to achieve victory. Its airlift and troop mobility were weak and inadequate, until it received the Boxcars in May 1953 and ultimately the Noratlas in June 1954. By then the new capacity was too late to influence the direction of the war. The lessons from the French experience in Indochina could be reduced to four factors:

- The need for a network of airfields for liaison, ground-support, and intra-theater airlift to enable effective air support of ground forces;
- The need for a solid, reliable, and simple ground-support aircraft capable of operating from forward airfields within range of ground engagements;
- The need for a capable intra-theater heavy-lift transport to supply the extended ground forces; and
- The need for helicopter capability beyond medical evacuation to enhance tactical troop mobility and support.

All of these would come to fruition in Algeria, where half a world away French officers were already applying the lessons of Indochina to a new insurgency.

66 Ibid.

Algeria

When war broke out in Algeria in 1954, *l'Armée de l'air* was in a much better position to wage it than it had been in Indochina. First, the logistical tail was a mere 600 miles south of France across the Mediterranean and not half a world away. Second, Algeria was a part of France and thus the NATO area of operations, a fact that made the French use of NATO equipment much easier in fighting a colonial insurgency. Algeria also had to a degree benefitted from a nascent air infrastructure remaining from World War II. These factors meant that it would be far easier to bring air support to a war in Algeria than it had been in Indochina. Last, Indochina had been a laboratory for air support in counterinsurgency, and the lessons gained from that experience would be actively applied to Algeria.

The employment of *l'Armée de l'air* in Indochina had often been marked by misunderstandings and mistakes; however, paradoxically it was the source of groundbreaking innovations in the use of aviation in fighting an insurgency. Beginning in the autumn of 1954, *l'Armée de l'air* began introducing innovations in Algeria, particularly the integration of the air and land forces in a united effort at hunting and destroying "rebel" bands. The first transfer of the Indochina experience to Algeria occurred with the establishment of ALAT, which had been heavily criticized earlier by *l'Armée de l'air* as a dilution of resources and thus effectiveness. In Indochina the ALAT had been extremely important in directly supporting ground forces with missions of observation, artillery spotting, liaison, and surveillance, and while army control of small aircraft had been experimental and controversial with *l'Armée de l'air*, it had proved so effective that it was refined and continued in Algeria, despite continuing objections.

Harvard. (Source: Archives de France)

The optimal situation for French forces was to adapt airpower to ground strategy, and it was for this reason that the tactical commands were created in Algeria in 1956 identically to the GATac in Indochina. This organizational structure allowed for the decentralization of decision-making to the local sector level and thus lent great flexibility in the employment of available local forces against an elusive target. This adaptability, allied with the acquisition of more modern equipment, was to prove a true melding of airpower as an effective support to ground forces, something that was weak in Indochina.

Another lesson from Indochina was the use of CAS, despite the fact that Algeria was quite different from Indochina. The need for a simple, robust, and stable aircraft that could deliver accurate fire was satisfied by the introduction of the T-6 Texan/Harvard in 1956, and this aircraft enabled *l'Armée de l'air* to become virtually omnipresent in the country.[67] In contrast to the adaption of such aircraft as the F8F Bearcat, F6F-5 Hellcat, and F4U-4 Corsair fighter-bombers or the B-26 Invader medium bomber to CAS, the T-6 represented the ideal embodiment of CAS during the initial years in Algeria. This aircraft was replaced by a similar one in the T-28S Fennec (Desert Fox) and finally by various versions of the Douglas AD-4 Skyraider.[68] These higher performing aircraft were, however, far more complicated with the attendant elevated maintenance and handling requirements. The T-6 represented the best compromise between simplicity, ease of maintenance, durability, armament load, and availability, and would become the CAS aircraft of choice in the *ultramar* for these reasons.

The final lesson from Indochina was the potential of the helicopter as a troop transporter and gun platform. In May 1954, *l'Armée de l'air* created in Indochina the 65th Mixed Helicopter Wing (*65ème Escadre Mixte d'Hélicoptères*), which had numerous detachments serving various commands. Put in place too late to be of much use in the Far East, the structure was resurrected in Algeria in April 1955 as Light Helicopter Squadron 57 (*l'Escadrille d'Hélicoptères Legers No 57*) stationed at Boufarik and became the Mixed Helicopter Group 57 (*Groupe Mixte d'Hélicoptères 57*) in the autumn of 1957. The army likewise acquired helicopters, and these were assigned to two helicopter groups. Those of the navy were initially organized in a single squadron, and then, in November 1957, reorganized as Naval Helicopter Group 1 (*Groupe d'Hélicoptères de l'Aéronautique Navale No 1* or GHAN-1), formed at Lartigue near Oran to oversee this and future squadrons. The approximately 250 helicopters employed in Algeria from 1959 became the true engines of combat and responded well to the requirements of counterinsurgency.[69] By the time of the cease-fire in 1962, this number had reached

67 Patrick Facon, "L'adaptation de l'armée de l'air à la guerre d'Algérie: la lute antiguerrilla" in Service Historique de l'Armée de l'Air, *Regards sur L'Aviation Militaire Française en Algérie, 1954–1962* (Château de Vincennes: Service Historique de l'Armée de l'Air, 2002), 32.

68 A *fennec* is a carnivore of the Sahara and is often called a desert fox (*renard des sables*).

69 Gras, 580.

600, a testimony to rotary-wing success.[70] Restricted solely to medical evacuation in Indochina, the helicopter reached its stride in Algeria and proved to be highly flexible and adaptable to the fighting there. Because it combined an inherent speed of response with a new dimension for the application of firepower, infantry tactics were built around its capabilities, and Algeria became a testing ground from which others beyond France would learn much in helicopter adaptation to combat.

When long-simmering resentments erupted in open rebellion in 1954, there were few resources in place. Indeed, when the new Governor-General Jacques Soustelle arrived in February 1955, the army possessed a solitary Bell 47G helicopter, which was rented from a private firm. Soustelle pressed the government into persuading the U.S. Air Force in Germany to transfer eight of its Sikorsky H-19s to *l'Armée de l'air.* Following this transfer, the government then executed an agreement to purchase additional helicopters in the United States.[71]

Soustelle worried that the military was still altogether a NATO-style force both in equipment and technique, for despite the lessons of Indochina, not everyone seemed to have benefitted.[72] Indeed, only a small percentage of the *l'Armée de l'Air* had been engaged in Indochina. Indicative of this situation was the initial deployment of the best and latest equipment in defense of Algeria: the new Sud-Est SE-535 Mistral jet fighter, which proved woefully ineffective.[73] The aircraft lacked endurance and proved difficult to maintain in the harsh conditions of North Africa, where sand and dust penetrate everything. Worse, the aircraft was too fast for effective counterinsurgency, as it is all but impossible for the lone pilot to spot, identify, and attack small groups of guerillas at jet aircraft speeds.[74] As James Eliot Cross so ably describes the problem:

> The speed of even the slowest fixed-wing aircraft is so great that the pilot has little chance of positively identifying an enemy who is not wearing a distinctive uniform, unless the latter obligingly waves a rifle or shoots at him. Even if he does this it is usually too late for the pilot to do more that turn for another pass, by which time the identified enemy, if he is prudent, will have disappeared.[75]

There were, however, two squadrons of Republic F-47D Thunderbolts, each with 18 aircraft, which were effective but ageing fighter-bombers. Parts for them were increasingly difficult to obtain. In September 1957, the surviving 24 airframes were consolidated into a single squadron and continued in service until replaced with Skyraiders. Since there was no equivalent replacement available for the Thunderbolt

70 Fall, 265.
71 Alistair Horne, *A Savage War of Peace: Algeria 1954–1962* (New York: Penguin Books, 1979), 113.
72 Ibid., 112.
73 The Sud-Est SE 535 Mistral was a license-built DeHavilland DH-100 Vampire VB Mk5.
74 Cooper.
75 Cross, 77.

in 1955, local French commanders began to arm light transports and trainers to fill the gap. Dassault MD-315 Flamants, Morane-Saulnier MS-500s Criquets and MS-733 Alcyons (Halcyon, a mythical bird identified with the kingfisher), and the Société Industrielle Pour l'Aéronautique (SIPA) S-111s and S-211s were all fitted with machine guns and 37mm rockets. In June 1955 this assortment of light aircraft was formed into the first of four Light Support Aviation Squadrons (*Escadrilles d'Aviation Légère d'Appui* or EALA) and used to good effect. With an observer spotting targets from the rear seat of one of these aircraft, it was twice as likely to locate and correctly identify a target as a single-seat fighter-bomber. Further, given the relative lack of sophistication in the adversary, the light armament of these transports and trainers was adequate at the time. These converted aircraft, however, were too light and fragile for a sustained combat role, and as the war matured, and the enemy became more cunning, the French had to look for sturdier substitutes.

Early in 1956, *l'Armée de l'air* began to acquire what would be nearly 700 North American T-6G Texan/Harvard aircraft, 294 of which would eventually be deployed to Algeria.[76] This World War II-era trainer was widely used by the Allies and was known as the Harvard outside of the United States. It suited the French purposes well, for it could sustain ground fire ably, had relatively good endurance, and was still available in quantity. The Harvard proved a robust and easily maintainable aircraft and was capable of carrying two twin 7.5mm machinegun pods, napalm canisters, bombs, and rockets. The French were the first to use the aircraft in a ground support role, and it became the mainstay of their operations. Four squadrons or *escadrilles* of T-6Gs were formed in 1955, and by 1958 there were more than 30 squadrons. Many of the aircraft were assigned to additional EALAs, as these commands were established, and there were eventually 38 EALAs. The EALAs were now under the new Light Support Aviation Group 70 (*Groupe d'Aviation Légère d'Appui* or GALA). GALA 70 was later split into three GALAs, one within each GATac. The T-6G remained in service until it was replaced by various versions of the Douglas AD-4 Skyraider in December 1959 and the North American T-28S Fennec in 1960. The Harvard would likewise become the mainstay of the FAP in the *ultramar*.

By 1958, the T-6G needed replacement, as the spare parts and engines were no longer easily available. Likewise, the airframes were reaching the end of their useful lives. Once again, the French sought to repeat their T-6G success and found it in another trainer, the North American T-28 Trojan. At the time the U.S. Navy had the rights to all new T-28B production, so the French found an alternate source. They were able to locate some 148 former U.S. Air Force T-28A aircraft at the Davis-Monthan Air Force Base storage facility and acquired these as a basis for their new CAS concept.

76 Capitaine Claude Pfirrmann, "Ces hommes du 1er Étranger de Cavalerie," *Historia Magazine* 293 (5 March 1973), 2120.

A private California firm, Pacific Airmotive, had developed a powerful, civilianized T-28A, which it named the Nomad. In place of the 825-horsepower standard Curtiss-Wright R-1300 Cyclone 7 engine in the T-28A, the Nomad mounted a 1,425-horsepower Curtiss-Wright R1820-76A Cyclone 9 salvaged from Boeing B-17G Flying Fortress stocks. It was an impressive aircraft, but there was no civil market for it. France employed Pacific Airmotive to design a Nomad variant, the T-28S, fitted with armor plate, reinforced wings, and increased cockpit ventilation for hot climates, and under the agreement it supplied design drawings, engines, and airframes for shipment to France. Here, Sud-Aviation would perform the conversion and install French equipment, armament, and sand filters. The T-28S was strengthened to take four underwing weapon pylons. The two inner pylons usually carried French-designed pods for twin .50-caliber machine guns. The outer pylons could carry shortened 440-pound bombs, rocket pods, napalm canisters, or single 105mm or 120mm rockets. By 1961, over 100 Fennecs were operating with four EALAs in Algeria, and the balance were either in reserve or final assembly. Their time in service, however, was brief.

At the same time, France sought a heavier piston-engined fighter-bomber to complement the Fennecs. While the converted trainers had served well, none had proved entirely capable of replacing the long retired Thunderbolts. Fortunately, the U.S. Navy was transitioning to newer aircraft and making quantities of piston-engine attack airframes available following the Korean War. Among these was the Douglas AD-4 Skyraider, with its powerful Wright Aeronautical R-3350-26WA *Duplex-Cyclone* engine delivering 2,800 horsepower. *L'Armée de l'Air* ordered 113 Skyraiders in three different versions: 20 of the AD-4, 88 of the AD-4N (3-seat night attack), and five of the AD-4NA (The AD-4N without the night attack equipment but fitted with four 20mm cannon). All were shipped to the workshops of a Sud-Aviation subsidiary, the Société Française d'Entretien et de Réparation de Matériel Aéronautique (SFERMA), which performed aviation maintenance and repair in Bordeaux. Here the radars and arresting hooks were removed and the air brakes welded in place, making them traditional, daylight close-support aircraft. With drop tanks, the Skyraiders enjoyed a long loiter time over the target area and were very maneuverable at low altitude and low airspeed, even with their four 20mm cannon and 8,000 pounds of ordnance on 15 external hardpoints. The type proved very effective, as by 1960 air-ground coordination was better than ever.[77] Like the Fennecs, the Skyraiders enjoyed a short career in Algeria, but they nonetheless proved to be the most successful of all the ground-support aircraft deployed by the French.

The successful experience with the B-26 in Indochina made the French anxious to obtain an inventory for use in Algeria. The Indochina B-26s had been loaned by the United States and returned following the conclusion of hostilities. Hence in 1956, French authorities began purchasing B-26s from different sources in the United States to fulfill their requirement. The 63 aircraft acquired were a mixture of B-26Bs,

77 Cooper.

Skyraider. (Source: Archives de France)

B-26Cs, and RB-26s, and were overhauled by Fleetway in California and Fairey Aviation in Great Britain.[78] Delivered to France and then to Algeria, these "flying artillery pieces" were used in level bombing, dive bombing, and strafing, and were usually directed by forward air controller (FAC) aircraft. *L'Armée de l'air* lost a total of 19 B-26s during the war with few of these apparently to enemy fire. The 44 surviving aircraft returned to France at the conclusion of the war, all with high airframe hours. The French experience with the B-26 in both Indochina and Algeria was exceptionally positive and would prompt the Portuguese to follow a similar acquisition path for the *ultramar.*

Transport presented fewer problems in Algeria than CAS, and the solid experience in Indochina with the venerable C-47 and the Noratlas proved applicable in North Africa. Nevertheless, the French soon discovered a requirement for a small transport with STOL capability but larger than the then available STOL types. In the early 1950s, Avions Max Holste developed a military version of its commercially unsuccessful design for a light observation aircraft. This bush aircraft, the MH-1521M, was named Broussard (one who lives in the bush) and could be used for airborne command, observation, FAC, medical evacuation, and commando transport. With space for six seats or two litters, it proved an enormous success, and it began arriving in Algeria in 1958 to replace the Criquet and the venerable Piper L4 Grasshopper, both of which were used for observation and FACs. Approximately 335 Broussards

78 Robert Lahalle, "La vie de dingue en plein ciel…," *Historia Magazine* 327 (9 July 1973), 2611.

Broussard. (Source: Archives de France)

were purchased by the French military for use with *l'Armée de l'air* and ALAT. With its 450 horsepower Pratt & Whitney R-985 Wasp Junior radial engine, the Broussard could take off in 200 meters and had a 700-mile range.[79]

As imaginative as the French use of fixed-wing aircraft was, the most lasting innovation from the Algerian experience was the heavy use of the helicopter and the development of associated tactics. In Indochina, a small number of Hiller H-23 Ravens and Sikorsky H-19s were available for casualty evacuation. The technology of the time did not allow for much more, but by the beginning of war in Algeria, larger and more powerful machines were becoming available in quantity. The Piasecki H-21 Workhorse or "flying banana" and the Sud-Aviation-built Sikorsky H-34 helicopters rapidly displaced fixed-wing aircraft for the transport of specialized troops and quick-reaction commando teams. By 1960, the French had developed airborne assault operations in which large numbers of troops were moved to enemy contact by well-orchestrated formations of helicopters.

It was soon found, however, that helicopters were highly vulnerable during the assault landing phase of an airborne operation. As Bernard Fall notes, "The Algerian nationalists soon learned about the foibles of the lumbering and noisy craft and quickly developed effective techniques for helicopter baiting and trapping."[80] Fixed-wing

79 José Munkelt Gonçalves, "Broussard, Bush Aircraft," *Mais Alto* 340 (November–December 2002): 9.
80 Fall, 265.

aircraft helped with CAS, but a disciplined enemy machinegun squad often successfully survived heavy air attack to ambush the troop-carrying helicopters that followed. The problem of relative vulnerability was soon painfully acute. The slow-moving and clearly outlined helicopter sat like a "tin duck in a shooting gallery," and its pilot knew that a hard hit on himself, the engine, the transmission, the controls, or the rotors would likely to bring him down.[81] The French, therefore, began to arm their helicopters.

In 1955 in the heart of the Aures, a group of peaks in the Atala Mountains south of Algiers, Colonel Félix Brunet, commander of the 2nd Helicopter Wing (*2ᵉ Escadre d'Hélicoptères*) made the first experiment in arming a helicopter. His troops were pinned by rebel fire from a hillside above their position and could not flank the enemy before dark. Brunet armed two men with automatic weapons and put one in each of the two casualty panniers mounted outside the Bell H-47. The enemy was soon routed by the helicopter fire, and Brunet's men gained their objective.[82]

Clearly the *ad hoc* Bell was not the answer, but it achieved its purpose and led to further experimentation and imaginative thinking. The first helicopter to be armed properly was the Sikorsky H-19, which carried a 20mm cannon and two rocket launchers as well as another 20mm cannon, two .50-caliber machineguns, and a 7.5mm machinegun mounted in the cabin. Many other combinations were tried, but the H-19 lacked the necessary power to fly with its needed weapons. Regardless of the various ways in which its armament and ammunition load were adjusted, it was unable to develop an effective balance between agility and firepower. It remained sluggish with maneuverability and performance characteristics too poor for offensive action. Essentially, it could not meet the basic requirement for its own defense, the ability to return enemy fire with overwhelming volume. The Piasecki H-21 was also tried, as it was an immediately available alternative, but it had the same deficiencies.

Finally in 1956, the Sikorsky H-34 appeared and proved able to carry the necessary armament to the fight and direct it effectively. Standard weaponry for the H-34 was the German-designed MG 151/20, a rubber-cushioned 20mm cannon on a French-developed flexible mount designed to fire from the cabin door, two .50-caliber machine guns firing from the cabin windows to port, and batteries of 37mm or 68mm rockets.[83] These gunships, known as Pirates, were increasingly requested to support ground troops, for after protecting a landing, they could remain on the scene, as they still had adequate fuel and ammunition for CAS. This situation led to the formation of a coherent doctrine for use of the helicopter and the establishment of Heliborne Intervention Detachments (*Détachements d'Intervention Héliportés* or DIH) composed of a *Pirate* and six troop transport helicopters.[84] This approach greatly increased the

81 Cross, 78.
82 Facon, "L'adaptation de l'armée de l'air à la guerre d'Algérie: la lute antiguerrilla," 37.
83 Ibid., 38.
84 Ibid, 37–38.

value of an infantry battalion through its integral mobility and fire-support; however, there was still a flaw in that a heavily loaded gunship found it difficult to maintain station with the lighter troop-carrier helicopters. Several weapon and helicopter developments were nearing fruition that would affect the combat performance and hence the lethality of the new helicopter gunship concept.

One of the most important air units in the Algerian war was the Overseas Group 86 (*Le Groupe d'Outre-Mer 86* or GOM-86). Stationed at Air Base 140 at Blida, about 45 kilometers southwest of Algiers, it participated in operations from the beginning of the insurrection in 1954 until the cease fire in 1962, maintaining a number of detachments in Constantine, Algiers, Oran, and the departments of the south. The unit was equipped entirely with the twin-engine Dassault Flamant light transport in three versions: MD-311, MD-312 and MD-315. The transports were converted into makeshift bombers with six hardpoints installed under the wings for bombs or rocket pods. In 1956 the Nord SS-11 ground-to-ground (*sol-sol*), wire-guided, anti-tank missile appeared with the French army and was quickly recognized by aviators for its potential in counterinsurgency. One particular Flamant model, the MD-311, was modified to accommodate the SS-11, which apparently would work from a slow-flying aircraft. Redesignated as the AS-11, the air-to-ground (*air-sol*) adaptation of the SS-11, it was first used in Algeria in a successful attack against insurgents who had taken refuge in heavily fortified caves along steep mountain gorges. The traditional attacks by T-6s and B-26s with rockets and napalm had failed, so four of the AS-11s were mounted on a MS-311 and fired successfully against the caves. In the hands of a seasoned operator, the AS-11 was very precise, and the bombardier with practice could actually put the missile through the specific window of a house. The weapon system had been a purpose-designed adaptation and had proved successful. The next step would be to try it from a hovering helicopter that would have certain obvious advantages over a moving fixed-wing aircraft. The H-34 was not ideal for this new weapon; however, French aircraft builders were now offering new, powerful, jet-engine helicopters that could easily handle the weight of the AS-11 missiles and their guidance equipment over and above the machinegun and cannon suite.

The first generation of these jet turbine-powered helicopters was the Sud-Aviation Alouette II (Lark), which went into production in 1957. Its sturdy design translated into reliability, easy maintenance, and adaptability. It was reminiscent of the early Bell models, in that it had a Plexiglas cockpit and open-framework fuselage, and was used for observation, liaison, search and rescue, training, casualty evacuation, and above all, offensive operations armed with the AS-11. It could also be fitted with a wide variety of rockets and guns. The next generation was the Alouette III, which was a significant improvement over its predecessor with its enclosed cabin that could accommodate up to seven, advanced equipment, and more powerful turbine engine. The Alouette IIs and IIIs could carry four AS-11s each, the latter often operating in mixed formations with gun-armed Alouette IIIs carrying a 20mm MG-151 cannon on a mounting in the port cabin door. Series production began in 1961, too late to have any significant impact in Algeria.

The Portuguese Approach

The clear lesson from the two French campaigns is that airpower in counterinsurgency does not necessarily require a high technology solution. Advanced aircraft and their weapon systems generally are unsuited for the austere support environment of such a conflict and are usually ineffective operationally, difficult to maintain, and not cost-effective. The most successful aircraft in the French experience were the simpler training and civil ones modified to military needs and able to perform the core missions of counterinsurgency well with their long loiter times, substantial armament, and relatively uncomplicated design. Above all else the French sought an air force that could be sustained and operated on a restricted budget with a limited aviation infrastructure. This was important in enabling a decentralized approach to operational control from relatively isolated airfields with the attendant local flexibility, as embodied in the EALA, ALAT, and DIH structures.

The more specific lessons can be divided into air strategy and operations. In the area of strategy, the air force must be able to blanket the theater of operations with an air umbrella in support of ground operations. This translates into the establishment of a system of airfield infrastructure that will enable timely air support to ground forces and will generally involve a central, fully capable air base supporting outlying, more spartan airfields in a spoke wheel-like concept. This infrastructure hierarchy likewise translates into support of a decentralized approach to counterinsurgency, one in which the local commander at the end of each "spoke" can respond readily to conditions in his area and use his local knowledge, experience, and imagination to tailor air support and deliver it effectively to sustain a ground offensive.

At the operational level, consideration must be given to the condition of the airmen and their aircraft. The aircrews must be well trained in the operating environment and acclimated to the severe weather and flying conditions. Few Westerners are, so this preparation is important. Aircraft will invariably pose a maintenance headache, as it is unlikely that they are designed for tropical or desert environments, and hence heightened maintenance cycles will affect operational capability and the readiness to do battle. The heat, whether tropical or desert, will severely limit the mechanics' ability to repair aircraft, as the sun in these climes can turn aircraft metal fiery hot and hence the aircraft themselves into ovens. Maintenance can be done only when the sun is low or set. Adequate numbers of the proper aircraft for counterinsurgency are critical in giving aviators the right tools to do their job and respond to ground force needs properly. The French air force never had more than 10 percent of its personnel engaged in Indochina, and the results reflect this low priority. Lastly, the enemy also gets a vote. He can be depended upon to show a surprising and skillful competence in air defense, whether it be by passively camouflaging his activities or actively engaging with anti-air artillery and missiles. In summary, do not enter the conflict blindly or halfheartedly when bringing airpower to the fight.

From the FAP perspective, the Clostermann paper helped it as a service to focus on counterinsurgency. While the FAP was close to the French experiences, and the

Clostermann paper was an ideal vehicle to generate service-wide interest in the use of airpower in counterinsurgency, in truth Arriaga and his air staff were developing a full awareness of what was required to defend the *ultramar* from the aviation perspective. They had studied these experiences of Indochina and the early lessons of Algeria along with the British ones in Malaya and Kenya and other post-World War II insurgencies and, as in all of the case studies, faced the classic dilemma of projecting airpower over vast reaches of forbidding terrain with limited resources.

Over time, the FAP developed its own concept of extensive operational decentralization and dispersal of infrastructure and aircraft as its solution to providing support to ground operations in the vast and difficult conditions of the *ultramar*. This concept was refined in various writings over the course of the conflict, although there was never any formal air doctrine published as such.[85] Shared guidance and experience appeared informally from time to time in published journals, but in reality the only formal document was the Standard Operating Procedures (*Normas de Execução Permanente* or NEP) produced by the air commands in each of the three theaters for a select set of circumstances.[86] Beyond these procedures there was a great deal taught, some of which has surfaced long after the war in memoirs and articles, but in truth much of the doctrinal thinking from this rich aviation experience has been lost. Perhaps this is so because of the decentralization that worked so well. Local commanders understood their situation intimately and were in the best position to craft successful solutions informally to the local problems. These commanders, ranging from the larger bases to the more isolated installations, appreciated the great operational freedom in supporting ground forces in their areas of responsibility, something that stimulated their incentive, imagination, and creativity in crafting solutions to difficult missions.[87] Unfortunately, their records on this thinking were largely either destroyed or abandoned at the local airfields when Portugal left Africa.

The excessive dispersion created some problems with flight indiscipline, resupply, and maintenance; however, these were relatively insignificant compared with the advantages that such freedom of action lent in the support of ground forces. In establishing such a doctrine that relied on independent action, it was imperative that the FAP recruit and retain individuals of the highest caliber. In fact, its recruits in both the officer and enlisted ranks were highly motivated and tended to remain in the force. They received excellent training and were consequently well qualified to implement the decentralized doctrine with their initiative, skills, and self-confidence. This approach to warfare was supported by a carefully considered and planned air-ground integration of forces that proved successful in countering an insurgency with its undefined lines and often unanticipated enemy contact. The decentralization of decision-making

85 Alves de Fraga, 44.
86 Ibid., 44–45.
87 General Jorge M. Brochado de Miranda, former Chief of Staff of the Portuguese Air Force, interview by the author, Alfragide, Portugal, 16 June 2011.

permitted opportune engagement based on limited intelligence and simple, easily understood plans. It also allowed local units to perform their primary ground support role by moving supplies, gathering aerial intelligence, evacuating wounded, transporting troops, and performing visual and photographic reconnaissance, all based on responding to immediate army requirements.[88] In contrast to others, the FAP embarked on the African Campaigns with a clear vision of how it would accomplish its mission, and as we shall see, this decentralized approach to counterinsurgency was enormously successful and holds valuable lessons for those who would craft an air doctrine for a similar conflict.

88 Ibid., 22.

4

Into Africa

In July 1955, when Kaúlza de Arriaga assumed the duties of Subsecretary of State for Aeronautics, he faced the enormous responsibility of preparing the FAP for war in the *ultramar.* Even as the French were conducting their laboratory of airpower in counterinsurgency, and many benefits would accrue to the Portuguese from these experiences in Indochina and Algeria, there remained the need to develop an air strategy for the *ultramar* and put it into effect. This was indeed the preoccupation of Arriaga, as everything would have to be done under the imminent threat of conflict. This produced an enormous sense of urgency that affected FAP organization, construction of infrastructure, acquisition of aircraft, and the recruitment, training, and deployment of personnel to realize an air strategy—one that was at the same time being creatively conceived in light of the limited Portuguese resources, the vast reaches of austral Africa, and allies willing to help. Because a military solution would not be the exclusive answer to a political counterinsurgency, Arriaga faced a number of stubborn obstacles in designing and executing FAP strategy. In examining his dilemma, we will quickly review the basis for Portuguese colonial unrest, next, look at the lessons of early diplomatic missions to identify allies, and last, examine the specific actions of FAP to develop its strategy and position itself to execute it against the headwinds of a shifting political environment and an increasingly agile enemy.

The Long View

The unconventional ideas of General José Norton de Matos, governor-general of Angola from 1912 to 1915 and later its high commissioner from 1921 to 1923, and other respected authorities on the large and rich colony swam against the tide of traditional colonial mercantile policy in advocating economic development. Development in the eyes of these progressives entailed liberating the populations of the *ultramar* politically and economically, that is, making them citizens with the full rights of those in the *metrópole* and allowing them to retain the fruits of their labors in a free economy. Under such a policy, the government would support rather than manipulate the largely agricultural colonial economy by building infrastructure to assist in bringing crops to market and by facilitating the establishment and functioning of those markets. This was seen as a better way over the mercantilist approach favored at the time and based on the notion of an industrialized colonial power using the resources of its colonies to feed its domestic economic engine and dictating the jobs to be performed in the *ultramar*.

The flaw in the mercantilist approach was that Portugal had no significant economic engine that needed to be fed and was thus unable to employ the resources of its colonies to its advantage. The Norton de Matos school of thought saw a different path to the release of latent colonial economic potential and its use as the driving engine for imperial prosperity. Subsequently, these ideas would be embraced with the abolishment in 1961 of compulsory labor and the several tiers of citizenship, yielding rapid growth of the Angolan and Mozambican economies at 11 percent and 9 percent respectively between 1961 and 1974.[1] This was not, however, the happy situation the FAP leadership faced in the late-1950s, as it sought to devise and implement an air strategy sensitive to the political battleground.

This lack of a coherent national approach to the *ultramar* led the government over the years to commit a number of unfortunate errors primarily because of its persistent short-term vision. There was no long-range plan that considered development and eventual independence on the model of Brazil or the integration of the people of the colonies into an overarching Portuguese political and economic entity.[2] Indeed, the reverse was true, and this incoherence in national policy and the domestic and colonial unrest that it spawned was seen as an opportunity by the nationalist movements.

The natives of the colonies were second class citizens in every way, having to carry identity cards that signaled their inferior status and lack of the normal rights of Portuguese citizenship. They were subjected to forced labor and egregious taxation, and the slave-like conditions under which they worked for the concession companies were the source of continued unrest, whether it was cotton in Angola, rice in Mozambique, or cocoa in São Tomé.[3] This government distain for its citizens included those in the *metrópole* as well, and the hardships that were the result of bad government policy prompted European as well as colonial emigration in search of a better life.

Metropolitan Portugal from the earliest times remained economically underdeveloped and was dependent on overseas commerce and colonial wealth to maintain more than a subsistence standard of living. Lisbon in the first half of the sixteenth century was a spectacularly opulent city, yet Portugal as a whole did not produce enough goods to feed and clothe its population at the time, and staples had to be purchased abroad.

1 John P. Cann, *Counterinsurgency in Africa: The Portuguese Way of War, 1961–1974* (Westport, Conn.: Greenwood Press, 1997), 15.

2 Fernando P. Resende, "A Força Aérea em Angola (1960/1963)" [The Air Force in Angola (1960/1963)], *Mais Alto* 228 (March–April 1984): 21.

3 Ibid.

Slaves were imported from the Guinean Coast to supply labor while rural Portuguese emigrated to western Spain in search of employment. The wealth of Lisbon, indeed, seemed useless to the population at large. During these years of plenty, Portugal failed to develop a domestic economy of any consequence, and hence there was no alternative to the stagnation at home when the wealth from abroad finally evaporated with the independence of Brazil in 1822. The Portuguese economy persisted at subsistence levels, and because it was so weak, it failed to participate significantly in the industrial revolution of the nineteenth century. As Portugal was not developing its economic engine, it was unable to employ the resources of its colonies on any appreciable scale and to exploit any relative trading advantage in international commerce. This sad economic state persisted and led to Salazar's emergence in the 1930s in an effort simply to preserve the government fiscally. As Portugal quietly dug itself out of its depression-era financial crisis, this formula for failure remained unchanged. There clearly was no coherent approach to the development of the colonial economies or any apparent idea of what their contribution to the metropolitan economy should be.[4]

Within this dysfunctional and often chaotic economic environment there was a widespread lack of individual opportunity. This void extended from the *metrópole* to the *ultramar* and persisted even in the wake of post-World War II economic expansion. It manifested itself in significant emigration throughout the Empire, something that Henrique Galvão termed vividly as "demographic hemorrhage."[5] In Angola, for instance, the positive economic developments across the northern and western borders in the Belgian Congo and Northern Rhodesia led to large numbers of economic migrants moving to both countries. Holden Roberto was but one example. In Mozambique, 10 percent of the population on average at any one time was working in the South African mines, where a laborer could earn five times the wages he could earn at home.[6] Southern Rhodesia also represented an attractive destination for Mozambicans, for the average migrant felt that he enjoyed a better life there. His wages were higher and paid promptly. There was no forced or contract labor. The articles that he liked were not only available but fairly priced. Women paid no tax, and there were leisure activities in the form of tea houses and bars scattered throughout the Rhodesian bush.[7] Indeed Rhodesia encouraged this migration with stores on the border that sold necessities at bargain prices and with lorries that took the migrants to their destinations inland.

Concurrently, even Portuguese citizens of the *metrópole* found it unattractive to emigrate to the colonies and held no faith that the *ultramar* led the way to any overall

4 Malyn Newitt, *Portugal in Africa: The Last Hundred Years* (London: C. Hurst & Co., 1981), 121.

5 Henrique Galvão and Carlos Selvagem, *Império Ultramarino Português*, IV Volume [Portuguese Overseas Empire, Volume IV] (Lisbon: Empresa Nacional de Publicidade, 1953), 132.

6 Newitt, *Portugal in Africa*, 115 and 119.

7 Malyn Newitt, *A History of Mozambique* (London: C. Hurst & Co., 1995), 514–515.

economic development. Indeed, it was more attractive in the mind of a Portuguese citizen to emigrate to Brazil or to another European country than it was to the colonies, and in 1961 alone the total number of Portuguese emigrating to these countries was more or less equivalent to the sum of the entire white population in all of the colonial possessions.[8]

Illustrative of this was an experience of General Fernando Resende, commander of the 2nd Air Region, during a 1961 air evacuation of wounded from Negage to the hospital in Luanda. There were with him four soldiers who had been wounded in an ambush at Santa Cruz. One of them, who came from Mesão Frio, a poor region of the Douro in northern Portugal, was only lightly wounded and was thus able to talk. During his conversation with the general he asked, "They say that Angola is ours, is this true?"[9] The general was astonished at such doubt, and his response in an attempt to justify the soldier's risk of life in the defense of Angola prompted another incredulous comment, "If so, then why is my family sacrificing to cultivate the furrows of our land instead of all of us coming to enjoy the immense fertility of Angola?"[10] Daily, in Resende's experience, one encountered such examples of this ignorance and economic dysfunction. Because of the mixed messages the government had historically sent to its people through flawed policy, the FAP leadership would have to work hard to overcome some very fixed attitudes and craft an effective air strategy that would support a successful military and political campaign to win the loyalty of the population, native and European.

Mission to Dakar

Reflective of Portuguese colonial policy was the experience of its delegation to a conference in Dakar in 1954. Portugal was invited to send representatives to this conference on colonial developments in Africa, which was sponsored and hosted by the French government. Other colonial powers present, in addition to France and Portugal, were the United Kingdom, Belgium, and The Netherlands. The purpose of the gathering was to develop approaches to the problems that the "new winds" blowing through the continent had created for these countries.[11]

In the absence of any precise guidance for the Portuguese delegation, it pursued colonial interests as understood by its members, who were predominately government and military officials and included General Resende. These naturally centered on security concerns and methods of coordinating available forces to maintain order in perpetuating the 400-year Portuguese rule. In the many forums that occurred, this

8 Resende, 21.
9 Ibid.
10 Ibid.
11 Ibid. Other attendees were Commodore Vasco Lopes Alves, the leader, Dr. Franco Nogueira, Captain Armando Roboredo, various customs and immigration officials, and the chiefs-of-staff of the military regions of Angola and Mozambique.

emphasis was out of step with the other delegations, which were exploring ways to acknowledge the independence movements and pave the way towards an orderly transition of their colonial possessions into new, independent nations. Not surprisingly, the Portuguese position produced many awkward moments. Indeed, the Portuguese representatives were surprised at the criticism they received on their stubbornly determined ownership theme and were forced to swallow a bitter pill when there was little collaboration forthcoming from their fellow colonial powers. This resulted in rather more negative than positive conference experiences for the Portuguese but served as an alert to the changes the others saw occurring in their own territories as well as neighboring states and certainly a forewarning of what lay ahead for Portugal.[12] Perhaps this was a rude awakening to the geopolitical environment facing the FAP, something that would affect its strategy and ability to wage war, but also helped to identify those who might support Portugal in the *ultramar*, such as France and Germany, and those who would not.

Five years later in 1959, when Portugal pressed the United States for the Lockheed C-130 Hercules tactical transports and Bell UH-1 Iroquois helicopters, which were proving themselves in Vietnam, an ambiguous reply was initially given. Later, a negative one was delivered from the new Kennedy administration, which had assumed an anti-colonial stance during the *Santa Maria* crisis. This shift in attitude not only with the United States but also with Great Britain forced Portugal to seek alternative supplies in France and the Federal Republic of Germany, more colloquially known as West Germany. There were interesting reasons for the new alignments.

In the case of France, its physical proximity with Portugal and the fact that it was currently fighting an insurgency in Algeria gave the two a common touchstone. Further France under President Charles de Gaulle's leadership was at the time following an independent line regarding NATO and its own defense. It was consequently happy to align itself with those European countries that were also at odds with the United States and NATO leadership. This liaison resulted in France becoming a substantial supplier of war matériel under generous credit terms alongside a general rise in the commercial links between the two countries. For the FAP, it meant that France would help supply the war effort with surplus T-6G Harvards, Nord Aviation Noratlases, and the Sud-Aviation (Aérospatiale) Alouette series of helicopters.

The liaison with West Germany was made for completely different reasons. In any European conflict with the Soviets, it was obvious that West Germany was going to be the primary battlefield. In reaction to this reality, the country undertook to develop during the late 1950s a conventional arms manufacturing capability, infrastructure, and training facilities to support its forces. The single limiting resource in this development was land, and consequently a provision needed to be made for a rear support area for these activities in the event of war. Because of French withdrawal from its military role in NATO, nations nearby that might serve this purpose were limited to

12 Ibid.

Portugal and Italy. Portugal proved ideal with its Atlantic exposure, natural ports, and relatively mild climate. Accordingly, negotiations were opened, and in 1959 a defense agreement was signed. Between 1960 and 1962 there were more than twenty accords between the two countries initiating the construction of infrastructure, the supply of technologically advanced armaments, and the modernization of defense industries under favorable financial terms to Portugal.[13] For the FAP this development was very significant in that Germany would supply surplus Fiat G-91 Ginas, Nord Aviation Noratlases, and Dornier Do-27s. Moreover, whenever there was a shortage of munitions in the African campaign, Germany promptly and without question supplied them to Portugal.[14]

Organizing for War

Arriaga was appointed subsecretary the year following the Dakar conference, as African security concerns assumed an increasingly elevated profile. During the closing months of 1956, he put in place organizational reform, first, consolidating the aviation support services from the army and navy into the FAP equivalent and, second, apportioning the Empire into Air Regions.

The new FAP aviation support was organized into six directorates to perform the needed sustainment functions, and these were:

1. Communications and Air Traffic
2. Recruitment and Instruction
3. Health
4. Matériel
5. Infrastructure
6. Management and Accounting.

While these performed vital support functions, and each would be represented at most aviation installations to sustain the assigned forces, there were two very critical ones in projecting power in the *ultramar*, the Directorate of Infrastructure (*Direcção de Serviço de Infra-Estruturas),* which we will examine shortly, and the Directorate of Management and Accounting (*Direcção do Serviço de Intendência e Contabilidade* or DSIC). This was the logistics arm and was responsible for subsistence, uniforms, fuel and lubricants, necessities of hygiene and comfort, and other consumables not attributed to Matériel.[15] Its most central aspect and the one that warrants comment is the

13 Newitt, *A History of Mozambique*, 524.

14 Ibid., 525. Given the politics of the time, none of these bore German marks of origin in order to prevent attribution and consequent accusations by the guerrilla organizations that Germany was taking sides.

15 Eduardo Velho da Costa, "Tempos Difíceis mas Estimulantes (1961–1975)" [Difficult but Stimulating Times] in *50 Anos de Administração Aeronáutica* [50 Years of Aeronautical

acquisition and distribution of aviation fuel, indispensable in the day-to-day operation of the FAP. Aircraft simply could not fly without the correct fuel delivered in proper condition. This was so important in projecting airpower that the DSIC maintained a presence on all major staffs and particularly those of the operational commands in the *ultramar*.[16] The portfolio of DISC responsibilities was complicated by the variety of products required, their fragility, and the need to deliver them to some very remote sites under harsh climatic and combat conditions. The distribution of kerosene, automobile gasoline, aviation gasoline in three different grades (80/87, 100/130, 115/145), and jet fuel in two (JP1 and JP4) was just the beginning. The list of products also extended to a vast number of specialty lubricants to cover the many and varied needs of aviation machinery.[17]

These products moved by ship to the various key entry ports, such as Bissau, Luanda, and Beira, and, in the case of fuels, were transferred to bulk storage tanks. The lubricants were generally in standard drums or even smaller containers, and the parts were in crates. From bulk storage the products were moved to rail car, tank truck, or drums that could be loaded on army flatbed trucks or navy landing craft for further shipment by road or river, depending on the ultimate delivery location. The army was the willing partner in this distribution, as it organized and ran the ground transport system and furnished the needed security for its operation. It also provided security for the many airfields and landing sites, particularly for helicopter support. Often the routes to supply the remote installations followed unimproved roads through some of the most primitive and daunting territory in the world, and hence travel by truck could be a slow and grueling process in most areas. Paved roads were easy routes but tended to connect only the principal towns, not the villages and the more remote sites, airstrips that were important in supporting the ground forces and the population. Delivering fuel, lubricants, and aviation spares to these more isolated areas required a fleet of supply vehicles negotiating not only the paved highways but poor roads, tracks, and trails. In many cases these latter *pistes* could degenerate into rutted tracks through deep sand. Vehicles could and did become stuck in these rudimentary, indeed vestigial, roads, requiring hours to extricate. During the rainy season unpaved roads became quagmires of soft mud and sand, and were virtually impassable. There was also the enemy to consider, and his activities were a further impediment in the more inaccessible areas.

These petroleum products were easily contaminated by dust and water. The oscillating temperatures and high humidity in the tropical theaters heightened this vulnerability and meant that the fuel handlers had to be careful at every turn, particularly at the more spartan locations where fuel was pumped by hand from drums. The filters

Administration], ed. Manuel de Campos Almeida (Lisbon: Força Aérea Portuguesa, 2007), 93.

16 Ibid.

17 Ibid.

and storage containers had to be kept meticulously clean under even the most environmentally hostile conditions. Fuel was tested regularly, and such care was reflective of a continuing preoccupation with quality assurance.[18] Indeed, it was almost miraculous that aviation fuel arrived in a timely way at its needed destination after often long and precarious trips either on flatbed trucks over the most unimaginable roads or by river transport under equally miserable conditions. Meeting this requirement with regularity and reliability remains one of the great and yet largely unsung aviation logistic achievements of the conflict. As Field Marshal A.C.P. Wavell so ably explained it:

> The more I have seen of war the more I realize how it all depends on administration and transportation (what our American allies call logistics). It takes little skill or imagination to see *where* you would like your army to be and *when*; it takes much knowledge and hard work to know where you can place your forces and whether you can maintain them there.[19]

The second reform was the apportionment of the Empire into three Air Regions (*Região Aérea* or RA). These were administratively and geographically manageable divisions based on geopolitical considerations:

- The 1st Air Region encompassed continental Portugal, and the adjacent Atlantic archipelagoes of Madeira, the Azores, and Cape Verde, with its headquarters in Lisbon.
- The 2nd Air Region included the territories of Angola, Guiné, and São Tomé e Príncipe, with its headquarters in Luanda.
- The 3rd Air Region extended to Mozambique, Portuguese India, Timor, and Macau, with its headquarters initially in Lourenço Marques.

The regions were further subdivided, each containing an area with peculiarly distinguishing local operating characteristics. In the 1st Air Region, for example, the subdivisions were called Air Zones (*Zona Aérea* or ZA) and there were three:

- 1st Air Zone: Continental Portugal and the Madeira Islands
- 2nd Air Zone: The Azores archipelago
- 3rd Air Zone: The Cape Verde archipelago

In 1958 Guiné was transferred from the 2nd Air Region to the 1st Air Region and combined with Cape Verde to form the Air Zone of Cape Verde and Guiné (*Zona Aérea*

18 Ibid., 94–95.
19 Archibald C.P. Wavell, *Speaking Generally* (London: Macmillan, 1946), 78.

de Cabo Verde e Guiné or ZACVG). In 1961, Portuguese India was removed from the 3rd Air Region, but otherwise this alignment continued until 1975.[20]

This designation of Air Regions and Zones was but the first of two phases in establishing an air support network. These were based on establishing a hub air base in the combat theater of each Air Region and in building the needed spokes from each of these bases to bring the umbrella of air support to the ground forces and population and enable decentralized operations. Both phases were targeted for completion by 1960. The ZACVG and the two *ultramar* Air Regions were virtually paper constructs at the time, and their command organizations and infrastructure would be established initially in Bissau (ZACVG), Luanda (2nd AR), and Beira (3rd AR). There would be at the start a medium-sized aerodrome based on a civilian field established at these locations, and later in the second phase these fields would be expanded to become wholly capable air bases. The definitions of the installation types within this array of facilities were formally established in a February 1960 dispatch from Arriaga and are described below:

- Air Base (*Base Aérea* or BA): A permanent air base that has the full array of support facilities and can handle all types of aircraft in significant volume.
- Aerodrome Base (*Aeródromo Base* or AB): A permanent air base that has essential primary support facilities and can handle most types of aircraft in limited volume.
- Logistical Aerodrome (*Aeródromo Logístico* or AL): A permanent air base designed to handle transport aircraft in the execution of primarily logistic missions.
- Transit Aerodrome (*Aeródromo de Trânsito* or AT): A permanent aerodrome designed to support transit flights with aircraft refueling and replenishment, aids to navigation, limited maintenance, and crew and passenger rest facilities.
- Alternate Aerodrome (*Aeródromo de Recurso* or AR): A permanent aerodrome with limited aids to navigation for use in special missions and emergencies.
- Training Aerodrome (*Aeródromo de Instrução* or AI): A permanent or temporary aerodrome to complement an Air Base or Aerodrome Base that supports flight training.
- Maneuver Aerodrome (*Aeródromo de Manobra* or AM): A permanent or temporary aerodrome, complementing an Air Base or Aerodromo Base, designed to provide the necessary support for air operations as dictated by the local tactical situation.[21]

20 Luís Alves de Fraga, *A Força Aérea na Guerra em África: Angola, Guiné e Moçambique, 1961–1974* [The Air Force in the War in Africa: Angola, Guiné and Mozambique, 1961–1974] (Lisbon: Prefácio, 2004), 40.

21 Eduardo Koll de Carvalho, "As Infra-Estruturas Aeronáuticas em África: Antecedentes e Implantação das Delegações em África" [Aviation Infrastructure in Africa: Its Background and Establishment in Africa], *Infra-Estruturas Aeronáuticas em África* – Colóquio, VI Encontro da Associação da Força Aérea Portuguesa [Aviation Infrastructure in Africa – Colloquium, 6th Meeting of the Portuguese Air Force Association], proceedings of

As a part of the due diligence effort at installing the FAP in Africa and because of its nearly complete absence there, Arriaga authorized an exploratory mission in the early months of 1958.[22] In January of that year, the Chief of Staff of the Air Force, General Costa Macedo, directed a survey team of seven hand-picked officers to visit key areas of the *ultramar* with the primary objective of studying the existing civil airfield infrastructure as to its suitability for military operations and identifying the needed infrastructure expansion for air defense of the *ultramar*.[23]

The task, led by his deputy, General Venâncio Augusto Deslandes, departed Lisbon in a Dakota on 12 May for the 39-day trip.[24] During that time the Dakota flew 215 hours, covered approximately 16,000 miles, and visited Cape Verde, Bissau, São Tomé, Luanda, Lobito, Luso, Lourenço Marques, and Beira. When the group returned on 19 June, it had developed significant intelligence on the *ultramar* and various other countries along the west coast of Africa, as well as South Africa and the Rhodesias.

The focus of the mission had been Angola, and its members recommended a triangular disposition of key airfields at Luanda, Luso, and Moçâmedes, which would enable complete coverage of the colony. However, by 1960 the threat was identified as likely to come from the former Belgian Congo, which had just been granted its independence, and this altered the initial recommendation. The first three fields were consequently established in Luanda (BA 9), Negage (AB 3) and Henrique de Carvalho (AB 4). Much later in the war it became apparent that Luso was of greater operational importance than Henrique de Carvalho. Curiously, today Moçâmedes has become an important base of the Angolan Air Force (*Força Aérea Angolana*), and hence, we see a certain coherence and thus justification in the original airfield disposition offered by the mission.

In the ZACVG, the primary airfield was established at Bissau as Aerodrome Base 2 (AB 2) in 1961 and later in 1965 upgraded to Air Base 12 (BA 12). It served as the hub in Guiné without the usual developed hierarchy of fields, as the slowest aircraft could be at any of the frontiers within 30 minutes of takeoff. While there were AMs in Nova Lamego, Aldeia Formosa, and Cufar, BA 12 had some forty regularly used extension airfields that in some cases could accommodate tactical transports as heavy as the Noratlas and Dakota but were mostly serviced by light aircraft. The primary

a conference held at the Academia da Força Aérea, Granja do Marquês, on 18 October 1997, 29–31.

22 José Krus Abecasis, "A Génese dos 'Transportes Aéreos Militares': As limitações, o ordem lógica e a competência técnica, 1947–1962" [The Genesis of the "Military Air Transports": The limitations, the logical order and the technical competence, 1947–1962], *Mais Alto* 285 (September/October 1993): 17.

23 Manuel Diogo Neto, "Recordando a Esquadra 92 da BA9 – Luanda" [Remembering Squadron 92 of BA9 – Luanda], *Mais Alto* 264 (March–April 1990): 38–39.

24 Other members of the mission were Lieutenants Colonel (Pilots) João Calhau and Ivo Ferreira, Major (Medical Corps) João Varela, Captains (Pilots) Manuel Diogo Neto and António Quintino, and Captain (Engineer) Teixeira de Almeida.

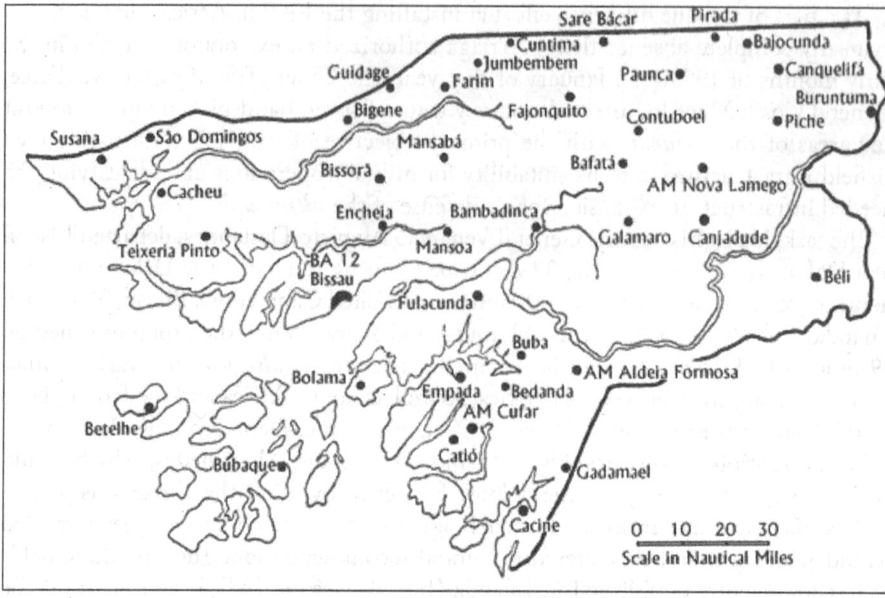

Map 4.1 Primary and Secondary Airfields of Guiné. (Drawn by the author)

ones are shown on the adjacent map. In fact, because of the poor roads and qualified reach of coastal and riverine vessels, almost every village of any size had at least a primitive airfield, although admittedly many of these were of limited capability.

In the 2nd Air Region, Angola was divided into three Air Sectors of Operations (*Sector Aéreo de Operações* or SEC AR), each identified with its primary airfield. The boundaries of the SEC ARs overlaid those of the ground forces, as follows:

- SEC AR AB 3 (Negage) – Northern Military Zone,
- SEC AR AB 4 (Luso) – Eastern Military Zone,
- SEC AR BA 9 (Luanda) – Central and Southern Military Zones, and Cabinda.

The commander of the region was located in Luanda throughout the conflict. The three SEC ARs and their airfields are depicted in the nearby map.

The 2nd Air Region also included the relatively isolated archipelago of São Tomé e Príncipe in the Gulf of Guiné, some 675 nautical miles northwest of Angola. The island chain, which contained the vital transit airdrome of São Tomé (AT 2), was a propitiously located waypoint between the transit airdrome (AT 1) on the Island of Sal in the Cape Verde archipelago and Air Base 9 (BA 9) in Luanda, and served as a vital link in the air route between the *metrópole* and the *ultramar.* Both transit aerodromes received lengthened runways, maintenance facilities, and rest amenities for crew and passengers.

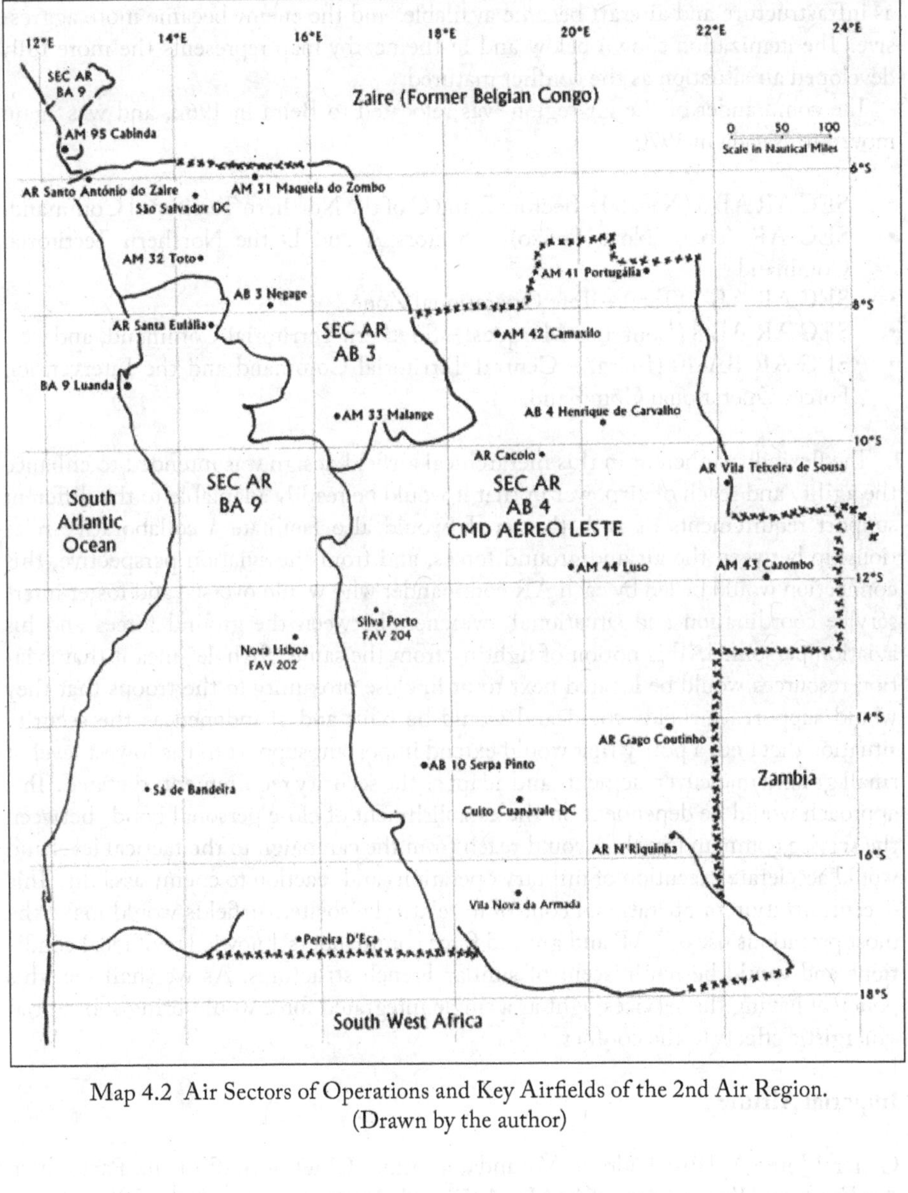

Map 4.2 Air Sectors of Operations and Key Airfields of the 2nd Air Region. (Drawn by the author)

The 3rd Air Region comprised Mozambique and a small airdrome with a helicopter detachment in Díli, East Timor. The SEC ARs in Mozambique, like those in Angola, were identified with a primary airfield, and their boundaries overlaid those of the ground forces. They began with a simple approach and became more elaborate, as infrastructure and aircraft became available, and the enemy became more aggressive. The itemization shown below and in the nearby map represents the more fully developed air situation as the conflict matured.

The commander of the air region was relocated to Beira in 1962, and was again moved to Nacala in 1970.

- SEC AR AB 5 (Nacala) – Sectors B and C of the Northern Territorial Command,
- SEC AR AB 6 (Nova Freixo) – Sectors A and D the Northern Territorial Command,
- SEC AR AB 7 (Tete) – Tete Operational Zone,
- SEC AR AB 8 (Lourenço Marques) – Southern Territorial Command, and
- SEC AR BA 10 (Beira) – Central Territorial Command and the Intervention Forces Operational Command.

The flexibility inherent in this hierarchical airfield design was intended to enhance the agility and reach of airpower in that it would be readily adaptable to the different support requirements of each theater. It would also facilitate a collaborative relationship between the air and ground forces, and from the aviation perspective, this connection would be led by each AR commander who would oversee and foster interservice coordination and situational awareness between the ground forces and his aviation personnel. This notion of fighting from the same "foxhole" meant that aviation resources would be located next to or in close proximity to the troops that they would support. Airfields could and would be built and abandoned as the security situation dictated, a policy that would extend important support to the lowest level of small ground maneuver elements and adapt as the security requirements dictated. This approach would be dependent on the establishment of close personal bonds between the service commanders that would reach from the campaign to the tactical level and would accelerate execution of military operations and reaction to enemy assaults. This decentralization of operational control to relatively isolated airfields would make the most propitious use of FAP and ground force commanders' knowledge of local conditions and would be reminiscent of similar French structures. As we shall see, this policy of having the services fight as a single integrated force would bring substantial synergistic effects to the conflict.

Imperial Airline

General Jorge M. Brochado de Miranda, a former Chief of Staff of the Portuguese Air Force, recalled the date of 18 May 1957 as the moment in which the full attention of the FAP shifted 180 degrees from its concentration on the defense of Europe under

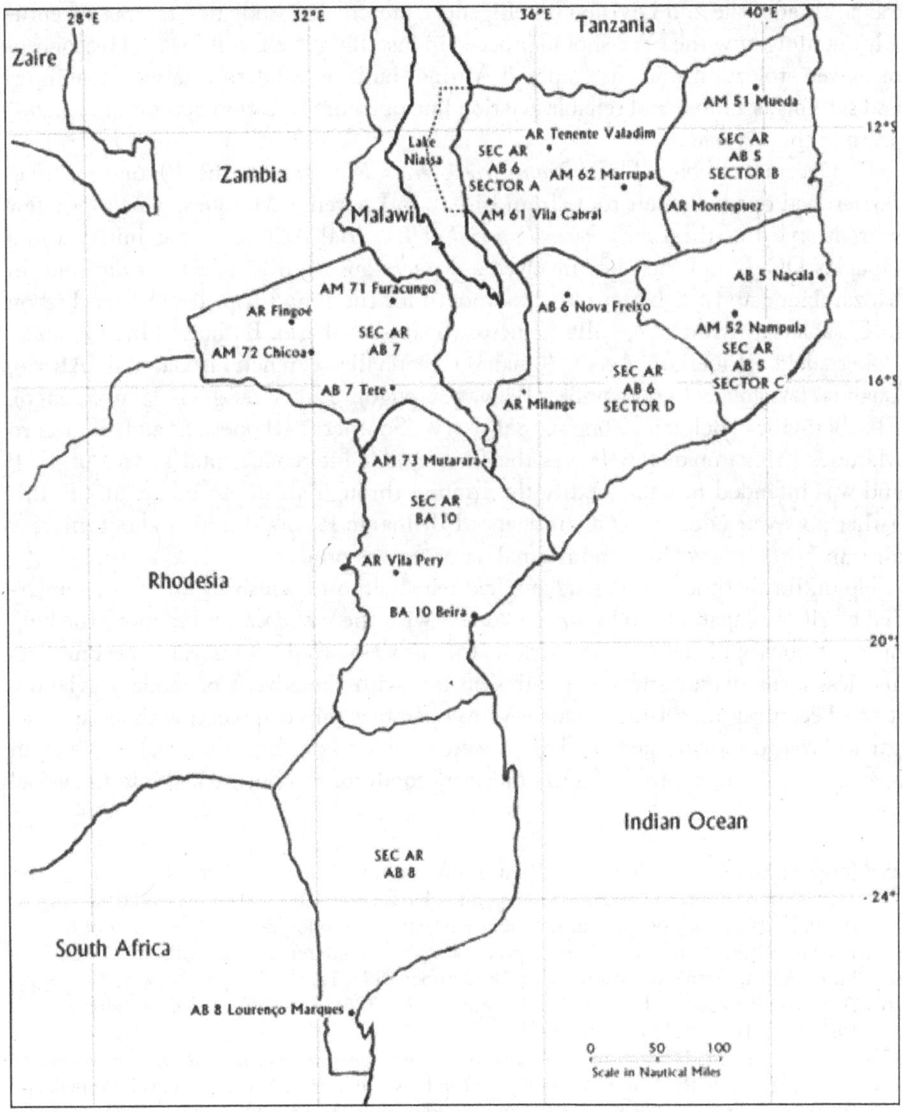

Map 4.3 Air Sectors of Operations and Airfields of the 3rd Air Region. (Drawn by the author)

the NATO umbrella to its preparation for defense of the *ultramar*.[25] On this particular date, Arriaga penned note No 195 directing the FAP Chief of Staff to prepare and deploy units to the 2nd and 3rd Air Regions, Angola and Mozambique respectively, and specifically to the cities of Luanda, Lobito, Beira, and Lourenço Marques.

The execution of Arriaga's directive prompted critical thinking from his talented and imaginative staff, and in November 1957, Lieutenant Colonel Joaquim Brilhante Paiva, chief of the 2nd Division (intelligence), presented a study to Arriaga conceptually detailing how the FAP should proceed in installing itself in Africa.[26] His concept of power projection was an "Imperial Airline" built on modern aviation capabilities and serving as a rapid and reliable aviation line of communication between the *metrópole* and the *ultramar*.

In December 1946, TAP (*Transportes Aéreos Portugueses*), the Portuguese Flag carrier, had opened an air route from Lisbon to Lourenço Marques, which for a few years it styled as the *Linha Aérea Imperial* or Imperial Airline. It was initiated in a Douglas DC-3, and the route involved a dozen stops from Lisbon to its terminus in Mozambique and the better part of a month for the round trip. It ran from Lisbon to Casablanca (Morocco), Villa Cisneros (Spanish Sahara), Bathurst (The Gambia), Robertsfield (Liberia), Accra (Ghana), Libreville (French Equatorial Africa), Luanda (Angola), Léopoldville (Belgian Congo), Luluabourg (Belgian Congo), Elizabethville (Belgian Congo), Salisbury (Southern Rhodesia), and Lourenço Marques (Mozambique).[27] It was the first regular intercontinental service of TAP and was intended to help solidify the Empire through an improved air link.[28] This earlier post-war effort not only influenced Brilhante Paiva's thinking about military air transport but gave him an informal name for the project.

Up until this time, the Portuguese had relied almost exclusively on their commercial maritime capability to link the *ultramar* with the *metrópole* and to move the bulk of the troops and matériel to Africa. This process would foreseeably become less and less attractive as any war progressed, for with the advent of modern aviation, it was becoming increasingly expensive to operate a ship compared with an aircraft, particularly in moving people. Troops were unable to perform their duties while on a voyage and, indeed, travelled in unhealthy conditions, often packed into the vessel

25 Jorge M. Brochado de Miranda, "Palavras de Abertura" [Opening Remarks], *Infra-Estruturas Aeronáuticas em África* – Colóquio, VI Encontro da Associação da Força Aérea Portuguesa [Aviation Infrastructure in Africa – Colloquium, 6th Meeting of the Portuguese Air Force Association], proceedings of a conference held at the Academia da Força Aérea, Granja do Marquês, on 18 October 1997, 12.

26 Arquivo Histórico da Força Aérea, Informação No 226/57 in *A Força Aérea em África*, Lisbon, EMFA, 2ª Repartição, April 1959.

27 António Jorge da Silva Soares, *Hisrórias com Asas: Memórias dos Pilotos da TAP dos tempos fo Hélice* [Stories with Wings: Memories of the Pilots of TAP in the Time of the Propeller] (Lisbon: Associação dos Pilotos Portugueses de Linha Aérea, 1992), 105.

28 José Manuel Correia, "A Linha Aérea Imperial" [The Imperial Airline] *Mais Alto* 372 (March/April 2008): 38–42.

like sardines in a tin.[29] For a nation pressed for manpower and cash, it made less and less sense to have an unusable manpower pool hostage to an expensive transit process. An aircraft could transport troops and priority cargo far more efficiently when all associated costs were considered.

The thesis of the report was that in order to project power throughout the Empire, there needed to be a communications network supported by a fleet of military air transports. This versatile network would have many potential benefits in that it could rapidly reinforce the *ultramar* with troops and urgently needed supplies, and transfer resources from one theater to another, as required. It could also evacuate the sick and wounded to the *metrópole* for advanced care, an enormous factor in troop morale. At the time, military air transport was limited to metropolitan Portugal, and Brilhante Paiva advocated vastly increasing this capability and extending it throughout the *ultramar* in support of both the political and military dimensions of security. There was, however, a drawback in that the FAP barely had the technical means to cover the vast spaces of Portuguese Africa and satisfy the critical requirement of having to use its own airfields exclusively, as in a continent hostile to colonial empires, it could obtain few landing rights outside of its territories.

The basis for Brilhante Paiva's concept was the appearance during World War II and immediately afterwards of large, reliable, four-engine military transports, such as the Douglas C-54 Skymaster, the Boeing C-97 Stratofreighter, and the Lockheed C-121 Constellation.[30] Portugal in 1947 had just begun to receive deliveries of the Skymaster and possessed only two in its inventory.[31] It would require many more such aircraft with its 3,650-mile range and useful cargo capacity of 28,000 pounds to execute Brilhante Paiva's concept successfully. Likewise, an accelerated upgrade of the Bissau airfield would be needed to accommodate the anticipated aircraft traffic and handling requirements, as it would be an important refueling stop for aircraft on their way to Portuguese possessions further south and east. Despite the lack of wherewithal, Arriaga would use this thinking as a springboard to build a robust air network with the improvement of existing airfields, the creation of new ones, and the acquisition of new, more capable aircraft. The Brilhante Paiva study led to the obvious conclusion that even with thin transport resources and an incomplete infrastructure,

29 Rui Tavares Monteiro, "Transporte Aéreo na FAP, Do zero aos 707" [Air Transport in PoAF, from Zero to the 707], *Mais Alto* 283 (May–June 1993): 5.

30 Mário Canongia Lopes, "Os Últimos Skymasters – Aviões Quadrimotores de Transporte da Aviação Militar Portuguesa (1952-1976) – II Parte" [The Last Skymasters – The Four-Engined Air Transports of Portuguese Military Aviation (1943–1951), Part II], *Mais Alto* 325 (May–June 2000): 20–23.

31 Mário Canongia Lopes, "Os Liberators e os Primeiros Skymasters, Aviões Quadrimotores de Transporte a Aviação Militar Portuguesa (1943–1951), I Parte" [The Liberators and the First Skymasters – The Four-Engine Air Transports of Portuguese Military Aviation (1943–1951), Part I], *Mais Alto* 324 (March–April 2000): 34–40.

the FAP had the capacity to begin immediate air operations to link the *metrópole* with the widely scattered overseas provinces.

Portugal in 1936 had acquired ten export versions of the Junkers Ju-52 transport aircraft, which could be adapted to bombing missions.[32] During the 1940s some of the aircraft had the armament capability removed, while others continued to serve in this role, but by the 1950s all of these transports had been relegated to training navigators, radio operators, and parachutists, and to domestic transport duties. While it had the capability of dropping a modest stick of 14 paratroopers, its relatively small cargo door limited its ability for both airlift and cargo drops by parachute. The Portuguese substituted the Pratt & Whitney R-1340-AN-1 Wasp powerplants with their increased reliability for the original BMW 132A-2 engines, and went on to acquire two additional Ju-52s of German manufacture and 15 of the French version in the closing months of 1950. Even with the engine upgrade, however, the Ju-52 was outmoded with its fixed narrow-track main landing gear and 132-knot cruise speed.[33] While the fleet had come cheaply, the Ju-52 was no bargain, as neither its 590 mile range nor its 4,067-pound payload would help realize the Imperial Airline concept, and the next generation of transport was needed.[34]

During 1943 and 1944, six Consolidated B-24D Liberators were forced to land in Portugal and were interned, as Portugal was a neutral at the time. Because they were of no use to the Portuguese as bombers, they were employed experimentally by the Transport Section of the Army Air Arm (*Secção de Transportes de Aeronáutica Militar*) as makeshift long-range transports. In April and May 1944, some long flights were made to Sal in the Cape Verde archipelago and Lajes in the Azores. In truth, the state of the Liberators and the lack of aircrew training in heavy aircraft restricted their utility. Nevertheless, this limited exposure opened a window on the sophisticated maintenance support system, infrastructure, and training needed to operate a fleet of heavy, four-engine aircraft.[35]

When the war ended in 1945, Portugal had only three interned C-47 Dakotas and four flyable Liberators of the original six for heavy air transport and hence no true capability. Nevertheless, it clearly saw the need for such a fleet and made the first tentative steps at acquiring one. Between the end of 1946 and the middle of 1947, it began

32 Mário Diniz, and Luís Proença, "Junkers Ju-52 No 6304: Da Alemanha ao Museu do Ar um Percurso de 73 Anos" [Junkers Ju-52 No 6304: From Germany to the Air Museum a journey of 73 years], *Mais Alto* 385 (May/June 2010): 39–44.

33 The fixed narrow-track main landing gear made the Ju-52 prone to bouncing from side to side on landing. Originally it was fitted with large spats to streamline the main wheels, but these were removed in the operational environment, as they quickly became clogged with sand and mud.

34 Mário Canongia Lopes, *Os Aviões da Cruz de Cristo* [Airplanes of the Cross of Christ] (Lisbon: Dinalivro, 2001), 298 and 333.

35 Lopes, "Os Liberators e os Primeiros Skymasters, Aviões Quadrimotores de Transporte a Aviação Militar Portuguesa (1943–1951), I Parte," 34–36.

to increase its competence in operating and supporting heavy aircraft with a mission to the United States to receive instruction in the flying and maintenance of the Boeing B-17 Flying Fortress and the Douglas C-54 Skymaster. The five crews under training took delivery of three B-17s modified for search and rescue (SAR) and two C-54s, and flew them from Kansas to Lajes Field in the Azores via New York, Boston, and Nova Scotia. This represented a major move forward in not only the acquisition of new technology but also the professional development of pilots, aircrew, and maintenance personnel. Indeed, it marked the beginning of modern military aviation in Portugal.[36] In the years immediately following, two additional SAR B-17s and six more C-54s were added to inventory, and these were lodged at Lajes in the Transport Squadron (*Esquadrilha de Transporte*), which had been formed in 1947 to accommodate the earlier acquisitions and represented the first step in establishing the Military Air Transport (*Transporte Aéreo Militar* or TAM) of FAP.[37] Subsequently, four additional C-54s were added in 1958 and 1959, providing a total inventory of twelve.[38]

In the mid-1960s, five SAR versions of the C-54 were added to the fleet, bringing the total C-54 family to 18.[39] These last aircraft were part of a joint effort with the United States at maintaining a SAR capability along the North Atlantic air routes, and their operations widened the competencies within the FAP of skilled pilots, navigators, aircrew, and maintenance personnel, and laid important groundwork toward the acquisition of more capable transport aircraft. By agreement with the United States, the six SAR C-54s were restricted to the North Atlantic, and this situation left eleven available for the Imperial Airline. The C-54, however, was a World War II-vintage machine, and as such, it was limited in its capability. It could just barely fly the 4,200 nautical mile distance from Lisbon to Luanda following the Lisbon–Bissau–São Tomé–Luanda route exclusively through Portuguese territory. This entailed two en route stops to refuel and change crews. The C-54 was slow with a cruise speed of 165 knots and took effectively two days to fly from Lisbon to Luanda.[40] Despite its load capacity, eleven aircraft would have no appreciable consequence in supplying the

36 José Krus Abecasis, "A Génese dos 'Transportes Aéreos Militares': As limitações, o ordem lógica e a competência técnica, 1947–1962" [The Genesis of the "Military Air Transports": The limitations, the logical order and the technical competence, 1947–1962], *Mais Alto* 285 (September/October 1993): 13–19. See also José Krus Abecasis, *Bordo de Ataque, Memórias de uma Caderneta de Voo e um Contributo para a História* [Attack Heading, Memories from an Aviation Log Book and a Contribution to History] (Coimbra: Coimbra Editora, 1985), 101.

37 Lopes, "Os Últimos Skymasters – II Parte" [The Last Skymasters – Part II], 20.

38 Lopes, *Os Aviões da Cruz de Cristo*, 323.

39 Mário Canongia Lopes, "Os Últimos Skymasters e os Searchmasters – Aviões Quadrimotores de Transporte da Aviação Militar Portuguesa (1952-1976) – III Parte" [The Last Skymasters and the Searchmasters – The Four-Engine Air Transports of Portuguese Military Aviation (1943–1951), Part II], *Mais Alto* 326 (July–August 2000): 20–23.

40 Monteiro, "Transporte Aéreo na FAP, Do zero aos 707," 5.

ultramar in a major emergency. Further, the ageing C-54s were complicated and difficult to maintain and thus uneconomical to operate alongside more modern aircraft. The Imperial Airline concept would require something else to be successful.

In 1961, Pan American World Airways was facing a decision about whether to conduct a major overhaul on two of its high-flight-hour Douglas DC-6A (cargo) and eight of its DC-6B (passenger) Cloudmaster transport aircraft or simply to retire them. Coincidently, Portugal was looking for a more capable aircraft to augment the C-54, and these were made available. The DC-6 Cloudmaster and its military version, the C-118 Liftmaster, were developed as faster, larger, pressurized versions of the C-54, and in fact, the initial model was a cargo transport. When the powerful Pratt & Whitney R-2800-CB-17 Double Wasp radial engine became available, Douglas engineers decided that they could extend the fuselage of the DC-6A by 4.5 feet to produce the DC-6B, a major advance in passenger aircraft. The Double Wasps were the most efficient piston engines available at the time and gave the aircraft the best operating economies of any large contemporary piston airliner. Likewise, passengers and crew appreciated its relative quietness, smoothness, and general comfort over others in its generation.

Portugal undertook to overhaul and modify the ten Pan American DC-6s for its use and put them to work on the *metrópole-ultramar* route. Round trips to the theaters were flown two or three times per week from Lisbon to Mozambique and took two days to perform the entire mission, about half the time as the C-54. Like the C-54, however, these missions always required relief crews stationed en route to maintain the schedule. The aircraft were operated safely at very high utilization rates and never had an accident. While laudable, this safety record should not be surprising, as with the acquisition of the DC-6, the experienced C-54 pilots were all assigned to the new aircraft. This left the C-54 community short of experienced crew, and its safety record suffered accordingly with the loss of four aircraft within the year following the reassignments.[41]

While a significant advance, the DC-6 was still not the perfect answer. Because of the mechanical nature of piston-engine aircraft and their consequent high maintenance requirement, DC-6 availability never exceeded six aircraft at any one time. The DC-6 series could carry only about 50,000 pounds of cargo, and thus this small fleet together with the C-54s still had a very limited impact on the overall logistic picture. The Portuguese were always short of spare parts to maintain the aircraft, and while there were measures taken to skirt this problem, such as chrome-plating the engine cylinders to retard wear, the situation never improved significantly. Cannibalization within the fleet ensued, and by August 1967 only a single DC-6 was flyable.[42] By April 1968 the fleet had recovered to the level of seven operable aircraft, and the crisis

41 Lopes, "Os Últimos Skymasters e os Searchmasters – Aviões Quadrimotores de Transporte da Aviação Militar Portuguesa (1952-1976) – III Parte," 20–23.

42 Monteiro, "Transporte Aéreo na FAP, Do zero aos 707," 5.

passed; however, severe limitations remained. By now the entire fleet needed a major overhaul, and because of this requirement, many DC-6s were limited in their cargo loads, able to carry little more than a C-54.[43] The ageing of both the DC-6 and C-54 fleets threatened to turn the TAM into a spent force, as one by one the aircraft were parked in front of the maintenance hangar at Alverca awaiting their major overhauls.[44] Even the seemingly advanced DC-6 was now considered outmoded by airlift standards of the day.[45]

In 1967, TAP had shifted completely to a jet aircraft fleet and operated the Boeing B-707.[46] This experience prompted the FAP in 1969 to consider buying three similar aircraft to resuscitate its transport capability. This was a controversial decision because of the expense of acquiring a new airframe and training its dedicated flight and ground support personnel. Ultimately, however, when the dust settled, two aircraft were authorized in February 1970.[47] These Boeing B-707-3F5C models were the last two of this type produced and were delivered in a relatively quick 18 months. They were purpose-built aircraft with a strengthened undercarriage and a mixed cargo and passenger cabin. Flown according to Pan American operating standards, they ultimately realized in Portuguese hands the highest utilization of any B-707 worldwide, a fact confirmed by Boeing. One of the two aircraft was always on a mission, and the reliability and flexibility of the plane made a dramatic difference in the logistic picture. In 1973 the two B-707s together flew 299 missions alongside 209 missions for the remaining fleet of DC-6s.[48] The DC-6s had by then been relegated to the shorter missions to Guiné, the Azores, Madeira, and Europe, and the B-707s were serving Angola and Mozambique. A B-707 typically left Lisbon at 1130, flew to Luanda, was unloaded, serviced, and loaded, and returned to the *metrópole* by 2230 the next day.[49] According to FAP records for 1972, transporting passengers and their luggage by B-707 instead of by ship saved the armed forces 266,633 Contos or about $10.1 million. In 1973 comparable savings were 267,633 Contos or about $10.6 million.

43 Ibid., 6.

44 Depoimento do Gen. Tavares Monteiro em 05 Abril de 1995 Sobre a Ponte Aérea Metrópole – Ultramar [Deposition of Gen. Tavares Monteiro on 05 April of 1995 on the Air Bridge *Metrópole – Ultramar*], Livro de Actos 1997, item 6, page 35, annex III, Archivo Histórico de Força Aérea, Alfragide.

45 General Tomás George Conceição Silva, interview by the author, 3 April 1995, Lisbon. General Conceição Silva was Chief of Staff of the Portuguese Air Force from 1988 to 1991, and had extensive airlift experience in Africa.

46 The military version is the C-135 Stratolifter.

47 Monteiro, "Transporte Aéreo na FAP, Do zero aos 707," 7.

48 Commander, 1st Air Region, "Resumo Estatístico das Missões Realizadas e Passageiros Transportados (Ida e Volta), 1972 e 1973" [Statistical Résumé of Missions Flown and Passengers Transported (Round Trip), 1972 and 1973], 30 September 1974.

49 Monteiro, "Transporte Aéreo na FAP, Do zero aos 707," 7.

These savings over two years paid for the two B-707s.[50] Finally, with this breakthrough in technology, the Imperial Airline was truly realized.

While the Portuguese began their counterinsurgency operations some twelve years after the Berlin Airlift (June 1948–September 1949), the lessons in this application of logistic power were an integral part of Brilhante Paiva's thinking. The Berlin Airlift had demonstrated that, given the machines and bases, air transport was more than a prop for short-term logistic emergencies. The intensive operational techniques developed there showed the way to a sure alternative to surface transport, providing the procedures and infrastructure were in place. Portugal sought to embrace these lessons in its airlift operations.[51] Except for U.S. operations in South Vietnam, other governments fighting comparable insurgencies seemed less attuned to airlift calculations. Malaya was the most comparable in distances and troop numbers, but there the British made no such effort. There was an endless stream of British troopships arriving in Singapore to support the war, and one newly commissioned subaltern described this expensive and leisurely process, in which "We spent hours eating huge P. and O. meals, and then more hours after them recovering in deckchairs, watching waves sliding past the ship, endlessly, without pause. …"[52] This process was neither a focused preparation for counterinsurgency nor an efficient troop insertion into the local environment, and was all the more surprising and inconsistent for a country that had in 1950 developed the first propeller-jet airliner, the Vickers Viscount, and the first turbojet airliner, the De Havilland Comet, two years later. The Portuguese had sought the rapid and cost-effective avenue of air transport to move troops and critical supplies, a choice that improved the sustainability of their war effort. Likewise, the shift from high-profile troopships arriving and departing with fanfare to the more subtle aircraft reinforced the low-key nature of their fight. For the Portuguese, routine aircraft traffic providing timely deliveries and a rapid response to needs in the theaters better fit the tenor of counterinsurgency.

Intra-Theater Extension

The "Imperial Airline" simply could not terminate at the primary theater air base and be effective. Perforce, there needed to be an air network that extended the transoceanic line of communications put in place by the intercontinental air bridge. Each of

50 Commander, 1st Air Region, "Resumo Estatístico das Missões Realizadas e Passageiros Transportados (Ida e Volta), 1972 e 1973."

51 Kaúlza de Arriaga, *Sínteses* [Synthesis] (Lisbon: Privately published, 1992), 190. General Arriaga served as the Subsecretary of State for Aeronautics from 1955 to 1962.

52 Oliver Crawford, *The Door Marked Malaya* (London: Rupert Hart-Davis, 1958), 13–14. The term "P and O" refers to the Peninsular and Oriental Steam Navigation Company, a British shipping and logistics company that dated from the early 19th century, which is usually known as P&O. In March 2006, it was sold to Dubai Ports World; however, the P&O brand has been retained.

these intra-theater systems would combine a web of variably capable airfields with an array of aircraft able to operate across the range from developed airfields to the very rudimentary strips in bringing air support to the entire theater. This capability would be built with two complementary efforts, the acquisition of the proper aircraft and the construction of the needed airfields.

Using the three Dakotas acquired during World War II, the FAP began in 1958 to build this fleet by acquiring three additional aircraft and adding to this inventory through 1972, as suitable Dakotas became available. The FAP would ultimately acquire a total of 28 Dakotas, mainly for use in the *ultramar* for short- and medium-range tasks. These came from diverse sources, primarily from the air forces of the United States, Australia, Israel, the United Kingdom, and South Africa, all of which were shedding their obsolete aircraft at bargain prices.[53]

The decision to pursue the acquisition of Dakotas, a proven and capable aircraft, was controversial within the air staff, and there were two lines of thought on the issue. The first or "pro-Dakota" position held that the Dakota would be adequate for local air transport in the *ultramar*, as it could land on unimproved fields with its wide-stance retractable main landing gear and carry a useful cargo load of about 6,000 pounds. The drawback with the C-47 as well as the earlier Ju-52 was that, while both had given good service, their basic landing gear design with a tail wheel gave them a nose-up attitude when parked, and this characteristic complicated cargo handling. Adding to this constraint was the comparatively small and restrictive cargo side door that limited ability of each to make supply drops. There were more modern aircraft available that would lift a heavier payload and had been designed for more efficient cargo handling.

In the *ultramar*, the Dakotas would show their versatility and be used for many types of missions, such as aerial reconnaissance, paratroop drops, medical evacuation, and search and rescue. They were even used as makeshift bombers in Guiné, including night missions, and in psychological operations in Mozambique.[54] While the Dakota

53 César Rodrigues, "Douglas C-47 'Dakota' e C-54 'Skymaster'" [Douglas C-47 "Dakota" e C-54 "Skymaster"], *Mais Alto* 385 (May-June 2010): 35.

54 Ibid. See also Mário Canongia Lopes, "Dakotas: Aviões bimotores de transported a Força Aérea Portuguesa–1944–1977, II Parte" [Dakotas: Twin engine transport aircraft of the Portuguese Air Force–1944–1977, Part II], *Mais Alto* 311 (February–March 1998): 34–35. One C-47, for example, was due for its 300-hour inspection at the OGMA works in Alverca in mid-1965 and was converted at this time into an "intervention aircraft" or "*avião de invervenção.*" This transformation into a night bomber or "*bombardeiro nocturno*" was made by attaching racks for six 50-kilogram and two 200-kilogram general purpose bombs under the wings. Storage was also built into the cargo bay for twenty 15-kilogram fragmentary bombs and 50 magnesium illumination flares, both of which could be launched from a free-fall chute. The pilot seat was also equipped with a bombsight. When work was completed in October 1965, the aircraft was flown to Guiné and participated in operations in December over Cacine and Cantanhez. Later in August 1966, it flew in support of operations at Medina de Boé and in the region of Cassebeche. Such fire support operations continued for the C-47s in Guiné until 1975.

had proven itself time and again, it had severe limitations in that its load capacity was relatively modest, and its maintenance was becoming an increasing problem because of its age as a World War II-vintage machine. No matter its venerable history, the Dakota clearly had its limits.

The second line of thought argued that the FAP needed a more robust, capable, and modern aircraft along the lines of the Fairchild C-119 Flying Boxcars, which had proven themselves in Indochina. There were at the time medium-weight cargo aircraft that could carry multiples of the Dakota load, could land on unimproved runways, were similarly easy to maintain, and were available at affordable prices. The obtainable aircraft of choice at the time was the Nord Aviation Noratlas with its two powerful and reliable Bristol Hercules 2,040-horsepower, 14-cylinder, twin-row, radial engines, each driving a Rotol four-bladed propeller. It was similar to the C-119 with its twin-boom nacelle-empennage construction and a large, pod-like fuselage slung under the high wing connecting the nacelle booms. This design was well conceived for operations under spartan conditions in that the propellers were positioned to clear rough terrain on landing, and the landing gear were widely separated for stability. Further, its soft fat tires gave it ideal unimproved field capability. The cargo bay with its rear-opening clam-shell doors was positioned near ground level for easy cargo handling. Its useful cargo load was 18,647 pounds.

The French air force had begun taking delivery of the military version of the Noratlas, the N-2501, in July 1951 and had eventually acquired a fleet of 228.[55] The West German Luftwaffe was faced with a similar obsolescence in its cargo fleet of Ju-52s and C-47s, and it too chose the Noratlas. It eventually acquired a fleet of 187, 25 of which were manufactured in France, and 162 of the N-2501D model in West Germany.[56] The Luftwaffe began disposing of its Noratlases as surplus in 1964.

A civil version of the aircraft, the N-2502, was offered by Nord Aviation with a small Turbomeca Marboré II turbojet added at each wingtip. It was adapted to the specifications of its two customers, and seven of the N-2502/N-2502A version were used by the Union Aéromartime du Transport (UAT), and three of the N-2502B version by CGTA–Air Algérie. Neither operator found much success in this civil adaptation of a military transport, and thus only ten were built.

When Portugal decided to augment its C-47 fleet with a heavier, more capable aircraft, it was offered six of the N-2502As by UAT.[57] These aircraft were based in Douala, Cameroon, and had been operated in Francophone Africa under grueling and harsh conditions. They were flown to Le Bourget outside of Paris, where they were

55 Mário Canongia Lopes, "Noratlas 1960–1977: Aviões bimotores de transporte da Força Aérea Portuguesa, I Parte" [Noratlas 1960–1977: Twin engine transport aircraft of the Portuguese Air Force, Part I], *Mais Alto* 312 (April–May 1998): 3–4.

56 Ibid.

57 Of the seven purchased by UAT, one was destroyed in an accident near Boukpayanga, Central African Republic, in March 1959, and the remaining six were subsequently made available to Portugal.

Nord Aviation Noratlas dropping a Stick of Paratroops. (Source: Museu do Ar)

overhauled, and where the Portuguese acceptance pilots and maintenance personnel were trained during the first half of 1960. These six aircraft were deployed to Angola in November 1960 as part of Squadron 92 and began operations from Luanda.

In 1962, six more Noratlases were acquired from the Israeli Air Force and, after overhaul, were designated N-2502Fs. In the same year, the three N-2502Bs owned by Air Algérie were also added to the fleet. The primary drawback of the nine civil models was an absence of personnel doors in the rear of the fuselage for paratroop drops. This shortcoming was corrected with their first major overhaul at Alverca.

Over the years 1965–1970, thirteen N-2501Ds were acquired from the Luftwaffe, bringing the total acquisition to 28 Noratlases. On balance, the Noratlas delivered good service in the *ultramar* and was a tactical airlift mainstay. If it had a single weakness, it lay in its acquisition from four different sources and hence with four very different instrumentation and electronics suites. This caused significant difficulties for the aircrews and maintenance personnel in flying and maintaining the aircraft.[58]

The Imperial Airline from its inception in 1957 until its rudimentary but reliable functioning in 1960 was moved from concept to reality in three short years. It would be expanded and refined over the course of the war and would reach its optimum efficiency with the acquisition of the two B-707s in 1971. Deployment of the Imperial

58 Rui Tavares Monteiro, "Recordando os Noratlas" [Remembering the Noratlas], *Mais Alto* 279 (September–October 1992): 12.

Airline followed a classic step-down pattern in which long-range transoceanic air routes terminated at a major airfield in each theater, and these in turn were connected by tactical transports to intermediate airbases. These short-field capable aircraft could also land on unimproved strips cleared from jungle and manned by small detachments to provide close support of army units in the field. This network was built, adapted, improvised, and later expanded to address the widening threat as the conflict matured. Ultimately, there was an extensive air logistic system stretching from the *metrópole* to the *ultramar*, having been constructed at a pace that would appear to be beyond the capabilities of the relatively limited Portuguese resources. This system remains a testimony to "doing more with less" in extending the strategic reach of a nation and its armed force.

A Look at the Neighborhood

In October 1959, Arriaga initiated a visit to the French Congo and South Africa, two neighboring and friendly countries abutting Angola and Mozambique, to explore the depth and sincerity of their political and military support. Extending any sort of support to a colonial power in an anti-colonial era was bound to be fraught with problems, particularly in Africa. Nevertheless, countries act in their own self-interests, and it was hoped that the interests of select African neighbors would align with those of Portugal. Certainly, the trip laid the groundwork for the support that Angola would receive in the emergency of March 1961.

France, it was thought, would welcome an informal alliance, as it had just fought an insurgency in Indochina, was currently fighting one in Algeria, and was likewise feeling estranged from the United States. It was still firmly in possession of the French Congo, and even after independence on 15 August 1960, had no plans to leave and curtail its assistance, as the new leaders did not wish to preside over a breakdown of government administration and services and a stagnation of development.[59]

France operated two air bases there, Base Aérienne 170 (Brazzaville Maya Maya), the capital, and Base Aérienne 173 (Pointe Noire). The French air commander in Brazzaville, General Georges Labit, was a close friend of Resende, as they were classmates in the 1952 course at the newly established NATO Defense College in Paris. Labit was a great enthusiast for the defense of Africa to the benefit of the West and subsequently did many things to help the FAP either in anticipation of or simply on the spot without any official authorization. These would include such overtures as sending the first four T-6G aircraft as the nucleus of a new close air support squadron at Negage.[60] These pirated aircraft were flown from Pointe Noire to Luanda with no insignia or documentation. Later during the March 1961 crisis, Labit would send

59 Francis Terry McNamara, *France in Black Africa* (Washington: National Defense University, 1989), 96–97.
60 Resende, 21.

resupply flights of Noratlas aircraft into Negage. These risky missions were crewed by French volunteers whose only navigation aid was a radio beacon 100 kilometers distant.[61]

Even after independence, France maintained a base at Pointe Noire, one of its three strategic bases in Africa, and did so until 1963, when President Abbé Youlou was overthrown by a group of radical young army officers.[62] Until the base was moved to Libreville in Gabon, Labit continued to extend French support.

South Africa was likewise most receptive and remained supportive over the years, supplying military equipment and sharing Portuguese airfields in engaging its own insurgency from the South West Africa People's Organization (SWAPO) for the liberation of South West Africa, now known as Namibia.[63] The national strategy of South Africa was to support the counterinsurgency efforts in the buffer states of Angola, Rhodesia, and Mozambique, which would form a protective arc stretching from the Atlantic to the Indian Ocean. These states would shield South Africa from the active insurgent threat to the north and prevent direct South African involvement. South Africa would support the Portuguese in Angola and Mozambique and later the Rhodesian government of Ian Smith to this end, insulating itself from the newly independent sanctuary states of the Congo, Zambia, and Tanzania.

Arriaga's trip ended on a positive note in solidifying relationships with France and South Africa. The handwriting was on the wall, however, with the Belgian Congo, which would gain independence in June 1960, a mere eight months hence, and become the primary threat to Angola.

Paratroops

With trouble brewing in the Belgian Congo, it became evident that Portugal would need to augment its security structure in Angola and particularly along its northern border. Addressing likely cross-border threats in the vast territory would require shock troops with their characteristic mobility and thus the ability to move quickly to the scene of trouble. These would be light infantry and pave the way for a heavier force to follow. During World War II, the U.S. 82nd and 101st Airborne Divisions and the British Parachute Division had demonstrated the value of such a force, and both had continued to show their worth in the post-war era. Arriaga saw the need for Portugal to have a similar capability to help secure the *ultramar*, as such troops could ideally be deployed by airdrop to trouble spots with the Imperial Airline and its tactical transport extension and gain control of a situation with the help of close air

61 Ibid.

62 The three strategic French bases in Africa were Dakar, Fort Lamy (N'Djamena) and Pointe Noire. Other small air bases with logistics facilities were Port Bouet, Cote d'Ivoire; Douala, Cameroon; and Niamey, Niger.

63 Jeeps and other vehicles, for instance, manufactured in South Africa are easily identified by their right-hand drive.

support, until heavier forces could arrive. This was seen as an important part of power projection in the *ultramar*.

The first Portuguese experimentation with paratroops began in 1922 with the jump of two officers from a balloon 500 meters over Alverca.[64] Later in 1930, a jump from an aircraft was made from 800 meters over Tancos, but in truth, little was done until World War II, when Portuguese Timor was occupied by the Japanese in February 1942. The Australians subsequently sought to insert into Timor indigenous personnel who would be able to gather intelligence on enemy activity and report it by radio to Allied intelligence. A dozen civilian Portuguese were recruited for the mission and provided parachute instruction as part of their intense physical training. The project met with mixed success when the insertion occurred in August 1944, as all of those infiltrated were ultimately either captured or killed.[65]

Following World War II and the success of both the German and Allied airborne forces, it was thought that Portugal ought to explore the idea of developing its own paratroop capability by sending some men from the army to be trained abroad. Accordingly, in August 1951 two infantry captains were sent to France to the École des Troupes Aeroportées (ETAP) to undergo parachute training. At the conclusion of their training, they returned to their units and continued their duties as before with no ability to maintain their newly found qualifications. The next year when the FAP was formed, the relevant Decree Law No 2055 made reference to the paratroops and the need to integrate their training and operations into the FAP. Meanwhile, the army continued to debate over the creation of one or more parachute units. Small numbers of additional officers and sergeants continued to receive parachute training in France and, later in 1953, in Alcantarilla, Spain. Still by 1954, the trained parachutists remained scattered throughout the army and had not been formed into their own specialized unit. Their skills were atrophying through bureaucratic indecision, and Arriaga, citing Decree Law No 2055, argued in 1955 that the FAP should have the paratroop portfolio and develop it as an elite force of airborne shock troops.[66] Because these were to be light infantry, and the FAP was not the army, there was still much debate. Would these troops belong to the FAP until they were on the ground, and then come under army control? The question was settled by the Minister of National Defense, who lodged the paratroops with the FAP but limited the force to a single battalion.[67]

At the beginning of 1955, there was a call for volunteers for paratroop training to flesh out the designated battalion. About 1,500 volunteers responded from the three service branches, and from these, 232 were considered qualified and sent to the Spanish Air Force for training at Alcantarilla. Called the "Spanish Course," it

64 Miguel Machado and António Carmo, "Tropas Pára-Quedistas Portuguesas" [Portuguese Paratroops], *Boina Verde* 158 (July–September 1991): 7.

65 Ibid., 9.

66 Ibid.

67 Luís António Martinho Grão, *História das Tropas Pára-Quiedistas Portuguesas* [History of the Portuguese Paratroops] (Lisbon: Comando das Tropas Aerotransportadas, 1986), 36.

provided the foundation for the Portuguese parachute battalion, and this initial qualification was completed in July with a graduation ceremony in which 188 men were awarded their green berets.

Following the training in Spain, the paratroop battalion (*Batalhão de Caçadores Pára-quedistas* or BCP) was formed and presented to the nation on 14 August 1955 in a formal ceremony. Shortly thereafter on 15 October, the unit made its first jump over its new home, BA 3 in Tancos, from the Junkers Ju-52. This exercise represented the beginning of a center of instruction at Tancos for the continued proficiency and expansion of the paratroops. Over the next several years the BCP broadened its training and capabilities to the point that it was ready for deployment as a frontline unit of the Portuguese armed forces. In step with the augmentation of the FAP in the *ultramar*, Arriaga and his staff conceived of Exercise *Himba*. This was an air transport exercise in which the BCP would be loaded on transports and sent to Angola, where power projection would be demonstrated by an air drop of a portion of the new force. There were fourteen aircraft participating:

6 PV-2 Harpoons from BA 6 (Montijo)
3 SC-54 Skymasters from BA 4 (Lajes, Azores)
4 C-47 Dakotas from AB 1 (Lisbon)
1 C-47 Dakota from BA 3 (Tancos)
1 Company of the BCP embarked in a C-54.

Unfortunately, the beginning of the exercise was marred by a tragic accident in which a C-47 crashed shortly after takeoff from Lisbon and went into the Tagus River with the loss of all aboard and the payroll for the entire exercise. The remaining fleet of aircraft flew to Luanda by way of Guiné and São Tomé on 30 April, and once in Luanda, flew to Nova Lisboa and Sá da Bandeira, where in each case the company jumped and then conducted a parade. Returning to Luanda, the paratroops participated in an air festival and concluded the exercise.

The entire exhibition was an unqualified success despite the tragic loss of the C-47. The paratroops were received joyously by the people of Angola and were truly impressive both in action and on parade. Indeed, in looking back on the exercise, one journalist observed in a halo of emotion that the expanded presence of the FAP in Angola represented a modern Lusitanian link with the *metrópole*.[68]

Enthusiastic about the success of Exercise *Himba*, Arriaga in the spring of 1959 ordered the planning of a similar mission for Mozambique with the code name Exercise *Ibis*. It would require 20 aircraft and 800 paratroopers.[69] Over the next several months

68 Miguel Machado and António Carmo, "Tropas Pára-Quedistas Portuguesas," 27.

69 Pedro Manuel Ferreira, "A missão que ficou por realizar 'Ibis' ou a pré-génese da instalacão da Força Aérea em Moçambique" [The Mission to Execute *Ibis* or the Groundwork for the Installation of the Air Force in Mozambique], *Mais Alto* 350 (July–August 2004): 22.

it became apparent that Moẓambique was not prepared to accommodate an exercise the size of the envisioned *Ibis*. The airfield infrastructure was largely undeveloped, and it would consequently take time to improve the existing ones and to build new ones. The logistics infrastructure also was not in place for any large-scale deployment of air units, and it needed to be built. In light of these shortcomings, the exercise was postponed until September or October 1960.[70] Again, after much consideration, the exercise was targeted for April 1961, the beginning of the dry season. Airfield and logistics problems were not the only obstacles, as the relatively large number of aircraft required posed a further hurdle. To assemble these would necessitate borrowing aircraft from TAP and cancelling Portuguese participation in concurrent NATO exercises, both fairly extreme measures.

In June 1960, Arriaga sent a mission to Mozambique to study the local problems associated with the deployment of FAP there. Its assessment and recommendations ultimately led to the improvement of facilities and the building of new ones, as depicted in the nearby map. Developments in the North of Angola in the first quarter of 1961 precluded the execution of *Ibis* and the need for the deployment of FAP resources to support it; however, by the beginning of 1962, Arriaga was pleased with progress and was congratulating those involved in installing the FAP in Mozambique.

Kalúza de Arriaga in the short space of six years (July 1955–January 1962) had reoriented the FAP toward Africa and prepared it to fight a very different war from the conventional European conflict that NATO envisioned. This would be a war that relied on air support of the troops in the field and of the native population. It would require the use of unimproved airfields, close liaison with ground forces, and operations in a harsh environment. All of this would be based on simple, reliable aircraft and on aircrew with experience in such warfare. The FAP would continue to gain strength on this course, and its first and perhaps most important test would come with the events in the North of Angola in the early months of 1961.

70 Ibid., 23.

5

Assault on the North of Angola

The Deslandes mission in May 1958 had focused on Angola, and its members had recommended a triangular disposition of key airfields at Luanda, Luso, and Moçamedes, which would enable complete air defense coverage of the colony. However, by 1960 it was clear that the threat would not come from the air and would likely come on the ground from the north and the former Belgian Congo. This realization altered the initial recommendation, as it became increasingly obvious that the FAP would be primarily supporting ground troops and the threatened population in the northern districts. Hence the primary fields were realigned as Luanda (BA 9), Negage (AB 3) and Henrique de Carvalho (AB 4). Because of the unrest in the Congo, construction at Negage began promptly in mid-1960 and proceeded feverishly, or at least as feverishly as could be done in a relatively remote environment. At Luanda, improvements were likewise progressing rapidly, facilitated by the fact that the base was being built on the opposite side of the runway from the existing civil airport and that it was at the commercial center of the province. Construction at Henrique de Carvalho, however, was seen as less urgent, for it was not close to the immediately threatened border areas. Its construction would begin in 1963, and it would not be operational until 1967.

Events in the north would develop rapidly and severely test the abilities not only of the FAP but of Portugal itself to protect its territory and citizens and to drive the forces of the UPA back into the Congo. This latter effort to rescue the besieged and terrified population of northern Angola and destroy the invading force is a story of accelerated mobilization, *ad hoc* planning, inventive adaptation, extraordinary flying skill, and enormous courage that remains a clear example of how airpower is applied with meager and ageing equipment by a force that is learning by doing in counterinsurgency.

Congo Anarchy

The Belgian Congo had prospered, since its transfer to the Belgian government from the estate of its King Leopold II in 1908. It now operated under a colonial ministry that answered to a democratically elected parliament that had opened the heretofore crown domains to private ownership and had ended its monopoly on trade. The right of the Congolese to own and work their lands had been restored. The system of forced labor had been abolished. As a consequence of these new liberalizations came new opportunities. The European population moved to take advantage of the potential in the Congo, and its numbers there had risen from about 3,000 in 1908 to more than 100,000 in 1960.[1] There occurred over this period a great rush of development in agriculture, banking, mining, and infrastructure. Despite all of this apparent progress and its attendant social benefits, there remained a strong undercurrent of resentment within the local population. No civil rights for the Congolese were thought necessary, and racial segregation was standard practice. The highest rank a Congolese could hold in the colonial administration was one of a minor clerk in the civil service or a noncommissioned officer in the Congolese army, known as the *Force Publique*. Until 1955 the Congolese were forbidden to attend college and even after that could only attend designated institutions so as not to be "corrupted." When independence came in 1960, there were fewer than 30 African university graduates. Of the 5,000 management-level positions in the civil service, only three were filled by Africans.[2]

Despite the democratic government of Belgium, the Congo since 1908 had been run by decree from Brussels through a governor-general in its capital, Léopoldville. The colony was permitted no politicians, no political parties, and no elections.[3] No Congolese or Belgians residing there had a vote on any matter. Following World War II and the subsequent political developments in the neighboring French and British colonies, Belgium was obliged in 1956 to permit the same political freedom in the Congo. Prior to that time there had been social, religious, and cultural clubs formed along regional and tribal lines. Within this framework there were discussions on how the Congolese might improve their plight and ultimately gain independence. These clubs now became the political parties of the Congo. By late 1959 the activities of these various political movements had matured to a point of demanding independence. The MNC (*Mouvement National Congolais* or Congolese National Movement), founded in October 1958 by the magnetic postal clerk Patrice Lumumba, had grown out of a government workers' club in Stanleyville. ABAKO (*Alliance des Ba-Kongo* or Alliance of the Bacongo), the influential cultural-political association of the Bakongo people, was led by Joseph Kasabuvu, a civil servant who became its president in 1955. CONAKAT (*Confédération des Associations Tribales du Katanga* or Confederation of the Tribal Associations of Katanga) was a former club among the Luba-Lunda tribesmen founded by Moise Tshombe. The leader of a soldiers' club in the *Force Publique* garrison at Thysville was a young sergeant-major, Joseph-Désiré Mobutu.

Beginning in January 1959, these demands for immediate independence were reinforced by rallies and strikes, many violent and bloody. By October, the violence and

1 Peter Forbath, *The River Congo* (New York: Harper & Row, 1977), 384.
2 Adam Hochschild, *King Leopold's Ghost* (New York: Houghton Mifflin, 1998), 301.
3 Mike Hoar, *Congo Warriors* (London: Robert Hale, 1991), 10.

tension had reached a level not seen since the time of Livingstone some 90 years earlier. Tribal conflict with painted warriors and dismembered cadavers, displays of religious fanaticism, the airing of latent grievances, all of these added to the strikes and rallies and caused hundreds of dead and injured.[4] This mayhem created a sense of urgency within the Belgian government and caused it to rethink its colonial policy of a five-year transition culminating in 1964. Tribal rivalries and agitation persisted and prompted the Belgians to clarify the political evolution and consider an African government in nine months.[5] This reconsideration was shortened by a month and formalized at the February 1960 Round Table Conference in Brussels, at which it was resolved to grant the Congo its independence in June, a mere four months hence. This generous concession was an attempt to satisfy the Congolese political leaders, who wanted immediate independence and wished to ascend to power without having to consult the people.[6]

In this short period the Belgian government attempted to overcome the enormous obstacles of accelerating the formation of a new Congolese government, a patently impossible task. It left little time for the Congolese to organize themselves politically, hold a wide array of elections, and assume the colonial administration from the Belgians using the largely untrained and uneducated local population. The seeds of immediate trouble began during the election campaign, in which candidates made wild promises, took the opportunity to vent long-repressed hostilities toward Europeans, and further stirred long-standing tribal animosities. Peter Forbath describes the scene leading to Independence Day as one of nervous excitement with crowds roaming the streets of major cities, "impatient for the miracles of freedom to begin, nursing tribal grudges, eager to become lords over the whites and take vengeance for past injustices."[7] In the 23 June elections, Lumumba's party outpolled Kasabuvu's; however, neither side could form a parliamentary coalition. Consequently, as a compromise measure they formed an uneasy partnership with Kasabuvu as president and Lumumba as prime minister. This fragile government might have survived in a benign environment; however, it stood little chance with the legacy bequeathed to it by Brussels. The mood rapidly degenerated and turned ugly during the first week of independence. Public services deteriorated with the wholesale departure of Belgian civil servants and a lack of trained Congolese to replace them. When the new civil servants were not paid, there were strikes that turned into riots when the *Force Publique* attempted to intervene. All semblance of order, however, vanished when the *Force Publique* itself mutinied.

4 Hélio Augusto Esteves Felgas, *Relatório Pessoal Acerca do Congo Belga: Referente a Outubro 1959* [Personal Report Concerning the Belgian Congo: Referring to October 1959] (Carmona, Governor of the District of the Congo, 4 November 1959), Resende personal papers, Caixa 17, Proc° 02, Arquivo Histórico de Força Aérea Portuguesa, Alfragide, 3.

5 Ibid.

6 Ibid.

7 Forbath, 388.

Despite the desire to replace Belgians with Congolese at every turn, Lumumba had not done so in one important sphere. He had elected to retain the 1,000 Belgian officers who led the 24,000-man *Force Publique* as an insurance policy against the difficult times ahead. In the minds of the black noncommissioned officers and troops this policy smacked of discrimination, and on 1 July there was a demonstration against it at the Lépoldville garrison. When the demonstrators refused to return to barracks, the commanding general ordered the Thysville garrison to intervene. In such a situation it would have been unrealistic to expect sympathetic brother to attack brother, and so the Thysville garrison predictably mutinied too. In a rapid chain reaction, most garrisons around the country followed suit with the result that the soldiers went on a rampage, attacking their officers, raping officers' wives, and looting. On 6 July, Lumumba, in an attempt to quell disaffection in the *Force Publique,* replaced the Belgian officers with Congolese. This move did not stem the tide, and the mutinies spawned a widespread wave of violence. Blacks attacked whites and each other, as they vented their anti-European hatreds and intertribal rivalries.

In the midst of this crisis, Lumumba in an irresponsible statement accused Belgian army officers of instigating the mayhem in a plot to re-annex the Congo. This charge and the developing chaos prompted a mass exodus of some 40,000 Europeans. Essential services ground to a halt. Food became scarce. Epidemics threatened. The spread of disorder prompted Tshombe, the newly-elected premier of Katanga Province, to ask Lumumba to appeal to Belgium for troops to restore order. Fearing a return of Belgian rule, he refused. Tshombe then, in an attempt to address his executive responsibilities and preserve his province, declared mineral-rich Katanga an independent state and seceded from the Congo on 11 July. The continuing riots prompted Belgium to deploy paratroops to protect its nationals, and it seems, they landed principally in Katanga, where they helped to sustain the secessionist state by rounding up and expelling the mutinous elements of the *Force Publique.*

On 12 July, Lumumba asked the UN to expel the Belgian "forces of aggression" and to compel Katanga to end its secession.[8] Three days later the first of 25,000 UN troops arrived, Belgian soldiers began to depart, and Lumumba declared martial law. The political situation then degenerated completely. The people of the lower Congo, who were represented by ABAKO, took their cue from Tshombe and demanded the abolition of the central government in favor of a system of independent provincial governments. This demand was followed by a declaration of independence in Kasai Province. Lumumba's country was falling apart with good reason, and having no reliable way to enforce his will, he turned to the UN with a request to use its peace-keeping troops to unify the Congo by force. When it refused, he countered by appealing to the Soviet Union. It obliged with a flight of 19 Ilyushin transport aircraft, thereby supplying sufficient arms and airlift capability to enable Lumumba to march on Kasai and pour the proverbial gasoline onto an already incendiary situation. In what Mike

8 Hoar, 11.

Hoar describes as a one-sided battle, Lumumba's forces massacred Kasai tribesmen by the hundreds in what was essentially a venting of ancient tribal animosities.[9] Manuel Diogo Neto, a FAP lieutenant colonel and squadron commander in Luanda at the time, described this hate-filled ethnic conflict as "two million blacks filled with the primitive feeling of animals that kill the ancestral enemy they instinctively hate."[10]

Lumumba next turned his force toward Katanga, where it was stopped at the border. Frustrated by this stalemate, elements of the *Force Publique* turned to brutalizing the population. Hélio Felgas, a Portuguese Army major and Governor of the District of the Congo, foresaw the threat from the Belgian Congo in this mayhem and in a report dated November 1959 urgently warned Portuguese authorities to prepare for spillover trouble.[11]

At this point Lumumba was a spent force in any attempt to unify the Congo politically. While he was a persuasive orator, he proved to be an unstable diplomat and an incompetent head of state. On 5 September, Kasabuvu dismissed the prime minister. In turn Lumumba dismissed the president. Kasabuvu prevailed by placing Lumumba under house arrest with the support of Mobutu, now a colonel and head of the army. All pretense of a functioning government seems to have ended. On hearing of Lumumba's arrest and the ensuing confusion, Antoine Gizenga declared the secession of Oriental Province, Lumumba's power base with its capital in Stanleyville. Gizenga's forces attacked Mobutu's garrisons, and a new bout of the most brutal savagery raged through the rain forests. It involved the worst sort of tribal warfare marked by wholesale massacres and cannibalism.[12] In the midst of the anarchy, Mobutu formed a "college of administrators" on 14 September and assumed the day-to-day running of the government, "neutralizing" all politicians. One of his first actions was to close the embassies of the Soviet Union and Czechoslovakia and banish their presence from the Congo.[13]

The situation stabilized for a time; however, in November, Lumumba left Léopoldville, where he had enjoyed UN protection, to travel to Stanleyville, where his supporters had control. En route he was caught on 2 December by Mobutu's troops, and on 17 January 1961 was flown to Katanga, where he was reluctantly received by Tshombe in an apparent near-death condition.[14] Later in the month he was shot by Congolese or Katangese troops while allegedly trying to escape from a rather dreadful mud hut in which he had been held prisoner.[15]

9 Ibid., 12.
10 Manuel Diogo Neto, "Acontecimentos no Norte de Angola" [Events in the North of Angola], *Mais Alto* 30 (June–July 1961): 8.
11 Felgas, 3.
12 Forbath, 390.
13 Hoar, 12.
14 Ibid., 13.
15 Ibid.

Lumumba's death caused an international outrage and raised the man to martyr-hero status. Mobutu returned the government to Kasabuvu, as his suspected hand in Lumumba's death became known, and in turn Kasabuvu made him commander-in-chief of the armed forces. The agony of the Congo continued until a now better prepared Mobutu stepped in again in 1965 and ruled for better or for worse until his downfall and death in 1997. It is estimated that over 200,000 Congolese perished in the five years between independence and the establishment of his second and final regime.[16] While these developments have been summarized loosely, it was into the fringe of this chaos that the FAP was thrust.

Luanda and the 2nd Air Region

The initial elements of the FAP arrived in Angola at the beginning of July 1960 to form a nucleus of the 2nd Air Region and to establish BA 9 in Luanda as a functioning military airfield. Construction of aviation infrastructure in Angola had begun in the 1950s and, as the threat became clearer, had as its refined priority the establishment of airfields in a defensive, interior line oriented toward the Congo. Luanda was to serve as the primary base supporting this projection of airpower throughout the territory and in defense of the north. Because there was an existing runway at Luanda, operations from there were able to begin in theory as soon as squadrons could be formed in the *metrópole* and flown south; however, the lack of a completed base and its continuing construction made aviation ground support somewhat rough.

In the last days of October, Squadron 91 with its six Nord Aviation Noratlas transports landed in Luanda, and the arrival of Squadron 92 with its eight Lockheed PV-2 Harpoon aircraft was anticipated shortly.[17] The delay of the Harpoons was the result of an undercurrent of diplomatic friction between the United States and Portugal over the use of equipment given to Portugal under the U.S. Military Assistance Program (MAP). The Azores Archipelago contained a Portuguese airbase (BA 4) alongside the village of Lajes on the Island of Terceira, an ideal mid-Atlantic location through which transited some 70 percent of U.S. military air traffic bound to and from Europe.[18] This base was initially leased by the United States in 1951, and its lease was renewed in 1957 for a five-year term, which was set to expire on 31 December 1962. Unlike other base agreements in Spain, Morocco, and Libya, where the rent was paid in cash, payment to Portugal was made in kind through the transfer of military equipment to enable Portugal to participate in NATO. There was a secret codicil to the 1951 lease agreement, which said that U.S. military equipment given to Portugal

16 Forbath, 393.
17 Manuel Diogo Neto, "Recordando a Esquadra 92 da BA9 – Luanda" [Rembering Squadron 92 of BA9 – Luanda], *Mais Alto* 264 (March–April 1990): 39.
18 José Freire Antunes, "Kennedy, Portugal and the Azores Base" in *John F. Kennedy and Europe*, edited by Douglas Brinkley and Richard T. Griffiths (Baton Rouge: Louisiana State University Press, 1999), 151.

could be used in the *ultramar* in the case of an emergency. In this secret exchange of notes, the United States stated unequivocally that, while the diversion of equipment supplied under MAP could not be made to the *ultramar* without U.S. consent, such consent would be uncontested and promptly granted in an emergency.[19]

In October, just weeks before Brigadier Fernando Pinto de Resende, the prospective commander of the 2nd Air Region, embarked for Angola, General João Albuquerque de Freitas, the FAP Chief of Staff, ordered eight PV-2 Harpoons to be deployed to BA 9 in secret on 17 November. Later that same month, the U.S. Military Advisory Assistance Group (MAAG) learned of Freitas's intentions, and on 21 October, Rear Admiral Wallis F. Petersen, head of the MAAG, notified Kaúlza de Arriaga, the Subsecretary of State for Aeronautics, that the deployment of the PV-2s abridged the agreement governing them. From Petersen's point of view, Portugal lacked the authorization to move the aircraft to Africa, as he believed they were supposed to be used exclusively to participate in NATO.[20] Eventually U.S. authorities yielded in the face of the vociferous arguments by Freitas that the aircraft would be used in an exercise in Angola and Mozambique and did not represent a permanent transfer.[21] This was a plausible representation in light of Exercise *Himba* (April 1959) and the planned Exercise *Ibis*. On the other hand, the U.S. intransigence was a bit hollow in light of the 1951 protocols, the developing threat in the Congo, and the fact that the PV-2 was an obsolete, World War II-vintage aircraft. It was due to be deactivated and replaced by the newer, more technologically advanced Lockheed P2V-5 Neptune, which was in use throughout NATO and which was currently being delivered to the FAP.[22] Indeed, in 1960 the U.S. Navy was on the verge of accepting initial deliveries of the even more modern, state-of-the-art Lockheed P-3A Orion, a third generation open ocean reconnaissance aircraft, so U.S. agitation over the deployment of what was virtually a museum piece was driven exclusively by political differences.[23]

Major António Silva Cardoso was assigned the mission of ferrying the eight PV-2s from BA 6, which was located just across the Tagus River from Lisbon and alongside the town of Montijo, to Luanda. He departed the *metrópole* on Friday, 12 November, in great secrecy, as this was the beginning of a weekend when the U.S. embassy was closed. This meant that U.S. authorities would not be at work and could not react for several days. By then the deployment would have been effected. In his memoires, Silva Cardoso describes in great detail the trials of ferrying of the aircraft along the West

19 Ibid., 161.

20 José Manuel Correia, "A Guerra do Algodão" [The Cotton War], *Mais Alto* 359 (January–February 2006): 41.

21 Ibid., 42.

22 António da Silva Cardoso, *Angola, Anatomia de uma Tragédia* [Angola, Anatomy of a Tragedy] (Lisbon: Oficina do Livro, 2000), 34.

23 Initial flight of the P-3A Orion was November 1959, and its first delivery to the U.S. Navy occurred in August 1962. Its first operational deployment was the Cuban Missile Crisis in October 1962.

African coast, labeling it "the most difficult mission" of his entire career.[24] Despite a lengthy transit fraught with political, mechanical, and weather difficulties, seven of the eight aircraft arrived safely in Luanda on 24 November, while the eighth would follow shortly after its balky left engine was changed in São Tomé.[25]

When Resende arrived in late October, he would have little time to prepare his small force to address hostilities. Overall there were precious few forces available to defend Angola at the time. The army had about 8,000 troops in the territory, the rough equivalent proportionately of having a mere 570 soldiers to defend the *metrópole*.[26] The navy had virtually no presence, as it was engaged in oceanic operations in support of NATO. The FAP likewise had effectively zero presence.[27] The absence of adequate security forces on the ground and in the air left northern Angola particularly vulnerable. This was an area of difficult terrain with its many valleys, river courses, and steep mountains, and posed a particularly thorny problem in projecting military power. Nevertheless, Resende injected a purposefulness to preparations from the first day of his formal installation as commander in November. He correctly perceived the looming threat and felt a keen sense of urgency to project airpower into the northern two districts opposite the Congo with the meager resources at his disposal, despite the difficulties.

With the arrival of the two squadrons, they were formed into Operational Group 901 under the command of Lieutenant Colonel Manuel Diogo Neto, who had overseen the initial training and qualification of Noratlas pilots and maintenance personnel in France. After giving the Harpoon crews a short rest, Resende had them back at work with familiarization flights in the region north of Luanda and in maritime reconnaissance missions along the coast in search of suspicious vessels that were possibly smuggling arms to rebel factions. No arms trafficking was observed.[28]

On 13 December, Resende ordered the four Austers and their ground support personnel to the small airfield next to town of Carmona, the district capital, as the

24 Silva Cardoso, 31–62.

25 Ibid., 59. See also Francisco Amando da Cunha, "Eu 'VI' Mesmo o Adamastor" [I "saw" the same Northwind] in *Histórias com Asas: Memórias dos Pilotos da TAP dos tempos do Hélice* [Stories with Wings: Memories of the Pilots of TAP in the time of Propellers] edited by António Jorge da Silva Soares (Lisbon: Associação dos Pilotos Portugueses de Linha Aérea, 1992), 147. The entire length of the "Costa da Mina" in the Gulf of Guiné is filled with squall lines that must be negotiated and large zones of instability extending across the east-west route that must be crossed. Adding to all of this is the intertropical front to the north of the transit route. These various weather activities could make for a difficult transit flight.

26 Fernando Pinto de Resende, "A Força Aérea em Angola (1960/1963)" [The Air Force in Angola (1960/1963)], *Mais Alto* 228 (March–April 1984): 21–22.

27 Ibid.

28 Ibid.

air base at Negage was under construction and not yet usable.[29] The following day, the Austers began their familiarization flights of the local area to the delight of a curious population, and the following week the disturbances in Baxia do Cassange began.

Negage—Sentinel of the North

Beginning in 1957, the Air Force staff undertook a series of studies and conducted a number of missions to Angola on the feasibility of developing aviation infrastructure. Although the Deslandes mission was the most remembered, the most important was the one headed by General Viana Tavares, which focused on the north of Angola and specifically on defense of the districts of Zaire and Uíge, which comprise the old Portuguese Congo. These northwestern-most districts were bounded by the Atlantic Ocean to the west, the Zaire River and a remote land border running along the approximate parallel of Latitude 5°50' South to the north, and the Cuango River to the east. They protruded into the Congo, and their northern border was only about 100 miles south of Léopoldville. This proximity made them clearly vulnerable to any assault, particularly as their frontiers were quite open, and once an enemy was inside the territory, there was little to stop him. Likewise, the agricultural economies of the north were built on numerous relatively remote and widely scattered farms and villages that raised sisal and coffee and presented soft and easy targets for any invading force, as they were far from help.

From Viana Tavares's observations, he was disposed to recommend a base be strategically located in the area of Carmona, a primary town roughly centered in geographic area of these two districts and about an hour of flight time from Luanda.[30] After overflying the area and returning to Luanda, he continued his studies of the region, as there were difficulties with Carmona. The runway there was too short for heavy aircraft, and its expansion and the installation of a base would encroach on a significant amount of the fertile and valuable local coffee-growing land.[31] Because of this, there was enormous public resistance to this site. Prime coffee fields around Carmona had already been sacrificed for the small runway, and there was no enthusiasm for any expanded presence.

On the other hand, the people of Negage waxed enthusiastic about the project and had no coffee zones to be jeopardized. Accordingly, they approached Viana Tavares and made the case that there existed ideal terrain for a base next to their town. Further,

29 Ibid. Negage, pronounced correctly, as the locals did, was "N'Gage" or "Angage," and by the same token "N'Gola" became "Angola." It was the influx of Europeans in the 1960s who pronounced it "Negage" that little by little made it so.

30 Américo Vilarinho Alves, "Aeródromo Base 3 – Negage, Súmula Histórica" [Airdrome Base 3 – Negage, Historical Summula], interview by Aniceto Ferreira de Carvalho, 31 May 2008, http://aerodino.no.sapo.pt/negage1.hmtl (accessed 24 August 2012).

31 Augusto Cândido Soares de Mouro, "Testemunho Norte de Angola (1961–1962)" [Testimony North of Angola (1961–1962)], *Mais Alto* 228: 49.

the area was perfect for air operations, as the metrological conditions were favorable for flying, and the local people would welcome the FAP.[32] The general followed this offer with a visit by automobile to Negage and concluded from this survey that, in fact, the town actually was an ideal site, as it possessed all of the requisite conditions. Approval of the base was granted, and construction began on the runway in mid-1960.[33] By February 1961, the runway had been cleared and graded, and there were approximately 1,200 meters of compacted earth on which aircraft designed for unimproved airfields could land.[34] On 6 February, two Austers landed, and two days later the remaining two along with a Noratlas touched down to celebrate inauguration of AB 3.[35] Eventually Negage would have a 2,600 meter asphalt runway (7,874 feet, according to the U.S. Department of Defense *Flight Information Publication–Supplement Africa* of the time).

Negage was to have three dependences in what was initially styled as the Uíge Zone:

- AM 31 – Maquela do Zombo
- AM 32 – Toto
- AM 33 – Malange

Of these, only Toto and Maquela do Zombo exhibited the same progress as Negage. Toto on completion would have a 1,500 meter laterite runway (5,400 feet, according to the U.S. Department of Defense *Flight Information Publication–Supplement Africa* of the time) and operate all types of aircraft with the exception of the B-26.[36] However, like Negage, by the spring of 1961, it would have only a graded runway of compacted earth.

Maquela do Zombo was progressing along the same trajectory and on completion would have a 1,450 meter runway (4,757 feet, according to the U.S. Department of Defense *Flight Information Publication–Supplement Africa* of the time), which would

32 Victor Elias, "Batuques do Negage, Tanta História por Contar" [Drumbeats of Negage, So much History to recount], http://aerodromobase3.blogspot.com/2008/06/tanta-histria-por-contar.html (accessed 26 August 2012).

33 Vilarinho Alves interview.

34 Fernando Pinto de Resende (Brigadeiro), Commander, 2nd Air Region, to João Faustino de Albuquerque de Freitas (General), Chief of Staff Portuguese Air Force, 1 January 1961, personal papers archive, Vol. II, No Prov. 323, Arquivo Histórico de Força Aérea Portuguesa, Alfragide.

35 Vilarinho Alves interview. The pilots of the detachment were Lieutenant Colonel Augusto Soares de Moura, Lieutenant Corte Real Negrão, Sergeant Pedro Carvalhão, and Lieutenant Elio Mesquita. There were four mechanics: Sergeants Rúbio and Gusmão, and Airmen Paiágua and Antunes. Lieutenant Maia, two vehicles and their drivers, Airmen Guerra and Fernandes, completed the detachment.

36 Ricardo Cubas, "O Guardião do Toto" [The Guardian of Toto], *Mais Alto* 327 (September–October 2000): 36

accommodate most FAP aircraft. Ronald Waring, a retired colonel in the Royal Hampshire Regiment and an instructor at the Institute for Advanced Military Studies in Pedrouços, a suburb west of Lisbon, offers an interesting description of a landing while on an August 1961 fact-finding assignment in Angola:

> The airstrip at Maquella do Zombo [Maquela do Zombo] had been cut out of the elephant grass and resembled more a switch-back railway on a fairground than any landing ground for aircraft. We bounded up and down, coming to rest at the far end with the propellers of the Nord Atlas [Noratlas] twin-engine plane churning in the elephant grass.[37]

While subjective, this description reflects the conditions of the airfields from which high-tempo combat operations were launched. Malange would not be usable until much later.

Because of events in the north, it would be another year before the completed base at Negage would receive a visit from Kalúza de Arriaga for its official opening ceremony.[38] In the intervening months, the 15 March UPA attacks came, and despite the initial warnings, Portugal was caught unprepared, much the same as was Great Britain after the fall of France in 1940. Like the Royal Air Force then, it was the FAP, a new-born force "still in diapers," which rushed to help with its scanty resources. AB 3 Negage, AM 32 Toto, and AM 31 Maquela do Zombo would play critical roles in this action.[39]

Mayhem

The assault on the north occurred primarily in the districts of Zaire, Uíge, and Cuanza Norte. The first two are next to the border, and the third contains the impenetrable Dembos region, a mountainous jungle area where UPA elements hid and prepared for the internal assault. During the weeks leading to 15 March, the UPA carefully prepared its assault plan, infiltrating agitators from the Congo. It assembled its forces not only on the frontier and in the Dembos but also in many different locales within the targeted regions and was able to do so largely in secret because of the remoteness of the area and the uncontrolled borders.[40]

37 Duke of Valderano, *The Owl and the Pussycat* (London: Minerva Press, 1998), 203. Ronald Waring inherited from his mother on her death the title of an Italian dukedom. With this event, he changed his name to Valderano and assumed the title.

38 Elias.

39 Ibid.

40 António Lopes Pires Nunes, *Angola 1961: Da Baixa do Cassange a Nambuangongo* [Angola 1961: From Baixa do Cassange to Nambuangongo] (Lisbon: Prefácio, 2005), 86–87.

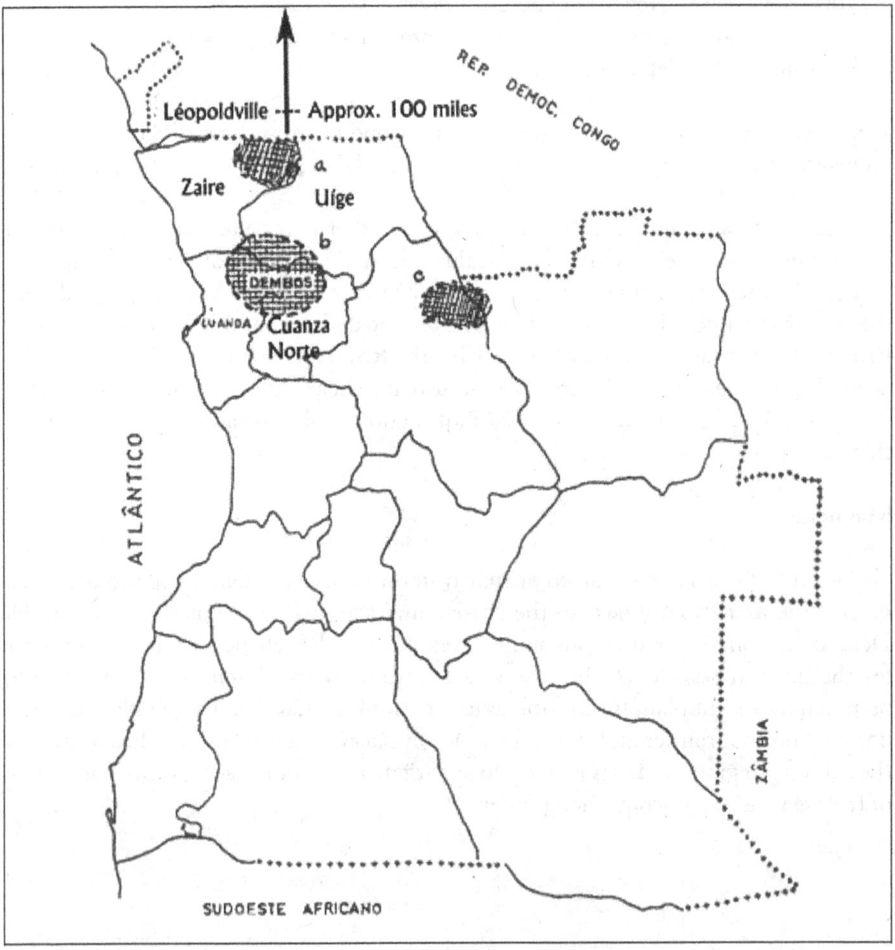

Map 5.1 Areas of Initial Subversion:
a. Primary area of frontier attacks on 15 March 1960
b. Area of internal attacks, Dembos Forest on 15 March 1960
c. Area of Baixa do Cassange uprising
(Source: António Lopes Pires Nunes, *Angola 1961: Da Baixa do Cassange a Nambuangongo* (Lisbon: Prefácio, 2005), 90)

The order to attack was described in a pamphlet widely disseminated to the population and marked the "beginning of the festival" as 15 March.[41] It cryptically ordered all those "followers to clean all *postos* thoroughly, and to treat the *chefs de posto* and their families and other whites well," and only some days later was the hostile intent of this message understood by the Portuguese authorities.[42] Another UPA document dated March 1961, Léopoldville, and titled "Rules to Follow for the Population of Angola in the Days of the Feast," reinforces the festival pamphlet and outlines specific procedures to be followed in the "celebrations."[43] It talks about women and children leaving their villages to participate in the celebrations, soldiers and sepoys allying with the people and joining the celebrations, and all people uniting in the success of the celebrations. The "celebrations," of course, were the intended acts of mayhem to destroy the agricultural commerce of the area and important communications infrastructure. These acts were described as "cleaning the farms" and the "destruction of bridges." The underlying logic in this plan was deeply flawed in that the population that was supposed to rise up and join the UPA forces depended for its survival on the prosperity of the farms and estates and the ability of these enterprises to get their crops to market by way of the interconnecting infrastructure. Thus despite their grievances, the resident population's loyalties could never ultimately lie with the UPA in this venture, and the people knew it.

The attacks were launched on signal, and the UPA attackers—an estimated 4,000 to 5,000 armed mob drugged with *quimbombo* or *cacharama*, a native alcoholic drink and hyperstimulant, and convinced that they were immune to the white man's bullets—swarmed from their redoubts in the Congo, the Dembos, and elsewhere in the early morning hours of 15 March.[44] This figure was expanded by an estimated 25,000 through forced recruiting.[45] From the Congo, they crossed the border at approximately seven points along a 300-kilometer front in the area of São Salvador and moved south. Approximately 700 European farms and estates, additional trading settlements, and over 100 government administrative posts were overwhelmed.[46] The mob acted with refined cruelty in massacring men, women, black, white, young, and old. At 1230 on that day, Radio Brazzaville announced that "the rebel forces of the

41 Ibid.
42 Ibid.
43 Ibid.
44 René Pélissier, *Le Naufrage des Caravelles: Etudes sur la Fin de l'Empire Portuguais (1961–1975)* [The Shipwreck of the Caravels: Studies on the End of the Portuguese Empire] (Orgeval: Editions Pélissier, 1979, 147; Tomás George Conceição Silva, "The Portuguese Air Force in the African Wars of 1961–1974" in *Memories of Portugal's African Wars, 1961–1974* edited by John P. Cann (Quantico: Marine Corps University Foundation, 1998), 113; and Alexandra Correia, "O Dia da Barbárie" [The Day of Barbary], *Visão História* 12 (June 2011): 59.
45 Willem S. van der Waals, *Portugal's War in Angola 1961–1974* (Rivonia: Ashanti Publishing, 1993), 64.
46 Pélissier, *Naufrage des Caravelles,* 147.

UPA had begun the fight for the independence of Angola, having decimated hundreds of whites."[47] At the same time that the sad news became known in Lisbon, there was great satisfaction in its receipt at UN headquarters in New York. The Brazzaville Delegate announced news of the "revolt," and immediately afterwards, the Liberia Delegate made the motion to condemn Portugal for defending itself and taking measures to restore order.[48]

All of Portugal was shocked at the horror. It has been estimated that 500 Europeans and about 20,000 local people died in this *jacquerie*.[49] The enemy was forced to act against the greater part of the native population that remained faithful to the Portuguese. Moreover, in this assault there were clear acts of intertribal rivalry. The UPA was based on the Bacongo people. In the north on the farms and estates there was a demand for labor to harvest crops, and these jobs were largely filled by Bailundo people from the south of Angola. While UPA forces were initially friendly to the Bailundos, after 7 April, when the enemy began burning the farms as well as the towns of Santa Isabel and Vista Alegre, these migrants began to fight the invaders.[50] As it turned out, the Bailundo workers were intensely loyal, and many died in an attempt to defend themselves, their employers, and Portugal against the Bacongo UPA.[51]

One of the best and most moving descriptions of the devastation wrought by the attacks is contained in an article by the journalist Sousa e Costa in the *Jornal do Congo*, published in Carmona and dated 20 April 1961, some 35 days after the slaughter.

> I returned to see little from an extensive flight over the valleys of [the rivers] Encoge and Loge. I saw below less than a few dozen [square] meters of soil, the spectacle left no room for anything but heartbreak, and tears came to my eyes. There is sadness in such horror. Dozens and dozens of farms in whose yards and woods were pursued day-by-day the painstaking work of coffee cultivation, abandoned and in such a way as to give the conclusion that life there had possibly existed at some time. Houses with doors broken open, automobiles abandoned just as they were before the terror, forests of coffee with grass encroaching and invading all, as if a tremendous wave of distress had inspired a malicious abandonment by proprietors of their property. And nothing can be imagined,

47 Edgar Pereira da Costa Cardoso, *Presença da Força Aérea em Angola* [Presence of the Air Force in Angola] (Lisbon: Secretaria de Estado de Aeronáutica, 1963), 90.

48 Ibid., 91.

49 René Pélissier, *La Colonie du Minotaure: Nationalismes et Révoltes en Angola (1926–1961)* [The Colony of the Minotaur: Nationalist Movements and Revolts in Angola (1926–1961)] (Orgeval: Editions Pélissier, 1978, 657–660; and Willem S. van der Waals, *Portugal's War in Angola 1961–1974* (Rivonia: Ashanti Publishing, 1981), 58–61.

50 Manuel Diogo Neto, "Acontecimentos no Norte de Angola" [Events in the North of Angola] *Mais Alto* 30 (June–July 1961): 9.

51 Ibid., 8.

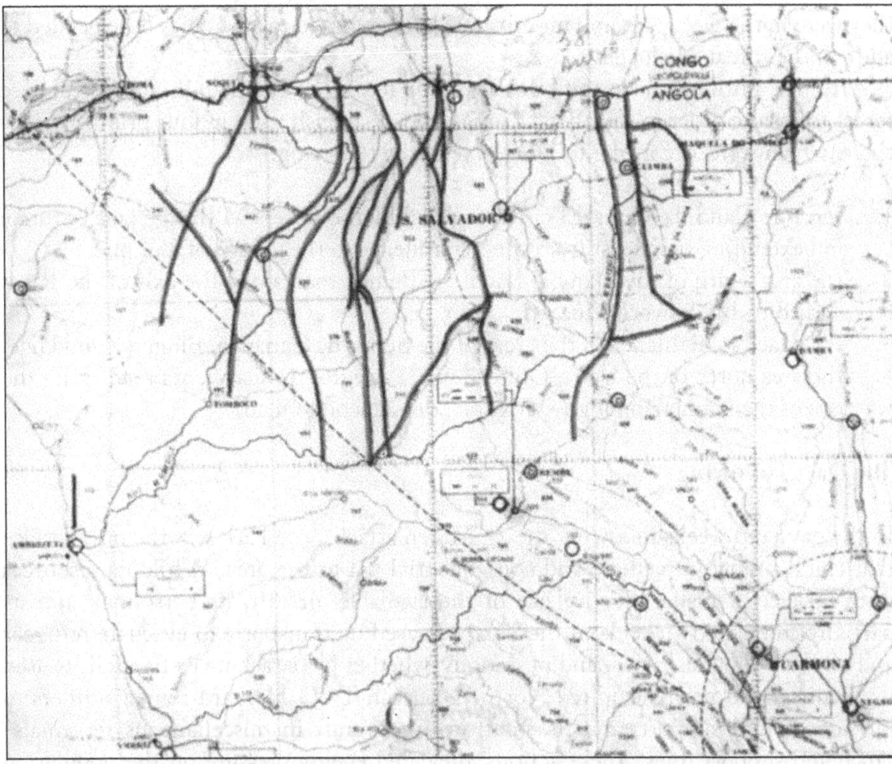

Map 5.2 Frontier Infiltration Routes in the Area of São Salvador, 15 March 1961. (Source: Personal papers of General Manuel Diogo Neto, Archivo Histórico do Força Aérea, Caxia 315-1)

> recognized, what shock these two extensive valleys hold with their small or great enterprises, all made with the heart, and now dead. What sadness![52]

Reaction by the authorities was swift, and in those 35 days, the FAP acted with every resource it could bring to bear in defense of northern Angola and the rescue of people who lived there. This initiative included the local Aero-Club of Angola and DTA (*Divisão dos Transportes Aéros*, the civil airline of Angola), both of which answered the call with a heroic effort. While the FAP acted quickly, there were unfortunately limits with six Noratlas, eight Harpoon, and four Auster aircraft. As far as land forces went,

52 As quoted in Pires Nunes, 88–89. Sousa e Costa was gravely wounded in reporting the war and was probably the first contemporary Portuguese war reporter to capture the essence of the conflict with journalistic success.

they were in the *metrópole*, and months would be required to assemble those needed to address the threat adequately.

After the initial attacks, the UPA was able to exploit the inability of Portuguese forces to restore order in the affected area and to establish itself in four primary zones there, which were, according to Diogo Neto, the following:

1. Serra de Candida—an area circumscribed by the Rivers M'Bridge and Lefunde and extending northeast from their confluence to the village of Cuimba.
2. The area south of the Damba–Lucunga–Bembe road along the axis of the River Zadi Bite-Bite toward Mucaba.
3. The Macisso highlands in the areas of the Serras de Cananga, Pingano, and Uíge.
4. The area north of the Úcua–Quitexe–Vista Alegre–Aideia Viçosa road along the axis of the Nambuangongo–Muxuluando–Mucondo road.

The Days Following

In the days and weeks following the 15 March attacks, the FAP was the only service that could supply immediate and consequential aid to the area. While its resources were meager, it made effective use of the available aircraft, its personnel, and its partially completed airfields in the north. It used its transports to evacuate refugees and air drop supplies, arms, and munitions, whether by parachute or free fall. It used its bombers and newly acquired North American T-6G Harvard armed trainers to provide ground support, and its small utility aircraft in miscellaneous reconnaissance and support roles. These actions alleviated enemy pressure on the population and provided what official resistance Portugal could muster in the initial days of the conflict.

In the early morning hours of 16 March, Lieutenant Colonel Augusto Soares de Moura, the commander of AB 3 at Negage, received a warning telephone call from Resende in which he was told to "be fully prepared."[53] This brief and cryptic message was necessary, as there were no secure military communications at the time between Luanda and AB 3. Soares de Moura left his quarters immediately and remembered, "I arrived at the field to see a Dakota land and load dozens of women and children fleeing the area."[54] In less than an hour it would return to transport an equal load of more refugees. Minutes later a Noratlas or "Nord" appeared and loaded a share of the expanding queue of refugees for Luanda. These operations continued throughout the day and in the days following, as the FAP had immediately initiated a rescue of the frightened and jeopardized population with its Noratlas and Dakota aircraft from Luanda. The Nords continued to operate into the night on the simple unimproved runway at AB 3 that was a packed earth affair. There was no formal illumination, and

53 Soares de Mouro, 49.
54 Ibid.

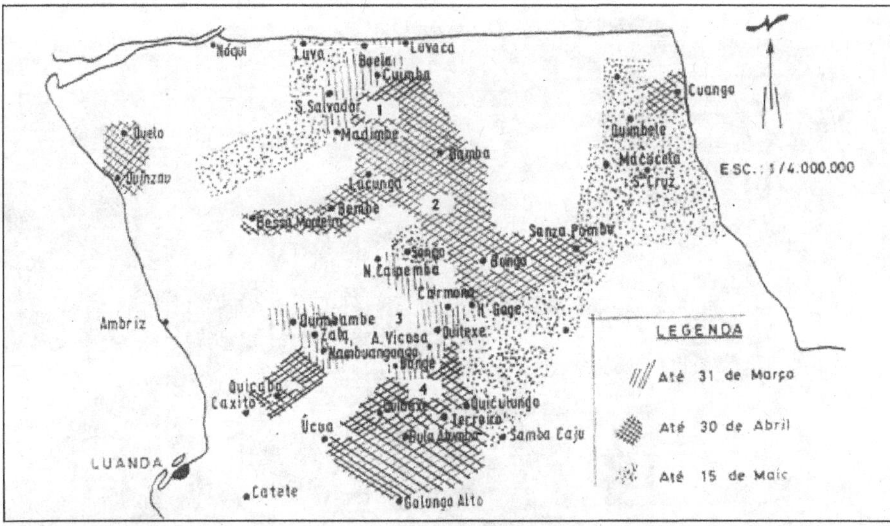

Map 5.3 UPA Insurgent Concentrations in the North of Angola. (Source: Luís Martinho Grão, *História das Tropas Pára-Quidistas Portuguesas*, *Vol. II-Angola* [History of the Portuguese Paratroops, Vol. II-Angola] (Lisbon: Comando das Tropas Aerotransportadas, 1987), 79)

runway lighting was provided by two jeeps parked opposite each other at the landing threshold. This rescue effort occurred during the rainy season between February and April, so periodically the field and its apron would be reduced to slime.[55]

Because of the washboard condition of the runway from rain, the pilots attempted to restrict the number of passengers to 30—the Nord had a seating capacity of 36. There was desperation in the embarkation queues, however, and the women in particular pushed and shoved their way on board until the Nord was packed. Passenger loads on the Nords began at 80–90 and advanced to 100–120. The record was 167, although it is certain that most of them were children.[56] The normal peacetime load limits of all aircraft were violated time and time again in removing refugees from danger. They seemed to be everywhere, and eventually the local cinema in Negage was used as a processing center in an attempt to bring order to the situation and prioritize the passengers. Locally owned automobiles shuttled the people between the cinema and the runway, as there were few military vehicles available.

The enemy now concentrated his activities on cutting the roads by felling trees across them. This had the effect of isolating a large number of small communities, and one of these blocked roads was only a kilometer from the Negage runway. Aircraft

55 Ibid.
56 Ibid.

Ground crew servicing a T-6G Harvard at Negage in 1961.
(Source: Private photograph collection of General Tomás G. Conceição Silva)

from all sources flew to the physical and psychological limits of air and ground crews to establish contact with the isolated populations and arrange for their rescue.

In addition to the larger transport efforts, the Austers of AB 3 worked hard to maintain a link with the anxious and isolated people by bringing mail and supplies over a period of many weeks. Offensively, the Auster pilots had only their pistols and were able to toss the occasional hand grenade, as in the "method Melange," so named from the procedure developed during the disturbances in Baxia do Cassange. Following the 28 operational missions that the PV-2s flew during that particular disturbance, Resende concluded that the actions of the four Austers in tossing hand grenades was relatively more effective that the PV-2s with their potentially greater capability.[57] It

57 José Manuel Correia, "Angola: Recordando 1961 e o Início da Actividade da Esquadra 93 'Magníficos'" [Angola: Regarding 1961 and the Initial Activity of Squadron 93 "The Magnificent Ones"], *Mais Alto* 395 (January–February 2012): 18.

was vastly simpler for the Auster pilot with his slow aircraft to be accurate with a hand grenade than it was for the heavier and faster PV-2, whose pilots were not trained in close air support and whose bombs were dropped inaccurately or, indeed, too low and failed to travel the required standoff arming distance.[58] The latter sank into the mud without exploding.

Nevertheless, Resende was worried about the obvious limitations of the Austers, and in a letter dated as early as 30 November 1960, he asked Freitas about the status of the replacement Broussard and Dornier aircraft and their scheduled shipment to Angola. As nothing was forthcoming in the immediate future, he was forced to rely on the fragile and fickle Austers. They were difficult to land when crabbing into the wind to maintain runway alignment, as the aircraft was extremely light and sensitive in crosswinds. On rollout, the brakes proved very erratic and were difficult to apply properly without experience. The pilots mastered these idiosyncrasies, but still such design flaws as having the flap handle on the opposite side of the cockpit from the throttle caused the pilot to have to switch hands to engage the flaps during the critical maneuvers of approach and landing. Experience would reveal that the fragile Auster was poorly equipped and inadequate for the hard operating conditions in the *ultramar* and was prone to mechanical failures.[59] When operated by inexperienced pilots, which was often the case, it was prone to accidents. Resende lamented the difficulties with the Austers in light of the rudimentary state of the outlying airfields, deteriorating conditions in the rainy season, and the lack of unimproved field capability in other aircraft: "Thus under the circumstances the pilots do not like them and neither do I….Moreover, I wish that the DO-27s would arrive so that we can transfer the Austers to the Aero-Clubs to good advantage…."[60] The bureaucratic "foot dragging" on sending the Broussards and solidifying negotiations on the Dornier DO-27s in the months leading up to 15 March had made Resende very anxious, particularly in light of his slender assets and the growing threat from the former Belgian Congo.[61] Now he had to depend on the Austers.

In these initial months, the primary mission was to defend or rescue the threatened population in the three districts around AB 3: Zaire, Cuanza Norte, and Uíge. This area was roughly the size of continental Portugal and was assigned exclusively to the

58 Aviation ordinance generally has an arming sensor driven by a vane that measures air travel to achieve a safe stand-off distance from the aircraft before detonation.

59 Aleixo Corbal, Oliveira Simões, Fidalgo Ferreira, and Heitor Almendra, "2° Região Aérea, Lições Aprendidas entre 1961 e 1975" [2nd Air Region, Lessons Learned between 1961 and 1975], proceedings of a conference "Almoços da AFAP," January 2011, 14.

60 Fernando Pinto de Resende (Brigadeiro), Commander, 2nd Air Region, to João Faustino de Albuquerque de Freitas (General), Chief of Staff Portuguese Air Force, 30 November 1960, personal papers archive, Vol. II, No Prov. 323, Arquivo Histórico de Força Aérea Portuguesa, Alfragide.

61 José Manuel Correia, "A Guerra do Algodão" [The Cotton War], *Mais Alto* 359 (January–February 2006): 43.

FAP and its paratroops, and indeed, the first troops to arrive in the area of Negage were a small detachment of "*páras*." They established their headquarters next to the Negage runway, and very soon thereafter the telephone rang. It was again Resende speaking in code and in English, "Be patient...," indicating more resources would soon be there.[62] Arrival of the troops gave the population confidence. Soon the paratroops took the offensive and began some probing actions, taking their first casualties.

Counter-offensive

When Soares de Moura arrived at the airfield on 20 March, he was stupefied to see four T-6G Harvards landing. These were not the training aircraft that he had piloted in the *metrópole* but the armed version with machineguns and rockets. The appearance of these aircraft was the result of Resende's negotiations with General Labit, the French air force commander at Pointe Noire in Congo Brazzaville, and they were described as a "gift" from him in their correspondence.[63] In light of subsequent events, however, it is unlikely that General Labit acted unilaterally. When U.S. President John F. Kennedy decided to renege on the understanding that MAP equipment could be used in the *ultramar* in an emergency, he threatened an arms embargo, unless all such equipment already deployed was returned to Portugal and used to support its NATO commitment. While the threatened arms embargo would not be made public as a matter of U.S. policy, private export licenses for war material would be revoked.[64] Two months after Labit's supportive action, General de Gaulle received Marcello Mathias, the new Portuguese ambassador to France, and told him, "Do not count on a change in the attitude of the United States. Look for new support. Resist. France will help you."[65]

At the end of World War II, the T-6 was in substantial surplus in the U.S. Navy and Air Force, as some 17,000 of them had been built.[66] Consequently, the surplus was made available to U.S. European allies at $100 per aircraft and $10 per spare engine, either new or rebuilt.[67] The T-6 had originally cost $30,000 to manufacture, so there was enormous value in its acquisition. France had purchased over 700 of this

62 Soares de Mouro, 49.

63 Fernando Pinto de Resende (Brigadeiro), Commander, 2nd Air Region, to João Faustino de Albuquerque de Freitas (General), Chief of Staff Portuguese Air Force, 10 March 1961, personal papers archive, Vol. II, No Prov. 323, Arquivo Histórico de Força Aérea Portuguesa, Alfragide.

64 Antunes, 162.

65 Ibid., 158.

66 Adelino Cardoso, "North American XAT-6E *Texan,* Un Avião-Protótipo Muito Pouco Conhecido" [North American XAT-6E *Texan,* A Prototype Aircraft Very Little Understood] *Mais Alto* 386 (July–August 2010): 45.

67 João Machado Vidal, "T-6/SNJ Harvard," Part I, *Mais Alto* 307 (June–July 1997): 3.

popular World War II trainer and modified them for close air support in Algeria.[68] The Portuguese found these armed trainers to be simple and effective in close air support and armed reconnaissance missions; however, their acquisition from so many sources resulted in cockpit and other variations that produced flying and maintenance complications and consequent adjustments. On delivery of the French T-6Gs, the four pilots told Soares de Moura that these aircraft were to be assigned to the base and that they would be returning to Luanda in a Nord.[69]

In fact, these four T-6Gs were actually struck from French inventory as surveyed aircraft, and their French markings removed before delivery to Luanda. They were flown there by French pilots without the markings, and it was only when they arrived at Negage that the Portuguese flag and Cross of Christ were painted on them prior to operations.[70] In correspondence between Freitas and Resende, Freitas asks about delivery of the four "toys" and a bag of "French almonds," codewords for the T-6Gs and their armament.[71] Resende replies, asking him about their "brothers," who were the additional T-6Gs promised from France.[72]

T-6G missions began immediately and were conducted at such a pace that shortages of munitions and aircrew fatigue became continuing problems.[73] Transport aircraft began bringing T-6G munitions to Negage from Luanda, and equally the French using volunteer pilots flew clandestine logistic missions from Pointe Noire to Negage directly, filing false flight plans for Fort Lamy. As French supplies became exhausted, two alternatives were discussed, approaching South Africa for munitions and manufacturing them locally. Likewise, the tempo of operations had an effect on the air and maintenance crews, and Resende points out to Freitas that there are only about 2,000 men defending Angola, the equivalent of defending the *metrópole* with two platoons. It is also at this point that Resende suggests to Freitas that a squadron

68 João Machado Vidal, "T-6/SNJ Harvard," Part II, *Mais Alto* 309 (October–November 1997): 34.
69 Soares de Moura, 50
70 Ibid. Soares de Moura noted that the national insignia were not to specifications but were affixed as best as could be done at Negage at the time.
71 João Faustino de Albuquerque de Freitas (General), Chief of Staff Portuguese Air Force, to Fernando Pinto de Resende (Brigadeiro), Commander, 2nd Air Region,18 February 1961, personal papers archive, Vol. II, No Prov. 323, Arquivo Histórico de Força Aérea Portuguesa, Alfragide.
72 Fernando Pinto de Resende (Brigadeiro), Commander, 2nd Air Region, to João Faustino de Albuquerque de Freitas (General), Chief of Staff Portuguese Air Force, 10 March 1961, personal papers archive, Vol. II, No Prov. 323, Arquivo Histórico de Força Aérea Portuguesa, Alfragide.
73 Fernando Pinto de Resende (Brigadeiro), Commander, 2nd Air Region, to João Faustino de Albuquerque de Freitas (General), Chief of Staff Portuguese Air Force, 29 March 1961, personal papers archive, Vol. II, No Prov. 323, Arquivo Histórico de Força Aérea Portuguesa, Alfragide.

of F-84s be sent as reinforcements, as there is a great deal of close air support that will be needed in the retaking of the north in the coming months.[74]

Most outposts and villages had a P-19 transceiver radio with which to communicate with the authorities, and from these came a deluge of requests for supply drops and fire support. Because these radio communications were not encrypted, there developed a simple, informal code in which each type of aircraft was designated an animal and each type of mission, a dance.[75] For instance, a village would ask for a lion to provide an urgent fox-trot—everything was urgent—meaning to send an attack aircraft for fire support. When the aircraft arrived at the village, ideally there would be a large log or white sheet pointing in the direction of the threat.[76]

Additional Austers and T-6Gs arrived from the *metrópole* during March and April, and initial deliveries of the DO-27 to Angola began in late April. The DO-27 was a West German, single-engine, short-takeoff-and-landing (STOL), utility aircraft manufactured by Dornier GmbH. Its configuration was the classic high-wing, tail-wheel aircraft with fixed landing gear, and it was notable for being the first mass-produced aircraft in West Germany after World War II. It was appreciated for its relatively wide and comfortable cabin and excellent short-field performance. Indeed, it was judged to be among the best of the FAP aircraft acquisitions. The West German *Luftwaffe* and West German Army had ordered a total of 428 of the aircraft, and Portugal initially purchased 40 new and later 106 surplus from West Germany.[77]

Arrival of the Dorniers was a welcomed development, as it was extraordinarily well adapted to the rudimentary airfields of the *ultramar* and could perform reconnaissance, transport of troops, passengers, and cargo, and evacuation of the wounded (MEDEVAC). The Dornier also had an offensive capability with its two pods or "nests" of eighteen 37mm rockets each, although this type of mission tended to be fulfilled by the T-6G. The ability of the Dornier to operate at night enhanced all of its capabilities, and almost immediately the Austers were overshadowed and flown only reluctantly.[78]

There were few landing fields that could accommodate the larger aircraft. Runways for the smaller ones were often the streets of villages. In some landing areas there was barely adequate wing clearance between the buildings lining the street for an Auster or Dornier DO-27. In the village of Mucaba, the runway was small and uneven, terminating in a building. In the village of Santa Cruz, the airfield consisted of two streets, one on which an aircraft landed, turned carefully and taxied to the next street for takeoff. As the winds were generally light, the orientation of the runways was rarely a problem.[79]

74 Ibid.
75 Ibid.
76 Ibid.
77 Mário Canongia Lopes, *Os Aviões da Cruz de Cristo* [Airplanes of the Cross of Christ] (Lisbon: Dinalivro, 2000): 184, 337–341.
78 Soares de Moura, 51.
79 Ibid.

Arming a Dornier at BA 9 with rockets, either smoke marking or anti-personnel fragmentation. (Source: Força Aérea Portuguesa)

Toward the end of March, sorties shifted from bombing missions to armed reconnaissance (*reconhecimento armado* or RCAR), and Soares de Moura flew his first one on 17 March.[80] The mission organization and targeting was discretionary, as there were no specific instructions for the pilots. Essentially they would survey isolated localities, the more remote farms, and secluded villages, and attempt to communicate with survivors. If any were found, then they were marked for a needed supply drop. If UPA groups were encountered, they were bombed or strafed. The only true guideline was to return from a mission with all munitions expended. After a few days of these patrols, all foot traffic ceased and the local population disappeared, as they were afraid of the aircraft.

The FAP goal was to seek and destroy an enemy bent on murderous activities and respond to the distress calls of settlers who refused to abandon their villages and who vigorously defended themselves in their desperate situation with the scanty resources available. Dramatic episodes followed one upon another. The FAP throughout its intervention and often in the nick of time saved the lives of entire populations faced with the threat of certain massacre. Ingenuity and close calls became everyday words in describing rescue efforts. In this vein, it was in the village of Mucaba that one of the most dramatic rescue episodes occurred.

80 Ibid., 50.

Village of Mucaba with the chapel in the foreground. (Source: Força Aérea Portuguesa)

Mucaba is located about 60 miles north of Negage in a densely forested mountainous setting. Horácio Caio, a journalist, describes the area from the air, "There below, the forest, dense and dark green, offered no clearing. Ravines and small plateaus were equally shrouded by a mist that began a few meters below the 'Nord Atlas'."[81] General Silva Cardoso elaborates similarly on the terrain in his memoires:

> We were going to learn to our cost and quickly arrive at the conclusion that the major enemy was not the armed guerrillas, but the terrain where war was made. The great density of the forest, the lack of lines of communication, the multiple water courses, and almost total ignorance of the potentialities of the enemy, all transformed the mission of the forces in the extraordinarily difficult terrain.[82]

Mucaba became increasingly threatened by UPA presence in the area. For almost a year the village had organized patrols of four men each night to raise alarm if the need arose. After the events of 15 March, there were terrifying accounts of subsequent enemy atrocities in Quitexe and many other neighboring locales, so the people

81 Horácio Caio, *Angola, os Dias do Despero* [Angola, the Days of Desperation] (Lisbon: 18ª Edição, 1966), 35.
82 Silva Cardoso, 161.

Chapel at Mucaba fortified by villagers in April 1961. (Source: Força Aérea Portuguesa)

of Mucaba were on heightened alert. The women and children had been evacuated. Road passage to the village became progressively irregular, and finally it was isolated. The village began to run low on fuel to generate electricity, a critical element in maintaining communication with the outside world. If the generator used its last fuel, then the P-19 transceiver would have no power to operate. On 29 April the remaining 30 civilians, one army quartermaster, and the *chefe de posto* Hermínio Sena gathered in the local chapel, which became a defensive redoubt, as the UPA entered the village. In the meantime, Sena succeeded in sending a dramatic call for help from the besieged chapel following wave after wave of UPA attacks, "We are running short of ammunition. By God, send us the Air Force."[83] With rebel trucks encircling the chapel and the mob, estimated at 2,000, only a little further away, a napalm strike appeared to be the most effective deterrent.[84] Unfortunately, a low-level dense fog developed during the evening.

BA 9 in Luanda had limited resources in three serviceable PV-2s and a stock of only six napalm bombs. The first PV-2 launched at 2100 hours with two of the 350-pound

83 "Um Nome para a História: Mucaba" [A Name for History: Mucaba], *Jornal do Congo*, 4 May 1961.

84 João José Brandão Ferreira, "A Heróica Resistência de Mucaba – Evocação" [The Heroic Resistance of Mucaba – Reminiscence], *Combatente* 347 (March 2009): 39

napalm canisters. Fog over Mucaba prevented any attack, and the aircraft was forced for safety reasons to jettison its precious bomb load and return to Luanda.[85] The second aircraft launched in another attempt but also encountered dense low-lying fog over the target. It too was forced to return to Luanda, but the pilot decided to accept the risk and landed with the precious bombs under his wings.[86]

The rebels felt that the end was close and that victory would soon be theirs. They proceeded to gather drums of gasoline from a nearby depot in obvious preparation for a gruesome barbecue and then congregated some distance away from the chapel, fortifying themselves with alcoholic libations in anticipation of the final assault.[87] The generator, which was critical to the radio, was within rifle range of the chapel, so its defenders were able to protect it from being shut off by the attackers. Increasingly pitiful and pleading distress calls came from the chapel, "We have no more ammunition. The enemy is firing through the chapel door. We can now only fight with bayonets. Please save us!"[88]

By 0500 on 30 April the situation was dire, as each man had less than 10 cartridges with which to defend the chapel. Yet at 0610, a Dornier DO-27 piloted by Captain Mário de Lemos de Mascarenhas, penetrated the low cloud cover and, flying close to the ground, landed his craft in the street leading to the chapel in a daring performance. This was an extremely dangerous maneuver, as while the street was straight, it was very narrow with garden walls along one side and a precipice descending through coffee plantings into a creek on the other.[89] Exploiting the element of surprise and capitalizing on rebel astonishment, he dropped ammunition and medicine boxes in front of the chapel, turned around, and took off with full throttle into the face of the mob that was yelling, shooting, and running towards the aircraft. This breath of life for the besieged chapel helped its occupants hold the terrorists at bay for a few more hours. As the sun rose, the fog suddenly lifted, and a pair of T-6Gs from Negage

85 Eduardo Alexandre Viegas Ferreira de Almeida, *Quarenta Anos de Aviação* [Forty Years of Aviation] (Lisbon: Privately published, 1995), 229–235. Eduardo Viegas was the chief pilot for DTA (Divisão dos Transportes Aéros, the civil airline of Angola), who, because of his intimate local knowledge of the territory, was asked by General Resende to fly copilot with Diogo Neto, whom he knew well, in a PV-2 to bomb the UPA attacking the chapel at Mucaba. The idea was that Viegas would be able to guide the attack through the fog cover and gain the area beneath it safely to execute the mission. The pilot at the last minute was changed to Captain Everedosa, who proved the reluctant warrior. After many disagreements between the two over the mission concept, the descent through the fog was never made, the bombs were eventually jettisoned in the forest, and the aircraft returned to Luanda after an unsuccessful mission. Everedosa deserted to Algeria a few months later. In reflecting on the experience, Viegas believes that Everedosa in truth never wanted to drop the bombs on his "friends."

86 Conceição Silva, 114.

87 Ibid.

88 Brandão Ferreira, 40.

89 Augusto Cândido Soares de Moura, "Mucaba, Uma Certidão Narrativa, I Parte" [Mucaba, An Attested Narrative, Part I], *Mais Alto* 329 (January–February 2001): 38.

Captain Mascarenhas and his Dornier DO-27. (Source: Força Aérea Portuguesa)

strafed in succession, followed by a third T-6G and the first of a series of PV-2s. The first PV-2 napalm attack hit the mob dead center, and the second finished the job. The population of Mucaba was saved from certain massacre; however, this episode marked the last opportunity to attack the enemy in the open.[90]

A further dramatic episode occurred in the last days of May, when the *chefe de posto* of Púri was alerted by a local native that murder of the loyal African population in the area was being planned by the UPA rebels. This warning was based on a pattern of events beginning on 17 May, when the *sanzala* or small native settlement of Soba Lengo, about 4 miles from Damba, was attacked by the UPA. The rebels cruelly murdered many of its inhabitants as they slept, and then pillaged and burned the *sanzala*. Next at the *sanzala* of Soba Tocoboco, near the village of 31 de Janeiro,

90 Conceição Silva, 114.

Survivors of the Chapel at Mucaba. (Source: Força Aérea Portuguesa)

various women and children were quartered. The same fate befell five prominent native chiefs in the region of Quibele, as they refused to join the revolt.[91] Púri was thought to be next, as the UPA leaders at the time were in the *sanzala* of Quibalundo, some four miles distant.

The *chefe de posto* immediately contacted the Negage administrator and the commander of AB 3, Soares de Moura. Quickly an imaginative plan was conceived. Twenty-five letters would be written in the local dialect of Quicombo with the forged signature of UPA officials in Léopoldville saying that Holden Roberto's UPA, with the support of Kasabuvu, was fighting for the independence of Angola. Holden Roberto or his representatives would fly to Púri to check on the performance of the local fighters. All of the leaders were to gather at the *curral dos bois* or cattle corral of the *sanzala* at a certain day and time.[92] The letters were delivered by local native volunteers. On the appointed day, an armed T-6G with no marks other than large UPA letters on its wings and fuselage and with the AB 3 pilot Soares de Moura at the

91 Cardoso, 183.

92 A *curral dos bois* or "corral of wood" consists of tree trunks closely planted in the ground in a circle. The trunks are inclined outward to make it very difficult for wild beasts to attack the cattle inside.

T-6 from AB 3 Negage used in a ruse to save the people of Púri.
(Source: Força Aérea Portuguesa)

controls and a black African in the back seat flew to the *sanzala*.[93] On arrival, the pilot lowered his seat and hid his head. He instructed his passenger to open the cockpit, raise his hand, and wave to the mob, which answered with obvious excitement and signs to land.[94] Suddenly two Negage T-6Gs appeared and strafed the crowd. The panic in the *curral* was indescribable. In seconds the area was deserted, and only pools of blood, the dying, and the dead remained.[95] This amazing deception helped to avoid another massacre, and the remaining enemy lost its appetite for intended assaults on Negage and Carmona.

General Tomás George Conceição Silva accords credit for the early FAP operations:

> The outcome of this first phase of the fight owed a large debt to the flying personnel and ground crews who worked around the clock with little or no rest. Flight crews covered an immense area through unforgiving tropical weather with few navigation aids and only the most rudimentary charts and in aging machines unsuitable for the tasks assigned. Ground crews rose to the most difficult aircraft maintenance challenge imaginable. They overcame the insidious tropical

93 Ibid., 184
94 Conceição Silva, 113.
95 Cardoso, 184.

> environment and the age of the aircraft in delivering serviceable machines to support a demanding operational tempo.[96]

The FAP during these early months of the conflict in truth made do with a paucity of resources, and its utility was probably best described by General José Lemos Ferreira in a deposition taken following the war:

> The Air Force was important in operations in which significant military action was not relevant but in which the psychological dimension was important. For instance, the resistance at Mucaba. Mucaba was besieged, and the people took refuge in the church. Support of this type was the Air Force through its PV-2….[97]

More than anything else, it was the spirit of the time that affected everyone in Angola, and the flying conditions and the risks involved are no better highlighted than in the accident report of the 17 June 1961 crash that led to the death of the junior pilot Jorge Eduardo Valadeiro at Negage. His T-6G experienced a crosswind in taking off on the unpaved runway that had a lateral incline. Because there was inadequate ramp parking space, the AB 3 aircraft were parked along each side of the runway, barely 3 to 4 meters from its edge. The crosswind and runway incline combined to carry the T-6G into the parked aircraft line, where it collided with a DO-27.[98] The accident report written by Captain Mascarenhas not only details the risks and contributing conditions that produced the accident but provides an accurate picture of the field operating environment. The details from "Section M—Violations" of the report Annex are most revealing and provide a window on these early operations:

> Maintenance of the aircraft was conducted in the open air, and in many months under clouds of dust, with inconsistencies in the maintenance manuals, with difficulties in obtaining tools, instruments, and verification tests. Some of the inspections were not made (for example, radios) for lack of equipment. There were no workshops. The pilots ultimately sent were very young and inexperienced (300 or less flight hours).[99] They were integrated into an operational unit (?), with terrorists only a few kilometers from the airfield, and consequently with severe emotional stress.

96 Conceição Silva, 113.

97 José Lemos Ferreira, "No céu da Guiné" [In the sky of Guiné] in *A Guerra de África 1961–1974, Volume 2* [The War for Africa, 1961–1974, Volume 2] edited by José Freire Antunes (Lisbon: Temas e Debates, 1996), 587.

98 Jorge M. Brochado de Miranda, "AB3, Negage, Início de actividade" [AB3, Negage, The beginning of operations], unpublished draft article delivered in correspondence to the author, 2 November 2012, Alfragide.

99 During this same period US Naval Aviators reported to their first squadron with about 450 flight hours. The 1,000-hour mark was generally considered the point at which a pilot knew the aircraft and how to fly it thoroughly.

There were no fire trucks; trusting that there would be no fire in the case of an accident (as there was not). The personnel knew that they would not be able to parachute because the chutes would not open [when tested] in the officers' mess, and there was no paraloft (and some of these [parachutes] were sent to the paratroops).

There was no capable radio: the D/F [direction finder] did not exist. The radio beacon, weak, functioned sometimes and only on demand; because of a lack of radio maintenance yet again the tower could not be heard and only began to function irregularly during the day.

No [paved] landing strip existed; the takeoffs and landings were made on the graded earth of the future runway; the lateral inclination was prohibitive and obstacles were not within the limits of security.

Personnel did not have adequate berthing, and the food was whatever was available.

There was no [hospital] emergency room; there were no ambulances.

[Flight] instruction, in the midst of high-tempo combat operations, was basic to takeoffs and without continuity or adequate preparation.

There was no metrology; there was no anemometer. The wind, in direction and intensity was estimated by "the eye." In the present "fogy" season, the wind is generally strong and always crosses the runway. The decision to fly or not was taken…by a majority vote.

There was no place to park aircraft. For this the "front line" was established on the runway itself; this was in the final analysis the cause of the accident, because it is certain that the pilot's eye was in the cockpit unaware of the direction of liftoff; on a normal runway nothing would happen, because the parked aircraft would not be located less than a few meters (3 to 4) from the runway for takeoff.

There was no ground equipment.

These factors were known by the Command and were well exposed. But the conditions of the Congo emergency created a need to take all the risks.[100]

100 Mário de Lemos de Mascarenhas, "Relatório ao acidente de que resultou a morte do Fur Mil Pil Jorge Eduardo Valadeira, no Aeródromo Base No 3, Negage, em 17 de Junho de 1961, Anexo, Secção M – Violações" [Report of the accident that resulted in the death of

A view of Negage in the spring 1961 showing a French T6G parked at the edge of the runway where Lieutenant Valadeira had his accident. (Source: Força Aérea Portuguesa)

Negage was born in the midst of nothing and began operations while still under construction. On top of this, it found itself the center of the dramatic events of 15 March and in the middle of a civil population with the wrath of terrorists visited upon it—terrorists whose aim was to incite hatred between whites and blacks and between the various black ethnic groups. These developments inspired the aircrews to commit themselves in the execution of their missions to save lives, evacuate wounded, and resupply isolated communities. They assumed inordinate risks, but these failed to deter them in their devotion to duty and spirit of sacrifice. The ultimate honed skills of flying in a counterinsurgency environment were born in this crucible.

While General Lemos Ferreira calls it a psychological dimension, it was more correctly perceived by those fighting as an epic struggle in which Portugal and the Portuguese would prevail against all odds. This spirit is no better captured that in the

Lieutenant Pilot Jorge Eduardo Valadeira, on Airdrome Base No 3, Negage, 17 June 1961, Annex, Section M – Violations], Negage, 8 July 1961, personal papers archive, Arquivo Histórico de Força Aérea Portuguesa, Alfragide.

battle of Magul in 1895, in which António, a civilian, remarked, "My captain, 6,500 blacks are many blacks...." To which Paiva Couceiro replied, "But 275 Portuguese also are many Portuguese!"[101] General Silva Cardoso in his memoires offers his thoughts in this regard:

> But for all of the difficulties that we faced in conducting the struggle, there was in every one of us an enormous force that was essential in this first phase: the belief, the will, the total commitment, the conquering of apparently insurmountable obstacles, these constituted our best and most powerful weapon. No one underestimated the struggle, no one hesitated, for the majority it was obvious that there were risks, including an awareness that death could come from an ambush at any moment.... There were no normal hours in this call for national unity. Every hour was one of service. There were no protests, but rather a great union among all to accomplish the mission demanded of the Air Force, and all without exception were understanding and appreciative.[102]

101 As quoted in Brandão Ferreira, 38.
102 Silva Cardoso, 162–163.

6

Retaking the North of Angola

Following the 15 March attacks, the FAP sought to accomplish two tasks as rapidly as possible. First, it needed to deploy the largest possible number of aircraft, flight and ground personnel, and paratroops in the shortest possible time, and second, it needed to acquire an inventory of aircraft adaptable to fighting the new war that it faced. Troops were on their way from the *metrópole* by ocean transport and airlift. Between the months of May and December 1961, troop strength was augmented to 33,477. These troops landed in Luanda and, as they became available over this eight-month period, moved to subdue the threat from the UPA and regain control in the north of the territory.[1] The transition from a small force aimed at reaction and defense to a large one of reoccupation and neutralization was gradual and occurred over a period of weeks characterized by the pace of reinforcements in both troops and aircraft arriving from the *metrópole*. Lieutenant Colonel Soares de Moura, commander of AB3 at Negage, noted that there began a new phase in May, as more and more troops passed through his airfield.[2] Instead of the RCAR, air operations now shifted and became predominately close air support in a close air-ground coordination.[3] An army sector command was established at Negage, and FAP officers assigned as liaison. There were daily meetings of the two services to work out coordinated operations for the following day, all conducted in the newly-built galvanized sheet metal buildings. The advance of the army and paratroop columns was accompanied by armed aircraft prepared to intervene when necessary. This fire support occurred alongside airlift evacuations, light logistics support, and reconnaissance, and was complementary to the grand operations of Nambuangongò, Quipedro, Serra de Canda, and Sacandica, all of which will be examined shortly.

1 René Pélissier, *La Colonie du Minotaure, Nationalismes et Révoltes en Angola (1926–1961)* [Colony of the Minotaur, Nationalism and Revolts in Angola] (Orgeval: Editions Pélissier, 1978), 658.

2 Augusto Cândido Soares de Mouro, "Testemunho Norte de Angola (1961–1962)" [Testimony North of Angola (1961–1962)], *Mais Alto* 228: 52–53.

3 Ibid.

On the second issue, the FAP had acquired a replacement for the Auster in its far more capable Dornier DO-27, which had proven itself at Mucaba, Santa Cruz, and elsewhere. The small squadron of Nords had also proven a worthy acquisition with its unimproved field capability, and further augmentation was planned. Addition of the T-6G Harvard brought excellent close air support capabilities to the counterinsurgency fight, and its initial deliveries from French inventory had been the ideal answer for the advancing columns of infantry. The T-6G was easy to maintain and fly, and more of these were to be added to FAP inventory. There was, however, the need for a heavier air-ground attack aircraft, and this would come in the form of the F-84G Thunderjet, an unlikely candidate at first blush. Its obsolescence as a cutting-edge fighter and hence its availability made it an attractive candidate. Here we will examine the imaginative marriage of developing FAP capabilities with the difficult operating environment to achieve strategic success in the north of Angola.

More Dangerous UPA

The UPA was changing too. By mid-May 1961, it controlled an enormous area in the north of Angola, an estimated 12,000 square kilometers.[4] Further, some 48 villages had been abandoned by the population. Since the March attacks, the UPA had continuously received instruction in fighting techniques and material aid in the form of modern weapons from abroad, and this external support began to alter the face of the conflict. The *canhangulos*, primitive firearms made with water pipes stolen from farms, were exchanged for automatic rifles. The frontal attacks were replaced with the refinement of ambushes, each more carefully planned, as UPA cadres gained experience. Mines began to appear in the main unpaved roads and their dirt trail access routes. As the enemy changed his methods, so did the Portuguese troops.

The bands of rebels were vulnerable to destruction from the air when massed, and hence the war of the guerrilla became increasingly prominent, particularly in the more remote areas away from the commercial and military centers. Air operations now lasted for days and weeks, and the army increasingly required permanent support from the FAP. As the war progressed, detachments became a routine requirement of AB 3, some temporary and others permanent. Quitexe, Nambuangongo, Toto, São Salvador, Cabinda, and Ambrizete, all required two aircraft, two pilots, and two mechanics, who came and went in constant rotation.[5] Reconnaissance, transport, fire support—some were successful and others disappointing. It was an exhausting war against a fluid enemy who retreated when attacked and advanced when the troops rested. UPA guerrillas destroyed crops and constantly threatened the more remote and isolated farming communities. Occasionally arms caches were discovered, and

4 Hélio Felgas, *Guerra em Angola* [War in Angola] (Lisbon: Livraria Clássica Editora, 1961), 109.
5 Soares de Moura, 52–53.

troops suffered losses. Pacification gained momentum, as little by little the enemy was driven back. Still, there were the all too frequent reversals, such as the one described by the journalist Hoácio Caio that occurred during the night of 10 April 1961:

> The farm of Maria José is one of the most important between Negage and Quitexe. Here whites and blacks worked, a total of a hundred of them. Over a period of about fifteen days there appeared here two blacks who came to tell the workers that the blacks were at war with the whites. That it would be better if the workers were also at war with the whites and thought that they should want to go with them into the bush. They would return in about a week to learn of the workers' decision.
>
> Thus it came to pass. About a week later a group of fifteen blacks came by night to hear the answer and to lead away those who wished to go to the bush. Those who remained, moreover, would be dead because they would be murdered. Thirty workers left that same night and in the nights following. Seventy-five remained, refusing to cede to the threats of the terrorists.
>
> Three nights later at about eleven o'clock, close to a thousand blacks came. First, they broke into all of the houses of the European employers, of whom there were none, as they had taken refuge in Negage. They plundered and destroyed all. Then with a great uproar, they destroyed the encampment of the native workers. Some awoke at the noise and came to the door. They were the first to be struck down. The rest died in their own beds. The two who were on guard failed to raise alarm, took to the bush, and fled. Others also had time to flee. The remaining were slaughtered savagely.[6]

The two guards, João Baptista and Maurício Jorge, appeared the next day in Negage, exhausted from having run the 30 kilometer distance from the farm. It took five hours for a detachment of troops from Negage to reach the farm and bury the bodies of the 33 Bailundos. One soldier made a cross with two sticks from the forest for each grave, and the entire detachment stood for some minutes in silent respect.[7] Portugal had much to do to protect its population.

Air Clubs and Volunteer Air Formations

In Angola, as in many of the territories, there were landing fields for light aircraft situated adjacent to the small towns and larger settlements, and these were used by air clubs and other voluntary air service organizations. Some of these were located immediately next to the town, such as Púri, and others some kilometers distant but

6 Horácio Caio, *Angola, Os Dias do Desespero* [Angola, The Days of Desperation] (Lisbon:18ª Edição, 1966), 141.
7 Ibid., 142.

located centrally among several small population clusters with a view to serving all of them. The air clubs performed a valuable and important role with their regular service to these isolated communities. Because of the nature of the air club mission, after 15 March much was demanded of their pilots and fragile aircraft. Whenever there was a call, they responded and were a lifeline to the population, bringing the bare necessities and evacuating them when danger threatened.

Because of the nature of their job, these pilots found themselves taking risks that the FAP pilots were not permitted to assume in operating their aircraft. They did this with a confidence based on an intimate personal knowledge of the area, weather, and other local conditions. The pilots knew the terrain over which they flew by heart and navigated visually, using the roads as their primary landmarks. Likewise they routinely flew at night in areas where there were no navigation aids. They flew in difficult weather and frequently landed in the streets of towns and villages. The emergency in the north occurred during the rainy season, so in this case the weather was often difficult for flying because of the clouds or mist and occasionally necessitated an unscheduled landing. In these cases the pilots knew of the sections of road where they could land in an emergency and often took advantage of these safe havens.

Fatigue was a major factor in air club operations, and when a pilot was far from his destination, and daylight was fading, he was forced to find a landing site for the night. When this occurred, he would seek to land next to an army bivouac, if possible, and pass the night with it. This practice eventually established a close bond between the air club pilots and army units operating in the north and led to a deeper understanding of the needs and capabilities of each. The pilots responded to army hospitality by bringing mail, drinks, cartons of shrimp, books, and recent newspapers.[8] Likewise FAP messes often had their menus enhanced with the addition of fresh game shot by one of the farmers in the isolated communities that the volunteers served.[9]

The Aeroclube de Carmona, for example, had two experienced pilots by the names of Albino Augusto Leite and Rui Moller Freira, who flew together and whom all in the north knew and respected greatly.[10] When the UPA began its attacks, these two wasted no time and, while the more primitive airfields remained free of UPA rebels and open, airlifted all manner of necessities—arms, munitions, medical supplies, and food—required for the isolated communities to survive under imminent threat. They worked around the clock at a debilitating pace and did the nearly impossible in supporting the population and complementing FAP efforts. When one of the Carmona aircraft literally became an unflyable mechanical wreck from wear, the FAP offered a replacement Auster so that Leite and Freira could continue to fly.[11]

8 António Lopes Pires Nunes, *Angola 1961, Da Baixa do Cassange a Nambuangongo* [Angola 1961, From Baixa do Cassange to Nambuangongo] (Lisbon: Prefácio, 2005), 105.

9 Ibid.

10 Ibid.

11 Ibid.

Sadly, on 4 June in the mid-afternoon, Leite and Freira departed in their Auster to deliver supplies and mail to the people of Quimbele and were forced to land 20 kilometers from Mucaba because of mechanical difficulties. From Negage a series of search missions were conducted in the Serra de Mucaba, but as the days passed, all hope of finding them alive was lost. Only years later was it discovered that Moller Freira had been killed instantly in the crash and that Leite had survived only to be murdered shortly afterwards by UPA rebels in the immediate area.[12] These two were the first heroes of the air war, and as Colonel Pires Nunes observes, "They risked their lives in working to save their friends, and in the Portuguese Congo, all were their friends."[13]

Other aviators answered the call, and probably the most colorful was Dr. Pequito Rebelo, who at age 69 purchased a Piper Super Cub and resumed his earlier war-time call sign "Viriathus of the Air" (*Viriato do Ar*) after the Lusitanian hero who led an uprising against Rome. Earlier he had flown in the Spanish Civil War on the side of the nationalists.[14] There were also other volunteers from a wide array of professions who were drawn fully to flying, making a part-time interest into a full-time commitment.

Later the air clubs were institutionalized, first, as the Flight of Air Volunteers (*Esquadrilha dos Voluntários do Ar* or EVA), and later, as the Volunteer Air Formations (*Formações Aéreas de Voluntários* or FAV) in which pilots imbued with the same spirit of total devotion were constantly in the air flying light aircraft that were long past retirement condition. These aviators, always ready to fly in an emergency, were the true miracle workers of 1961.

F-84G Arrival

Resende's earlier appeal for an aircraft capable of delivering heavy fire support was answered with the arrival by ship in Luanda of a dozen F-84G Thunderjets between April and June 1961. These aircraft were in the process of being retired, and many of them had been transferred to the aircraft rework facility at Alverca with cracked combustion chambers in their Allison J-35 jet engines, a development that threatened to ground the majority of the fleet.[15] Kaúlza de Arriaga was initially reluctant to authorize the transfer, but after the events of March, he relented.

12 Pedro Manuel Ferreira, "FAV Formações Aéreas Voluntárias, Uma Singularidade Portuguesa no Teatro de Guerra Africano, Parte I" [FAV Formations of Air Volunteers, A Portuguese Uniqueness in the African Theater of War, Part I], *Mais Alto* 348 (March–April 2004): 6.

13 Ibid.

14 Ibid.

15 José Manuel Correia, "Angola, Recordando 1961 e o Início da Actividade da Esquadra 93 'Magníficos'" [Angola, Rembering 1961 and the Beginning of Operations of Squadron 93 "The Magnificent Ones"], *Mais Alto* 395 (January–February 2012): 18–19.

The first F-84Gs arrived in Portugal in February 1953 and were based at Ota (BA 2) in Squadrons 20 and 21. When the F-84G entered service, it was immediately acknowledged to be obsolete for its primary mission of aerial combat with its straight-wing design. It was no match for the Soviet MIG-15 and MIG-17 swept-wing fighters. These were faster aircraft, were capable of a tighter turning radius than the F-84G, and hence could bring their guns to bear easily on the hapless Thunderjet. For

Servicing an F-84G Thunderjet at BA 9 in Luanda. (Source: Força Aérea Portuguesa)

Arming an F-84G Thunderjet at BA 9. (Source: Força Aérea Portuguesa)

this reason the various NATO air forces were anxious to acquire more sophisticated and capable aircraft as quickly as possible, and this came in the form of the F-86G Sabre, which began to enter service in Portugal in 1958 as well as other NATO countries. By 1961 the sturdy F-84G remained potentially valuable in certain roles, and the FAP was not yet ready to abandon it. Originally the aircraft was designed and built as a fighter-bomber and was still able to perform credibly in attack missions, particularly in a permissive environment with no appreciable anti-air threat.

Following the drama associated with the transfer of the PV-2s to Luanda, Portugal was anxious to avoid another confrontation with the United States through a perceived violation of the Mutual Defense Assistance Program (MDAP), which was a system of "renting" or lend-lease. Under MDAP the United States retained control in the use of its loaned or rented equipment. Portugal avoided this restriction by purchasing each of its F-84Gs outright for 6,000,000 Escudos (approximately $208,695 at the average 1961 Escudo-Dollar exchange rate), which released it from the MDAP restrictions and left it free to deploy the aircraft to Africa.[16] The F-84Gs, however, would be of little use in the *ultramar* without the needed engine replacements. In a remarkably inconsistent application of its anti-colonial policy, the United States was happy to authorize the private sale to Portugal of an abundant supply of new replacement J-35 engines with zero hours of operation, an item that was absolutely vital to the operation of the F-84G in Africa.[17]

When the Thunderjets arrived in Luanda, they were initially housed in the civil aircraft hangar and supported by the Angolan civil airline DTA.[18] Key operations to retake the north began in April and ran into the late summer of 1961, and as soon as the F-84Gs could be made flight-ready and their new Squadron 93, the *Magnificoes* (Magnificent Ones), stood up at BA 9, they would join the fight. The first mission of Squadron 93 occurred on 17 August.[19] Despite the late start, Squadron 93 completed 43 missions in August, of which 32 were solo attacks or, as the FAP pilots termed it, "localized destruction" (*destruição localizada* or DLOC).[20] On 25 August a formation of Thunderjets provided close air support for the paratroops retaking Canda, their baptism of fire. Again on 19 September, the Thunderjets attacked Scandica in preparation for the air drop of paratroops. Unfortunately the jets were armed with wholly inadequate 100-pound bombs, as that was all that was available. There was also a shortage of 350-liter napalm canisters, and these shortages for a time made the Thunderjet missions more psychological than dangerous to the enemy.[21] The number and intensity of combat missions increased with the addition of each airframe, and

16 Ibid.
17 Ibid.
18 DTA (*Divisão dos Transportes Aéreos*) was formed as a division of the Administration of Railways, Harbours, and Air Transport of Portuguese West Africa in September 1938.
19 Correia, 19.
20 Ibid., 21.
21 Ibid.

there developed a proportionate increase in the need for munitions, particularly 500-pound general purpose (GP) and 350-liter incendiary bombs. On 10 September, Resende notified Lisbon that the arsenal at BA 9 had only 110 350-liter napalm canisters and 90 500-pound GP bombs, enough for about 15 days of Harpoon and Thunderjet operations. The 100-pound bombs were being used as a last resort, and these were sufficient for only two days of Thunderjet operations.[22] The constant strain on the long supply chain to deliver the needed ordnance became the greatest preoccupation of Resende during this period, and the magazine at BA 9 would reach the lowest level during the entire war.[23] Resende noted that the problem always cast a long shadow over operations and the FAP ability to conduct them: "Our delivery of fire support was limited to 25 percent of its potential because of a severely reduced arsenal"[24]

Typical of Thunderjet operations was Operation *Onzo* launched to recover the settlement of Beira Baixa in which the pilots made precision drops of napalm canisters less than 300 meters from army troops in waiting columns, as there were no 500-pound bombs or 2.75-inch rockets. Nevertheless, the available ordnance delivered precisely produced the desired results and had a "persuasive effect" on the enemy.[25] Later on 10 October, various army columns advanced on Quicabo-Nambuangongo with the ultimate goal of Pedra Verde with support from the Thunderjets. Days afterward there were jet operations against UPA concentrations in Serra da Cananga, Camabatela, and Bembe.[26]

Major António Silva Cardoso, the command pilot in a Harpoon, describes the events of a mission on 10 October with the Thunderjets against Serra da Cananga, an enormous elevated plateau with great escarpments completely covered in dense vegetation. The plateau extends northwest from Quitexe to Nova Caipemba, and the UPA had taken refuge there and was intensely subverting the population. Patient and lengthy aerial reconnaissance had located the UPA "sanctuary" with its flag flying from a flagpole in a clearing. Accordingly an attack was planned in which a formation of four Thunderjets would deliver their ordnance from altitude in a steep angle of attack that would place the bombs with great precision. This would be followed by three Harpoons in sequence together dropping a salvo of 18 200-pound bombs. As the Harpoons approached the target at 3,000 feet, the Thunderjets passed overhead and commenced their bombing runs, each delivering two 750-pound bombs. One of the Thunderjet pilots, Lieutenant António Seabra Dias, developed target fixation in his dive and failed to exit successfully. It was a great tragedy for all in the aviation

22 Ibid., 22.
23 Ibid., 19.
24 Fernando Pinto de Resende, "A Força Aérea em Angola (1960/1963)." [The Air Force in Angola (1960/1963)]. *Mais Alto* 228 (March–April 1984): 26.
25 Correia, 22.
26 Ibid., 24.

F-84 on the ramp at BA 9 ready for a mission with two 350-liter napalm canisters. Note the new Fletcher-style tip tanks and the dive brake extended. (Source: Força Aérea Portuguesa)

community at AB 9—the loss of a comrade, a friend and an excellent pilot.[27] This was the first of five fatalities with the "Magnificent Ones" that marred the 13 years of service (1961–1973) in Angola.[28]

Another 13 Thunderjets arrived in Luanda in 1962, bringing the total delivered to 25. The squadron conducted missions nearly every day, and it flew over 2,000 sorties a year for the first several years. As the war progressed, more Thunderjets were lost, and by 1973 only 12 survived. Each Thunderjet averaged 100 combat sorties a year, and this high rate of availability was due solely to the skill and dedication of the squadron maintenance personnel. During the war the original tip tanks were replaced by Fletcher-style tanks from the T-33. This was a field modification performed in Luanda.

Nambuangongo

Nambuango was a thriving small town situated on the western edge of the mountainous Dembos forest about 180 kilometers northeast of Luanda and in the center of a region rich in coffee cultivation. It was remote enough to be vulnerable to the UPA attacks and was savagely assaulted on 16 March. Following the murder of its population of nearly 300 Europeans, the local UPA leadership established its headquarters there and for five months rained terror on the surrounding countryside. Because of this successful regional savaging, the UPA cadre chief Maneca Paca named

27 António Silva Cardoso, Angola: Anatomia de uma Tragédia [Angola: Anatomy of a Tragedy] (Lisbon: Oficina do Livro, 2000), 233–234.

28 Correia, 25.

Nambuangongo the "terrorism capital of Angola." Within the month he would die in an encounter with Special Infantry Battalion 96 (BCE 96).[29]

By the middle of July, Portugal had assembled enough manpower and equipment to launch an offensive against this stronghold and began a three-pronged attack known as Operation *Viriato,* as depicted in the nearby map.[30] The initial column, which was composed of Cavalry Detachment 149, departed on 15 July from Ambriz, some 154 kilometers from the objective, and proceeded east on the Quimbumbe – Zala route. The second column, composed of an infantry tactical unit reinforced with artillery and engineering components, departed Luanda on the same day and proceeded northeast on the Caxito – Quicabo axis across the Lifune River. The third column, composed of Special Infantry Battalion 96, departed Dange and followed the Vista Alegre – Mucondo (Mucando) – Zemba-Quiraundo (Quicunzo) – Muxaluando route. Throughout the advance of these three columns the FAP provided air cover and reconnaissance with PV-2s from Luanda and T-6Gs from Negage in a closely coordinated air-ground support effort.

Once the columns were advancing, FAP aircraft overflew Nambuangongo or "Nambu," as the troops called it, and confirmed that it was being abandoned and that UPA forces were taking to the bush. This development meant that ambushes from the tall grass bordering the roads and trails would pose the greatest danger. Thus, it became the primary FAP mission to scout ahead and around the advancing ground forces, to alert them to enemy presence, and to destroy any detected UPA elements with air attacks. The columns were constantly forced to halt to clear obstacles, and on these occasions the aircraft would provide a protective perimeter. Felled trees lay across the roads, and bridges had been destroyed right down to the foundation piers—all done with explosives stolen from salt mines in the nearby Quicabo region. The PV-2s spent hours and hours maneuvering over the columns, as they moved through difficult terrain and constant rains. This protective air cover was complemented by supply drops from cargo aircraft between Quicabo and Nambuangongo.[31]

When Battalion 96 reached the village of Mucondo, where sections of the UPA had taken refuge but fled when the aircraft appeared, its troops discovered signs of cannibalism. This was the work of the infamous Maneca Paca, among other cadre leaders, who had had a hand in the killing and mutilation of four of a section of five soldiers investigating the situation in the small settlement of Cólua on 2 April. The

29 Edgar Pereira da Costa Cardoso, *Presença da Força Aérea em Angola* [Presence of the Air Force in Angola] (Lisbon: Secretaria de Estado de Aeronáutica, 1963), 188.

30 Following Hannibal's defeat in 202 B.C., Carthage lay prostrate, and Rome pushed westward from Spain into Lusitania, which proved a difficult venture. A Lusitanian war chief by the name of Viriato or Viriathus, who was originally a simple shepherd, proved more than a match for the Romans and defeated them in battle after battle. Unable to defeat him in combat, in 139 B.C. Rome bribed some of his disloyal followers to murder him as he slept.

31 Correia, 21.

fifth escaped for a time but later was captured and killed. According to a Bailundo witness, Severino José:

> As to the massacre of Cólua, nothing is known over and above the cooking of the four soldiers whose fingers and testicles were cut off and hung on skewers to cure….Bomboco and especially Simão Lucas organized authentic spectacles of cannibalism beginning with the four white soldiers captured at Cólua, forcing the Bailundos to cook them and enter into the festivities….From the first hour all those who died, whether whites, Bailundos, or local natives were eaten and always buried afterwards….[32]

Ronald Waring notes that it was a routine practice: "[Portuguese] casualties were buried where they fell, but only too frequently terrorist groups which came on the scene several days later would dig up the bodies and eat the hands and fingers, being the tastiest parts."[33]

The gruesome murders of the white farmers and their families and the native workers on the estates and the further practice of cannibalism delivered a psychological advantage to the Portuguese in demonizing the enemy. The UPA was successfully presented as grotesque and even invisible alongside the Portuguese, who were visible and thus normal human beings.[34] The Portuguese also presented themselves as non-racist in contrast to the highly tribal Bacongo UPA. Such enemy racism and tribalism practices were successfully portrayed as primitive and negative to the elevated multicultural image of Portugal.

Nambuangongo was reclaimed at 1745 on 9 August.

Quipedro

In order to maintain pressure on the UPA, Operation *Nema* to retake Quipedro began promptly on 11 August following the reoccupation of Nambu. For the FAP paratroops, this would be the first of only ten jumps during the war between August 1961 and April 1963, and these were undertaken only because there were insufficient helicopters at the time for a helicopter assault. Later, when the helicopter squadron was formed in the late summer of 1962, the paratroops would increasingly become helicopter-transported air-assault troops. In the meantime the Dornier would serve the role of a light cargo helicopter.

The concept of operations was to have Cavalry Detachment 149 continue its advance through Nambu on the road to Quipedro. The detachment would link up with the 1st Company of Parachute Battalion 21, which would be dropped into a landing zone

32 Deposition of Severino José as quoted in Pires Nunes, 103.
33 Duke of Valderano, *The Owl and the Pussycat* (London: Minerva Press, 1998), 205–206.
34 Inge Brinkman, "War and Identity in Angola," *Lusotopie 2003*: 196–197.

adjacent to the village. The UPA cadres on leaving Nambu had proceeded north in the direction of Bembe and Caipenda, and thus it was hoped to establish a successful hammer and anvil maneuver in which the paratroops would land behind the UPA and act as a blocking force for the cavalry to drive the rebels into the waiting paratroops. PV-2s from Luanda and T-6Gs from Negage would support both forces.

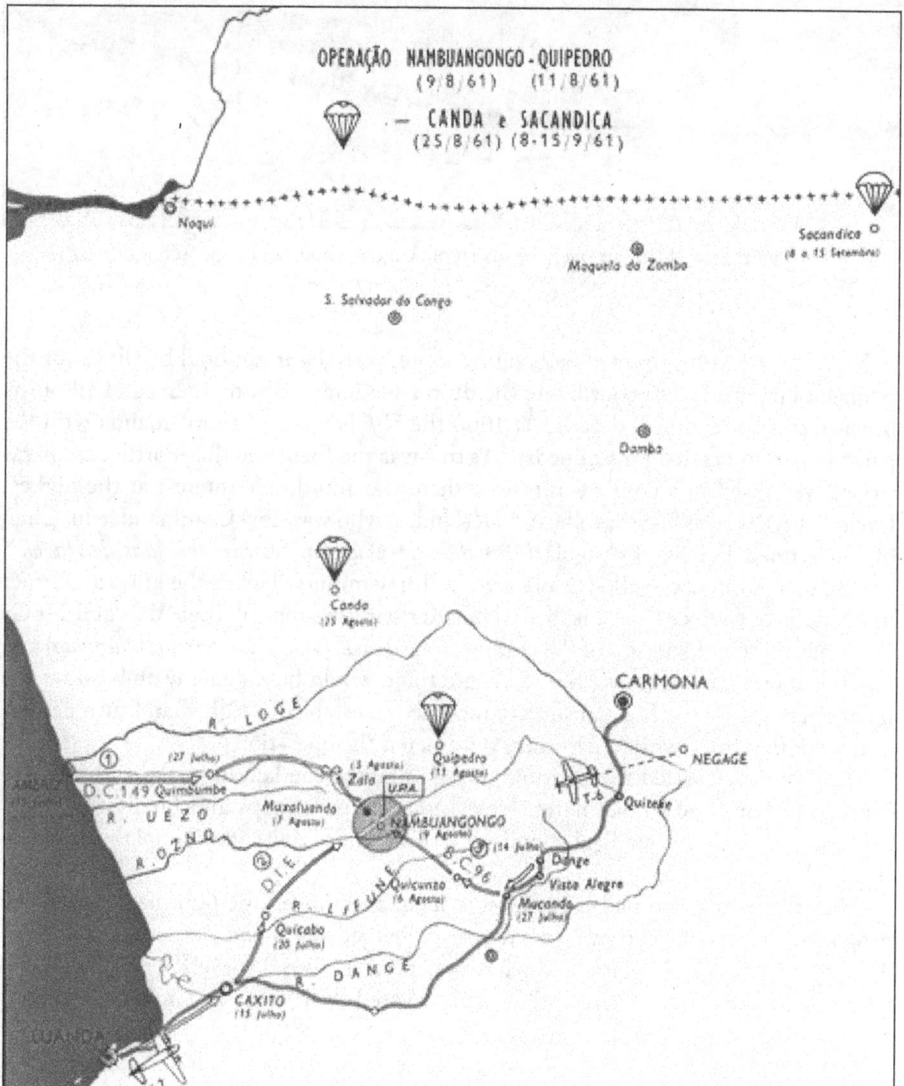

Map 6.1 Key Operations in Retaking the North of Angola 1961.
(Source: Força Aérea Portuguesa)

T-6G Harvard armed trainer over the north of Angola. (Source: Força Aérea Portuguesa)

Early on the morning of the operation there were the traditional briefings for the pilots and paratroops to coordinate the drop zone references, markers, and all of the other pertinent technical details. At 1000 the C-54s took off from Luanda with the paratroops and headed for Quipedro. As this was the first time that Portuguese paratroops were used in a combat insertion, there was inordinate interest at the highest levels of FAP. Generals Resende and Deslandes, who was now Commander-in-Chief of the Armed Forces of Angola (*Comandante-Chef das Forças Armadas de Angola*), among many others travelled as observers. Thirty minutes before the jump a Dornier marked the objectives to orient the PV-2s and their fire support. The PV-2s and T-6Gs then covered the vegetation of the landing zone or LZ (also ZL or *zona de largada*) and its surrounds with withering fire to ensure there would be no enemy ambush on the landing paratroopers. The transports proceeded to establish a right-hand turn pattern over it for the drops while a Dornier coordinated the operation.[35]

The first drop carried the pathfinders who marked the landing zone with fluorescent cloth strips and smoke flares. Next came the paratroops and finally the provisions, munitions, and medical supplies.[36] All of this activity gave ample warning to the UPA.

After the paratroops had landed, regrouped, and moved the 600 meters from the drop zone to the village, they remained unchallenged and found a deserted Quipedro. To avoid the suppressing fire and paratroopers from above, the UPA occupants had moved into the nearby ravines that were covered with trees and foliage and thus

35 Manuel Diogo Neto, "Operação da Serra da Canda," personal papers archive, Vol. I, No Prov. 315, Arquivo Histórico de Força Aérea Portuguesa, Alfragide, 4.
36 Costa Cardoso, 195.

Paratroops readying an improvised runway at Quipedro, 1961.
(Source: Força Aérea Portuguesa)

became invisible and safe from the advancing ground forces and the aircraft circling overhead.[37]

While the paratroops met no enemy, the cavalry detachment was not so fortunate. Rather than moving on foot, the detachment was mounted in vehicles and bound to the road bordered by high grass. As the column of vehicles proceeded, it encountered log barriers across the road that had to be cleared. When the column halted to remove the barriers, the detachment found itself in a classic ambush scenario and took intense fire from UPA gunmen hiding in the thick, tall grass 15 to 20 feet high growing right to the edge of the road. It was precisely as the men left the vehicles and were most vulnerable that the rebels attacked. This skillful setting of ambushes and the creation of killing zones were new refinements that clearly came from outside the UPA. Rebel ambushes increasingly made good use of the terrain, another newly imported skill. They also had abandoned their home-grown weapons of primitive shotguns and bush knives for modern automatic weapons, and during these attacks wore a quasi-uniform of blue trousers with white or red bands on them.[38] Still despite this new look, UPA

37 Ibid., 196.
38 Ibid., 196.

forces were not military professionals in that they continued their unrestrained acts of savagery and barbarism.

While the reoccupation of Quipedro was successful, the UPA was able both to escape the village and the paratroops and to lay effective ambushes on the attacking column. The necessary element of surprise was completely absent from Portuguese planning, and the UPA made the most of this deficiency.

Likewise at Quipedro the Portuguese experimented with air supply of its troops. The extended lines of communication from Luanda, the logistics hub some 245 kilometers distant, made resupply by poor roads tedious and lengthy, and an alternative was needed. Paratroops in this case cleared a landing site near the village and prepared it to receive a Dornier. While supplies had been dropped during the operation, it was considered an important capability to be able to land next to the troops to unload fragile cargo that could not survive a drop and evacuate the sick and wounded. On completion of the improvised landing strip, a Dornier nearly wrecked itself in landing because of the uneven surface. While it took off successfully, there would be no more attempted landings at this field, and only air drops would be used for resupply until the column advanced to a capable and proven site. This experimentation in the use of a Dornier was driven by the lack of helicopters. Ultimately their arrival numbers would change the circumstances, and the Dornier would no longer attempt such risky operations. The FAP was routinely asked to operate from improvised runways, straight sections of roadway, or the streets of a village in performing missions, and such requests often led to very risky maneuvers and consequently a high incident of Auster and Dornier accidents.[39] This pilot survived and was lucky, but it was not always so.

There were other examples of the versatile use of the Dornier in aiding the advance, particularly in helping ground forces bypass the blocked roads and destroyed bridges through airlift. As an example, at dawn on 7 July, the Dorniers of AB 3 created an air bridge from the village of Sanza Pombo to Santa Cruz, where they landed in one street and took off in another in a constant shuttle.[40] In this case, the Dorniers managed to airlift an entire platoon of some 30 men complete with arms and baggage.

On 15 July, following the Santa Cruz operation, The Dorniers supported the ground advance on Quimbele. Here the risks were considerably greater technically, as the landing strip was indescribably tight. It had been improvised by the population next to the entrance of the village and was so small and restricted that only the Dorniers could execute a landing fully loaded with cargo and make a successful take off. Use of the airfield, such as it was, was never again authorized, and it was closed following the

39 Aleixo Corbal, Oliveira Simões, Fidalgo Ferreira, and Heitor Almendra, "2º Região Aérea, Lições Aprendidas entre 1961 e 1975" [2nd Air Region, Lessons Learned between 1961 and 1975], proceedings of a conference "Almoços da AFAP," January 2011, 14.

40 Soares de Moura, 52–53.

single day of operations.[41] The risk was taken at Quimbele, as this was the first time since March that anyone had been in or out of the village.

Serra de Canda

Because of the slow and labored progress overland in retaking Nambuangongo and Quipedro, it was thought that another quick strike from the air would leapfrog the road problems. These were not to be taken lightly. For example, the Baptist missionary David Grenfell noted that between Bungu and 31 de Janeiro, a stretch of about 30 kilometers, the road was blocked by 800 trees and 200 trenches.[42] As there were still inadequate numbers of helicopters available in Angola, a paratroop drop at Quipedro had been the only viable alternative. This insertion had worked well there in putting troops where they were wanted, and while there were teething problems, the overall success of the operation appeared to merit a repeat. Certainly using the paratroops would avoid the trees felled across roads, covered pits dug in the roads, destroyed bridges, mines, and ambushes—each occurring with increasing frequency.

Exactly two weeks after Quipedro, the FAP launched Operation *Auta* as a clearing operation in the Serra da Canda. Aerial reconnaissance and captured documents indicated that this mountainous region had become a refuge for various armed elements and served as a waypoint for UPA cadres coming from the Congo and infiltrating into the areas of Carmona, Toto, Vale do Loge, and Dange.[43]

Canda was a difficult area, as it had a median altitude of 1,000 meters and covered some 1,200 square kilometers. It had the geographic outline similar to the face of a laundry pressing iron with its front pointing south and its rear nearly abutting the Congo frontier. In the west it was defined by the M'Bridge River, and in the east by the meandering Lefunde and its tributaries. The area was particularly difficult to reach during the rainy season when flying weather was irregular with clouds and mist and the roads often a quagmire.

In the north of the Serra there were two villages: Cuimba, which was abandoned, and Maquela do Zombo, which was the location of AM 31. Nearby were the rich copper mines Mavoio, and further to the south on the Lucunga River were Bembe and Toto (AM 32).

In the Serra the settlements generally remained active, and there was considerable traffic on the connecting trails. There were in the area palm groves, orange trees, and plantings of cassava, and at the higher elevations, zones of thick, dense elephant grass

41 Ibid.
42 Inge Brinkman, "Refugees on Routes, Congo/Zaire and the War in Northern Angola (1961–1974)," paper delivered at the international symposium *Angola on the Move: Transport Routes, Communications, and History*, Berlin, 24–26 September 2003, 3.
43 Diogo Neto, 2.

that made aerial reconnaissance difficult. Some aircraft flying over the area were taken under fire to no effect, but it did reveal the presence of armed elements hiding below.[44]

The concept of operations followed that of Quipedro with the collateral purpose of demonstrating Portuguese capabilities to a larger international audience. After extensive and discrete aerial reconnaissance, an LZ was identified next to the settlements of Pingo and Banza-Tadi located about 75 kilometers northeast of Toto in the plain between the Lefunde and M'Bridge Rivers and south of the 1,110-meter peak of Bundiungo. The LZ would be cleared using a large number of incendiary bombs to burn off the grass and facilitate paratroop landings by the 2nd Company of Parachute Battalion 21, a force of 120 men divided into four platoons, and preceded by a pathfinder team. Once on the ground, the paratroops would reconnoiter the surrounding area, set ambushes, advance into the Serra, eliminate the enemy, and join with two army columns, one advancing south from Maquela do Zombo and the other north from Lucunga. The southbound column was composed of a battery of artillery, an infantry company, and a reconnaissance platoon. The northbound column was a 40-man army equipment recovery detachment.

The paratroops would all carry the new, light 7.62 caliber ArmaLite AR-10 battle rifle, and their heavier equipment would be left behind to give them added mobility. There also would be two para-nurses on 5-minute standby in Toto (AM 32) with a forward-deployed Sud Aviation Alouette II helicopter to evacuate the wounded.[45]

The support aircraft were:

- Three C54 Skymasters and a Noratlas for air transport.
- Four PV-2 Harpoons and four T-6G Harvards for close air support.
- One Dornier DO-27 for local coordination.
- One Sud Aviation Alouette II helicopter for evacuation of wounded.[46]

Four paratroop-capable transport aircraft would be required to effect the drop, and as the 2nd Air Region did not have the needed aircraft, they would have to come from abroad. The three Skymasters came from BA 4 in Lajes, Azores, and AB 1 in Lisbon. The Noratlas was lent by the French from the air base at Pointe Noire, as the available Portuguese Noratlases were the civil version without the lateral doors for paratroop drops. Later the Portuguese Noratlases would be converted for paratroop drops.

The enemy was lightly armed, mobile, and avoided the open areas and roadways that would expose him to detection and fire by aerial reconnaissance. He was shy of direct confrontation with superior numbers and firepower and limited his activity to ambushes on isolated forces and proselytizing the population. Because of this behavior, the air staff felt that he would maintain his distance from the paratroops in

44 Ibid.
45 Ibid., 3.
46 Ibid.

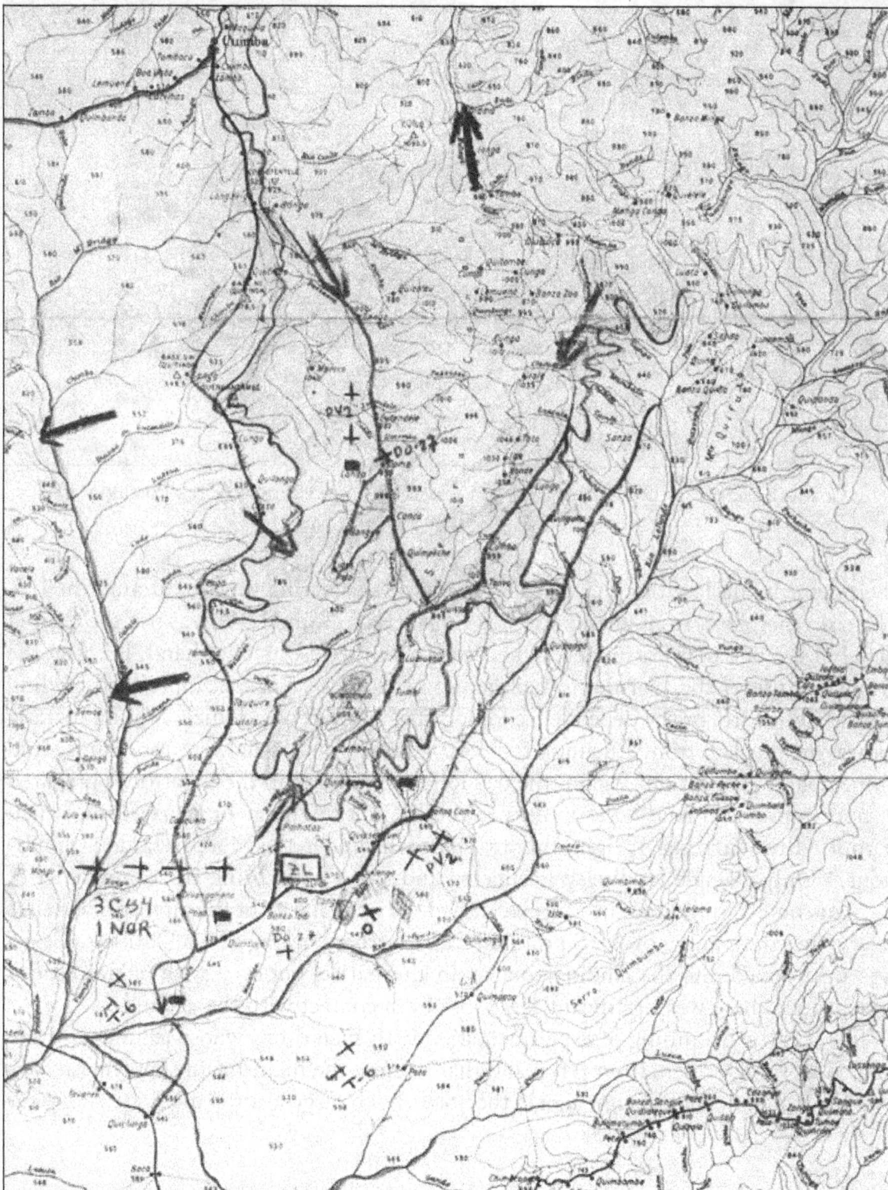

Map 6.2 Serra da Canda Operating Area. (Source: Força Aérea Portuguesa)

Loading a Dornier DO-27 for a TLCA mission. (Source: Força Aérea Portuguesa)

their descent and landing phases, as he would be evacuating the area around the LZ. The risk, therefore, from enemy fire was considered minimal.[47] It was not the desired plan but the one that Portugal could execute with the resources at hand. Further, the attempted use of the Dornier to replace the helicopter, as at the improvised landing strip at Quipedro, or the dependence for logistics support on a vehicle column making its way to the LZ from Lucunga some 50 kilometers distant were too clumsy and vulnerable to disruption. Again, the day following the jump the paratroops prepared a short improvised landing strip next to the LZ in the village of Banza-Tadi, which permitted the landing of light aircraft but again under the most hazardous conditions.[48] Only a single landing was effected and that was to bring the commander of the parachute battalion to the LZ. Because of the irregularity of the airstrip and hence safety, the commander of AB 3 prohibited any further landing of his Dorniers. Cargo was air dropped into the landing zone as an alternative, but even with reduced speed and altitude there was significant cargo loss from contact with the ground.

This was the beginning of experimentation with such drops whose techniques were refined by the Dorniers from AB 3 and that became the mainstay in supplying troops over the remainder of the campaign in the north. Such cargo airdrop missions (*transporte*

47 Ibid.
48 Ibid., 4.

ligeiro de carga aérea or TLCA) required extraordinary pilot skill and flying precision. The pilot would fly very low and slow, 35 to 40 knots at full flaps, over the objective. The side door of the Dornier was removed prior to the mission, and the mechanic, secured by a harness anchored to the aircraft, pushed 20-kilogram sacks through the door on the pilot's signal. The sacks were doubled, the inner one containing the goods and the slightly larger outer one containing the cargo wrapped in a protective padding of elephant grass. The procedure proved successful, and rarely did a sack burst.[49]

Following the paratroop jump, the Dornier returned to areas near the villages of Landa and Sanza Coma, where earlier armed enemy elements had been observed. Landa was well to the north of the LZ and is identified on the nearby map of the operating area by a black square representing the targeted kill zone. The two PV-2s and a Dornier are depicted by crosses. Sanza Coma, where a group of 15 to 20 rebels fired on a FAP aircraft, is located to the east of the LZ, and the black square representing the target kill zone is a short distance northwest of the village. The loitering area for the PV-2s is marked by two crosses and labeled "PV-2." The dark bold arrows pointing toward the edges of the map show the direction of fleeing UPA rebels, and the lighter arrows pointing towards the Serra, the approach and deployment of army troops coming from the north and south.

The paratroops regrouped in the settlement of Pingo, 400 meters south of the ZL, following the jump and found it deserted.[50] The company then split into two detachments of two platoons each to reconnoiter the area. One detachment headed northeast toward Zanza Coma, and the other toward the confluence of the M'Bridge and Lefunde Rivers. Neither detachment made contact with the enemy but recovered documents that revealed the UPA organization in the area. During 26 and 27 August the paratroops established ambush sites on the trails and only found three local Bembe natives fleeing the UPA.[51] The enemy had moved from the area to avoid contact with a superior force in classic guerrilla style.

On 28 August, as the company began probing the Serra, it was ambushed but suffered no casualties. It captured five enemy elements and a number of weapons. During the days that followed, clearing operations progressed without enemy contact. Indeed, the population had also fled from the potential conflict and vanished into the sea of elephant grass and dense vegetation. Many crossed the border into the Congo. The Serra da Canda ceased to be a safe refuge for the UPA, at least for the moment.

After eight days, the paratroops linked with the army column and then returned to Luanda, completing the operation. This type of search and destroy mission highlighted two aspects of airpower in counterinsurgency. First, troops with a high level of training

49 Tomás George Conceição e Silva, "The Portuguese Air Force in the African Wars of 1961–1974" in *Memories of Portugal's African Wars, 1961–1974,* ed. John P. Cann (Quantico: Marine Corps University Foundation, 1998), 117.

50 Diogo Neto, 5.

51 Ibid.

General Resende visiting the paratroops at the conclusion of Operation *Sara*. He is standing in front of the only significant building in Sacandica, the "Pousada da Felicidade". (Source: Força Aérea Portuguesa)

and cohesiveness proved very effective in moving against a difficult, dangerous, and evasive enemy. This lesson would lead to the expansion of elite forces in the FAP paratroops and navy *fuzileiros* (marines), and establishment of the army commandos.

Second, parachuting a force requires substantial preparation on the ground in readying the landing zone, marking it, inserting the pathfinders, and regrouping the forces once on the ground. This activity takes time and thus not only alerts the enemy but also gives him adequate notice to evacuate the area or make other military decisions. Without the element of surprise, it is nearly impossible to damage the enemy in counterinsurgency. In August of 1961, this type of operation was the best of the available alternatives, and more effective tactics would await the arrival of helicopters in numbers, the forming of additional elite units, and the development of refined tactics.

Sacandica

Sacandica was a *posto* located 25 kilometers south of the Congo frontier and 150 kilometers east of Maquela do Zombo (AM 31). On 15 March, it was overrun in the initial hours of the UPA assault and had been in UPA hands since then. Because of its distance from the current disposition of land forces, it was assigned to the FAP and its paratroops. In planning Operation *Sara*, it was initially believed that a Nord operating from AM 31 could simply land on the grass runway there and discharge the paratroops and their equipment, but when the Nord arrived over Sacandica, it became apparent that the short runway was no longer usable.

Returning to Maquela do Zombo, the paratroops lightened their packs and prepared to jump into the target. As there was no pathfinder team immediately available because the original plan did not require one, pathfinder duties were performed by a lone Dornier that dropped the needed flares to mark the LZ. Following insertion of the platoon, 1,800 kilograms of cargo were dropped. This was enough to last the platoon for 20 days, the time thought necessary to complete the mission objectives.[52] The *posto* was held by a strong, fanatical contingent and required some fierce fighting over 29 days, until it was finally freed on 14 October. During this action, the thirty men were resupplied entirely by air.

Fuessi

While Portuguese forces had reoccupied the key centers of the north, they had not eliminated the rebels in the least. UPA forces still controlled in various degrees an area approximately 150 miles wide 200 miles deep from the Congo border into the north of Angola.[53] The conflict had shifted from a fluid chaos into an organized insurgency. UPA lines of communication ran from the Congo over an intricate and constantly changing system of footpaths to a UPA headquarters known as Fuessi (Fuesse), which was about halfway between the frontier *postos* of Luvo and Buela. The trails continued south from Fuessi some 50 miles to the Bembe area, where the famous UPA commander João Batista ran a military command center.

UPA forces were organized into small "mobile groups" of 10 to 12 men suited to hit and run action.[54] They received weapons from various sources within the Congo, where several sympathetic nations had forces under a UN peacekeeping operation that was consistently losing its light arms. These weapons were carried to Batista in his jungle hideaway and distributed to some 40 odd mobile groups, which carried them on a trek that could last as long as two or three weeks and range as far south as Úcua.[55] These treks travelled trails, crossed bridges, and bivouacked at established rest areas with shelters. The rest areas also had other infrastructure and represented attractive targets should they be identified. From the UPA standpoint, the FAP was making increasingly effective use of airpower with its RCAR and DLOC missions.[56] The UPA countered with slit trenches as air raid shelters, mock villages as decoy targets, and a warning system of spotters. According to John Marcum and George Houser, who in January 1962 visited the UPA in the north, the air raids were devastating despite these countermeasures.[57] Fuessi was subsequently dismembered in an opera-

52 Costa Cardoso, 211.
53 John A. Marcum, *The Angolan Revolution, Volume I: The Anatomy of an Explosion (1950–1962)* (Cambridge: The MIT Press, 1969), 229.
54 Ibid.
55 Ibid.
56 Ibid.
57 Ibid., 230.

tion during the last week of April 1962, three months after Marcum and Houser's visit and Batista's death on 6 February.[58] Portuguese intelligence had identified the dangerous settlement and headquarters, from where mobile groups were regularly laying ambushes throughout the area, and sought to destroy the center and its ability to make mischief.[59]

On 20 April 1962, the 1st Company of Parachute Battalion 21 was airlifted from Luanda to Sao Sãlvador and began its operation to neutralize Fuessi. The objective was about 30 kilometers to the east of Luvo in an area that was difficult to access. The only route was by a footpath that had not been travelled in a year and was thought to be abandoned. The company of 120 men travelled by truck from São Salvador to Mamarrosa and after that followed a road next to the River Gunguege until reaching the River Lunguege. Here the troops forded and proceeded on foot, as the bridge had been destroyed, and the vehicles could not continue. The rough roadway leading away from the bridge degenerated into a path that was overgrown and difficult to follow. It was also crossed by others, and the route became very confusing. The guide, Pedro Tumissongo, was familiar with the routes around Luvo and south, but the company had left the area that he knew, and now it depended on a Dornier from Negage to provide orientation and drop supplies, as the troops travelled with light packs.[60] This help and the size of the force would have made enough of a disturbance to destroy any element of surprise, as UPA lookouts and sentries could hear and see the aircraft and detect the approaching column. It was the rainy season, and the heat and humidity made going difficult, even travelling with light loads. The progress was further impeded by the need to be on guard against ambushes. Still, there were four more rivers to cross before arriving at Fuessi.

The company reached the Rivers Mata-Matende, Cai, and Magina and forded them before finally reaching the Luvo, which was wide and swollen by the rains and had a swift, powerful current. It took time to rig a line and to cross carefully, one by one, but it was done without incident, and the march resumed at an accelerated pace. The company reached Fuessi at noon and quietly deployed, encircling the compound and establishing a counter-ambush formation. When the paratroops entered the compound, it was deserted, but there were warm fires, food, plates and utensils, and other abandoned property. There were even photographs of "*mademoiselles*" with inscriptions written across them in French and local idioms.[61] In the largest hut, which housed the UPA leaders, there was a notice posted on an exterior wall. Pedro Tumissongo read it for the paratroops, "It is possible to enter into this place

58 Ibid., 238. Batista was killed in a 6 February 1962 raid on a Portuguese camp near Bembe when UPA forces were surprised in finding unanticipated and prepared Portuguese forces far outnumbering their attackers. Batista's death represented at a minimum a failure of UPA intelligence.

59 Costa Cardoso, 265.

60 Ibid., 266–269.

61 Ibid., 269.

Supply drop over abandoned UPA headquarters at Fuessi, April 1962.
(Source: Força Aérea Portuguesa)

but difficult to leave; blacks who enter into this place must obey, thereafter, the orders of Holden Roberto."[62] The disappointment of the failed attack was extreme, but then this failure held lessons that would lead to more effective operations. Guides would become more competent, force sizes reduced, and power projection more comfortable with the bush. Dorniers alerting the enemy would become a thing of the past. Many operations would have to be conducted to eliminate the UPA in the north, but eventually success would come.

Arrival of the Helicopter

During 1958, Portugal acquired seven Sud-Aviation Alouette II helicopters, which represented the initial step in creating an effective helicopter force. The first Alouette II was wrecked in January 1960; however, the remaining six were sent to Angola at the end of 1962 and later, with the acquisition of twelve Alouette III helicopters in the spring and summer of 1963, formed the basis for Squadron 94, the *Moscas* or "Flies," at BA 9. This addition of the helicopter fleet to the resources of the 2nd Air Region

62 Marcum, 230.

would have a transformative effect on the war in the development of tactics for moving light infantry forces and supporting them with localized firepower.

Depositing troops exactly where they were desired was difficult with a parachute insertion, as there were so many variables and vulnerabilities. The wind was an enormous factor, as it could cause injuries on landing or even cause a miss of the LZ. The LZ may also be some distance from the objective and subject to ambush. On the other hand, the helicopter could deposit a force precisely next to an objective for its taking or destruction.[63] It was the perfect answer to mobility in the difficult terrain of the *ultramar*, and illustratively, a one-minute flight in a helicopter equaled about an hour on foot in the jungle. It was a constant challenge to make contact with the enemy. When contact was made, complete reconnaissance was often impossible or the opportunity for engagement would be lost. Consequently, the infantry required speed, mobility, and flexibility to capitalize on these infrequent opportunities, and the helicopter was the answer to profiting from these and gaining tactical successes.

Initially, the helicopter assault concept was built around the 20-man combat group. A flight of five helicopters would insert five 4-man sections on the ground quickly and judiciously to make the most of insurgent contact. If the flight was expected to encounter resistance and needed covering fire to protect it, then a helicopter gunship or *heli-canhão* with its 20mm cannon covered the insertion and recovery and provided support as required. The development of this gunship capability required several years and was done in consultations with Sud-Aviation.

Finding the solution to troop helicopter vulnerability during the landing and recovery phases took experimentation and time. Initially fire from the hand-held AR-10s (7.62mm caliber) carried by the troops was tried and found inadequate. A 12-gauge shotgun with buckshot loads was also tried with some success, but something more powerful was needed.[64] Next, the MG-42 light machinegun (7.62mm) was tried, but it was difficult to handle and could not be held on target with the movement of the helicopter. Looking for more serious firepower that could hold the target or sweep across the LZ, the FAP borrowed from the Thunderjet its Browning M3 (.50-caliber) machineguns that fired 1,200 rounds per minute and mounted them in a side-by-side configuration in the left door of the Alouette III.[65] Only the Alouette III could carry the weight. This dual M3 mount was slaved to a targeting visor borrowed from the T-6G, and the arrangement worked partially, providing more robust and controllable fire support. However, the weapon was bulky, and both its weight and that of the ammunition and other modifications made it cumbersome. It took about two years to find the desired solution in the French Matra MG-151, a 20mm cannon

63 Arménio Duarte Martins, "Cotova e Falcão" [Skylark and Hawk], *Mais Alto* 311 (February–March 1998): 8.

64 Ibid.

65 Ibid., 9.

Matra MG-151 20mm cannon mounted in an Alouette III. (Source: Força Aérea Portuguesa)

with little recoil and a firing rate of 700 rounds per minute. This proved ideal, particularly when used with high-explosive incendiary munitions.[66]

On the negative side, with the cannon mounted in the helicopter, its two boxes of ammunition aboard, the gunner in position, and 450 liters of jet fuel in the tank, the Alouette was heavy. The situation was further aggravated by the center of gravity now being shifted forward, making it a challenge to fly and requiring great concentration by the pilot. Experienced pilots worked best in the *heli-canhão*.[67]

The key to inserting the combat group lay with the lead pilot in the flight. He was invariably an aviator qualified in both helicopters and fixed-wing aircraft, and would spend much of his initial familiarization flying in the theater at low altitude in a Dornier over his assigned tactical area. In addition to learning the terrain, he would mark a map with suitable helicopter landing sites that he might use in future operations. In designating the landing sites, he would also mark approach routes to each. These were chosen for their natural terrain features that would mask the arrival of five noisy helicopters and potentially deliver surprise. Alongside the pilot's area familiarization, paratroops, army commandos, or other troops rehearsed operations as a team until the unit was proficient.[68]

Once prepared, the combat group was kept on alert at a primary base, such as Luanda, until an operation was imminent. At that point the group would relocate to

66 Ibid.
67 Alfredo Cruz, *O Voo do Falcão: Piloto de Combate, Moçambique* [Flight of the Falcon: Combat Pilot, Mozambique] (Lisbon: Fronteira do Caos, 2014), 125.
68 Duarte Martins, 9.

an airfield in the area of anticipated operations, where there might be a wait of several days before an opportunity occurred. The entire success or failure of the mission depended on surprise, and considerable pains were taken to achieve it. Reconnaissance of the target was kept to an absolute minimum to avoid flushing the guerrillas prematurely, and discussion of the mission was forbidden outside the secure briefing area. Once called, the combat group would be given its final briefing and then promptly launched, taking the most direct of several preselected routes to the target and its prescribed LZ.[69] Travelling with the troop transport helicopters in their nap of the earth approach, the helicopter gunship would circle and clear the LZ prior to landing the troops. It would continue to cover the area with its cannon to provide suppressing fire, if required, while the troop carriers were most vulnerable in debarking the assault force. The MG-151 was much feared by the guerrillas. For additional fire support there might be a Dornier with two nine-rocket canisters, one under each wing, loaded with versatile 37mm rockets developed by the French. They were used either for marking targets with smoke or making an attack on guerrilla sites. After depositing the combat group, the troop carriers would take off and loiter nearby awaiting conclusion of the action, which would generally last only about ten minutes. At its conclusion, the helicopters would land at the nearest suitable site, as the original may or may not be near the concluding action, and load the assault team and perhaps a highly valued prisoner. The team would then be airlifted to another objective.

The combat group with its helicopter transport would normally remain in the field throughout the day, pursuing the insurgents and coordinating operations with other ground forces. At the conclusion of daylight operations, the heli-borne combat group would rendezvous with a preselected, specially equipped ground unit with whom to bivouac for the evening. As only select units were equipped with the necessary helicopter field support in fuel, ammunition, and maintenance capability to host an air assault team, this choice required some considerable planning and coordination. During the evening the helicopters would be fueled and rearmed, and the men rested for further operations the following day. This process continued for three to four days before they returned to the original base.[70]

Upon occasion two air assault teams would be employed together, thereby doubling the force on the ground from twenty to forty troops. Unless the team was part of a larger operation and formed one of the closing forces in a "hammer and anvil" tactic or a similar maneuver, there was a risk that such a small body could be overrun by a larger insurgent force. Thus care had to be exercised not to insert a team into an untenable situation. On the other hand, use of armed helicopters was carefully controlled, so that any damage or casualties in the population were avoided through

69 Tomás George Conceicão e Silva, interview by the author, 3 April 1995, Lisbon.
70 Duke of Valderano, interview by the author, 17 March 1995, London.

Filling napalm canisters on a PV-2 at BA 9. (Source: Força Aérea Portuguesa)

the overenthusiastic or indiscriminate use of firepower. Helicopter assault operations were always executed away from populated areas.[71]

While the helicopter was seen as a high-technology tool essential to modern counterinsurgency operations and represented a major advance in this field, it also had a number of important disadvantages. Helicopters were relatively expensive, and their addition to a campaign increased costs significantly. Added to the initial cost of a helicopter were the extraordinarily high maintenance and operating expenses, particularly in a tropical environment. Helicopters required a great number of very specialized men to service and fly them, and their expense was a substantial and continuing burden, particularly as the war progressed. Nevertheless, the helicopter and its effective utilization was without question the biggest difference in military capability between that of Portuguese forces and the enemy.

71 Ibid.

Quivitas and CMIG-Zero

In the Dembos region and the mountain ranges between Uíge and Quitexe, with their dense vegetation, there persisted relatively secure hideouts for the rebels, as described by Houser and Marcum.[72] The Serra de Uíge range was an estimated 60 kilometers long by 5 kilometers wide and, with its difficult terrain, represented a true challenge for any ground force assault. Ground troops crawling through the mountainous jungle had to operate in the gloomy shadow of great mossy trees supporting a dark green canopy. The humidity and a web of hanging vines made progress exhausting, and it was easy to become disoriented in this suffocating environment. On two prominent, nearly impregnable summits rising from this grim and torrid landscape, there was a pair of strong UPA redoubts. The first was called "Quivitas" by the local people, and the second was known by its acronym CMIG-Zero, which stood for the Military Center for Instruction in Guerrilla Warfare, Number Zero (*Centre Militaire d'Instruction Guerrilla–Zero*). The latter boasted a parade ground, an elaborate infrastructure of huts, and the signature flagpole. Both enjoyed large, open, and sweeping approaches and were ably defended by heavy machine guns so positioned to defeat any advance against them. Various unsuccessful attempts had been made on a regular basis to destroy these centers, but they had had limited effect in reducing enemy capabilities in the area.

In September 1965, the army sector commander allied with the FAP to construct a campaign plan designed to eliminate these two sites permanently. The plan called for a concentrated and lengthy artillery and air bombardment prior to a ground assault. For 20 days army artillery and FAP sorties pounded the critical vulnerabilities of the installations. From BA 9 a steady stream of PV-2s and F-84Gs dropped their ordnance loads on the UPA centers. The bombing was so intense that during this period the magazine in Luanda was reduced to a mere two days supply of munitions.[73] The BA 9 offense was supplemented by the sixteen lighter but no less lethal T-6Gs from AB 3. During this bombing, the T-6Gs were a mere 20-minute flight from their objectives and so would fly three consecutive sorties without refueling. On the two return landings, they were hot rearmed by the ground crews in a few minutes and launched again for their targets.[74]

The Dorniers were integral to a systematic reconnaissance plan that identified the rebel-infested areas, and as a method of identifying targets within the vast complexes, subdivided them into smaller sectors. These were easily identifiable by their farming

72 George M. Houser and John A. Marcum, "Joint Press Statement," Overseas Press Club, New York, 1 February 1962; and George M. Houser, "A Report on a Journey through Rebel Angola," American Committee on Africa, New York, 1 February 1962. See also Marcum, *The Angolan Revolution*, 228–232.

73 Correia, 22.

74 Conceição e Silva, "The Portuguese Air Force in the African Wars of 1961–1974," 116–117.

patches, newly trod tracks or abandoned ones, huts hidden beneath the trees, and so on. The most useful reconnaissance hours were noon with the sun directly above, or late afternoon with the sun low in the horizon and penetrating laterally into the forest vegetation. The Dorniers also regularly supported the ground forces with TLCA missions that delivered fresh food and the all-important mail.[75]

Following the air and artillery offensive, several combat groups of paratroops and other infantry forces were inserted into the mountain and penetrated the redoubts, encountering little enemy resistance. They remained bivouacked on the two mountain summits after the action to ensure that they were not reoccupied. The enemy fled the Uíge range, and later at Quivitas a short runaway was built that simplified logistic support.

New Fronts

With the general neutralization of these strongholds, the UPA and MPLA began to look elsewhere for potential success. The first new front attempted by both was Cabinda. Negage sent a detachment in support of the paratroops and *fuzileiros* operating there.[76] Next, intelligence pointed to enemy infiltration in the northeast and east of Angola, and detachments were deployed to AM 33 at Malange to conduct reconnaissance missions in the districts of Lunda and Moxico. These were vast and inhospitable areas where any navigational error or mechanical failure would likely be fatal.[77] Reconnaissance missions in this area were increased and expanded in 1963.

As the war progressed, the life of improvisation at Negage shifted to one of relative comfort. The greatest evidence of this change was the presence of a vast new hangar. This and other buildings were part of a program to simplify all installations in the *ultramar* with standardized prefabricated modules. Together these improvements made one forget the earlier hardships, according to Soares de Moura:

> Our "slum" had been dismantled, and we were able to live much more comfortably. Gone were the packing crates for tables and the floor of loose bricks. In the town of Negage, the personnel of AB 3 developed an elegant neighborhood where the schools and cinema functioned well. Carmona expanded with new streets, modern buildings, businesses and restaurants. There was a new coffee exchange where farmers could sell their crop. The military presence lifted the economy, and the road improvements created reliable links between the more remote areas and the primary towns. Villages were improved with better sanitary conditions and facilities for health care and commerce, all assiduously patrolled by the military. Local militias were established to aid in self defense. Regular

75 Ibid.
76 Soares de Moura, 52–53
77 Ibid.

> patrolling by air and ground forces restricted enemy movement by day, forcing him to hide in the elephant grass; however, he posed a continuing threat.[78]

By the mid-1960s, neither the MPLA nor the UPA had met with any substantial success in a classic subversion of the population in the north or in truly displacing the Portuguese position there. This lack of insurgent progress occurred for several reasons. While the army lacked both the manpower and mobility to seal the land border completely in an area of poor roads, dissention between the MPLA and the UPA, Léopoldville's favoritism of the UPA, and Holden Roberto's failure of leadership served to limit any insurgent offensive. Friction between the two movements had its origins in the external recognition and choice of which one was to be the representative government of any independent Angola. As the UPA was comprised largely of Bakongo, a people prevalent in the north of Angola and that area of the Congo immediately north of the common border, it felt that it more accurately represented the population of Angola at large, and this view was backed by Léopoldville. However, the MPLA saw things differently and described the tribally-based UPA as "racist" and "foreign"—foreign because its leadership had no ties with Angola and the movement could only survive with Mobutu's support.[79] The MPLA on the other hand was comprised largely of the *mestiço* elite and was perceived to have less in common with its host and the people of Angola. As a consequence, there was competition and open friction between the two, and this misdirected energy diluted the attention of both in prosecuting the insurgency. Indeed, UPA troops took to ambushing MPLA troops in northern Angola, and afterwards Portuguese forces came upon these bloody fratricidal scenes.[80]

Léopoldville's favoritism of the UPA was unabashed, and it worked to give its chosen movement every advantage. As an example, Léopoldville because of its own security problems did not permit insurgents to bear arms within its borders. Both the MPLA and UPA insurgents would be transported by Congolese soldiers to the Angolan border, where they would be issued their weapons from depositories. On returning from operations, their weapons would be cantoned, and they would return to their training base with a Congolese escort.[81] Léopoldville employed this control measure to discriminate against and harass the MPLA. In 1962 the MPLA was often impeded from reaching the border, and its arms confiscated, whereas the UPA was accorded privileged and unimpeded access. Despite advantages in the international fora and on the battlefield, Roberto, whose eyes seemed permanently masked behind

78 Ibid.

79 Brinkman, "War and Identity in Angola," 198.

80 Ibid.

81 John A. Marcum, *The Angolan Revolution, Volume II: Exile Politics and Insurgent Warfare (1962–1976)* (Cambridge: MIT Press, 1978), 42.

dark glasses, proved unable to lead effectively in either the political or the military dimension, and Léopoldville sponsorship was unable to reverse this trend.

From the beginning the MPLA had been unwelcome in Léopoldville. In June 1963 following the UPA change in name to FNLA and its establishment of its exile government GRAE, the MPLA moved its headquarters to Brazzaville. Its ability to project itself militarily into Angola became almost completely circumscribed by November, and it looked more and more isolated. In January 1964 at its Conference of Cadres (*Conferência de Quadros*) in Brazzaville, it was decided to begin rebuilding the movement as a serious revolutionary force. Part of this plan involved opening a front in the east of Angola through Zambia. The MPLA had made overtures to Kenneth Kaunda, the first president of Zambia following its independence from Britain in 1964. It had also befriended Julius Nyerere, the first president of Tanzania, likewise following its independence from Britain in 1962. These relationships were to prove fortuitous. Beginning in 1965, Tanzania and Zambia permitted the transit of Chinese and Soviet weapons and MPLA cadres across their territories to the Angolan border. Because of this access, the MPLA was able to open a major offensive and to conduct an intensive insurgent campaign in the eastern Angolan districts of Moxico and Cuando Cubango. Preparations were begun accordingly, and by the opening months of 1966 proselytizing of the population in the east was evident. The first armed incursions occurred in April, and the war had now clearly moved to the east of Angola. Portuguese resources were redirected to the east and southeast of the territory in response to the new threat. Before moving eastward, however, we need to examine developments in Guiné and Mozambique.

7

Aerial Reconnaissance and Intelligence

Gaining intelligence on the guerrillas was vital in keeping them separated from the population, denying them shelter, food, recruits, and intelligence, and destroying them. Information was needed on their political leaders, their military commands, their forces in the field, their operational plans and intentions, and their sources of support both within the colonies and abroad. This information was obtained through reports of infantry patrols, interrogation of prisoners or turned insurgents, captured documents and equipment, signals intercepts, paid informers and agents, and aerial reconnaissance. Here we will examine the development of these sources, the perfection of their use, and the integration of their contributions into a complete intelligence picture at the strategic, campaign, and tactical levels of the conflict with the primary focus on aerial reconnaissance and its importance.

Enemy Behavior

Intelligence requirements always center on developing a sound understanding of the environment and the enemy. This was a difficult task in that throughout the theaters there was a general lack of insurgent organization and communication. Insurgent thrusts were loosely organized and reflected only a vague and general campaign plan. It was thus difficult to understand precisely what they were doing tactically that would support their strategic plan.

The insurgents in all cases frustrated the traditional intelligence processes, as there was not a firm relationship between their decisions at the political level and the actions by their military commanders in the field. This disconnect was due primarily to the vast physical distances involved, the lack of good radio communication, and some lack of discipline. Consequently, the political direction located in the neighboring sanctuary country often bore only casual resemblance to activities in the field. Intelligence gained on the intent at the top needed to be verified and refined by intelligence on local operations. At the local level, the insurgents demonstrated an ability to coordinate their efforts effectively.

Because of this disconnect, everything was tactical, and destroying insurgents depended on making successful contact. Effecting this contact was very complicated, as the areas to be covered by surveillance were extensive, difficult, and remote, and a group of twenty enemy could easily hide in the vastness of eastern Angola, the Téte district of Mozambique, or the jungles of Guiné. Gaining useful intelligence and moving to intercept incursions were based largely on educated guesswork that was often as imprecise as the insurgents' plans themselves. Nevertheless, the insurgents provided clues, and Portugal capitalized on these to develop the picture piecemeal through a number of contributing methods and to generate successful operations based on the mosaic assembled through these sources.

The normal modus operandi of the various insurgencies was to establish operational bases situated strategically along the border of Portuguese territory but safely within the sanctuary country. They would then project their power into the target possession and attempt to extend this penetration by planting camps in its interior. These would then be used to project power further by launching attacks against local villages, military convoys, and civil and military installations in the hope of securing a foothold for further advances.

Classic insurgent operations had no geographic objectives per se. The enemy occupied no permanent site, followed no set itinerary, and gave little importance to communication with his headquarters. He remained fluid, mobile, and frugal.[1] His operations were designed to cause losses to government troops, take prisoners, capture armament, and destroy military and commercial infrastructure. These required the proper conditions to assault a militarily stronger government, and thus it was necessary to establish a hidden, secure, and strategically located refuge to rest and prepare for this favorable opportunity. This refuge would be deep in the targeted colony, and from here the insurgent would conduct his reconnaissance to learn about Portuguese troop movements and practices and to identify potential weaknesses to be exploited. Such a refuge or encampment, which the insurgents called *quartéis* or barracks as an aggrandizement of the small huts that housed their troops, were by necessity established along his "corridor of transit" en route to population concentrations and thus sited in predictable areas. These *quartéis* represented his true Achilles heel. In order to have a reasonable chance of attacking these bases and successfully annihilating the insurgents and their equipment, the Portuguese needed to develop very favorable conditions themselves, for the insurgent well knew his vulnerability and could be easily flushed. Attacking him from the air would destroy some huts and perhaps kill some insurgents, but this only caused a temporary disorganization and some insecurity in his movements. Attempting to block his resupply routes for arms, munitions, and supplies

1 Comando da 2ª Região Aérea, "Observação Aérea" [Aerial Observation], supplement to Intelligence Bulletin No 9, Commander 2nd Air Region, Operations Directorate, 4 July 1966, 2.

was also difficult. Probably the most effective disruptor was the destruction of his local crops and other support infrastructure; however, this proved to be difficult, as he was easily spooked and would uproot at any undue government interest, especially targeted reconnaissance. A successful assault, therefore, had to be done by beating him in his "habitat."

In concert with this profile, there was the trademark ambush that was conducted using a minefield laid along a road or trail. When the Portuguese troops encountered the anti-tank and anti-personnel mines buried in the roadway and alongside it, the hidden insurgents would lay down blistering gunfire from their automatic weapons. The entire engagement would last perhaps ten minutes before the insurgents would retreat and regroup in safety, while Portuguese forces were recovering. For the insurgent who carried the 15-kilogram anti-tank mine from his sanctuary country to the ambush site, a distance that could easily be up to 50 miles, its weight was heavy and made heavier by the heat and humidity. In fact, in all of the theaters the climate was both hot and humid nearly year-round, and this combination made the insurgents' operations wearing and difficult. It made progress in the targeted territory slow and predictable for an ambush team, and raised its desire to lay the ambush and be rid of the heavy anti-vehicle mine that was its anchor.

The local population, whose loyalty was desired by both Portugal and the nationalist movements, was mostly of the great Bantu group. This may be subdivided into many ethnolinguistic groups, and each of these in turn into numerous tribes, making a very complicated mosaic. National borders in this part of Africa often divided these groups, leaving a portion in two or more countries. Such a situation proved to be a tailor-made opportunity in fostering subversive and insurgent activity, as people of the same tribe freely crossed and recrossed porous borders to engage their kinfolk or conduct commercial trade. Insurgents used this freedom of passage to infiltrate Portuguese territory and influence the political loyalties of the local population. This influence was effective in gaining support in the border areas and along routes of enemy travel; however, this enemy proselytizing was not particularly effective outside the areas of immediate fighting and rarely affected the political leanings of the Portuguese African population overall. This basic truth was a critical factor in maintaining stability within each theater.

Another consideration in gathering intelligence was the fierce competition between nationalist movements for international recognition and victories over the Portuguese and each other. These successes could be touted in important international forums to bolster the legitimacy of each in claiming preeminence as the recognized government-in-waiting that would assume authority over a territory when victory was finally realized. This competition generated enormous animosity within and between the various movements, and this extreme feeling led to deadly tactical exchanges of fire between them and direct military attacks on each other. It also led to deadly leadership struggles within the movements. Portugal was able to exploit these open divisions for intelligence and its own tactical successes in fracturing the already fragile and divided nationalist movements.

Intelligence Organization

In the Portuguese army there was an intelligence system at the beginning of the war. The officers and men serving in the *2ª Repartição* or *2ª Secção* (Second Division, or on a lower echelon, Second Section, which was responsible for the intelligence function) of the General Staff of the Army and its replications down to the battalion level were personnel with diverse warfare specialties (infantry, cavalry, artillery, and so forth) who were assigned to an intelligence section. The rationale for this policy was that officers "self-specialized" themselves, following a natural interest and developing a proclivity for the intelligence craft.[2] When an intelligence service was established, these officers gravitated to it. As early as 1958 the Secretary of National Defense attempted to draw the various services together in an intelligence-sharing effort and began training a cadre for this purpose at the Intelligence Center of the British Army, Maresfield Park Camp, United Kingdom.[3]

This effort was overcome by the events of 1961. As early as January of that year there had been a working group in Lisbon composed of representatives from the military commands, PIDE, the General Command of the Portuguese Legion, the Director General of the Political and Civil Administration, the Customs Police, and other government arms.[4] The working group recommended the creation of a local theater intelligence service that would draw together intelligence from all sources and disseminate it in a timely fashion to relevant users. This body would be known as the Intelligence Service of the Governor-General (*Serviço de Informações do Governo-Geral* or SIGG) and would be responsible for a host of duties in this vein, including psychological operations, intelligence, counterintelligence, electronic surveillance, prisoner interrogation, etc.[5] This path was followed in both Angola and later Mozambique, where the bodies were known respectively as SIGGA and SIGGM.

Following the March 1961 attacks in the north of Angola and the appointment of General Venâncio Deslandes as the Governor-General and Commander-in-Chief of the Armed Forces of Angola (CCFAA), he created a true Service of Centralization and Coordination of Intelligence (*Serviço de Centralização e Coordenação de Informações* or SCCI) on 29 June. It included not only local intelligence gathering and dissemination but was also the beneficiary of intelligence gathered through the Ministry of Foreign Affairs, the Ministry of the Overseas Provinces, and other national sources. These diverse avenues provided access to both tactical and strategic intelligence. Later an SCCI was created in Mozambique, in Guiné, and in the other colonies as and when

2 Colonel Luís Alberto Santiago Inocentes, interview by the author, 5 June 1995, London.

3 Pedro Alexandre Gomes Cardoso, *As Informações em Portugal* [Intelligence in Portugal] (Lisbon: Instituto da Defesa Nacional, 1980), 106.

4 Estado-Maior do Exército, *Resenha Histórico-Militar das Campanhas de África (1961–1974)* [Historical-Military Report of the African Campaigns (1961–1974)] (Lisbon: Estado-Maior do Exército, 1988), Vol. I, 360–363.

5 Cardoso, 107–108.

it was deemed appropriate. Oversight of the SCCI was the responsibility of the theater Intelligence Commission (*Comissão de Informações*), which set policy on intelligence operations, military security, counterintelligence, and the like. It was also responsible for the creation of the district, regional, and local Intelligence Commissions that held responsibility for the district, regional, and local sections of the SCCI. As with most undertakings of this magnitude, the SCCI required tuning to make it function properly across the various military and civil bodies. By February 1963 Deslandes affirmed the progress and its importance by acknowledging that each passing day continued to reinforce the need of moving quickly to coordinate and analyze raw intelligence, to organize its procurement, and to learn not only what the enemy wanted but especially what interested the Portuguese.[6]

It was never an easy task for the SCCI to reconcile the diverse viewpoints and motivations of the various civil agencies, the police, and the military commands in the interest of an accurate, overall intelligence picture. Professional rivalries between intelligence agencies are a service tradition virtually everywhere, and Portugal was no exception. The most notable difficulty arose with PIDE, normally considered a superior and effective intelligence service, disagreeing with the army. PIDE in a number of these instances withheld information and acted on it through the offensive operations of its own *flechas*, indigenous forces initially of Bushmen trackers. For the most part, however, PIDE did not feel threatened by the military intelligence sections and worked well with the army. Its conflict was usually with its fellow civil intelligence agencies.[7] The SCCIs did not have a legal "heavy hand" to ensure coordination. In 1967, the SCCI in Angola proposed to Lisbon that a Secretary of Security be created in Angola to provide national oversight and coordination, a sort of "Home Office." Lisbon did not agree. Notwithstanding these difficulties, cooperation was viewed as essential and was pursued by the leadership. It improved considerably when the Governors-General were also Commanders-in-Chief.[8]

In each theater Portuguese intelligence coordination was theoretically done at every level. Militarily it was effected from the smallest patrols using and gathering intelligence to the headquarters of the Commander-in-Chief. On the civil side it ran from the local *posto*'s contact with the population to the Governor-General. At every stage intelligence gathering, analysis, and dissemination coordination was sought between military and civil arms, and the mechanism for this joint process was the SCCI with all of its levels.[9] Contact with the population was considered paramount. Interpreters and translators were used extensively. Agents infiltrated the enemy's infrastructure. Troops moved freely and regularly among the population, and trackers accompanied

6 Ibid., 109–118.

7 Major General Renato F. Marques Pinto, interview by the author, 30 March 1995, Oeiras. Marques Pinto was Chief of Military Intelligence in Angola 1963–1965, and Director of the SCCIA 1965–1968.

8 Inocentes interview, 5 June 1995.

9 *Resenha Histórico-Militar das Campanhas de África (1961–1974)*, Vol. I, 368–370.

the patrols to interpret the signs of the forests and jungles and locate the enemy. Special militias were also formed for the self-defense of the population and, it was hoped, reported on enemy contact. Additional sources were ground reconnaissance through infantry patrols, signals intelligence, analysis and interpretation of captured documents and equipment, and aerial reconnaissance.[10] All of these activities formed a network or system of intelligence gathering that was coordinated, documented, and shared throughout the civil and military arms. Operations were planned and executed to exploit the products of this information flow on a regular basis.[11]

Intelligence Operations

In its counterinsurgency campaign Portugal sought to develop its intelligence picture piecemeal through a number of sources and methods. Each of these made an important contribution in solving the intelligence puzzle, and because of this insurgent operational vagueness, it was important that information from one reinforce that from the others to produce an accurate picture. Each is examined below with an emphasis on aerial reconnaissance, which remained the key in rapidly constructing a tactical picture.

Ground Reconnaissance

Ground reconnaissance was the most basic method of finding and fixing the enemy in the difficult terrain of Guiné or the vast expanses of Angola and Mozambique and was a tricky job. To address this problem at the field level, each battalion had a staff that included an Operations and Intelligence Section (*2ª Secção*). This section was comprised of two officers, two sergeants, and two enlisted men. Supporting this staff was a reconnaissance platoon of twenty-nine men commanded by an officer with three sergeants and twenty-five enlisted men. It was equipped with jeeps and radios to give it added mobility. In some instances, depending on availability, helicopters might be used; however, this practice was the exception rather than the rule. Thus overall, there were thirty-five officers and men, or about 5 percent of battalion strength, devoted to intelligence duties, a somewhat greater proportion than is normally found in a force structured for conventional warfare.[12]

10 Renato F. Marques Pinto, "Intelligence, The Key to Counterinsurgency," in *Memories of Portugal's African Wars, 1961–1974*, ed. John P. Cann (Quantico: Marine Corps University Foundation, 1998), 17–20.

11 Estado-Maior do Exército, *Subsídios para o Estudo da Doutrina Aplicada nas Campanhas de África (1961–1974)* [Aids for the Study of Doctrine Applicable in the Campaigns of Africa (1961–1974)] (Lisbon: Estado-Maior do Exército, 1990), 159–160.

12 Duke of Valderano, interview by the author, 17 March 1995, London; Marques Pinto, correspondence with the author, 18 July 1995, Oeiras.

Patrols by both reconnaissance platoons seeking information on the enemy and normal combat groups obtaining intelligence in the course of performing other missions proved among the most valuable and productive sources.[13] Both types of units were constantly attempting to make contact with the enemy and maintain the initiative, and thus all operations to one degree or another embodied a reconnaissance role. Experienced troops were able to deduce information through the simple skills of knowledgeable observation, for example, the length of time an insurgent encampment had been abandoned and the number of men in the enemy force. One of the clues in this reasoning process was the state of the latrines at the campsite. Attempts were always made to verify this type of information through contact with the population. Also the type, condition, and origin of any captured weapons reliably fingerprinted insurgents, as it was known who sponsored which nationalist movement with what armament.[14] This fingerprinting also indicated the capability and mission of the insurgents and their logistic needs, and this knowledge was valuable to a tactical commander in enabling him to maneuver against the enemy. These several examples of ground reconnaissance and its contribution to the puzzle illustrate the importance of "foot slogging," a very labor-intensive but valuable exercise.

The population represented a key source of information, and indeed it was the primary battlefield, or in Clausewitzian terms, "the center of gravity" of an insurgency. The Portuguese soldier was indoctrinated in this principle and instructed to engage the population on enemy particulars: times of sighting, nature of armament, and numbers of insurgent forces.[15] Not only must he be alert to intelligence on the enemy, but he must also protect its primary source from insurgent intimidation. Indeed, it was found that the higher the confidence level of the population in the Portuguese soldier's ability to protect it, the greater the information it would provide. Insurgent intimidation could easily build a *muro do silêncio* ("wall of silence") between the population and the soldiers, and it required enormous patience, time, and goodwill to penetrate this "wall," again, a labor-intensive process.[16] The rewards of doing so, however, were great.

Captured Insurgents

Captured enemy personnel potentially represented the best source of information and thus "a terrorist captured alive is much more valuable for intelligence purposes

13 José Emídio Pereira da Costa, "Informação e Contra-Informação Militar em Ambiente de Guerra Subversiva" [Military Intelligence and Counterintelligence in the Environment of Subversive War], *Revista de Artilharia* (September–October 1961): 149.

14 Ibid.

15 Estado-Maior do Exército, *O Exército na Guerra Subversiva* [The Army in Subversive War] (Lisbon: Estado-Maior do Exército, 1963), Vol. II, Pt. 1, Chap. V, 2.

16 *Subsídios para o Estudo da Doutrina Aplicada nas Campanhas de África (1961–1974)*, 157.

than a dead one."[17] Ground reconnaissance patrols were often assigned the secondary mission of capturing a prisoner, and this mission required careful preparation and special training. Upon capture, an insurgent was immediately interrogated to obtain current information about the local area, the presence of other insurgents, their camps and staging points, their equipment, their lines of advance and withdrawal, and so forth. This highly perishable information was generally tested for soundness and verified immediately, at least to the extent possible under field conditions.

The Portuguese were always concerned about mines and booby traps, and thus captured insurgents were also asked to unmask these devices for destruction. Some were cooperative and others less so, depending on their fear and the proximity of their insurgent compatriots. In these cases the prisoner was asked to lead the way to the devices, for as one Portuguese veteran commented, "After all, they laid them and they knew where they were."[18]

A second interrogation would occur at battalion or sector headquarters to obtain more detailed information on the insurgent himself, his training, background, companions, and instructors. While most of these interrogations were performed by Portuguese specialists, there were exceptions. In Zala, Angola, for instance, there was an insurgent captured in 1965 who was named Alfredo and who now worked for the Portuguese. He had developed a talent for interrogation, as he knew the language of the enemy who habitually infiltrated into the area and had formerly served an insurgent commander in interrogating captured Portuguese. Captain Ricardo Alcada from the garrison post at Nambuangongo explained that Alfredo "usually spent an hour or two with a man in a room. By the time he came out he knew the man's history, who his grandmother was, whether she was a communist, what the man was doing in the area, what his unit was, what he was going to do—in fact everything the Portuguese could wish to know."[19]

An assessment would then be made as to whether the man was wholeheartedly with the nationalist movement or whether he had been coerced into joining by fear, promises, or both. In these cases the prisoner was shown troops in training and demonstrations of firepower, and an attempt was made to convince him that the Portuguese would prevail in the struggle. The prisoner was also exposed to the new villages and the social work that was being done for the population, including meeting *apresentados*. These were Africans living in the bush due to fear either of the insurgents or of the Portuguese troops and who had presented themselves to the authorities for resettlement in the protected villages. They could also be civilians forced to work for the insurgents as porters, cooks, and the like, and had simply run from the enemy camp. With this exposure to an environment of law and order, organized administration,

17 Pereira da Costa, 150.
18 Colonel Luís Alberto Santiago Inocentes, interview by the author, 14 April 1994, London.
19 Captain Ricardo Alcada, interview by Al J. Venter in *The Terror Fighters: A Profile of Guerrilla Warfare in Southern Africa* (Cape Town: Purnell and Sons, 1969), 72.

and medical care, the prisoner was supposed to conclude that life under a Portuguese administration would be infinitely preferable to the hard and dangerous existence of an insurgent in the bush.[20]

Treatment of Africans by the insurgents also contrasted sharply with that of the Portuguese. Anything from a minor mistake, such as the loss of a weapon or kit, to a rational questioning of authority could draw a severe penalty from the insurgent leadership. It was thus very rare for a captured insurgent, *apresentado*, or deserter to rejoin the insurgents, although he had the opportunity to do so.[21] The alternative was one of the prisoner of war camps. These facilities were isolated, and all were similar to São Nicolau on the southern coast of Angola, which was bounded by shark-infested sea on one side and desert on the other. With the opportunity to flee foreclosed by natural obstacles, the atmosphere was relaxed. Frequently the former insurgents would take a new wife and with Portuguese help build a new life.[22]

Lieutenant Colonel João Barros e Cunha, garrison commander at Nambuangongo in 1968, described another aspect of the rehabilitation program: "We offer the hand of friendship to these people—partly because we feel we have to, from a Christian point of view, and partly because we know they can help us win the war."[23] In exchange for their lives the insurgents were required to disclose all from beginning to end: "names, places, codes, signals, dates of training, future and past programs and explain any documents and plans that may have been found in their possession."[24] Finally, in a moment of truth, the prisoner was required to lead the Portuguese to his former headquarters and provide a tour of his operational area, thereby compromising himself completely and providing a tacit death sentence to his former comrades.

Signals Intercepts

The Portuguese army had a good signals intelligence service with an excellent decrypting capability. The insurgents initially used no radios, so signals intelligence (SIGINT) was denied. This situation was due to three factors: radios were unavailable; the proliferation of languages made communication difficult; and the low educational level of the insurgents made it very difficult to operate and maintain higher-technology equipment. When they began to use radios in the latter stages of the war, however, the Portuguese did listen and break their codes.[25] In the meantime Portuguese analysts spent a good part of their time listening to, recording, and analyzing enemy

20 Valderano interview, 17 March 1995.
21 Inocentes correspondence, 5 June 1995, London.
22 Marques Pinto interview, 30 March 1995.
23 Lieutenant Colonel João Barros e Cunha, interview by Al J. Venter in *The Terror Fighters: A Profile of Guerrilla Warfare in Southern Africa* (Cape Town: Purnell and Sons, 1969), 55.
24 Ibid.
25 Inocentes, interview, 14 April 1994.

propaganda. Although it was grossly exaggerated and distorted in its view, it proved to be a good source of information.

Captured Documents and Equipment

Captured documents and equipment represented an important facet of the intelligence collection effort and were carefully handled and preserved. Normally documents were written in Portuguese, as it was often the only common language available to the insurgents, although French was sometimes used.[26] For the most part the local languages of Portuguese Africa are spoken only, so there were no documents in native tongues. When a prisoner of any importance was captured, documents were usually found on him or with his equipment. These documents were used to verify his story during interrogation and provided a valuable index to his credibility. In many cases the circumstances of capture proved to be as important as the document itself. For example, any document listing guerrilla units was particularly important if connected with the precise time and place of capture. Battlefield searches also often uncovered official papers, as dead insurgents frequently carried documents that revealed their intentions and were of immediate tactical value. These were always processed as rapidly as possible through channels, for such information was invariably highly perishable.[27]

In one instance captured documents provided the Portuguese with confirmation that their operations were effective in denying the PAIGC permanent camps in Guiné. A field message captured in July 1971 near Gadamael on the southern frontier was written in Portuguese and reiterated a decree to the effect that no PAIGC unit was to remain in a local area for more than two days. It also reminded *bi-grupo* leaders and commissars of the consequences of disobeying this order both from the PAIGC command and from the Portuguese Air Force once the unit had been located, as they would be punished by both.[28]

The study of captured equipment, such as weapons, spent or unspent ammunition, mines, booby traps, radios, and the like provided technical intelligence. It provided the Portuguese with an indication on the type and degree of support that the guerrillas were receiving from external sources and helped define the enemy order of battle and combat capability. The taking of such equipment confirmed or complemented the strategic intelligence provided by Lisbon. The Portuguese published booklets describing the equipment and its capabilities, cataloging all types of armament used by the enemy and their location of capture. By these means the troops knew the

26 Marques Pinto interview, 30 March 1995.
27 Pereira da Costa, 149.
28 Lieutenant Colonel Lemos Pires, interview by Al J. Venter in *Portugal's War in Guiné-Bissau* (Pasadena: California Institute of Technology, 1973), 45.

enemy's order of battle and what to expect in combat. The information on mines and booby traps was particularly helpful.[29]

Agents and Informers

For the Portuguese the use of agents and informers was a normal process in obtaining information. In some cases within the theaters such agents worked with the military directly, and in others they sought the police, depending on which authority was present in the area. It was for this reason that a close liaison between the two was so vital and competition and jealousy in intelligence gathering were so counterproductive. More often than not, the police were in closer contact with the population than the army and possessed better facilities for centralizing this activity.[30]

In penetrating the sanctuary countries where the nationalist movements resided, border agents were generally handled through the army and agents on missions further afield were handled by PIDE.[31] Competition between the various factions within the insurgent movements provided a fissure that bred informers and fostered agents. Disaffected members of all the movements proved a fertile source of recruits and an opportunity for PIDE to sow the seeds of dissention.[32] In Guiné, the PAIGC "suffered from internal frictions between the mulatto Cape Verdean leadership and the African Guineans; and this was naturally exploited by the Portuguese."[33]

In Mozambique the main nationalist movement, FRELIMO, was in open competition with rival nationalist forces until 1972 not only for external sponsorship but also for nationalist movement dominance. Additionally, within FRELIMO there was a split between those committed to revolutionary socialism and the conservatives. The conservatives found themselves isolated and followed their leader, Lazaro Kavandame, in joining the Portuguese.

Another difficulty, particularly acute in the early stages of development, when many of the movement's members know little about one another, is the danger of infiltration by Portuguese agents. The complexity of motives behind divisive conduct makes it the more difficult to guard against individual neuroses, personal ambitions, and genuine

29 Marques Pinto correspondence, 26 June 1995, Lagos.
30 Pereira da Costa, 148–149.
31 Marques Pinto correspondence, 26 June 1995, Oeiras.
32 James Eliot Cross, *Conflict in the Shadows: The Nature and Politics of Guerrilla War* (London: Constable & Co., 1964), 34. The author argues that "first-rate counterespionage or counterintelligence is critical for all guerrilla and counterguerrilla operations. Treason, or even the fear or danger of it, can stop an insurrectionist movement dead in its tracks. There is no defense against it, for any important information that reaches the authorities gives them the very advantage they need to trap and destroy the militarily weaker guerrilla." The late President Magsaysay of the Philippines once told the author that one of the most effective counterinsurgency measures was judiciously applied bribery.
33 Peter Janke, "Southern Africa: End of an Empire," *Conflict Studies*, no. 52 (December 1974): 3.

ideological differences, all of which are muddled up with the tactics of the enemy secret service.[34]

The Angolan nationalist movements spent a substantial amount of time fighting among themselves and thus had little united war effort. The Portuguese were so successful in exploiting these differences and causing defections that it prompted René Pélissier to comment in 1971 that "the PIDE networks and Portuguese informants in Congo-Kinshasa are equal in effect to a division of parachutists on the ground."[35]

The Portuguese consistently sought to undermine the organizational structure of the nationalist movements during the war. The existence of many and varied ethnolinguistic groups throughout the theaters made the task of maintaining order among their constituencies far more difficult for the insurgents than for the Portuguese. Any action, normally brutal, against the population by either side was an unpardonable offense and an unbearable transgression against a group, particularly if it were inflicted by insurgents of a different group. An undercurrent of vendettas was a continuing condition of the battlefield.[36]

Guides and Translators

The Portuguese routinely used guides and translators to identify and track insurgents and to communicate with the population. Normally such guides were simple, unsophisticated people whose confidence and cooperation were vital to Portuguese success.[37] In Angola, for instance, the best trackers were Bushmen from the Kalahari Desert in the southeast, who were also initially employed as PIDE agents. The Bushmen were intelligent and gifted people who had lost none of their age-old talent for tracking in the bush.

With the plethora of local languages in each of the theaters, translators were attached to every unit to accommodate those who might be encountered in its operating area. In Guiné, for example, each company had between twenty and twenty-three translators attached, as there were some twenty different languages that they might use in the course of their operations. The importance of being able to communicate with both the population and the captured enemy cannot be overemphasized in the political atmosphere of an insurgency. These translators provided a bridge between the population and the soldiers in the field, for they knew the social form, the proper introduction, and the appropriate humor.[38] The Portuguese soldiers recruited in Africa knew one or more local tongues and seemed to have a facility for acquiring them.[39] Translators took great pride in their responsibilities for the simple reward of appre-

34 Eduardo Mondlane, *The Struggle for Mozambique* (London: Zed Press, 1969), 132.
35 Douglas L. Wheeler and René Pélissier, *Angola* (London: Pall Mall Press, 1971), 216.
36 Inocentes correspondence, 5 June 1995.
37 Pereira da Costa, 149.
38 Robert J. O'Neill, *Vietnam Task* (Melbourne: Cassell Australia, 1968), 75–76.
39 Marques Pinto interview, 30 March 1995.

ciation and the satisfaction of their physical needs.[40] Their pay, while normally equal in value to that of a Portuguese soldier, was a mixture of kind and cash, and often payment was divided and made not only to the translator but to his wife or wives.[41]

Aerial Reconnaissance

Aerial visual reconnaissance (*reconhecimento visual* or RVIS) was the complementary method to these other sources and, of all of the aviation missions flown daily, was one of the most important. Initially the insurgents assaulting Angola took refuge in the Dembos region and the mountain ranges between Uíge and Quitexe with their dense vegetation. The Uíge Range was approximately 5 by 60 kilometers (3 by 36 miles) of dense tropical undergrowth and difficult terrain and represented a dangerous problem as a waypoint along a "corridor of transit." In order to locate and destroy the insurgents hiding there, a "systematic reconnaissance plan" was developed that divided the insurgent-infested region into small manageable sectors that could be easily identified from the air by their obvious terrain features.[42] Each of these sectors or areas invariably had a conspicuous point of reference, such as a prominent rock outcropping or the confluence of two rivers, to orient a pilot's map and aid in fixing his observations, and indeed, one of the most notable outgrowths of the conflict was the development of excellent charts and maps of the *ultramar*. Each sector was assigned to designated pilots who were responsible for flying there regularly and would readily notice changes in the pattern of activity on the ground, such as expanded farming, newly-cut or abandoned tracks and trails, or recently built huts hidden beneath the trees. Typical clues to insurgent presence were such things as a new footbridge, a new trail, new and heavy use of an old trail, boats concealed alongside a stream, shelters built at the edge of the forest rather than in the open, and the presence of crops, clothes drying, or domestic animals around huts.[43]

RVIS was a difficult skill for a pilot to acquire, as it took many months for him to become acclimated to the flying environment in any of the theaters, develop a deep reservoir of experience and skill in understanding what he was seeing, and learn what the various signs and clues on the ground revealed to him. He needed to know his assigned area well, so that he could easily detect the subtle changes in it and identify precisely on an aviation chart the location of any suspicious site. This precise identification was important in planning and conducting a raid with ground forces.[44]

40 Ibid.

41 Óscar Cardoso, interview by the author, 1 April 1995, Azaruja, Portugal. Sr. Cardoso is a former Inspector with the PIDE/DGS and was instrumental in founding the *flechas*.

42 General Tomás George Conceição Silva, "The Portuguese Air Force in the African Wars of 1961–1974" in Memories of Portugal's African African wars, 1961–1974, ed. John P. Cann (Quantico: Marine Corps University, 1998), 116.

43 Joaquim Vito Corte-Real Negrão, "Subsídios para o Reconhecimento Aéreo Visual" [Aid to Visual Air Reconnaissance], *Boletim do Estado-Maior da Força Aerea* (December 1962): 34.

44 *Normas Gerais para a Execução de Reconechimentos Visuais,* 18.

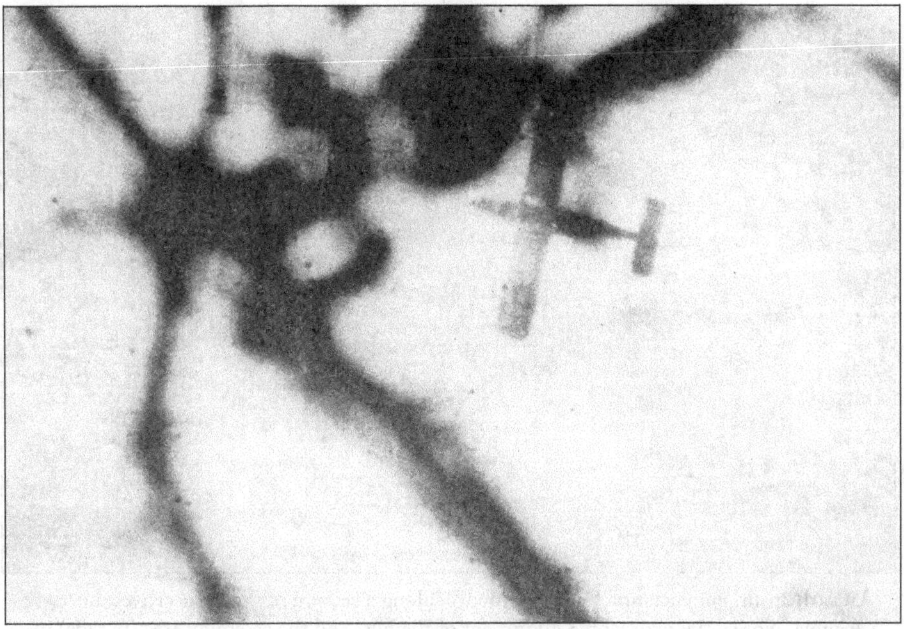

A Dornier DO-27 on a RVIS operation in Guiné, as seen from below by a PAIGC force in hiding. (Source: Tadahiro Ogawa)

This RVIS skill required a competence patiently acquired by flying over the assigned sector routinely in a Dornier DO-27 or Auster observation aircraft initially at between 800 and 1,000 meters (2,500 to 3,000 feet) and later at the relatively low altitude of 200 meters (600 feet) for many hours.[45] These missions were flown at or near noon, ideally between 1000 and 1400 hours local, so that the sun would be directly overhead and make visibility through the foliage easier than at other times. Flying in the late afternoons was also an important technique, as the sun would at this time penetrate laterally into the vegetation revealing clues that might otherwise be hidden from the midday sortie.

In observing a suspicious site, the pilot would attempt to avoid attention on the ground by continuing to fly his aircraft on a straight course, as if nothing curious had been noted. He would make no maneuver for at least three minutes or 8 to 10 kilometers before passing out of the ground audio-visual envelope and then turning.[46]

45 Conceição Silva, 116.
46 *Normas Gerais para a Execução de Reconechimentos Visuais* [General Principles for the Execution of Visual Reconnaissance] (Luanda: Comando da 2ª Região Aérea, June 1966), 22.

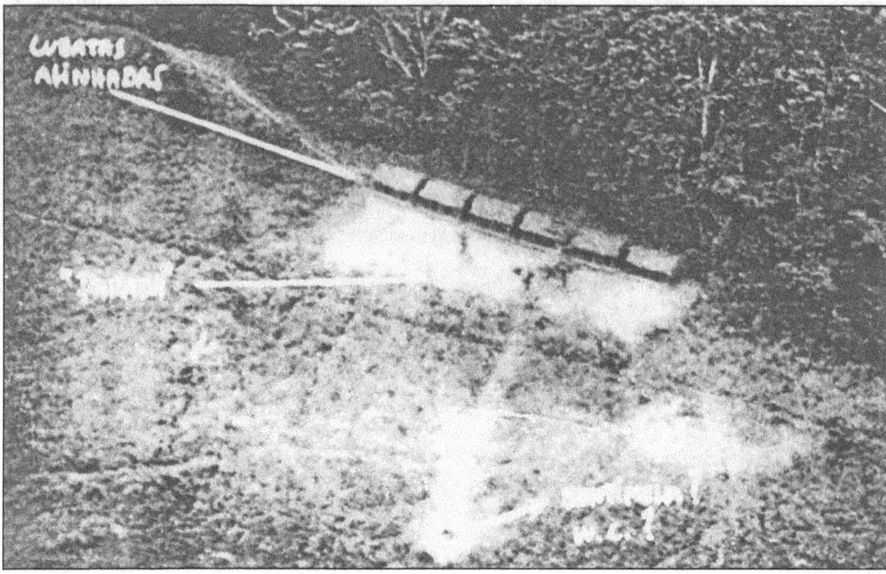

An early insurgent encampment at Quipedro, along a section of the Nuno River, north of Angola. Notice the regimented alignment of the huts and the assembly area (parade) in front of them. Such formal and visible alignment acquiesced to the need of preservation through deception and camouflage with a more random appearance of the huts as the conflict matured. (Source: Força Aérea Portuguesa)

This was important, as even when the aircraft had passed from view, the sound of its piston engine indicated whether it was continuing in a straight line or turning. The sound in a turn varied from its highest frequency when approaching to its lowest frequency when departing. It was hoped that this routine would not alarm the insurgents. Showing any interest, such as altering course and losing altitude for a closer look, would risk alarming them and causing their flight. In the case of a bridge, a boat, a trail, or another object of transit for insurgent use, however, a close look at no less than 50 meters (150 feet) was generally warranted.[47] In these latter cases, there was rarely anyone on the ground in the vicinity to be alarmed at the low pass of an aircraft. The insurgents worked at night, and aircraft flew RVIS sorties in daylight, so with this limitation, deductions had to be made from available signs. There were other complications in quartering a sector, such as the torrential rains during the rainy season and the morning fog or mist during the *cacimbo* period following the rainy season.

Overhead imagery with hand-held cameras proved a useful tool, and its analysis was helpful in revealing the presence or absence of insurgents and changes in their

47 Negrão, 42.

operations. These photographs were generally made of hamlets or other remote shelters, particularly any construction. Other objects of interest were transit signs, such as a footbridge, a boat, new crops, newly-cut trees, or a recently used trail. A series of photographs over time provided clues to insurgent activity and revealed changes in detail not easily visible to the naked eye in the brief time available to the pilot for his observation. For instance, the construction of normal huts included a door and windows and possibly a zinc roof. Insurgent huts had two or three large doors, no windows, and were much larger than the normal hut. There might also be the occasional "porch." In some instances these would be for local functions, but normally these would indicate insurgence presence.[48] The general orientation of the huts was linear and distinctly geometrical with a parade ground and flagpole visible. The insurgent encampment was also always separate from the huts of the population that supported it. Insurgent huts tended to be located at the confluence of two rivers or at the headwaters of a river, as the rivers were used for orientation and landmarks for enemy travel. These subtleties might elude the naked eye in flight but could be discerned easily from postflight photographic analysis, particularly through a comparative analysis of sequential missions.

Before the 1961 attacks in the north of Angola, a local community generally had two villages or settlements: an official one next to the road or primary trail where crops were grown and taxes paid on them; and a clandestine one in an inaccessible locale where crops were grown to evade tax and where the community habitually lived. These second, clandestine villages and their inhabitants were vulnerable to insurgent cooption in supporting a nearby enemy camp. One of the obvious signs of this development was a sudden increase in crop acreage around a clandestine village to support the hidden camp.[49]

The insurgents would locate their camp nearby, perhaps a kilometer away. Initially these *quartéis* were poorly hidden, but by the end of 1962 the insurgents had developed a sensitivity to air observation and began to hide their camps better. For the Portuguese aviator, finding these encampments became increasingly difficult, as they were located in areas where there was no cultivation or visible human activity to alert him. This gave the appearance of an area empty and uninhabited, but there were other clues that it was neither. The presence and condition of trails was one of the most important, particularly if they appeared well-used and seemed to disappear into canopied vegetation and reappeared on the other side. This obscured area interrupting the trail warranted a closer look, as the enemy trails were frequently designed to appear as a simple and direct route between population centers that merely passed through a thicket. The enemy continued to perfect these deceptions, and the real trails to the encampments were often lightly marked and appeared only as small bumps in the grass. Even with repeated passes by the observer, it was not possible to see the huts

48 Ibid., 9.
49 Comando da 2ª Região Aérea, "Observação Aérea," 6.

A local clandestine village in the north of Angola. (Source: Força Aérea Portuguesa)

Enemy encampment in the north of Angola in early 1962. The enemy was beginning to understand the danger posed by an aircraft and starting to conceal his bases, although not effectively just yet. (Source: Força Aérea Portuguesa)

beneath the foliage, although it was an almost certainty that they were there. If the vegetation was thick and copious, it was impossible to see anything.[50]

If an aircraft approached a clandestine settlement or camp too closely or appeared too curious, the insurgents immediately abandoned their camp and the local population that had been supporting them and fled to safety across a national frontier or, if that was not possible, reestablished themselves about 10 kilometers or more distant from the original site and in an area of more difficult access. The abandoned encampment soon acquired an unrefined appearance, as the huts became warped, and the grass untrimmed. Birds and herds of wild animals soon appeared nearby and indicated the absence of people.

The new encampment was invariably more difficult to identify and could be visible in part for only a brief moment through a hole in the foliage, as the pilot passed overhead. It was likely to be well under the canopy of treetops or overhanging vegetation and frequently hidden at the base of a cliff between large blocks of stone.[51] Quite often a pilot after many hours of patient and exhaustive observation was astonished to catch a glimpse of a hut in a sudden opening in the treetops. He would then recognize actual camouflage nets constructed of living vines and foliage. Such an encampment likely held other clues for its identification through its needed support infrastructure: the use of burned-over areas, fishing structures along the rivers, suspension bridges, float bridges, trails, rafts, dugout canoes, caves, and so forth. Burned out areas had various origins and were normally the initial phase of clearing a field to cultivate crops. A clue as to their ultimate use was the examination of the trails that led to and from the burnt area. Fishing structures along waterways were often constructed by natives in small streams. They were simple, rough affairs formed by driving stakes into the river bed and were common in small tributaries. Their sudden appearance was a clue to enemy presence.[52] Native bridges were only built when local traffic reached a level to justify the expenditure of labor capital. A bridge built to accommodate insurgents would likely be built by impressed local labor, be sited along a relatively straight portion of the river, and concealed from above by large amounts of vegetation projecting from both banks.

Trails were complicated in that they were made variously by animals, insurgents, refugees, local people, and smugglers.[53] These trails all invariably disappeared beneath vegetation, and the pilot constantly sought their reemergence on the other side of a river, bush, road, or other obstacle. Rarely did a trail take the most direct route between points, as its users had their particular reasons for following that peculiar route. Enemy trails followed such routes so as to be invisible from the air

50 Ibid., 7.

51 Negrão, 10.

52 Ibid., 11.

53 Inge Brinkman, "Refugees on Routes, Congo/Zaire and the War in Northern Angola (1961–1974)," paper delivered at the international symposium *Angola on the Move: Transport Routes, Communication, and History*, Berlin, 24–26 September 2003, 9.

Huts of an enemy encampment in mid-1962. Increasingly the camp infrastructure is hidden from above, as the insurgents learned about aerial reconnaissance. (Source: Força Aérea Portuguesa)

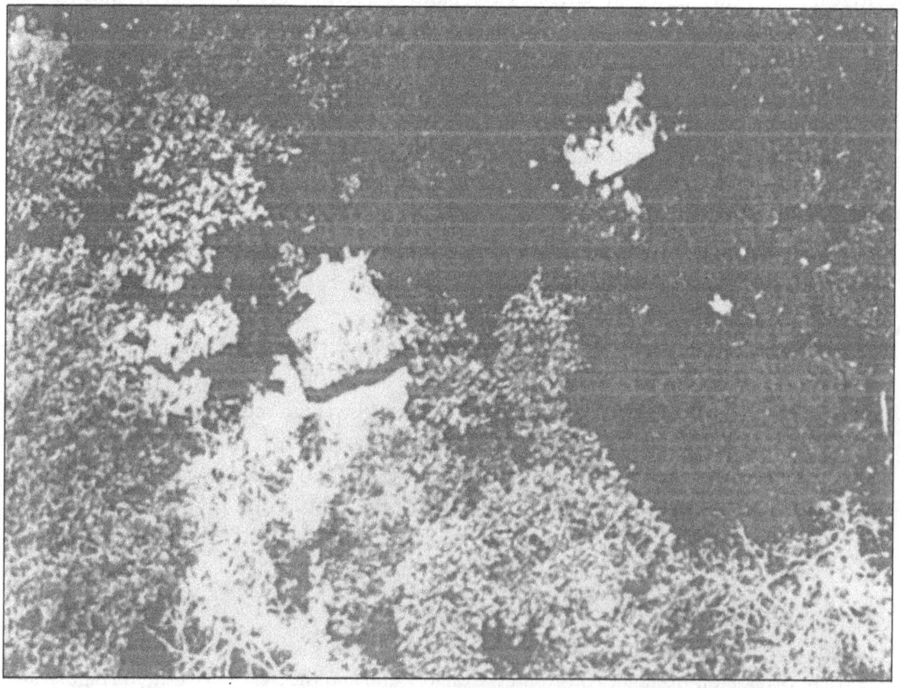

Huts of an enemy encampment seen through a break in the foliage during a noon reconnaissance in late 1962 in the north of Angola. (Source: Força Aérea Portuguesa)

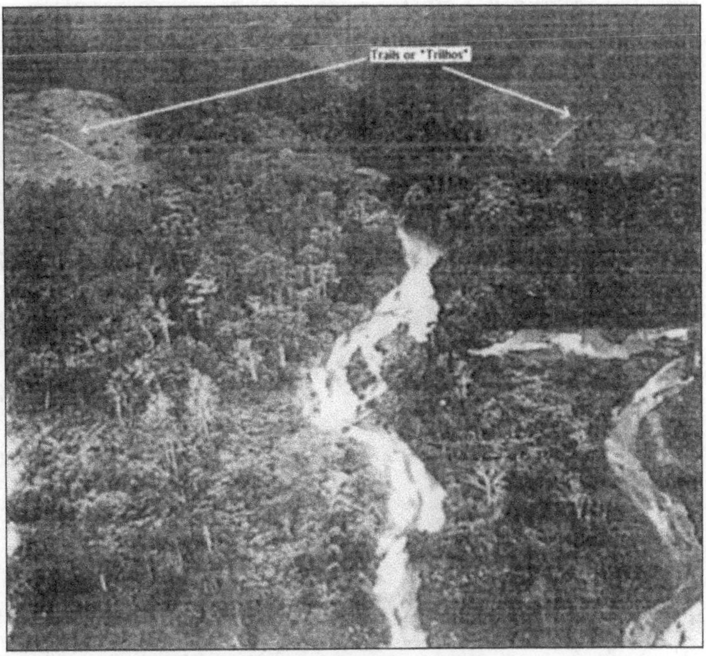

A trail leading into and hidden by luxurious vegetation bordering a river in the north of Angola. (Source: Força Aérea Portuguesa)

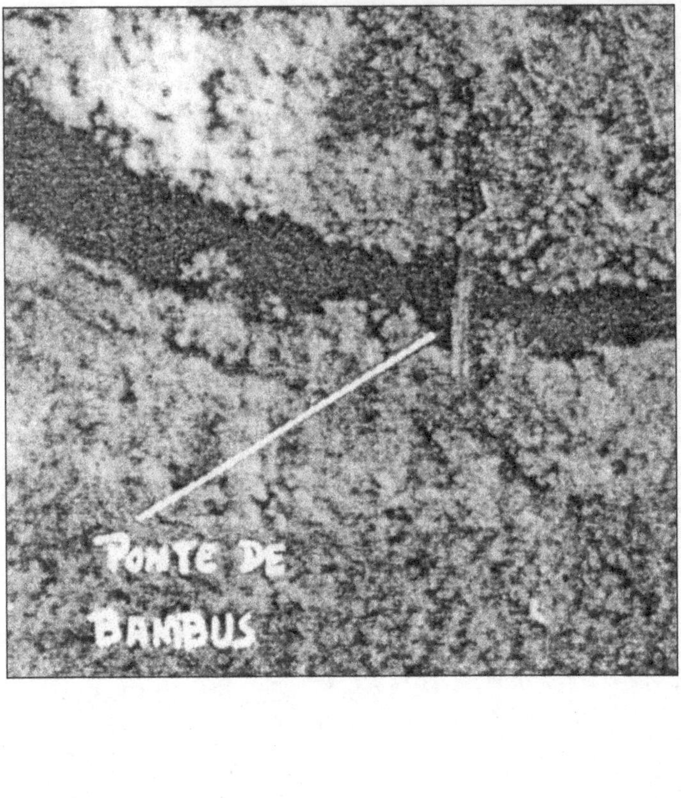

Overhead photograph of a bamboo footbridge, Lulovo River, Angola. (Source: Força Aérea Portuguesa)

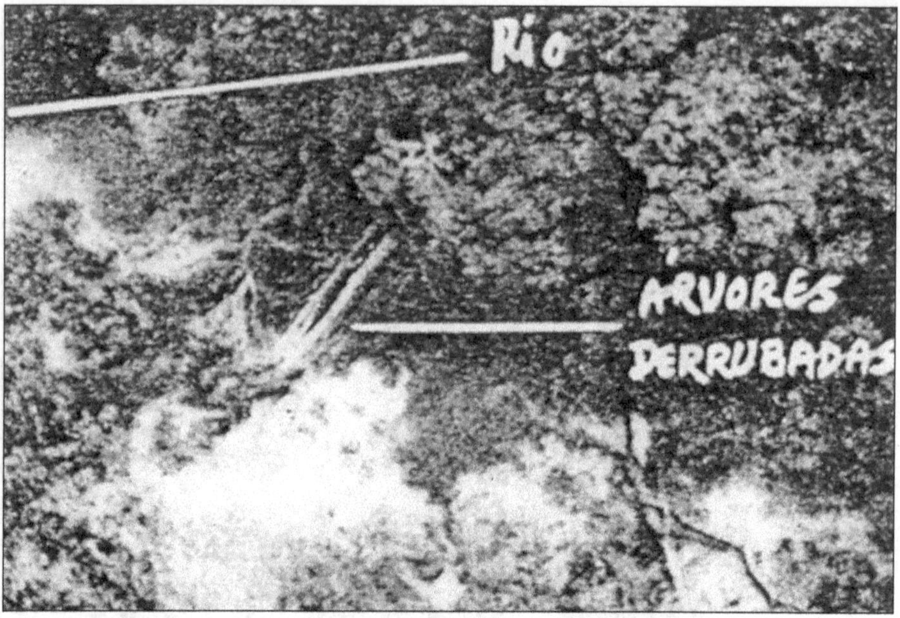

A cut log bridge across the Calambinga River, Angola. (Source: Força Aérea Portuguesa)

and were veritable tunnels through the vegetation. Within them the enemy marched single file. Often the trail branched into similar parallel trails so that a large group of infiltrators could travel single file and yet remain roughly together and concentrated rather than being strung out. Frequently the trail became a dry river bed or a shallow stream overhung with vegetation. Often also the insurgent trails paralleled a road, a watercourse, or a well-used trail, and the insurgents crossed these landmarks only when it was judged to be safe and then preferably when there was a hard, stony surface available that would reveal no tracks. Rafts, *pirogas* (native log canoes), and more traditional fabric-covered frame canoes were used by the insurgents to cross the larger rivers and were always hidden in the overhanging vegetation along a riverbank. Sometimes there was a platform covering the vessels, and this structure alerted the pilot.

Information from aerial reconnaissance was fed to the army on a continuous basis, as an enemy beaten in one place today would tomorrow establish new camps at another with extraordinary speed. The enemy's tactics were also in a constant state of flux, and it was vital to detect these changes promptly and to modify operations accordingly.[54]

54 Negrão, 34–43.

All of this was complicated by the enemy's awareness of the danger posed by RVIS and aviation in general, and consequently he used both active and passive defensive measures. He became an expert in the techniques of disguising, camouflaging, and dispersing. He developed alternative sites and adapted to and became expert in deception and vigilance that took advantage of the dense vegetation and configuration of the terrain. Indeed, the enemy lived by deception, constructing shelters beneath the roots of great trees or living in huts half buried. He developed an understanding of flying habits and routines and knew which aircraft posed a danger and which did not. When caught in the open, he and his compatriots would appear in work clothes that were a part of their uniform, hold up their hats, and wave at the aircraft as if they were ordinary field workers.[55]

The enemy only launched his operations when he could anticipate the support of a substantial part of the population in his zone of action. This was important, as many extra hands were needed as porters, cooks, and guides to achieve the common goal of a successful operation contributing ultimately to "independence." This labor would quickly be evident from the air, as a new settlement of huts would suddenly appear with domestic animals, *mandioca* (cassava) patches, irrigated land, cooking utensils, clothes lines, etc. The insurgents lived in an alternative camp well-hidden and perhaps up to a few hundred meters away, but their support infrastructure would be visible. Their huts would be half-buried or alternatively natural caves or caverns, depending on the terrain. Destruction of the insurgent and support villages was always a temptation for the pilots; however, the best, most fruitful course was always to conduct visual reconnaissance by being as invisible as possible and let the ground forces make contact.[56] Indeed, there was much care taken not to spook the insurgents and always to leave them in substantial doubt about having been observed. So that they would not connect the overflight of an aircraft with a subsequent attack, there was always an interval of days or weeks before a ground assault or helicopter envelopment was executed.[57] Indeed, overflights of a particular target were done no less than a month apart in order not to alarm the enemy, as he was very shrewd. Interestingly, a pilot would hear through the native grapevine one or two days later about his overflight.[58]

The enemy manipulated the people to his ends by instilling widespread fear of any government rehabilitation efforts and thus making it very difficult for the Portuguese to pierce the protective population shield and find and destroy the armed bands. In locating them, a pilot must have wisdom, intuition, and a measure of good luck, for if he alerts the enemy, all will flee and hide with a rare mastery.[59]

55 *Normas Gerais para a Execução de Reconechimentos Visuais,* 3.
56 Ibid., 4.
57 Ibid., 5.
58 Ibid., 17.
59 Comando da 2ª Região Aérea, "Observação Aérea," 7.

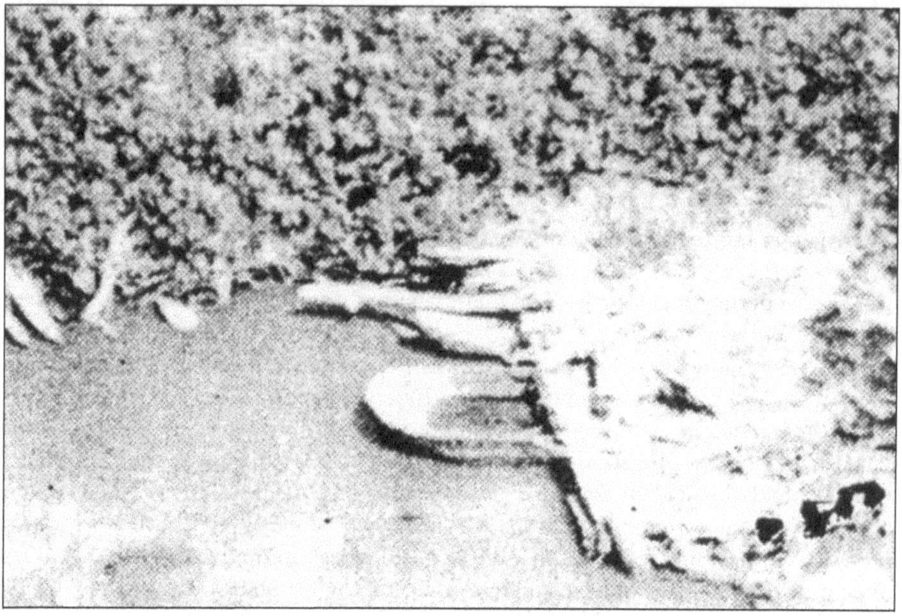

Canoe hidden beneath a platform, M'Bridge River, Angola.
(Source: Força Aérea Portuguesa)

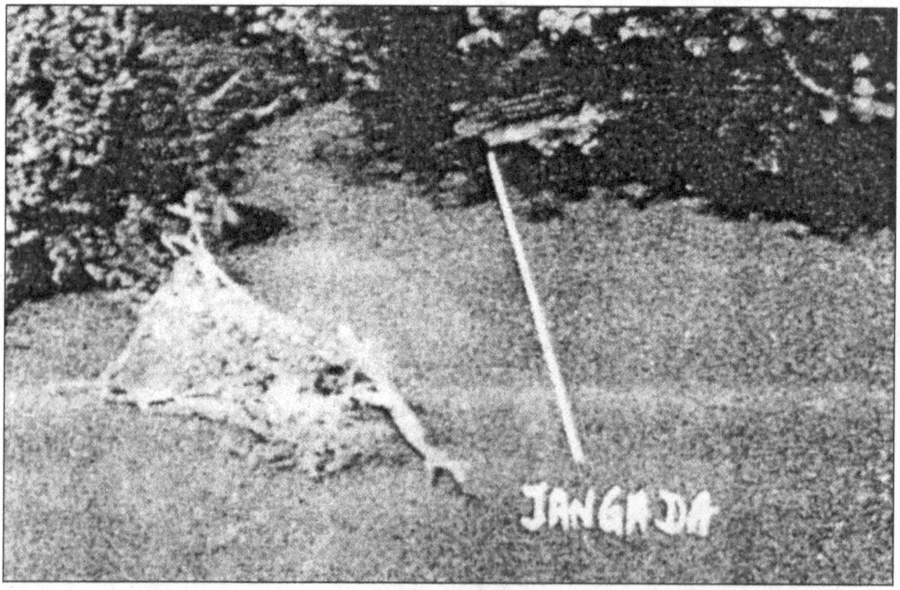

Raft hidden beneath riverbank foliage, M'Bridge River, Angola.
(Source: Força Aérea Portuguesa)

The sense of urgency in reacting to the results of air reconnaissance depended on many factors: the activity of the insurgents, their proximity to or remoteness from the population, and the availability of local troops. Normally insurgents did not concentrate in large numbers, so locating them required quartering the large expanses of Angola and Mozambique in search of their telltale signs. Once the insurgents were located; however, the Portuguese intelligence network began to develop a picture of their intentions. As they moved on foot, their progress was normally slow and gave the Portuguese time to prepare a trap. They were easily spooked, however, and great care had to be taken in planning any operation. After a few years of fighting, they had been hunted so relentlessly that at any hint of danger they quickly melted away. At that point the intelligence picture would then have to be developed anew.

New Tools

In addition to RVIS, the nature of the conflict in Guiné and Mozambique with the advent of hand-held surface-to-air missiles changed the equation, and other means of air surveillance were required. Overhead photography debuted in 1973 and was used against the insurgent bases in Tanzania, Malawi and Zambia. The aircraft camera, a Fairchild K-20 aerial camera, was enclosed in a pod fashioned from a 200-liter Napalm canister, as is shown in the nearby photograph.[60] This improvised camera pod was mounted beneath the right wing of a Fiat G-91 and was effective from an altitude of 15,000 feet, a height well beyond the range of the SA-7 *Strella* surface-to-air missile used by PAIGC and FRELIMO forces and the *Rapier* installations in Zambia.[61] The system was used to photograph the important FRELIMO bases of Nashingwea and Mtwara in Tanzania, of Chipota in Zambia, and of those in Malawi. Nashingwea was an important FRELIMO base and the command post for Samora Machel, and Mtwara was the logistical and recruitment center for FRELIMO. Photographic reconnaissance of the Zambia-Mozambique frontier further revealed the FRELIMO bases and arms dumps supporting enemy operations in Mozambique.

While the system provided excellent targeting, the drawback with the advent of this technology was the need for the reconnaissance aircraft to return to Nampula to

60 The aerial camera K-20 was manufactured by the Fairchild Aerial Camera Corporation and used a 5.25"x20-foot to 5.25"x200-foot roll film, with an image size of 4x5 inches. The lens was a 161mm f/4.5, with shutter speeds of 125, 250, and 500. The aperture could be turned down to f/22. Focus was fixed at infinity, something that made sense. Unless you are flying at low altitude, everything is at infinity. The shutter is built into the lens, which was made by either Kodak, Llex, or Bausch and Lomb, as available at time of order. An interesting feature is the use of a vacuum to keep the film flat in G maneuvers.

61 José Armando Vizela Cardoso, "A Guerra Untramarinaem Moçambique, Operações Militares" [War in the *Ultramar*: Military Operations], presentation given at the Instituto de Estudos Superiores Militares, Lisbon, 20 June 2012.

Napalm canister modified to accommodate a Fairchild K-20 aerial camera. (Source: Força Aérea Portuguesa)

Improvised overhead camera pod under right wing of a Fiat G-91. (Source: Força Aérea Portuguesa)

deliver the film to be developed and interpreted and to reload the camera after each flight. For missions along the Téte-Zambia border, the return flight was nearly 400 nautical miles with few navigational aids.[62]

Toward the end of 1973 the FAP acquired airborne infrared equipment capable of registering images from the heat of human bodies and thus revealing the presence of hidden FRELIMO bases in Mozambique. The equipment was mounted in a modified Douglas C-47 Dakota and was very effective. Once the camps were identified, their location was passed to the Fiat G-91 squadrons, which destroyed them promptly. The system was so devastating to the insurgent camps that FRELIMO began to face a severe recruitment problem.[63]

The Key to Counterinsurgency

The Portuguese clearly understood that the centralized flow of intelligence was the key to counterinsurgency and that this flow would come primarily from the population. They consequently designed their intelligence-collecting machinery to work in this special environment. It was here again that their entire effort was reoriented from conventional war with its focus on the enemy as the primary source of information to counterinsurgency, in which the operating area and the civil population there were primary. The understanding of this principle was well articulated by a Portuguese junior officer writing in a military journal of the time:

> The primary objective of the terrorists is to destroy the population's confidence in the guarantees of protection that can be provided by local government. With this objective achieved, the local population is reluctant to provide intelligence to the authorities for fear of reprisals. Only with well conducted operations will it be possible to reestablish this loss of confidence. And these depend upon the timely utilization of reliable intelligence. …[64]

The search for information was the key to success, and in this Portugal excelled.

As intelligence networks operate largely in secret, the details of their organization and operations are not often visible to outsiders. The litmus test for the effectiveness of an intelligence apparatus is the frequency with which it is taken by surprise. In the war the Portuguese were seldom caught unawares, but when they were, the results were severe. The initial events in Angola in 1961 represented more a failure to act on intelligence than poor intelligence itself. Thereafter, the system in all three theaters appeared to work relatively well, and it was more often the insurgents who were taken by surprise in operations that were planned and executed regularly to

62 Ibid.
63 Ibid.
64 Pereira da Costa, 147.

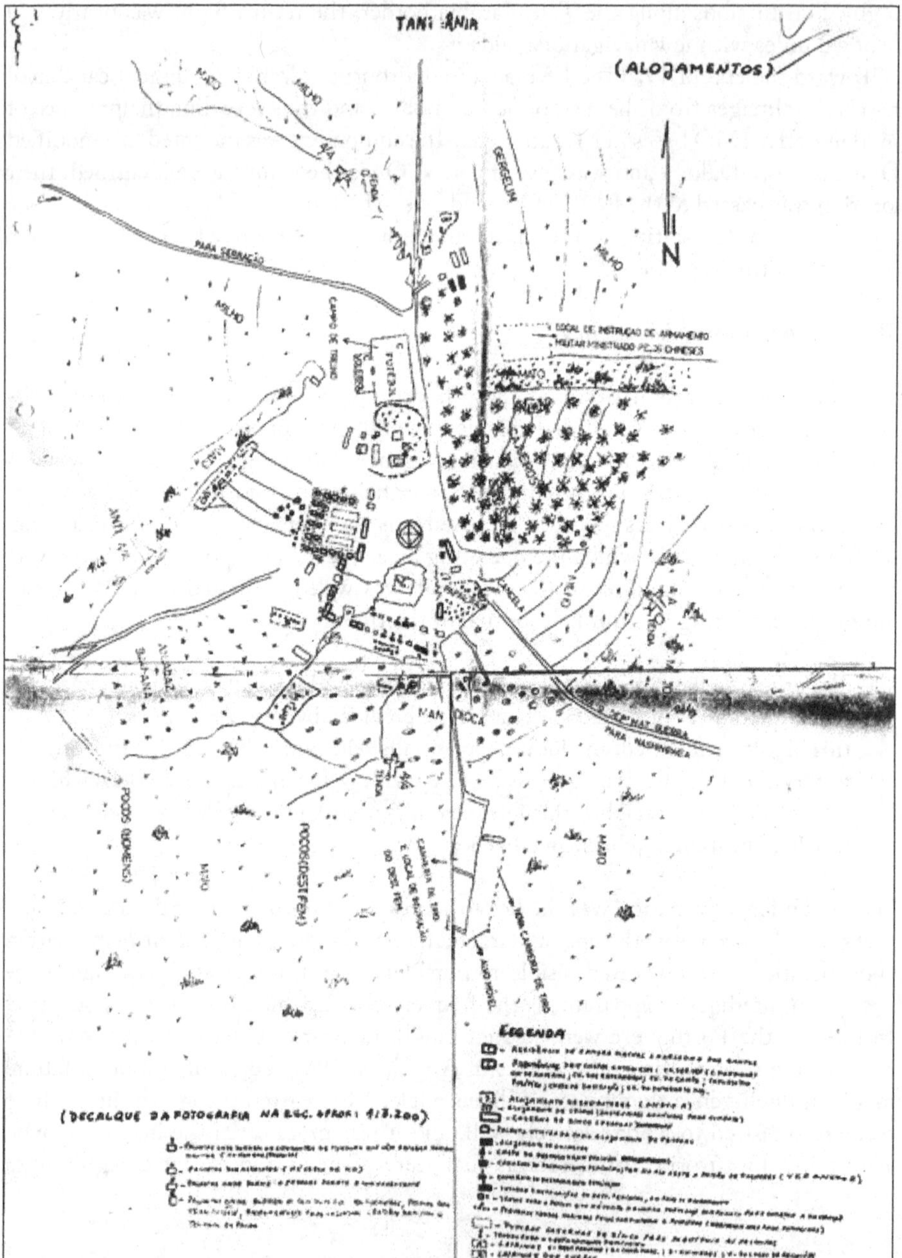

Layout produced from an overhead photographic mosaic of Nachingwea, Tanzania. (Source: Força Aérea Portuguesa)

Interior of a C-47 Dakota showing the layout of the airborne infrared system used in the north of Mozambique. (Source: Força Aérea Portuguesa)

exploit information. The contribution of a good intelligence network was one of the decisive factors in the Portuguese ability to sustain the conflict for thirteen years and to employ its limited resources in effectively controlling the insurgent threat. It was able to anticipate the enemy well and largely defeat him over this prolonged period through consistently thorough intelligence work.

8

Guiné: A Beastly Place

Guiné is a complicated land in which to project force, and it is thus not easy to secure it against threats from either within or abroad. Its climate is unappealing and hostile. Its topography and geography are difficult, and its population is highly fragmented and unevenly distributed. Its economy is built on subsistence agriculture and is thus devoid of any reasonable economic potential. In summary, its viability as a Portuguese colony was questionable under any circumstance. The greater mystery was why Portugal felt compelled to devote a disproportionate measure of its treasure to defending such a worthless possession and tie its national strategy to a land of such little economic or strategic value. The rationale offered was that its loss would jeopardize the enormous investments in Angola and Mozambique, a political and economic illusion.[1] The defense of this tiny enclave against an insurgency with inviolate sanctuaries nestled immediately opposite its frontiers and from which a resourceful enemy could project its own power proved a supreme challenge and looked increasingly like a Pyrrhic victory, if indeed victory could be achieved by any definition.

A Complicated Theater

For Portugal the situation was complicated by a number of factors. First, the country is tropical and was described by Al Venter in 1973 as a "grim, torrid stretch of swamp and jungle."[2] It is a wet and steaming land under a burning sun. The intense humidity and enervating heat make for an oppressive atmosphere. As José Talhadas, a *fuzileiro*, described it: "Hardly had I put my foot on the ground, then discomfort permeated my body: a scorching heat and harsh humidity seemed to explode my chest, sweat

1 Marcello Caetano, *Depoimento* [Deposition] (Rio de Janeiro: Distribuidora Record, 1974), 191–192.

2 Al J. Venter, *Portugal's Insurgent War, The Campaign for Africa* (Cape Town: John Malherbe, 1973), 13.

followed, forever, from those first minutes in the land."[3] While there is a dry season that provides some relief from November to May, it too is not without discomfort. As it nears its end in the late spring months, the sun approaches its seasonal zenith over the colony, and tropical weather systems develop. These are characterized by low pressure areas and strong ascending warm air currents. These updrafts are fed from the continental air mass over the Sahara and create what is termed locally as the "east wind." The population experiences it as a hot, dry airstream that comes from the desert to the east, absorbs moisture from the soil and vegetation, and carries a fine dust that affects those with respiratory problems. This is accompanied by temperatures rising to 100°F (35°C–40°C) on average.

The rainy season runs from June to October and ensures a high humidity. It is characterized by frequent monsoons that develop from the disequilibrium between the oceanic and continental air masses. These storms are magnified by the shock of the "east wind," and are characterized by violent electrical activity with the attendant thunder and lightning, tornados, strong winds reaching 60 knots, and heavy rainfall. In the coastal region annual rainfall averages 3 meters, and in the interior, 1.25 meters.[4] These conditions greatly inhibit air operations.

Because of this pervasive dampness and heat, the staying power of any force becomes difficult. Ammunition, spare parts, uniforms, maps, rations, medical supplies, and all of the other military matériel and varied paraphernalia of modern war not protected from the dampness or not used immediately will rot or rust. Equipment maintenance problems multiply by several factors over normal wear and tear in a temperate zone, and all machinery requires constant attention if it is to be reliable. For the FAP it was indeed a place that defeated machinery with careless ease. It is also a land humming with mosquitoes and many other kinds of insect life. Again, Talhadas describes it as a place "where flies and mosquitoes teemed and seemed to form dangerous 'aerial squadrons' that attacked us, principally at night. Before long we felt their devastating effects, although the mosquito nets managed to limit the damage. The next day, my face and other parts of my body were swollen with bites on all sides. Curses and unspeakable words were more than many on awakening at daybreak."[5] Debilitating tropical diseases abound, and are transmitted by these *petites bêtes*. They also lodge in aircraft, and no Pitot tube, gun port, or wheel well is immune from their interest. Their potential presence demands an attentive preflight inspection on the part of every pilot prior to his sortie, or unwelcome surprises could await. Venomous snakes thrive in such a habitat. Rapid rates of decay denigrate the quality of front-line medical care.

3 José Talhadas, *Memórias de um Guerreiro Colonial* [Memories of a Colonial War] (Lisbon: Âncora Editora, 2010), 71.

4 Estado-Maior do Exército, *Resenha Histórico-Militar das Campanhas de África, Vol. III, Dispositivo das Nossas Forças Guiné* [Historical Military Report of the African Campaigns, Vol. III, Disposition of Our Forces in Guiné] (Lisbon: Estado-Maior do Exército, 1989), 20–21.

5 Talhadas, 72.

In short, the climate becomes a very important component of Clausewitz's friction of war.

Second, the topography of Guiné is difficult. It can be divided roughly into two distinct geographical areas, the west (littoral) and the north and east (interior). The most important is the western one, which on a tactical level is characterized by a forbidding and inhospitable stretch of mangrove and swamp forests covering the coastal inlets and deltas of half a dozen rivers. Tidal action floods twice daily, and as the water rises, it fills these estuaries and creates vast tracts of impenetrable swamp.[6] The banks of these sinuous rivers are covered with dense vegetation projecting over the water, a condition that is ripe for the ambush of any vessel. In travelling along a waterway, naval vessels make slow, tedious, and noisy progress that alerts any enemy. Currents in the rivers during flooding and ebbing can reach 7 knots, enough to overwhelm the small engines in the majority of river transport. This tidal action particularly affects navigation in certain areas, such as Gadamael on the upper reaches of the Cacine River, where it is only possible for larger vessels to operate on the four or five days of high tide during a full moon.[7] At low tide vast expanses of dreary mud flats (*bolanhas*) are exposed and impede the projection of naval power and commerce through the river system. Vessels must carefully travel their courses in the center of the channels, for they could easily become grounded and mired in the mud during ebb tide. The low water level would then leave them completely stranded, isolated, and vulnerable to insurgent attack. This necessary caution and the added requisite planning for each trip or patrol was an important component of wartime friction. Ground forces had to cross these rivers and mud flats to engage any enemy, and this entailed either travelling long distances overland by truck, or boarding a naval landing craft, or accepting a lift by helicopter. Each method had its limitations. In the case of the roads, they were often poor and impassable in the rainy season and were frequently mined as part of an enemy ambush. Indeed, many of the roads during the rainy season become rivers themselves, making passage all but impossible. The noise of the trucks easily alerted the insurgents. Landing craft readily moved the needed volume of forces but, like land travel, were noisy and vulnerable to ambush. Movement by helicopter was rapid, but the limited space available meant that only a small force could be airlifted even with several rotorcraft, and these too were noisy.

Further, the thousands of miles of Guinean rivers and tributaries are obscured from the air by mangroves and thick foliage, making clandestine insurgent movement simple and its interdiction a difficult military problem. Cameras aloft rarely see the forest floor. These rivers are navigable by medium and small craft deep into

6 John P. Cann, *Contra-Subversão em África: Como os Portugueses Fizeram a Guerra,1961–1974* [Counter-Subversion in Africa: The Portuguese Way of War, 1961–1974:], 2nd ed., rev. Major General Renato F. Marques Pinto (Lisbon: Prefácio, 2005), 25.

7 António José Telo, *Homens, Doutrinas e Organização 1824–1974* [Men, Doctrine, and Organization 1824–1974] (Lisbon: Academia de Marinha, 1999), 576.

the country and provide vital lines of communication that cannot be matched by the primitive road system. The coast also has many small islands, the most important of which form the Bijagós Islets. The land rises in the northern and eastern interior areas, where the coastal growth gradually disappears, and the land becomes forested, as the terrain transitions into the sub-Saharan savanna plain of grasslands and scattered, scrawny trees. Elevation never exceeds 300 meters.[8]

Third, Guinean geography also plays an important role. Officially Guiné has a land area of 36,125 square kilometers, but its twice daily inundation covers as much as 22 percent of the country, and reduces the surface area to an estimated 28,000 square kilometers. From this figure an additional 3,200 square kilometers can be deducted for the area periodically flooded by rainwater.[9] Under these considerations, the habitable land mass is about 24,800 square kilometers, or about the size of Switzerland, or about one-quarter the size of the 95,440 square kilometers of the Congo District of Angola, the scene of earlier action.[10] Its land frontier is about 680 kilometers, of which 300 comprise the northern border with Senegal, and 380 the eastern and southern borders with the Republic of Guinée.[11] Guiné is thus so small that the entire country became essentially a frontier and battle zone, making it extremely difficult to defend. Perhaps the best characteristic of this small area was that there was no part of it that could not be reached by an aircraft in 30 minutes.[12]

As if these impediments of climate, terrain, and geography were not enough of a challenge, the population represented a fourth obstacle as much for the insurgents as for the Portuguese. In 1960 the population was 525,437, and was concentrated in the western coastal delta with about 100 people per square kilometer.[13] The arid eastern half was virtually empty with about one person per square kilometer, and it was here that insurgent infiltration occurred with the least opposition. Over 99 percent of the population was black and fragmented into two primary groupings covering 28 ethno-linguistic groups, each exhibiting various degrees of loyalty to Portugal or the insurgents. Indicative of early exposure to the Portuguese, the coastal population is Christian or animist. The interior peoples are Muslim, reflecting their contact with the ancient interior trade routes from the Gulf of Guiné through Timbuktu to The Maghreb. Because Guiné was an economic

8 Cann, 25.
9 António Afonso and Carlos de Matos Gomes, *Guerra Colonial* [Colonial War] (Lisbon: Notícias, 2001), 88.
10 Hélio A. Esteves Felgas, *O Clima do Congo Português* [The Climate of Portuguese Congo] (Carmona: Imprensa Angolana, 1959), 9.
11 Estado-Maior do Exército, 17.
12 José Manuel Correia, "'Rotina' de Guerra: Os F-86F 'Sabre' na Guiné, 2° Parte" ["Route" to War: the F-86F Sabre in Guiné], *Mais Alto* 358 (November–December 2005: 40.
13 Ibid., 27.

The PAIGC logistic system relied on a fleet of these canoes plying the Inxanche and the other river systems of the Quitafine and Cantanhez peninsulas to supply its forces in southwest Guiné. (Source: PAIGC archives)

backwater, it was never a country of European colonization, and there were less than 3,000 whites in the colony.[14]

Defense of this territory was consequently difficult under the best of circumstances. The enemy did not have far to travel from his sanctuaries in Senegal and the Republic of Guinée to reach population centers. With such a short logistics tail that led over only partially challenged waterways from the south and overland routes in remote areas in the east, insurgent troops operated in concentrations almost anywhere in the territory. The Portuguese consequently found themselves in a defensive position, where Bissau, the capital and main port and city, was the center through which all power flowed to support the interior line represented by the frontier or border areas. Because the nature of the battlefield characterized the height of Clausewitzian friction, the absolute key to mobility was airpower. It would move troops to battle, support them in their fight, and bring them home afterwards. This support was accomplished with a relatively small cadre of specialized and highly trained aviators and support personnel and a varied assortment of obsolete aircraft.

14 René Pélissier, *Le Naufrage des Caravelles, Etudes sur la Fin de l'Empire Portugais (1961–1975)* (Orgeval: Editions Pélissier, 1979), 208.

The use of airpower in this theater began in 1963 as part of a military offensive and was successful as far as it went. However, it had its limitations for a number of reasons. Initially leadership of the war was divided between Brigadier Fernando Louro de Sousa, the commander-in-chief, and *Comandante* Vasco da Gama Rodrigues, the governor-general. While Rodrigues, a naval officer, was a forthright individual with a wealth of operational experience, Louro de Sousa was a staff officer with little practical experience in the field. Because of these two contrasting perspectives, the men clashed constantly, and the campaign against the PAIGC suffered accordingly. Rodrigues attempted to be practical, and Louro de Sousa clung to his theories of conventional war.[15] This division of leadership created a fragmented approach that could not address the war effectively or successfully.

Brigadier Arnaldo Shultz, the commander-in-chief and governor-general during the period 1964 to 1968 decided to strike a defensive posture and to counterattack the PAIGC as it advanced in an attempt to regain control of lost areas. This strategy was doomed to failure, as it allowed the PAIGC a toehold within national territory and a relatively free hand with the population. Shultz did not see the population as a primary battleground, and while the PAIGC was establishing medical clinics, schools, village markets, and the like, Shultz ignored it. In fact, it was a key factor in the equation. Shultz pursued a military solution, indeed gambled on it, and the strategy proved a failure. The small territory was increasingly being overrun and controlled by the PAIGC in the face of his offensives. Early on the PAIGC was able to extend its influence into the northern and southern areas and deep into the east of Guiné in the *chão manjaco* (Manjaco ground). In 1965 it had begun to receive substantial military support from the communist bloc. The momentum was in its favor, and the FAP would try to reverse this.

The FAP was an important factor in Shultz's plan, as it was deeply feared by the PAIGC. The cumulative effect of FAP action during the Shultz years, blunted the PAIGC military initiative and ultimately forced it into a strategic defense. Indeed the PAIGC was virtually paralyzed by the weight of the bombing, as the terrified villagers in its areas of influence abandoned their homes for safer ground. This was, however, hardly a "hearts and minds" program, and it would prove counterproductive, as the subverted population continued to welcome the PAIGC in the face of its continued setbacks. Nevertheless, the FAP executed Shultz's plan with faithfulness and imagination, and this is the story of how it did so despite the substantial obstacles it had to overcome.

15 Alpoim Calvão, "Ataque a Conakry" [Attack on Conakry] in *A Guerra de África 1961–1974* [The African War 1961–1974] (Lisbon: Temas e Debates, 1996), 508.

The Air Force Arrives

The war in Guiné was a very different insurgency from most not only because of the physical nature of the battlefield, as has been noted, but because the PAIGC proved resilient and resourceful. This extended to a willingness to threaten the livelihood of the population, if it suited PAIGC purposes. Selective destruction of the dike system on which much of the littoral population depended to raise rice was used to punish those loyal to Portugal. Likewise, the destruction of the tourist infrastructure on the beaches at Ponta Varela in the northwest caused the abandonment of its development and the loss of local jobs.[16] A year into the conflict the rice and ground nut receiving and processing facilities and lumber mills were forced to close, and the local jobs connected with growing, harvesting, and processing rice, cashews, peanuts, and timber wound down.[17] These hardships meant little to Cabral in his greater strategic view of the conflict, which was to damage the Guinean economy and make waging war more expensive for the Portuguese. Portugal would now be earning no foreign exchange from vanishing exports and indeed would have to import foodstuffs to feed the population. Further, the abandoned facilities that were remote in many cases were taken over by the PAIGC and used as logistics depots.[18]

The first PAIGC attack occurred on 21 July 1961 on the village of São Domingos in the northwest and resulted in four wounded. Other actions followed within days at the neighboring villages of Varela and Susana further westward and adjacent to the Senegalese border. The PAIGC strategy was to divide and exhaust the small number of Portuguese forces by first initiating attacks in the north that drew them toward the Senegalese border and then to infiltrate its forces into the southern areas of the province. Portuguese army troops in Guiné at the time numbered about 3,000.[19] Little actually happened that was visible after these first several attacks, as the PAIGC spent its time organizing the population, recruiting fighters, training its new troops, forming them into cadres, and attempting to isolate selective parts of the territory

16 José de Jesus Bispo, "A Participação da Força Aérea na Guerra de África" [The Participation of the Air Force in the African War (1961–1975], *Revista Militar* 2507 (December 2010): 1391.

17 Mustafah Dhada, "The Liberation War in Guinea-Bissau Reconsidered," *Journal of Military History* 62 (July 1998): 578. The PAIGC government never saw fit to restart these industries after the war and left it to outsiders to help. Please see Rui Hortelão, Luís Sanches de Baêna, and Abel Melo e Sousa, *Alpoim Calvão, Honra e Dever* [Alpoim Calvão, Honor and Duty] (Porto: Caminhos Romanos, 2012), 425–443.

18 José Nico, "Pedaços das Nossas Vidas, III – O Rosto do Inimigo" [Pieces of Our Lives, III – The Face of the Enemy], *Mais Alto* 363 (September-October 2006): 21-22.

19 Comissão para o Estudo das Campanhas de África, *Resenha Histórico-Militar das Campanhas de África (1961–1974), 1º Volume, Enquadramento Geral* [Historical-Military Review of the African Campaigns (1961–1974), Volume 1, General Situation] (Lisbon: Estado-Maior do Exército, 1988), 260.

Rice-growing dikes and a PAIGC canoe on the Quitafine Peninsula; the enemy mixing with the population, as seen from a Dornier DO-27 observation aircraft. (Source: Personal archive of José Krus Abecasis)

Partial view of Bissalanca in 1961. (Source: Força Aérea Portuguesa)

by planting land mines and felling trees in the roads.[20] As trained troops returned from communist China and weapons arrived from the communist bloc countries via Conakry, strengthened cadres moved into the southern districts of Catió and Fulacunda. By late 1962, there were sufficient numbers of both troops and arms to begin the PAIGC offensive.

Meanwhile because of PAIGC developments and the Portuguese unpreparedness to address them locally, Portugal decided in July 1961 to reinforce Guiné with a flight of eight North American F-86F Sabre jet fighters from the *metrópole*. These were to fly directly to Africa with refueling stops in the Canary Islands and at Sal (AT 1) in the Cape Verde Archipelago before proceeding directly to Bissalanca (AB 2, later BA 12). This transit, codenamed *Atlas*, pressed the range limits for the F-86F and was impressive in its planning and execution.[21] All eight aircraft were safely on the ground in Guiné by mid-August; however, their deployment outside of the NATO area of operations, contravened the agreement on their use with the United States, which had supplied them. This controversy would dog their presence in Africa until their pressured return to Europe in late 1964.

This choice of F-86Fs was a strange one in that the aircraft was ill-suited to either counterinsurgency or operations from a remote and spartan environment, conditions that characterized Bissalanca and its surrounds in classic fashion. The fact of the matter was that little else was immediately available to meet the urgent need for air support. The F-86F could carry a far heavier bomb load with higher performance than the handful of T-6Gs at Bissalanca, and the FAP needed the more substantial aircraft to execute its mission.

In the spring of 1961, Bissalanca was the only important aviation infrastructure in Guiné. It had a decrepit runway of almost 6,000 feet, and the FAP occupied a modest hangar next to the civil terminal. Prior to the arrival of the F-86Fs, there were at Bissalanca two C-47 Dakotas, two Austers, and four T-6G Harvards; however, the operational activity of these aircraft was for all practical purposes symbolic. The arrival of the Sabres to this backwater highlighted the enormous deficiencies in the airfield. The first order of business was to resurface the landing strip, as it had been degraded by the constant landings of the DC-6s, C-54s, and Noratlas heavy transports. The F-86Fs were consequently ingesting pieces of cement and loose pebbles, and this caused foreign object damage (FOD) to the precious J-47 jet engines. In addition to FOD sweep-downs of the runway and taxi areas, the ultimate answer was a small mechanical broom to maintain clean jet intakes. While paving the runway and taxiways, the runway was extended 1,000 feet and lighting added. By October night landings could be made; however, the runway lights would on occasion operate

20 Jesus Bispo, 1391.

21 José Manuel Correia, "'Atlas' Nome de Código: Os F-86F 'Sabre' na Guiné, 1° Parte" [Codename "Atlas": the F-86F Sabre in Guiné], *Mais Alto* 357 (September–October 2005): 31–38.

intermittently because of the effects of humidity on the electrical system, and the subsequent darkness would heighten pilot anxiety. On the positive side, the paved taxiway gave the pilots an alternative to the main runway in an emergency landing, if the runway were fouled.[22]

The terminal radio aids to navigation were sparse and unreliable. There was a single radio beacon in poor condition with its antenna exposed and vulnerable to enemy action. Metrological conditions often interfered with its operation, and it could cease broadcasting with no warning. Instrument flight (IFR) was thus impossible. Air traffic control was haphazard, and local air traffic might or might not contact the tower. Hence it often came as a surprise to the F-86F pilots on approach to discover an unannounced intruder in the pattern with potentially disastrous consequences. Not infrequently the jets had to abort landings because of these. There were other problems in runway safety, as in one instance a native woman wandered onto the runway, and several days later a herd of cattle appeared on its downwind apron. Both of these incidences caused aborted landings. A complete review of the air traffic situation resulted in the addition of signal flares and new UHF and VHF transceivers for the tower controller to organize and direct air traffic.[23] Unfortunately the VHF tower radio could not be heard beyond 60 nautical miles, and all communications were frequently interrupted by equipment failure due to the high heat and humidity.[24]

In flying the initial F-86F familiarization sorties in late August, pilots were frustrated by the severe metrological conditions that limited their training. During these flights it was also discovered that the air navigation charts were inaccurate, and as a result, it was easy to violate Senegalese airspace along the northern border. Other operating problems centered on a series of bird strikes, a lack of oxygen to supply the aircraft systems, and other minor but debilitating failures that disrupted operations. These were solved as the deployment progressed, and by the end of the rainy season in October aircraft readiness stood at 100 percent of the eight F-86Fs.[25]

The radius of action of the F-86F was such that it was able to undertake reconnaissance and frontier patrol missions throughout the province. Both visual (RVIS) and armed reconnaissance (*reconhecimento armado,* RCAR or RECA) fell into the category of "sovereignty flights" (*voos de soberania*), and these comprised the majority of missions flown by Detachment 52 throughout 1962.[26] Rotation of pilots between the parent squadron in the *metrópole* and the detachment occurred on a regular basis.[27]

22 Ibid.

23 Military UHF or Ultra High Frequency radios operate in a bandwidth of 225–400 MHz. Military VHF or Very High Frequency radios operate in a bandwidth of 138–144 MHz.

24 Correia, "'Atlas' Nome de Código," 31–38..

25 Ibid.

26 Armed reconnaissance or *reconhecimento armado* appears to have been known by its acronym RCAR in Angola and RECA in Guiné.

27 Correia, "'Rotina' de Guerra," 34–43.

An F-86F Sabre being armed on the ramp at Bissalanca for a mission. (Source: Força Aérea Portuguesa)

The F-86F, as used initially in the "sovereignty flights," soon transitioned into pre-targeted raids on known enemy concentrations. Its use in close air support was limited at first, as its radios were incompatible with those carried by ground forces. At the time only the T-6G could communicate with ground forces, which carried the ubiquitous PRC-10, a 10-tube UHF FM transceiver and the primary radio of an infantryman. Consequently, the F-86F could do only preplanned missions and had very limited flexibility after launching. This situation deeply affected the morale of the F-86F pilots, who were never able to understand clearly how effective their actions were in advancing the ground force objectives or to make adjustments in flight.[28] Nevertheless, the F-86F packed a big punch. Its normal armament configuration consisted of six .50-caliber machineguns, two bombs up to 1,000 pounds, and four 2.75-inch rockets. The machineguns might carry incendiary or fragmentation rounds. The bombs might be general purpose (GP) or napalm.[29]

The T-6G was used in the close air support of both naval patrols and ground forces. In the case of the navy, the T-6G would fly along the river route scouting ahead of the vessel to identify a potential ambush and destroy it. The Harvard was armed with two Browning 7.7mm machineguns with 150 rounds each, one mounted beneath each wing. These had been taken from the Spitfires and Hurricanes that had been struck

28 Ibid.
29 Ibid., 42–43.

from the FAP inventory and adapted to the Harvard by the aircraft rework center at Alverca.[30] Later when heavier firepower was needed, the T-6G carried pods of 37mm rockets with fragmentation warheads. It was able to deliver well-directed firepower in support of land forces and conduct armed reconnaissance against PAIGC bases and encampments. It was lighter and far less potent than the F-86F but very flexible and could remain aloft for upwards of three hours with careful fuel management.

The first aircraft loss in Guiné occurred on 29 May 1962, when a T-6G with two pilots flew low over the Corubal River to inspect some canoes. Later on 17 August, as an F-86F returned from a mission and encountered a violent rainstorm over Bissalanca, the pilot attempted several landings without success. When his fuel was down to 200 pounds (about 33 gallons), he tried a final time, landing halfway down the runway and overrunning the barrier at the end to find himself inverted. He scrambled from the cockpit before the remaining fuel caught fire; however, the detachment was now reduced to seven aircraft. Of those remaining, maintenance crews were able to have only three to four ready for daily missions at any one time. This prompted a serious look at maintenance problems under tropical operating conditions at a forward airfield, and adjustments were made.[31]

In January 1963, the PAIGC launched its first serious and successful military operation against Portuguese troops in attacking Tite, a town and military stronghold on the south side of the Geba River opposite the capital of Bissau. A similar attack occurred on Catió, a major town and garrison on the Tombali Peninsula in the south. The PAIGC military successes were rapid and in the following months were characterized by well-executed ambushes coupled with vehicle mines planted in roadways and antipersonnel ones along their margins. The tactic was to stop a column of vehicles by disabling the lead one, whereupon the other halted vehicles would have to be abandoned by their drivers and passengers under the blistering fire of the hidden ambushers. These fleeing troops would seek refuge in the mined shoulders of the road to avoid the PAIGC gunfire and would suffer the effects of the antipersonnel explosives there. The conflict spread in the south fuelled by an elusive logistics tail emanating from Sansalé and Kandiafera, two PAIGC bases close to the border in Guinée.

As the PAIGC worked to consolidate its position in the southern littorals, the FAP made its first air attack in force on 4 April 1963 in a firepower demonstration on a strip of land next to a village that had been isolated by the enemy. The attack had been prompted by small arms fire taken by an Auster from the tiny settlement of Darsalame on the Cumbijã River in the south, which had boasted a flagpole that flew the PAIGC flag.[32]

30 Aleixo Corbal, Oliveira Simões, Fidalgo Ferreira, and Heitor Almendra, "2° Região Aérea, Lições Aprendidas entre 1961 e 1975" [2nd Air Region, Lessons Learned between 1961 and 1975], proceedings of a conference "Almoços da AFAP," January 2011, 9.

31 Correia, "'Rotina' de Guerra," 34–43.

32 Jesus Bispo, 1392.

At this time the PAIGC had no credible air defense capability, although this would eventually change, and had established relatively large bases within Guiné that were located next to native villages. Many of these were in PAIGC "liberated zones" in the more remote and inaccessible areas of the territory where Portuguese presence had been traditionally absent. Having such forces concentrated in prominent and easily identifiable bases was begging for targeting and air attack. Indeed, this strategy was a substantial PAIGC vulnerability and was heavily exploited in air operations during the early years of the war.

As air operations became increasingly aggressive, there were further aircraft losses. On 22 May 1963 Sergeant António Lobato and his wingman Sergeant Eduardo Casals departed Bissalanca in T-6Gs for their bombing objective on Como Island. With Como in view, they used hand signals to establish their intentions and then separated to operate independently against the target.[33] Lobato acquired the target, a large hillock in the center of the island covered with trees, and centered it in his gunsight. After a burst of machinegun fire, he recovered from the attack in a *chandelle* and felt the impact of anti-aircraft machinegun fire on the bottom of the wings. Even as all appeared normal, he requested his wingman to have a look and advise him of any damage. Casals approached the underside of Lobato's aircraft and, misjudging the relative speed of the two, contacted Lobato's propeller with his cockpit canopy. Casals became inverted with this collision and crashed, while Lobato experienced severe vibration from a bent propeller. He headed northeast toward Tombali, as he searched for a suitable site to ditch his damaged and failing machine. This he found in a *bolanha* that had been prepared for rice cultivation, and as he landed, the under-wing machinegun pods created such drag that the wings separated from the fuselage, and the fuselage began to roll over and over in the *bolanha*. When it stopped, all was silent. As Lobato extracted himself from the cockpit, he came face to face with a group of local Africans stupefied at the sight of a European emerging from a wrecked aircraft. The PAIGC in the area quickly found Lobato and took him first to Como, then Bedanda, and then south through the Quitafine Peninsula to Guinée, where he remained a prisoner for some seven years until rescued in 1970 by *Comandante* Guilherme Almor de Alpoim Calvão in Operation *Mar Verde*.[34]

On 31 May, an F-86F performing a bombing run in Bedanda was hit six times with antiaircraft fire. Jettisoning the external fuel tanks to improve performance, the pilot, Captain Barros Valla, attempted to reach Bissalanca only to experience a flameout over the Geba River at 500 feet altitude. He ejected successfully about two kilometers from the field, and the aircraft crashed into the river. The wreckage was recovered and deposited on the base waste tip.[35]

33 António Lobato, *Liberdade ou Evasão: O Mais Longo Cativeiro da Guerra* [Liberty or Evasion: The Longest Captivity of the War] (Amadora: Erasmos Editoria, 1996), 37–45.

34 Ibid.

35 Correia, "'Rotina' de Guerra" 36.

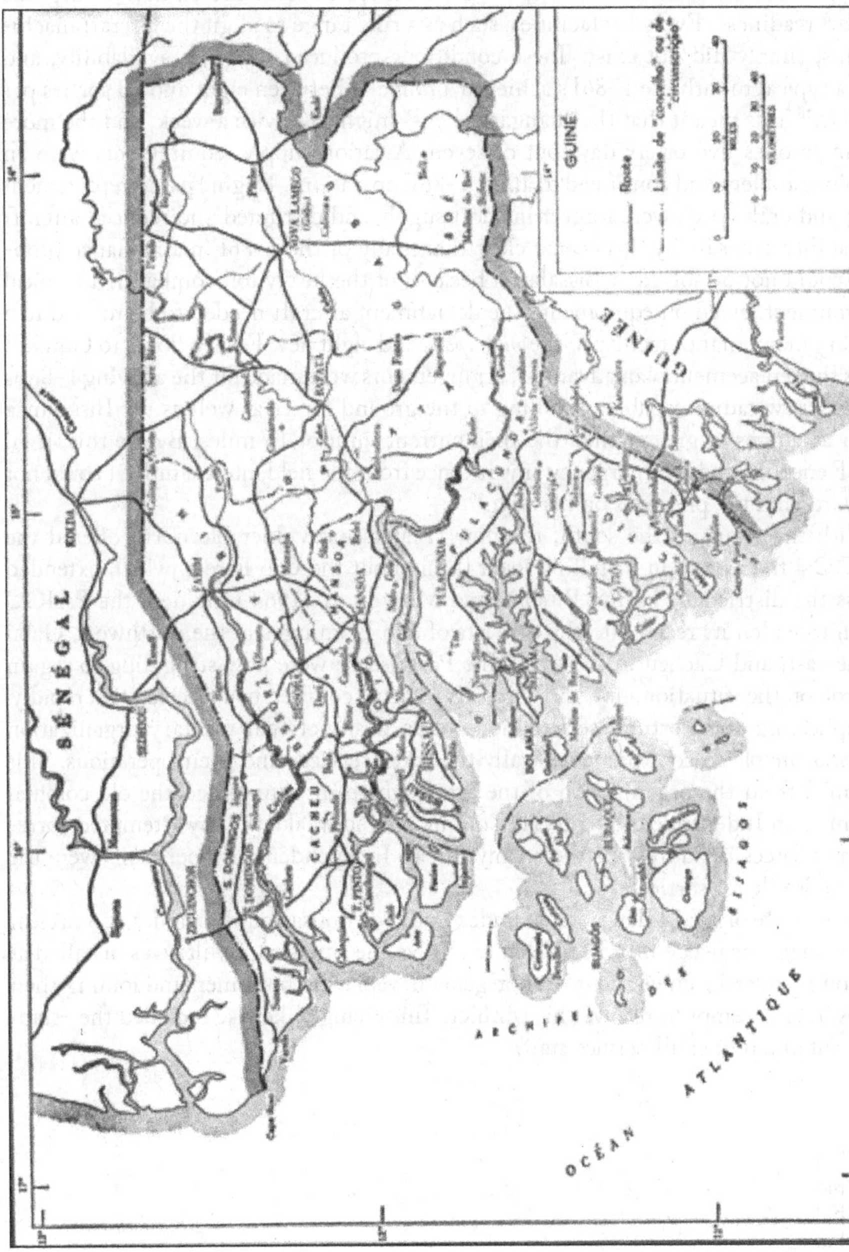

Map 8.1 Map of Guiné showing its various districts (*concelho* or *circunscrição*) (Source: Força Aérea Portuguesa)

The F-86F was a relatively sophisticated aircraft for the time and consequently required a great deal of maintenance attention to fly and fight to its full potential. Such maintenance was difficult in Bissalanca despite the best efforts of the ground crews, as a lack of facilities and spare parts conspired with the climate to degrade aircraft readiness. Everyday facilities, such as a rifle range to align the aircraft machineguns, simply did not exist. These conditions produced irregular availability, and over a typical month the F-86Fs achieved a range of between eight and 23 sorties per aircraft.[36] This meant that the "hangar queens" might fly twice a week, and the more fortunate ones five or six days out of seven. Aviation supply requirements were an ongoing problem and continued to affect F-86F operations. Engine parts, replacement tires, and brake pads were forever in short supply and disrupted and reduced aircraft availability accordingly. It became clear that many of the F-86F maintenance problems could not be solved at Bissalanca because of the heavy toll combat in a tropical environment levied on equipment. The detachment aircraft needed to be rotated to a primary maintenance facility in the *metrópole,* and eight new F-86Fs flown to Guiné.[37]

As the replacement plan advanced, arrangements were made for the arriving F-86Fs to carry new radios capable of talking to the ground forces as well as the Bissalanca tower at ranges of greater than the then current limit of 15 miles. Before this, if an F-86F encountered an emergency any distance from the field, quite simply it could not be heard, and the pilot was on his own.[38]

With the focus on the south, the now vacated and vulnerable north offered the PAIGC a freer hand in installing itself throughout the Oió Forest, which extended across the districts of Farim, Bissorã, and Mansoa. As 1964 unfolded, the PAIGC began to widen its reach into the districts of São Domingos in the northwest, Gabú in the east, and Cacheu in the west. The Portuguese were now struggling to regain control of the situation, for the intensity of the conflict had accelerated rapidly. Complicating the Portuguese situation was a dysfunctional military organization that was unable to coordinate centrally the three services and their operations. This stemmed from the organization of the army, which in Guiné used the old colonial system of an Independent Territorial Command, and weakened any attempted correlation of forces in addressing the enemy threat. Independent air operations were the normal profile for sorties.[39]

As the state of affairs deteriorated under Louro de Sousa and Rodrigues, it was clear that change was needed. In May 1964 to resolve the situation, Shultz was installed as the commander-in-chief and governor-general, replacing both men and joining their duties in an attempt to resolve this conflict. This change likewise included the establishment of a proper all-service staff.

36 Ibid.
37 Ibid.
38 Ibid.
39 Ibid.

Shultz too had his weaknesses, as while he was a man of great intelligence and good sense, he was very prudent and cautious. This approach lacked the needed forcefulness to fight successfully against an aggressive enemy. He felt comfortable in seeking a compromise on every decision and avoiding any conflict with his commanders. Decisions were made as a team, and this consensus leadership invariably fell short.[40] According to Calvão, the starkest example of Shultz's lack of aggressiveness was the capture and subsequent imprisonment of Lobato, who was held two to three months in southern Guiné before being taken across the border to Conakry. During this opportune time an operation to free him while he was still known to be in Portuguese territory was not even considered. There simply was no urgency to rescue any prisoners.[41] One can only imagine how this affected morale. The pilot community was small at the time, perhaps 20 pilots. The lack of a search and rescue component and the will to use it on top of the several deaths was a poor message to deliver to those on whom so much depended.

In 1963, as the PAIGC began its serious fighting, Air Force Intervention Zones (*Zonas de Livre Intervenção da Força Aérea* or ZLIFA) were established. These were essentially "free fire zones" in which aircraft were able to conduct operations independently of ground or naval forces and without their participation.[42] Before such operations were approved, the target areas were "cleaned" or "sanitized" by the army to prevent any accidental death or injury to innocent civilians. This entailed removal of known native village populations and their possessions, and these precautions were aimed at honoring the just war principle of non-combatant immunity.[43]

Even with its limitations in communication and the cumbersomeness of ZLIFA coordination leading to an alerted enemy, the F-86Fs proved extremely effective. This was particularly so during August, as they executed various assignments against PAIGC installations on the islands of Como and Caiar and in the bush adjacent to Tombali and Catió. The PAIGC had introduced heavy antiaircraft machineguns to project its encampments in these areas, and these needed to be eliminated. On 21 August the F-86Fs attacked and silenced two antiaircraft redoubts next to Incassol, and at the end of the month in the area of Cachaque Balanta, attacked various PAIGC encampments with napalm. Both of the latter were on the Quitafine Peninsula. The PAIGC complained about these successes and was able to reignite U.S. pressure to return the F-86Fs to Europe. George Ball, the Undersecretary of State, made this point on his August visit to Lisbon.[44]

40 Ibid.

41 Alpoim Calvão, "Ataque a Conakry," 509.

42 Neto, 316.

43 Luís Alves de Fraga, *A Força Aérea na Guerra em África: Angola, Guiné e Moçambique, 1961–1974* [The Air Force in the War in Africa: Angola, Guinea and Mozambique, 1961–1974] (Lisbon: Prefácio, 2004), 106.

44 Correia, "'Rotina' de Guerra," 37.

In the middle of September, F-86Fs attacked bases along the banks of the Corubal River in the area of Xitole, in Campeane, and in Caulaca, and at the end of the month, they provided close air support in Exercise Sabre, an operation in which they attacked PAIGC encampments in the region of Enxalé and Malafo and again on Como Island. In this case a Broussard acted as the airborne forward command post (*posto de comando avançado* or PCA) to overcome the lack of an air-ground radio in the F-86F and coordinated air-ground operations.[45]

Also during September and in response to the widening war, the FAP Chief of Staff conducted an inspection tour and determined that Bissalanca needed augmentation to form a more balanced and enhanced force structure. He recommended the following changes:

- 10-15 Alouette III helicopters (3 Alouette IIs currently assigned)
- 12 F-86F Sabre fighter jets (8 currently assigned)
- 20 T-6G Harvard trainer-fighters (8 currently assigned)
- 12 DO-27 observation aircraft (4 currently assigned)
- 3 C-47 Dakota transports (3 currently assigned).[46]

Additionally, he recommended an urgent resolution to the weak photographic reconnaissance capability, an expansion of the runway, and exterior lighting of the buildings to improve security. Finally, he assigned Lieutenant Colonel Manuel Barbeitos de Sousa as commander of Bissalanca. Barbeitos was an experienced F-86F pilot who had a comprehensive understanding of its gun and ordnance systems and their tactical employment. He became intimately involved in Detachment 52 operations from November onward and was an important factor in revitalizing the mission profiles of both the F-86Fs and T-6Gs and the adoption of new procedures for close air support.[47]

The shortage of spare parts for the F-86F engines and brake systems continued to plague operations. Even with these impediments, in mid-December the Sabres undertook a series of missions against targets in and around Darsalame, which defiantly continued to fly the PAIGC flag.[48] These operations in the closing months of 1963 were precursors to Operation *Tridente.*

Operation *Trident*

By early 1964 the southern border area with Guinée and the coastal areas around Catió and Cacine had become progressively infected with PAIGC insurgents, so

45 Ibid.
46 Ibid.
47 Ibid., 39.
48 Ibid.

much so that a serious Portuguese challenge was long overdue. The navy in particular worried that without such action its access to Catió would be closed.[49] Accordingly a jointly executed Operation *Tridente* was planned to combine army, navy, and air force resources to isolate and sanitize the familiar islands of Caiar, Como, and Catunco, which were bounded by the Atlantic Ocean and the Cumbijã, Caiar, and Cobade Rivers. Caiar Island had been the scene of an earlier December 1962 action, which had been purely an infantry sweep and had had no lasting effect. The expanded area of operations for *Tridente* now covered nominally some 210 square kilometers of land, about 170 of which were actually usable because of the tides.

The three islands were characterized by large tracts of rice patties with their intricate dike systems and alternatively by vast areas of very dense and impenetrable grasslands and swamp forests. These latter areas provided ideal hiding places for insurgents, as troops found passage through such terrain very difficult going. Clearly the operation would break new ground, for it was to be the first such full all-service operation in any of the theaters.[50] It would last for more than two months and would be supported by a substantial portion of military resources in Guiné.[51] This would also be the first and largest of a number of grand operations that pre-staged the Shultz years with their heavy military focus.

Como and its adjacent islands were considered important both to the insurgents and to the Portuguese command for obvious reasons. From the point of view of the insurgents, it was a perfect staging point, for it was remote, and this isolation brought with it security. It could be supplied by sea from the sanctuary of Guinée only about 25 miles distant, a few hours at most by boat, and it provided a springboard for PAIGC subversive operations further into Portuguese territory. From the Portuguese point of view it was perceived as a festering sore of insurgent activity and an unwelcome subversion of its population in a prime rice-growing and a very modest cattle-grazing region. The insurgents were diverting the rice crop and livestock to their own ends and establishing themselves as a relatively permanent fixture within the population. Portuguese intelligence estimated the PAIGC force on the three islands to be about 300, which number included about 15 military advisors from Guinée.[52]

The operation had five goals:

- To secure the Portuguese riverine lines of communication in the south.
- To protect the population from the PAIGC subversion and intimidation.
- To reinforce Portuguese administrative authority over the island.

49 Guilherme Almor de Alpoim Calvão, correspondence with the author, 9 May 2005, Cascais.

50 Aniceto Afonso and Carlos de Matos Gomes, *Guerra Colonial* [Colonial War] (Lisbon: Notícias, 2000), 79.

51 José Alberto Lopes Carvalheira, "Acção da Marinha em Águas Interiors (Guiné)," 7; Telo, 585.

52 Ibid.

- To improve the economic well-being of the population and the area.
- To inflict a grave loss of credibility and prestige to the enemy.[53]

The concept of operations was, first, to isolate the area by securing the surrounding waterways of the Caiar, Cobade, and Cumbijã Rivers, next, to disembark and support the land forces with naval and air firepower in their sweep for enemy troops, and last, to secure the maritime and riverine lines of communication so that a permanent presence could be maintained in the area. To this end the navy established a *cordon sanitaire* around the islands and, with its forces and the aid of aerial reconnaissance in the form of aged but capable P2V-5 Neptunes from Sal, would prevent insurgent escape. The land forces would sweep the three islands and kill or capture the fleeing insurgents, while the navy destroyed their boats.

The operation was divided into three phases. The first began on 15 January 1964 with an amphibious landing at various sites on the islands of between 1,100 and 1,200 troops, a substantial commitment of available troops. The landings were given close air support by the T-6Gs, which attacked the bases of Cauane and São Nicolau with repeated rocket runs, and by army artillery from its support base in Catió some 5 to 10 miles distant, depending on where one was on the island group. As one might suspect, there was a challenge to the assault, and the patrolling Neptune identified a number of canoes, which were destroyed by the navy.

Armed natives appeared opposite the village of Darsalame on the left bank of the Cumbijã and were dispersed by naval fire. The Neptune later identified a substantial group of insurgents near the village of Cametonco on Catunco Island about 400 meters inland from the Cumbijã riverbank. This group fired at both the Neptune and later a patrol boat before fleeing. And so it went over the several days of the first phase and into the second, with the landings encountering sporadic resistance from the insurgents and their local recruits while they tried to flee the islands in canoes or on foot or to hide in the swamp forests. The second phase occurred between 17 and 24 January and was centered in patrolling the three islands to establish a wide presence, to flush out insurgents, and to gather intelligence.

Following the landing and establishment phases, the third phase lasted from 24 January to 24 March and involved periods of intense fighting throughout the island of Como. The primary fighting occurred in the area of Uncomené near the river of the same name that separates the islands of Como and Caiar and in the Cachil swamp forest between the islands. The F-86Fs flew daily sorties, and by mid-February the lack of tires had grounded four jets. Before the conclusion of Como operations, one F-86F had inoperable guns and another cracks in the nuts fastening its right wing to the fuselage. Even so, the third phase saw dozens of F-86F sorties in support of ground operations, a tribute to the maintenance crews.[54]

53 Ibid., 6.
54 Correia, "'Rotina' de Guerra," 40.

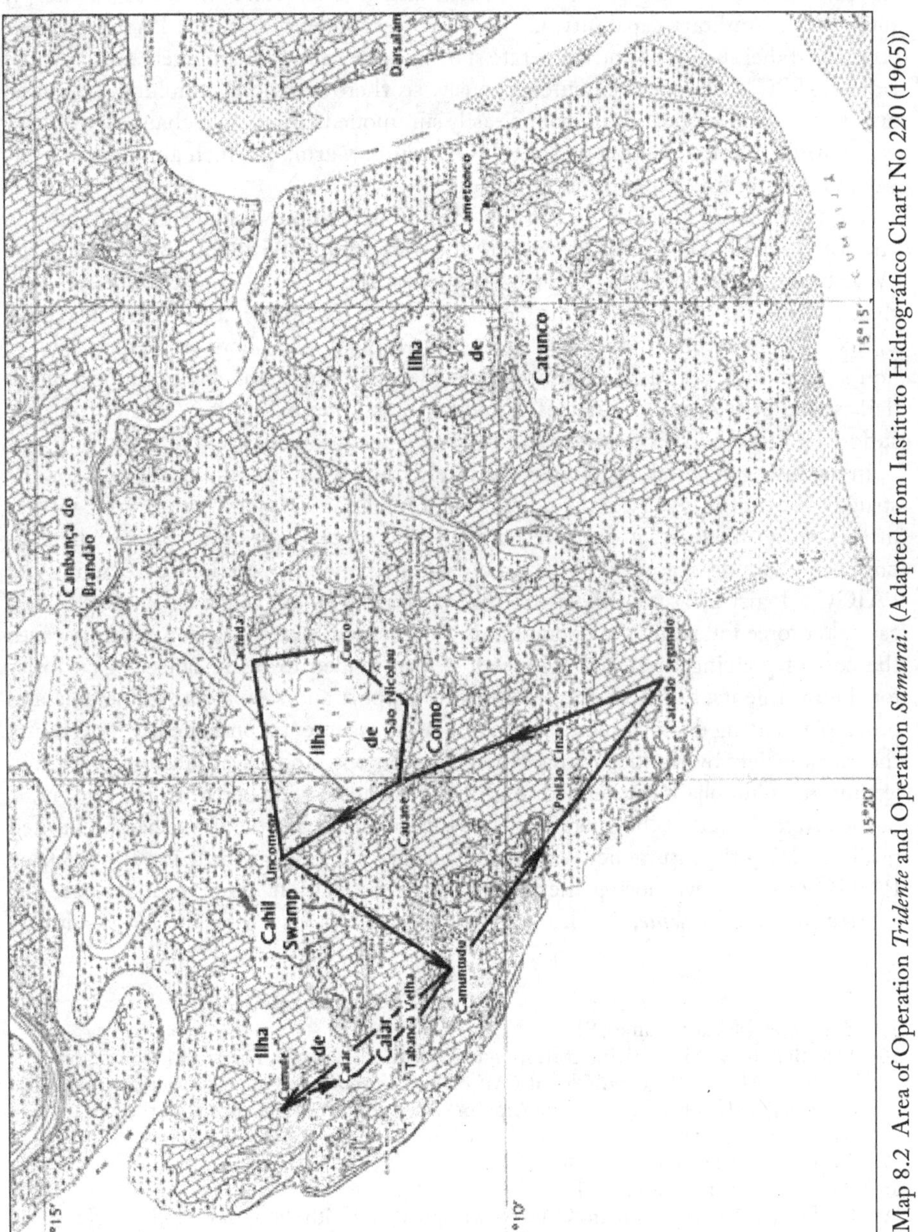

Map 8.2 Area of Operation *Tridente* and Operation *Samurai*. (Adapted from Instituto Hidrográfico Chart No 220 (1965))

The insurgents suffered substantial losses in 76 killed, 15 wounded, and 9 taken prisoner, a casualty rate of around 30 percent.[55] These numbers do not, however, consider unknown casualties from artillery fire and close air support. These PAIGC troops were "well armed, battle-hardened, and trained," and their loss cut deeply into PAIGC military capability. Captured correspondence between PAIGC leaders confirmed their besieged and desperate state and the need for reinforcements.[56] At the time PAIGC units operated autonomously, so there was no overarching command and control, and support could not be easily summoned. This would change within the year with the reorganization of the PAIGC military arm, of which antiaircraft units would be a prominent and important feature.

The FAP played no small part in this PAIGC reduction. By the end of March the aircraft from Bissalanca had flown 851 missions and 1,105 flight hours during the "War of Como." The T-6Gs logged 141 sorties; the F-86Fs, 73; the Dorniers, 180; the Austers, 46; the Alouette IIs, 323; the P2V-5 Neptune, 16; and the Dakotas, 2. Six were hit by enemy fire, and one T-6G was lost. The pilots on average flew 40 hours per month, and the maintenance crews worked night and day in support of the daily strikes, performing "miracles of good will and improvisation."[57] The Portuguese suffered 9 killed and 47 wounded in the 71-day operation, all ground forces.[58]

In an insurgency, taking territory is rarely critical and usually irrelevant, for the struggle is over the loyalty of the population. Operation *Tridente* was a Pyrrhic victory in this sense; however, it did assert Portuguese sovereignty over an area where its presence had long ceased to be felt and disrupted enemy operations by denying the PAIGC a logistic base. While a contingent of army troops remained in Cachil as a caretaker force following the operation, it was eventually removed as not being worth the cost of policing the terrain. Indeed, it also tended to be an object of curiosity for the insurgents, as it was isolated from its support and seen as the vulnerable and proverbial "sitting duck."[59] Increasing numbers of insurgents consequently moved into the vacuum left by the departing Portuguese troops, and naval craft entering and leaving the Cumbijã River tended to give the islands a wide berth. In the end, none of the original goals of the operation were fully realized. Portugal was having limited results treating the insurgency with a conventional solution, as while it was punishing PAIGC troops, it was losing the loyalty of the population. As if to underscore the shortcomings of *Tridente*, Amílcar Cabral held his famous Cassacá Congress during

55 Afonso and Matos Gomes, 81.

56 João Bernardo "Nino" Vieira, correspondence with Rui Demba Djassi and Domingos Ramos, 3 March 1964, undisclosed PAIGC base in southern Guiné, typewritten transcript furnished by Guilherme Almor de Alpoim Calvão in correspondence with the author, 9 May 2005, Cascais.

57 Correia, "'Rotina' de Guerra," 40.

58 Afonso and Matos Gomes, 81.

59 Guilherme Almor de Alpoim Calvão, correspondence with the author, 22 June 2005, Cascais.

February 1964 at his base near the settlement of that same name on the southern border in what he declared to be a "liberated zone." Its purpose was to reorganize the PAIGC war effort and consolidate the independent commanders into a national army, the FARP.

While the Como Island area would remain important as a PAIGC springboard into the center of Guiné for the next several years, it would eventually lose its importance because of its peripheral location. Alpoim Calvão confirmed this when he returned to the island five years later in Operation *Torpedo* and found only weak resistance.[60] The PAIGC persistently moved onto the peninsulas of Quitafine and Cantanhez adjacent to the Guinée border and proceeded to isolate the local market towns of Catió and Bedanda.[61] The ports of these besieged centers, however, remained the secure link to Bissau via the waterways. These moves positioned the PAIGC advantageously to subvert the interior of Guiné, as its lines of communication from its Guinean sanctuary through these two peninsulas became increasingly safe from Portuguese interdiction. For the Portuguese, this proved to be one of the grand operations that from time to time punctuated what might have been an otherwise sound and patient strategy of counterinsurgency. This type of large conventional maneuver seemed to hold a fatal attraction for local commanders, as it appeared to be a means to end the war quickly. In fact it was just the reverse. It inevitably fell short, for while many arms were generally found, insurgents captured or killed, and bases destroyed, the enemy tended to melt into the terrain or into the local population. The operation could not be concealed because of its large scale, and other areas of Guiné were denuded of troops to support it, a situation that left the population elsewhere vulnerable. It was not a sound counterinsurgency operation and invariably yielded disappointing results at great expense. The constant pressure of small patrols as part of a *tâche d'huile*-style strategy engaging the population, for instance, would have been a far more effective approach in its low-key, low-cost methods.[62] True, the PAIGC would require substantial firepower to stop, but it had also successfully infiltrated the population, something the Portuguese at present were neglecting.

Following *Tridente*, operation of the F-86Fs became increasingly difficult, as need for their constant use, the consequent heavy maintenance demands, and the scarcity of spare parts conspired against their availability. Faced with underutilized F-86F pilots, the commander of Bissalanca decided that the ten of them must be able to fly one of the other types of aircraft assigned to the base. This would have been the T-6G or the

60 Guilherme Almor de Alpoim Calvão, correspondence with the author, 9 May 2005, Cascais

61 Afonso and Matos Gomes, 81.

62 The *tache d'huile* technique was perfected by Louis-Hubert Lyautey during his Moroccan campaigns of the early Twentieth Century in which he progressively reoccupied hostile territory by offering Arab tribes French protection and access to commerce and social help. His expansion on a broad social-military front with the notion of building an Arab nation succeeded where his predecessors had relied on subjugation by the military column.

DO-27.[63] The situation with the F-86F and the T-6G, which proved slow and vulnerable in combat, also prompted thinking on their replacements. Available alternatives were the North American T-28 Trojan (series), the Chance Vought F4U-4 Corsair, and the Douglas AD-4 Skyraider, all of which were available from French surplus stocks with the 1962 conclusion of the Algerian war. Another alternative was the purchase of 65 Sabres manufactured in Canada and sold to the West German Luftwaffe.[64] As Germany had replaced these with the Lockheed F-104 Starfighter, Portugal became a willing buyer for the now surplus aircraft. However, complications arose with the sale, as both the United States and Canada sought to honor the UN arms embargo on the African conflict. The search for a replacement aircraft continued, and ultimately Germany offered to sell Portugal 40 of its Fiat G-91/R4 attack aircraft, which would be free of UN embargo complications.[65] These Fiats had been retired in 1965 and would be available from storage for deployment, as soon as they were refurbished, and Portuguese pilots trained.[66]

As 1964 advanced, F-86F operations began to wind down, and its final mission was flown on 20 October. The aircraft were subsequently dismantled and put aboard ships for transport to the *metrópole*, the last one departing on 22 December. For the next 18 months the intrepid T-6G would be the primary fire support platform, until the Fiats arrived in May 1966.[67]

PAIGC Antiaircraft Developments

From the beginning of the conflict and for its first several years, the FAP enjoyed unhindered freedom of action. There was little or no anti-air threat, and aircraft losses were largely those of pilot error. For the PAIGC insurgent, this near helplessness in the face of Portuguese airpower was dangerous both to its military operations and to the population it was attempting to subvert. A stream of invective on the "injustices" perpetuated on the "innocents" from the air emanated from PAIGC mouthpieces, as if such acts were somehow illegal in war. There was certainly a conscientious effort on the part of the Portuguese to discriminate between the enemy and the non-combatant and to honor the latter's immunity; however, the PAIGC was unquestionably engaging in impermissible deception when it came to the laws of war and then claiming that "atrocities" were being committed.[68] This obscuring of conditions that enable moral

63 Correia, "'Rotina' de Guerra," 40.

64 José Matos, "A Saga do Sabre Mk 6, 1ª Parte" [The Saga of the Sabre Mk 6, Part 1], *Mais Alto* 394 (November–December 2011): 37–41; and "Saga do Sabre Mk 6, 2ª Parte" [The Saga of the Sabre Mk 6, Part 2], *Mais Alto* 395 (January–February 2012): 37–43

65 Correia, "'Rotina' de Guerra," 41.

66 Matos, Part 2, 43.

67 Correia, "'Rotina' de Guerra," 43.

68 Alberto R. Coll, James S. Ord and Stephen A. Rose, eds., *Legal and Moral Constraints on Low-Intensity Conflict* (Newport: Naval War College, 1992), 153.

restraints on the use of force undermines those same restraints. The Guinean native became the true sacrificial lamb when PAIGC fighters intermixed with him, and it became virtually impossible to fight without causing substantial civilian casualties and damage. This intermixing was a well-established tactic that Cabral used mercilessly and deliberately to alienate the population from the Portuguese and taint the reputation of any intervening allies.[69] Certainly the PAIGC reaction reflected its respect for airpower, and it would have to do something about FAP success, if it were to succeed.

During the 1961–1963 period there were methods that the PAIGC used with varying degrees of effectiveness to frustrate the dominance of airpower. The weather in Guiné, for instance, was predictably deplorable for flying in the rainy season, and this factor offered opportunity for relatively unhindered PAIGC operations. PAIGC operatives learned camouflage techniques and concealment practices that could only be penetrated by clever and alert aerial reconnaissance. The PAIGC difficulty with these passive methods was that the FAP reconnaissance pilots were learning to recognize such measures as rapidly as the PAIGC developed its skills. Such a development was occurring in Angola, and there was every reason to expect similar success in Guiné, and later Mozambique. For the PAIGC, respect for the aircraft remained vital to survival.

Passive measures were but a small menu of potential options, and the PAIGC went through several phases of antiaircraft defense. In the beginning there was only indiscriminate firing of individual small arms. Next the Soviet 7.62mm machinegun appeared on a wheeled mount, but this was only effective against aircraft flying at low altitudes. Cabral was soon open to new methods, particularly from the Chinese. By the close of 1963, his forces had received a more potent weapon of Soviet design and Chinese manufacture: the Degtyarev-Shpagin (DShK) 12.7mm heavy antiaircraft machinegun. This gun had the capability of producing major effects but was easily seen from the air, as it was habitually situated in the open. The DShK was quickly followed by the ZPU-1 14.5mm heavy machinegun. Both of these weapons were wheeled to enable mobility, and the transport mechanisms converted to stable support systems for use. The ZPU-1 could be broken down into several 80-kilogram sections for transport over rough ground, and it was soon followed by the less mobile but more powerful twin- and quad-barrel models, the ZPU-2 and ZPU-4 respectively. The latter weapons were ponderous and not particularly mobile. In fact, they were only used in readily identifiable open emplacements ringed with earthworks and sited in zones where the PAIGC considered itself impregnable, areas straddling the southern border.[70] Conversely, the DShK and ZPU-1 were found at sites well into the interior because of their mobility.

Despite PAIGC propaganda that it was downing FAP aircraft on a regular basis, such was not the case. While the PAIGC gunners attained many hits, none were fatal

69 Ibid., 194.
70 Jesus Bispo, 1394.

to either the aircraft or their crews. For example, in 1966 hits were recorded on eight DO-27s, six T-6Gs, one Alouette II, two G-91s, and one Dakota. In 1967, hits were sustained on five DO-27s, five T-6Gs, three Alouette IIIs, and two G-91s. The most serious case did not occur until five years into the fighting in 1968, when a G-91 was hit, and the pilot successfully ejected.[71]

From the beginning the Chinese and Soviet advisors taught the PAIGC gunners well, and they proved to be quite fearless in using these new weapons. They were able to hold their fire from concealment until the last moment, when the chances of a hit on a passing aircraft were increased, and their position would not be betrayed. The PAIGC gunners proved quite gifted and earned respect, particularly among the T-6G pilots, who were now at increased risk below 8,000 feet. Colonel José Krus Abecasis, Commander of the ZACVG, observed that he had "on occasion admired the courage with which the antiaircraft gunners stood their ground, maintaining their positions and shooting continually, when we were enveloping them in metal, indifferent to the risks that they ran."[72]

In November 1965, Abecasis faced a substantial threat in a rapidly growing PAIGC anti-air capability. He reached this conclusion on personally observing the intensity of fire and the luminosity of the projectiles from grouped gun emplacements on the southern border shooting at his aircraft and thought that the positions must be equipped with 20mm guns or their equivalent.[73] This observation was likely the result of seeing the ZPUs in action, as the 14.5mm ZPU-1 had a rate of fire of 600 rounds per minute per barrel or 2,400 rounds, if it were a quad-mount. This would have been enough to impress Abecasis, particularly as he was the target. The larger problem now was that the PAIGC begin to protect its forces successfully from airpower and deny Portugal its largely unfettered hand in the air.

Operation *Resgate*

Operation *Resgate* (Rescue) was in the words of Abecasis instituted to "rescue our freedom of action in the skies of Guiné" and was the first of a series of all-resource air operations.[74] In the final analysis, the enemy had no aviation, and yet he was proclaiming the Cantanhez Peninsula impregnable. Cantanhez was the equivalent of the Barrio de Alvalade, a neighborhood in Lisbon, and such a boast was viewed as a gauntlet thrown before the FAP.[75] The dense vegetation in which the enemy hid with his antiaircraft weapons was not impossible. It simply required an imaginative plan to

71 Ibid.

72 José Krus Abecasis, *Bordo de Ataque, Memórias de uma Caderneta de Voo e um Contributo para a História.* [Leading Edge: Memories from an Aviation Log Book and a Contribution to History] (Coimbra: Coimbra Editora, 1985), Volume II, 610.

73 Ibid., 487.

74 Ibid, 490.

75 Ibid.

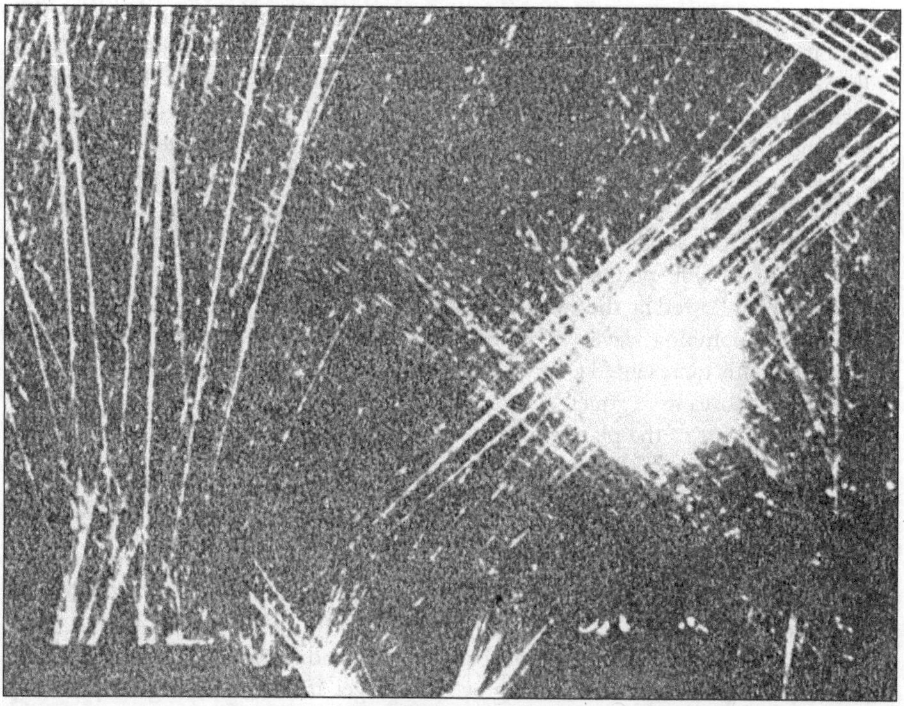

Night antiaircraft fire at a Colonel Abeçasis's aircraft from several ZPUs along the southern border of Guiné. The large bright spot is a magnesium parachute flare.
(Source: Personal archive of José Krus Abecasis)

neutralize the threat, and the FAP responded to the challenge. In Abecasis's words, "Moreover, we would respond this time with all of our firepower, and the FAP would appear with an offensive capacity of which the enemy never dreamed....We would rescue our freedom of action using a powerful deception."[76] As we shall see, the term "all of our firepower" was quite limited and required considerable inventiveness to employ with success.

The initial phase of the FAP plan was to provoke the PAIGC antiaircraft artillery with a night test flight by a C-47 Dakota, and on 6 December the decoy with its running lights lit flew over the target. The PAIGC gunners fell for the ruse and responded. This enabled the planners to pinpoint the enemy gun emplacements and develop a scenario to destroy them.

The plan called for the two C-47s at Bissalanca, which had been modified to carry 50-kilogram fragmentation bombs and 60-liter canisters of napalm for horizontal night bombing, to lead the way. The Dakotas and their crews had developed

76 Ibid.

a proficiency in using an adapted bombsight and magnesium parachute flares to illuminate targets and drop their ordnance with deadly accuracy.[77] A single Dakota was to fly over the earlier identified targets, again with its running lights lit, and drop a series of illuminating flares that would burn for two minutes. The intense light from the magnesium flares would not only reveal the guns but also ruin the night vision of the PAIGC gunners. When the flares were spent, immediately two Lockheed P2V-5 Neptune patrol bombers from the detachment at Sal were to follow in close sequence to drop their 750-pound bombs selectively. Each Neptune would continue to make passes, until its 8,000-pound capacity bomb bay with its ten bombs was exhausted. This would be followed by the two Dakotas with their napalm canisters. Targeting for the successive bombing waves would be facilitated by the fires sparked in the initial explosions.[78] This represented a very plausible scenario in light of the available aircraft now that the Sabres were gone.

Abecasis, author of the plan, worried enormously about collateral damage and the fact that punishment of the enemy could well translate into numerous losses in the civilian population under PAIGC control. As he observed, "It was a war with an inhuman and cruel face."[79] It was also a price that the enemy was willing to pay for the usurpation of Portuguese territory—an enemy that had had no hesitation in slaughtering the resisting youth of Farim, a cruel act of terror in the interests of PAIGC designs on the country.[80]

"H" hour was fixed at 2000 on 17 December 1965, and the operation went exactly as planned with the lead Dakota arriving over the target at 2100, drawing the predicted fire, and releasing its flares. The entire sky was now lit with antiaircraft fire as well. The magnitude of this reaction was quite a surprise to the pilots, as it was two or three times the volume of fire that had been observed in the earlier probing sortie. Indeed the entire area defined by the triangle of Cafine–Santa Clara–Catesse vomited high-caliber metal. It was clear that in the short period of 11 days the PAIGC had built additional antiaircraft capacity. As darkness fell after two minutes of magnesium daylight, the Neptunes hit their initial point (IP) at 2105 for their first bomb run. Following this they circled for 50 minutes, as they made successful run after run. When the Neptune loads were exhausted, the Dakotas delivered the *coup de grace* with their napalm drops. The enemy was given no rest during the entire night, for the aircraft returned to Bissalanca to rearm and make additional strikes at midnight, 0200, and 0500. In all some 30 tons of bombs were dropped during the sequence of

77 Mário Diniz and Luís Proença, "Dakota No 6157: Douglas C-47A Skytrain, Três Décadas de História nos Céus de Três Continentes, 2ª Parte" [Dakota No 6157: Douglas C-47A Skytrain, Three Decades of History in the Skies of Three Continents, Part 2], *Mais Alto* 390 (March–April 2011), 40.

78 Krus Abecasis, 492.

79 Ibid., 491.

80 Ibid.

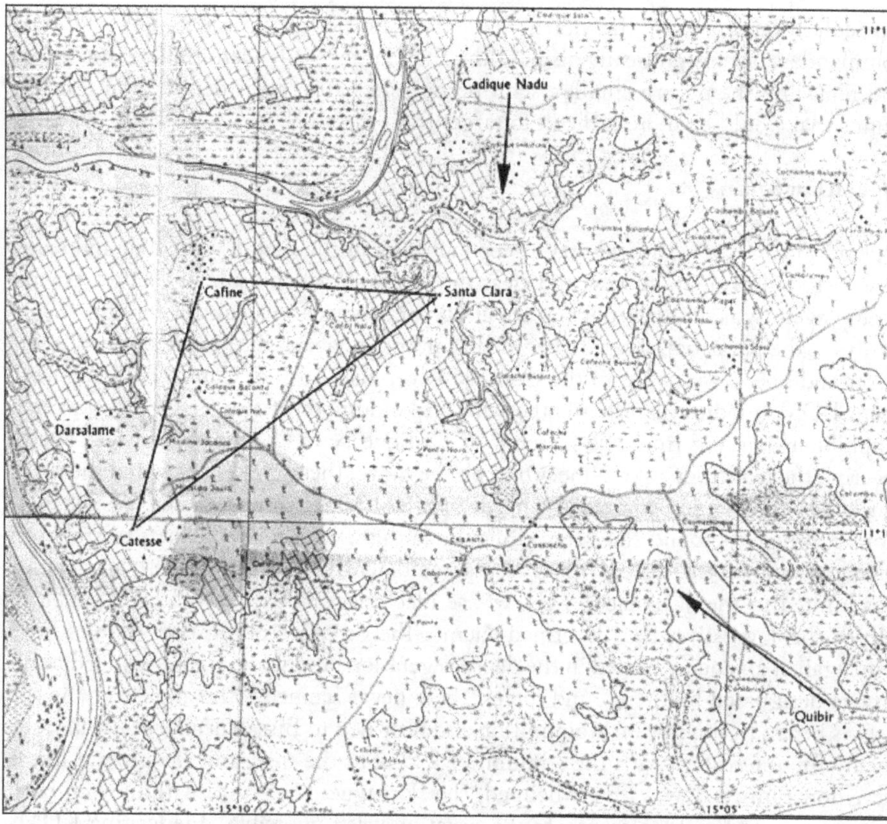

Map 8.3 Area of Operations *Resgate* and *Safari* on Cantanhez Peninsula. (Adapted from Instituto Hidrográfico Chart No 220 (1965))

bombing waves that evening.[81] The ground crews at Bissalanca worked through the night to rearm the bombing force and prepare the aircraft for their serial missions. Bissalanca was now truly at war. At dawn a flight of T-6Gs flew over the site on an armed reconnaissance mission, and this would be followed by an infantry sweep with close air support to rid the Cantanhez of enemy forces and exploit the success of Operation *Resgate*.

Resgate and the new FAP initiative to identify and destroy PAIGC bases systematically forced changes in enemy behavior. While much of the target identification was done by scouts and informers, the information from these sources was confirmed through aerial reconnaissance, again a threat from the air. FAP operations were

81 Ibid., 494.

particularly effective in the north.[82] As the destruction proceeded, the PAIGC commanders began frantically dismantling their large, easily identifiable installations and replacing them with smaller, semi-permanent bases, arms depots, and training camps that would be more difficult to target from the air.[83]

Operation *Safari*

Operation *Safari* was the infantry sweep launched in the first days of 1966 following *Resgate* to follow on its success and clear the Cantanhez Peninsula of PAIGC troops. It employed two combat groups or about 60 paratroops and a contingent of *fuzileiros* to do so. The paratroops were to begin at Quibir (Canábria) in the east on the Cacine River and sweep westward to the Cumbijã River just south of Catesse. The *fuzileiros* in an LDM (medium landing craft) with a like number of troops were to make an amphibious landing from the Cumbijã at Cadique Nalú and work their way south, meeting the paratroops at Cafine. Once on the move, these two forces found the roads blocked with cut trees to delay their advance and establish enemy ambush settings. Both contingents found the going difficult, as there were many pockets of stubborn enemy resistance that forced them to assume defensive positions. Close air support was limited to the available T-6Gs, whose machineguns packed too little punch to silence the aggressive, heavily-armed enemy. Their many strafing runs were of little avail against an enemy buried in the dense vegetation, and while they continued to hit targets designated by the ground forces, there never seemed to be any effect. As the T-6Gs began to run low on ammunition and returned to Bissalanca for rearming, the enemy fire intensified with the addition of RPGs to the mix, and the ground forces were now pinned down. Overwhelmed, the Portuguese began to withdraw on both the eastern and northern fronts, as the situation was impossible. The waiting LDMs were unable to beach because of the low tide, and according to Abecasis, it was every man for himself, as many attempted to swim to the waiting craft.[84]

The defeat constituted an enormous disaster in that the prestige of Portuguese troops suffered enormously, while that of the enemy was justifiably enhanced. It was another classic example of underestimating the enemy in an insurgency, whether through a lack of intelligence or a simple dismissal of his mythical inferiority. Clearly his capabilities were misjudged and hence not taken seriously enough. Portuguese planners seemed to have no notion of enemy ground strength and had sent an inadequate force that was quickly isolated and made vulnerable with the adverse terrain and tides. There was no reserve, no means of reinforcement, and inadequate fire support. Small wonder that *Safari* ended in a shameful disaster, one for which much blame was

82 Dhada, 580.
83 Ibid.
84 Krus Abecasis, 505–506.

generated among the participants.[85] This was certainly a poor sequel to the success of *Resgate.*

In the Cantanhez and Quitafine peninsulas the enemy was well established and continued to better his military and political organization. He was firmly entrenched with well-hidden shelters and earthworks, well-established antiaircraft weapons, excellent light and heavy arms, and an effective logistics system supporting it all. His command structure was efficient and responsive, and his troops were well trained and equipped. He understood how to deploy them and employ his reserves, and the population in the area was completely controlled. Under these conditions a satisfactory solution to Cantanhez was virtually impossible with the thin means available to the Portuguese. The evolution of the PAIGC in this area was taking a very dangerous turn and becoming increasingly difficult to reverse. Portuguese towns and military installations on Cantanhez were becoming isolated, and with the approach of the rainy season the roads there would be impassable. The only communication route would then be by water.[86] The enemy had indeed achieved a significant victory and solidified his position in Portuguese national territory. The FAP alone would have difficulty dislodging him.

Fiat G-91 Arrives

Little happened until the end of March, when enormous and heavy crates arrived by sea—the first of the Fiats. Between October 1964 and May 1966, almost 19 months, the T-6Gs had shouldered the fire support mission. While these aircraft and their pilots did their best, major operations, such as *Resgate*, had to be patched together with the few heavier aircraft available. During this period the enemy continued to build his war machine in the south on Cantanhez and Quitafine and in the Oió forest between Bissau and Senegal. From nightfall until the break of day well organized, disciplined columns of porters marched undetected through the *bolanhas* and forests carrying supplies, as the PAIGC operatives steadily widened their subversion of the population. The enemy had reestablished himself solidly in the south following *Resgate* and had built the necessary infrastructure and accommodation with the population to sustain and expand his presence. All of this was protected by an anti-air net of artillery. Whenever Portuguese troops encountered the enemy, they found an imaginative and aggressive foe that followed the classic insurgent tactics of knowing when to fight and when to flee. The Fiats would have their work cut out for them in rolling back this PAIGC presence.

The Fiats brought the FAP a major operational capability throughout the territory, principally in the dangerous region of the south and southwest, which was slightly more than 15 minutes flight time from Bissalanca. While the Fiat brought a boost

85 Ibid.
86 Ibid., 507.

Fiat G-91 landing at Bissalanca. (Source: Força Aérea Portuguesa)

to airpower, it had limited offensive capability alongside the Sabre. It was, however, an excellent photo-reconnaissance platform with its vertical and oblique cameras that could identify clearly the antiaircraft threat as well as other enemy developments. These would include installations in the neighboring sanctuary countries and infiltration routes. The Fiats were armed with a combination of external fuel tanks on the wing root stations, eight 2.75-inch rockets on the outboard ones, and internal .50-caliber machineguns. In a full bomb load, the tanks were replaced with two 200-kilogram bombs and the rockets by four 50-kilogram ones. Napalm canisters could replace the 200-kilogram bombs on the wing root stations as an alternative.[87]

The nine Fiats arrived in several shipments spread over April (4), May (4), and November (1), and were assigned to the newly established "Fiat Squadron," Squadron 121, the "Tigers." The early months of flying were precarious, as there were no explosive charges for the Martin-Baker ejection seats. While this was corrected in September, the logistic system in the early months of Fiat deployment was adjusting to the new aircraft.[88]

The principal objective of the "Tigers" initially was to attack and destroy the PAIGC anti-air defenses installed in the south. As described by Abecasis, they were broadly to attack:

- The enemy's critical vulnerabilities in the area of operations,
- All enemy movement on roads, trails, and rivers, particularly those coming from Guinée, and
- Enemy infrastructure to include his depots, encampments, bases, and any other installations relative to his operations, political, economic, or military.[89]

87 José Manuel Correia, "Fiat G-91 na Guiné 1966-68, Os Dois Primeiros Anos de Combate, 1ª Parte" [Fiat G-91 in Guiné 1966-68, The First Two Years of Combat, Part I], *Mais Alto* 398 (July-August 2012): 31-35.

88 Ibid.

89 Krus Abecasis, 556–557.

As the pilots began to fly over the area, they experienced barrages of fire from 12.7mm single- and 14.5mm single- and multiple-barreled guns with each pass over a target site. The focus then became the elimination of the anti-air threat. In making a pass over an antiaircraft gun, the pilots would begin with a high altitude reconnaissance to identify the target, which was generally in a cleared area surrounded by an earthwork defense. These shielded the enemy gunners from the bomb blasts of near misses and yet provided them with an uninterrupted field of fire. Lookouts were posted around each site, sometimes in trees, and some distance away to provide advance warning to the gunners. There may well have also been agents watching Bissalanca to alert the gunners to a strike.[90]

The first missions to the south carried the maximum bomb load. The pilots' inexperience and fear in attacking a hostile target led to some very real difficulties in placing the bombs as desired, for they tended to fall short, when the pilots anxiously sought to avoid the antiaircraft fire and made their drops early to begin their exit. The result was a series of enormous holes around the gun emplacements, and indeed these pits became known jokingly as "the caves of Vasquez" after a nervous pilot, José Vasconcelos e Sá, who made persistently spectacular misses.[91]

It was rapidly becoming clear that suppressing the anti-air threat in the zones of Cassebeche and Quitafine would be an arduous mission and would require the utmost concentration in its daily execution. The multiplying complex of gun emplacements and their increasingly accurate fire made simply completing a photo reconnaissance sortie difficult.

Operation *Estoque*

As soon as the Fiats became available, they were put into action with Operation *Estoque* (Rapier) between 9 and 11 August. Primary air resources would be the Fiat squadron for daylight operations and the three Dakotas modified for night bombing. The operation was designed to reduce the enemy redoubt defensively established in the border area and would be conducted by the FAP independently in two phases:

- The first phase would begin at 1900 on the 9th and exclusively involve the heavy FAP aircraft (Fiats for daylight and Dakotas for night), which would attempt to destroy the maximum number of targets before the second phase began.
- The second phase would employ the T-6Gs, which would attempt to seal the border and destroy any reinforcements coming into Guiné or troops withdrawing to the sanctuary south of the border. The T-6Gs were to maintain a safe standoff distance of at least 5 kilometers from known antiaircraft sites.[92]

90 Nico, "Pedaços das Nossas Vidas, III – O Rosto do Inimigo," 21.
91 Ibid.
92 Krus Abecasis, 556–558.

Intake and nose of a Fiat showing the impact of two DShK 12.7mm rounds. Note the optical glass port for the overhead camera. (Source: Força Aérea Portuguesa)

The Dakota missions on the night of the 9th did not go well because of the weather and heavy cloud cover, and despite the fact that three aircraft dropped 800 kilograms of bombs, any successful hits were random luck. By the following day the weather had improved, and the Fiats were able to attack enemy antiaircraft positions at Cabonepo, Cassacá, Cassebeche, and Camissorã. On the 11th enemy reaction was intense, and one of the Fiats suffered hits and sustained damage.[93] On the nights of the 11th and 12th Dakota bombardments continued with great risk to the crews, as the aircraft drew substantial fire from the area of Cassebeche. One suffered hits in its tail with 14.5mm projectiles. Intelligence subsequently learned that in this attack a white Soviet advisor was killed when an antiaircraft gun was destroyed. Many enemy were also killed around Cassumba, and those who survived the raid were disoriented and took refuge in the *bolanhas* along the banks of the nearby rivers, their morale shattered. There still remained a large enemy concentration around the village of Tade next to the border. In the final raid a newly emplaced ZPU-4 was destroyed near Cassacá.

From all the intelligence developed following the operation, it was apparent that the enemy had been severely damaged in the border area. He was not destroyed, however, and his organization remained intact. The FAP had surprised him with its

93 Ibid.

prolonged three-day air offensive and severely reduced his antiaircraft capabilities. He was weakened, at least for the present, and any sense of security in his "liberated zone" no longer existed.[94]

Alouette III Arrives

Portugal was the first foreign military to use the Alouette III, which was a natural outgrowth of helicopter development by Sud Aviation for the French conflict in Algeria. When it first appeared in 1959, it was a substantial improvement in transport capability over its predecessor, the Alouette II. Its larger cargo area, streamlined aerodynamics, and more powerful engine gave it excellent altitude and tropical operating characteristics. Portugal began taking deliveries of the Alouette III in June 1963, and Bissalanca received its first six machines in November 1966, some three years after it had been designated for an upgrade. In March 1967 an additional three arrived, raising the total inventory to nine. Two of these were equipped with the French Matra MG-151 machine cannon. These gunship helicopters were each known as the *Lobo Mau* or the "Wicked Wolf" and proved to be one of the more fearsome and accurate weapons in the FAP arsenal. The Alouettes were assigned to Squadron 122, the *Canibais* (Cannibals), based at Bissalanca.[95] Hopefully this development meant that there would be no more failed operations, such as *Safari*, where the troops were stranded because of the tides. It also meant that troops could be inserted and protected, and those who were wounded could be evacuated immediately. Finally, to the pilots it meant that on ejection or ditching they would be rescued promptly. This new capability was to prove extremely effective against the enemy.

Operation *Samurai*

In October, Abecasis created the Center for Photographic Reconnaissance at Bissalanca, which specialized in aerial photography acquisition and interpretation. By November the Center had created a complete mosaic of Como Island with overhead, panorama, and oblique photographs.[96] Key to this work was the Fiat and its four Vinten F95 MK3 (70mm) high resolution cameras mounted in its nose: one frontal oblique, two lateral oblique, and an overhead. The arrangement was designed to operate at high subsonic speeds (450 knots) and low altitude (600 feet). The vertical arrangement was used for high altitude photography. The great clarity and detail of the Vinten images, particularly of the PAIGC base at Cauane and other encampments

94 Ibid., 562.

95 Mário Diniz and Luís Proença, "AL III: História Breve de uma Vida Longa, 1963–2013" [AL III: Brief History of a Long Life, 1963–2013], *Mais Alto* 404 (July–August 2013: 20–21.

96 Correia, "Fiat G-91 na Guiné 1966-68," 36.

Map 8.4 Border area in southern Guiné. (Adapted from Instituto Hidrográfico Chart No 220 (1965))

in the center of the island, were the essential to analysis and directly attributable to the pilots who flew the risky missions to obtain these photographs. While the relatively low altitude made the Fiat vulnerable, its speed made achieving a hit with an antiaircraft gun difficult. The images gathered from these flights were an integral part of the planning for Operations *Samurai* and *Valquiria* (Valkyrie) launched in the region of Cafine on 13 November and 19 December respectively.[97]

In the fall of 1966, some 34 months after the conclusion of *Tridente,* the islands of Caiar and Como were again invested with the PAIGC and now being reaffirmed as a "liberated zone." Intelligence estimated that there were 50 "Popular Army" troops on the two islands and that these had formed and trained an armed militia of between 150 and 200 men. The principal PAIGC base was at Cauane with secondary bases at Caiar, Cachida, and Catabão Segundo. The FAP armed reconnaissance on Como was centered on the hillock that Lobato had attacked in the center of the island and was bounded by the villages of Cachida, Uncomené, Cauane, São Nicolau, and Curco, forming a pentagon. The PAIGC was well aware of this targeted area and avoided it. It was also very wary of FAP capabilities and moved its headquarters on a weekly basis in a predictable rotation between the villages of Cauane, Uncomené, Camuntudu, Cumule, Caiar, Camuntudu, Catabão Segundo, and Cauane again. To protect this new base infrastructure from the FAP and maintain the solid hold it had built over the local people and its "liberated zone," the PAIGC had placed numerous antiaircraft guns throughout the two islands. Operation *Samurai* was designed to put an end to this party.[98]

The concept of operations mirrored that of *Resgate* in that there would be a three-day period of continuous bombing using the Dakotas at night and the Fiats and T-6Gs during daylight hours. Once the enemy air defenses were shattered, ground force units would envelope the two islands and sweep them of enemy infestation, killing or capturing the PAIGC and militia troops. This time the ground forces would come in strength and have generous close air support with the Alouette III gunships, Fiats, and T-6Gs.

The operation opened at 2200 hours on 13 November, the night following the new moon, as total darkness was an important defensive factor for the slow-moving Dakotas. No Dakota pilot wanted a moonbeam reflection from his aircraft to give the PAIGC a target. The three modified Dakotas made serial trips to the target area for each of the three nights. During daylight the Fiats and T-6Gs flew in continuous waves against the antiaircraft sites and PAIGC base areas. When the bombing was completed, the intelligence assessment developed through aerial photography and interviews of prisoners gave a picture of immense destruction and psychological damage to the enemy. All of the targeted gun emplacements were destroyed as well as about a third each of the two bases at São Nicolau and Cauane. The enemy was

97 Ibid.
98 Krus Abecasis, 580–581

thoroughly demoralized over the large losses in both equipment and personnel. Of the approximately 50 "Popular Army" troops, there were six killed and 20 missing and presumed dead. According to prisoner interviews, the most feared attacks were those of the Fiats with their napalm canisters, and these were closely followed by the T-6Gs with their strafing.

On the 16th some 223 troops, of which 97 were paratroops, were inserted into Caiar and Como by air from the Dakotas. The remaining 126 were airlifted by the seven transport Alouettes in several trips and protected by the two gunships. After ten days of operations, the enemy was either killed or forced to flee the islands. Firefights with PAIGC troops had partially destroyed the villages of Tabanca Velha and Poilão, and fires from the fighting continued to burn in parts of Cauane, Cassacá and São Nicolau.

Subsequent intelligence disclosed that the enemy had been divided between his primary base of Cauane and secondary encampments when *Samurai* began and immediately organized a retreat. The exodus of the population with the PAIGC elements provided a strong indication of enemy control over it. By mingling with the fleeing population, a traditional insurgent practice, many of the PAIGC were able to escape and live to fight another day.

Operation *Valquíria*

Predictably the enemy returned to Cantanhez to rebuild its presence, and by December 1966, only a year after *Resgate*, he was again a sizable threat. His attacks on naval traffic in the Cumbijã River threatened the riverine communications link with Bedanda. Further, he had reactivated his antiaircraft capability and raised it to a dangerous threat level. The goal of Operation *Valquíria* (Valkyrie) was similar to that of *Resgate*: force him to reveal his defensive positions so that they would be open to destruction from the air, particularly those interrupting riverine navigation in the area of Cantanhez.

The operation occurred between 19 and 21 December and was comprised of four Dakota night sorties, eight Fiat sorties, and 12 T-6G sorties for a total of 74 flight hours. Some 5,930 kilograms (13,073 pounds or 6.5 tons) of bombs and rockets were expended.[99] The action was apparently successful, as far as it went, for the enemy was forced to alter his strategic plans. His efforts to establish himself in the south had been at enormous cost at the hands of the FAP counteroffensives over the past year. In both Cantanhez and Como he had repeatedly suffered substantial troop and equipment losses. He was down, but not out, as the expression goes, and had not abandoned his long-term strategy. He would simply make adjustments. The north now became a threat area. PAIGC operatives began an endless series of ambushes and attacks on military columns along the roads connecting Mansoa, Mansabá, Bissorã, and Bissau,

99 Ibid., 609–610.

and troops were tied down simply escorting convoys. It was also obvious that the Portuguese strategy of a heavy reliance on air strikes was not working, as there was the missing human element, and the real battle was for the soul of the population. The people would appear to cooperate in the day but would work for the PAIGC at night. In the capital itself security was weak, and terror had become a permanent threat. The answer was more polarization. While the military was hitting the enemy, so were the police. PIDE was a specter that held a certain dissuasion in its questioning of undocumented natives, sympathizers, and collaborators; however, as the detentions and interrogations of suspects intensified, there was a perverse turn. The suicide of one suspect who was a respected professor in Bissau produced a general atmosphere of unease in the white community and simmering resentment among the natives.[100] This mistrust became palpable and would be obvious to outsiders, such as Admiral Américo Tomás and Brigadier António Spínola, when they visited Bissau in early 1968 for an appraisal of campaign progress.

Operation *Barracuda*

Valquíria may have appeared to be successful on its surface; however, in certain regions the enemy was so thoroughly entrenched with his encampments and their a vast infrastructure of crop cultivation, field hospitals, arms magazines, and living quarters, all beneath a vast complex of powerful antiaircraft artillery, that it was impossible for air power alone to eject him. His antiaircraft defense became increasingly dangerous, as its capability and armament became heavier. These weapons were protected by extensive sand earthworks and positioned tactically to provide mutual protection with interlocking fields of fire while defending PAIGC installations. The gunners were well-disciplined and unusually warlike, uncharacteristic traits for the typical insurgent army. This sophisticated antiaircraft organization could only have been achieved with the help of outside advisors. At various times these were Soviet, Chinese, or Cuban, and on occasion their bodies would be discovered by ground troops following an air raid. Suppressing such an entrenched foe and his antiaircraft capability in the areas of Cassebeche and Quitafine was always going to be an arduous air mission—one that required full concentration of the pilots in their daily missions to attack each new enemy gun emplacement, and these now appeared with alarming frequency.[101]

In January 1967, Abecasis was relieved as the ZACVG commander, and his replacement was Colonel Rui da Costa Cesário, who immediately initiated a rare event: an operation to capture intact one of the PAIGC antiaircraft guns. There was a particular site opposite Bissau and a short distance upstream on the Geba River near the village

100 Ibid., 612.

101 José Manuel Correia, "Operação Barracuda: Os Fiat G-91 ao Ataque na Guiné" [Operation Barracuda: The Fiat G-91 on the Attack in Guiné], *Mais Alto* 367 (May–June 2007): 23.

of Gã Formoso. It threatened aircraft in the traffic pattern at Bissalanca and could not be permitted to continue. Based on photographic reconnaissance, Cesário planned a multiple force assault on the bush around Gã Formoso. It envisioned the use of the large navy landing craft *Alfange* to land six combat groups of paratroops at Ganduá Porto, just west of and downstream from Gã Formoso, and two combat groups of paratroops brought in two waves aboard six Alouette III helicopters from Bissalanca to envelope the site. The force would be covered by the two Alouette gunships, and the ninth Alouette would remain on alert at Bissalanca for medical evacuation. The remainder of the FAP resources would also be fully engaged. The Fiats would lead the initial air attack wave at dawn to maximize surprise, and these would be followed by T-6Gs in close air support of the deployed land forces.[102]

The operation opened on 4 February with the six Fiats rolling down the runway at Bissalanca at 0600 hours. These were followed by eleven T-6Gs, seven Alouettes and a DO-27. One Alouette gunship was not available. In ten minutes the entire aviation attack force was airborne, and at 0610 the six Fiats attacked their objectives. The first Fiat dropped two bombs that failed to explode; however, the enemy was unprepared, and there was no reaction. Three Fiats acquired an adjacent target but dropped ineffectively because of poor visibility. Finally, the enemy antiaircraft machinegun at the initial site began firing and hit one of the Fiats, as it made a bombing run against it.

At 0615 eleven T-6Gs attacked the bush areas adjacent to Gã Formoso with 15- and 50-kilogram bombs, 37mm rockets, and 7.62mm machineguns to destroy the artillery support installations, such as magazines and personnel shelters. At 0620 the six Alouettes arrived with the first wave of paratroops protected by the gunship. At 0630 the second wave arrived. At 0640 the landing craft disgorged its six combat groups in Ganduá Porto under the air umbrella of a gunship and two T-6Gs. By 0900 the advancing ground forces had overrun Gã Formoso and captured PAIGC antiaircraft weapons and ammunition, as the enemy fled. Those in flight were attacked by the Dornier and the gunship. By 1300 the paratroops had occupied the entire enemy encampment without resistance. At 1800 the troops returned to the landing craft and were in Bissau by 2100 that evening.[103]

This was a perfectly executed operation in that FAP put the enemy to flight in the area of Gã Formoso, destroyed his infrastructure, and eliminated the threat of the DShK 12.7mm site on the Geba riverbank. Other light arms and hundreds of rounds of heavy machinegun ammunition were taken. Enemy losses were unconfirmed, but it was thought to be about a dozen, while there were no Portuguese casualties. On the other hand, many aircraft had sustained hits.

This action had demonstrated the utility of low altitude photographic reconnaissance, as the images obtained by the Fiat were in such detail that the analysts were able to put together a complete picture of the threat and its support complex of huts,

102 Ibid., 24.
103 Ibid., 25.

magazines, caves, foxholes, and trenches for shelter from air attack. The photographs even showed the furrows made in the earth by the gunners dragging the machinegun tripod, a valuable clue that enabled analysts to measure the distance between the legs and identify the precise weapon.[104]

While it was a successful military operation, the enemy abandoned his equipment and fled to fight another day. Missing from the equation was any people-to-people program. When the local population passively allowed the enemy to install a dangerous weapon nearly opposite the capital, alarm bells should have sounded. This deafness would continue for the remainder of 1967, with Portugal following its exclusively military solution throughout the balance of Shultz's tenure.

Cumbijã River Action

The PAIGC soon returned to the southern border peninsula and by mid-1967 had strengthened its position around Cafal and Cafine with new antiaircraft weapons protected by ramparts of sand arranged in the form of a snail. Installed next to the Cumbijã River was the bigroup of Sana Naiana, which was harassing the passing naval vessels with raking fire, as had been done before.[105] The transit of launches in this zone had become a veritable headache and required escort for protection. These were not completely effective, as they took fire as well. Intelligence had identified the bigroup and two antiaircraft guns well installed with fortifications and foxholes for the gun crews. A sunset raid was planned to eliminate these two weapons, as it was thought that the gun crews would have retired from their positions for the night.

Just prior to sundown on the designated day two Fiats taxied to the approach end of runway 03 at Bissalanca, each with two 350-liter canisters of napalm. Lieutenant Vasconcelos e Sá and his wingman Sergeant António Mota Cardoso, pushed the throttles of their Fiats forward and heard the Orpheus Mk803 D11 engines spool to takeoff power. As each Fiat with its gross weight of 12,500 pounds accelerated down the runway, it quickly reached 140 knots and rotated, lifting off after about 4,000 feet. At 160 knots, the pilots raised the gear and flaps and continued to accelerate to 300 knots and climb at this speed toward their cruise altitude of 10,000 feet. The pair turned south toward their targets and, on reaching the town of Catió, reduced power to flight idle to descend. They separated to approach their two assigned targets from different angles in the hope of confusing the gunners. With airbrakes deployed, they entered a steep dive angle to achieve maximum surprise. As Mota Cardoso reached release altitude with the treetops sliding rapidly beneath the nose of his aircraft, he locked onto the target and advanced the throttle to reach his escape speed of 400

104 Ibid., 26.

105 José Nico, "Pedaços das Nossas Vidas, II – O G-91 que Não Trazia Correio" [Pieces of Our Lives, II – The G-91 that Did Not Bring the Mail], *Mais Alto* 360 (March–April 2006): 31.

knots. The engine did not respond, and a chill went up his spine. Just then he noticed that his target was visible and growing in his bombsight. He made a small correction and "dropped the eggs" slightly short, and with the aircraft now lighter and climbing, he turned toward Catió. He continued to trade airspeed for altitude and eventually reached 1,500 feet, but his efforts to revive the engine became an anxious and fruitless exercise. As the instinct for survival dominated, he considered ejecting; however, he would be in enemy-dominated terrain, and it was doubtful that a search party from Catió could reach him before the PAIGC.[106] If he were captured, he would certainly meet Lobato's fate.

He was now drenched in sweat because of the tension and his failing engine which powered the cockpit air conditioner. Warning lights began to appear, and as the engine wound down, its temperature rose. It was now impossible to maintain altitude. He would try to reach Catió and eject. At that moment the nose of the aircraft passed the runway at Cufar next to the Cumbijã River. Mota Cardoso knew immediately that he had found the solution. He would head there and eject if he ran out of altitude.[107]

In reality the field at Cufar was not intended for an aircraft of this type. It could only accommodate those designed to land on unimproved runways, such as the Dornier, the Auster, the T-6G, or the Dakota, for it was only 3,000 feet of beaten and rough earth. Luck, however, was with Mota Cardoso, and his level of stress bred a great confidence of the possible. As he aligned with the runway and lowered his landing gear, the situation deteriorated. Violent vibrations ran through the aircraft, and the hiss of the engine turned into a howl of screeching that came from metal rubbing against metal. He immediately closed the throttle and shut off the fuel to the engine. The vibrations stopped immediately, and the remainder of the flight was conducted in silence except for the buzz of the inverters driving the now battery-powered instruments. He aimed the aircraft for the approach end of the runway, expecting to sink gently into a sea of grass on its apron, but then felt a violent contact with the ground. The aircraft bounced short of the runway, and Mota Cardoso popped the drag chute. It seemed like forever before it took hold, as the barbed wire surrounding the military compound rushed toward him. He swept through the wire and made his second contact with the ground on the runway. The aircraft began to shake from the rough surface and was difficult to control through normal breaking. His speed was now 80 knots, and he was standing on the brakes fearful of blowing his tires. As his aircraft sped past the army buildings with its nose now decorated with the mouth of a tiger and many strands of barbed wire, there was little notice from the soldiers. A game of football continued alongside the runway, while Mota Cardoso struggled to save his life. As he overran the landing strip and came to rest in the tall grass beyond, the company commander approached the aircraft and activated the electric motor that opened the canopy. After removing his seat harness, Mota Cardoso was helped trembling from the aircraft. As he was

106 Ibid., 31–32.
107 Ibid., 32.

about to speak, a soldier arrived and, not understanding the gravity of the situation in which an aircraft had "fallen from the sky," asked a reflexive question, an expression of hope, that left Mota Cardoso dumbfounded: "Did you bring the mail?"[108]

In defense of the soldier, mail was considered among the most essential air cargo duties of FAP in its support of ground forces because of its importance to morale. It carried the highest priority, and aircrews never passed the opportunity to load it onboard their aircraft. Mail not only brought news of dear ones at home but was tangible evidence that a soldier serving in the field had not been forgotten. Consequently, mail distribution was always an important event for every unit.[109]

More on the Cumbijã River

Nine months later in reaction to the reintroduction of antiaircraft guns in the border peninsula, particularly at the villages of Dameol, Tanene, Banir, Cassebeche, Canefaque, and Cassequelã, a flight of five Fiats was dispatched to address this development. The preceding day, 6 March 1968, two Fiats piloted by Lieutenants José Nico and José Vasconcelos e Sá had conducted a patrol the length of the southern frontier. The aircraft each carried external fuel tanks and eight 2.75" rockets. During their patrol they were to check the enemy infiltration routes between Kandiafara south of the border and Guilege as well as the road that linked Aldeia Formosa to Gadamael Porto. The pilots concentrated on the roads and trails over which they flew. While there were army units below, there was no radio chatter, as all was calm. There were no clouds in the sky, and the surfaces of the rivers beneath them were glass-like. The truth was, however, that the PAIGC insurgents did not dare show themselves during daylight for fear of being seen by aircraft. Traffic on the rivers and trails always began when darkness came.[110]

As the two aircraft flew over the southern border, it was easy to see the quantity of new earthworks and other construction surrounding the empty emplacements for heavy antiaircraft machineguns. The sand bunkers and double revetments were easy to identify, as they contrasted with the green foliage surrounding them. The clearings around the emplacements were extensive to provide a good field of fire. This also enabled the pilots to identify the sites from a considerable distance. A new danger was the threat of antiaircraft weapons hidden beneath the tree foliage. Several months earlier the PAIGC had reaffirmed the status of the area as a "liberated zone," and it

108 Ibid., 33.

109 José Montalvão, Depoitmento do Cor Pilav José Montalvão que ilustra a importância do transporte aéreo durante a guerra do ultramar. [Deposition of Colonel Pilot-aviator José Montalvão, which illustrates the importance of air transport during the war of the *ultramar*]. Livro de Actos 1995, item 3, pages 135–138. Archivo Histórico de Força Aérea, Alfragide.

110 José Nico, "Pedaços das Nossas Vidas, III – O Rosto do Inimigo": 21.

was easy to understand the confidence in this declaration on seeing the formidable defenses.[111]

New developments were recorded through photographic reconnaissance missions over a targeted area, and the intelligence harvested from the photographs enabled attacks to eliminate the new sites. The results initially were very mixed, as the pilots needed to acquire experience and learn the limitations of their aircraft as well as the capabilities of the enemy. While the Fiat had speed and agility, its guns were relatively light for the job of destroying these sites, and hence bomb aiming precision was vital. Putting a bomb on target was very difficult, for if it exploded more than 10 to 15 meters wide, the gun and its crew would be protected by the revetments, and the only results would be a loud noise and a cloud of dust. The poor aim of the pilots was also in part because of the courage of the PAIGC gunners, who stood their ground and remained by their guns in the face of daunting attacks when they might have fled to safety.[112]

Also because the availability of bombs was reduced during the last several months of 1967, attacks were made bomb by bomb against the munitions inside the holes where the gunners and their arms hid. In most cases these openings were less than two meters in diameter, and dropping one "into the hole" was as much an art as it was a science. Such bombing runs required a pilot to take heightened risk, and all understood that this would eventually lead to the loss of the first Fiat. This feeling was nowhere more evident than in the missions to Quitafine and along the frontier northeast from there towards Gandembel, where a trail crosses the border from the PAIGC base of Kandiafara. The practice of single-bomb passes, which increased risk with each pass, was only changed at the beginning of 1968, when a significant supply of 200- and 50-kilogram bombs arrived in Bissau. From then on the Fiats used an "overloaded" configuration of two 200-kilogram bombs on the interior racks next to the wing roots and four 50-kilogram ones on the exterior racks. This entire 600-kilogram (1,322-pound) load was dropped in salvo on a single high-speed pass. The arrival of the six bombs nearly simultaneously on the target was much more effective in terms of increasing the probability of a hit while reducing the risk considerably to the aircraft. In a fight that was delivering increasing losses to the enemy, this type of bombing had a great effect in causing his gun crews to abandon their weapons. In most cases the enemy could not endure such strikes, and after the first attack in a flight of five, the remaining four were unopposed.[113]

On this day Nico and Vasconcelos e Sá observed the construction of several new sites. Some were concentric circle, double revetment types with a hollowed-out gun pit in the center. Others were very different in that they were not concentrated together in a mutually defensive pattern with interlocking fields of fire but were largely open and

111 Ibid.
112 Ibid.
113 Ibid., 22–23.

had no excavated shelter. These appeared to be hastily constructed earthworks that were clearly destined for wheeled arms, as there was a clear trail of tire marks from their use. No weapons were observed in any of the epaulements. The settlements of Tanene, Banir, and Dameal were all deserted. Soon the two Fiats arrived opposite Kadigné, a small "port" on the south (left) bank of the Inxanche River, which marked the border. This village was a major logistics center for the PAIGC, and from it vessels of various types shuttled to and from Guiné with enemy supplies. A number of these vessels had been stolen from the commercial companies in Guiné and were now used by the PAIGC.[114]

There was an area around Cassebeche, where the FAP regularly felt the sting of antiaircraft fire between October 1967 and February 1968. The gun emplacements in this area had proliferated during this period, so it was natural that the two Fiats turned toward Cassebeche to observe the new construction. What they found was indeed new: a revetment that spiraled outward from the central gun pit to encircle it twice and hence to form two barriers to bomb blasts. From the air it appeared to resemble a snail's shell. Whatever weapon would be installed in its center was large and powerful, as its tire tracks were very visible. The pilots saw little else. Whatever was to be placed in the snail's center would be easily seen at considerable distance, for there was a wide clearing of foliage from around the site. After making three passes over the fortifications, the pilots decided to cut their mission short and return to Bissalanca to report their observations. Before departing, they circled the new site to establish the best approach for an anticipated attack and to identify visual references. There was indeed a large and prominent tree to the north of Cassebeche, and if an aircraft passed over its top on a heading of 210°, it would arrive directly over the target.[115]

On returning to Bissalanca, the two pilots recounted their findings, and their squadron commander, Lieutentat Colonel Francisco Costa Gomes, agreed that in the morning five aircraft would attack with the "maximum load." The intention was to arrive over the area near noon when the sun would be high, and the pilots could execute a dive with the sun at their backs. The bombs would be dropped in salvo, as usual. Afterwards, they would return to Bissalanca to rearm and attack again. Additionally there would be a SAR helicopter standing by on the Island of Melo in case a pilot needed to eject. The early experience in this type of mission, in which various approaches in sequencing the attack actions were explored, had yielded no real conclusion. Pilot opinion was divided. Some preferred the treetop approach, and others a high angle from altitude. The weapons load was also open to debate, as some were convinced that a combination of napalm and fragmentation was best, while others preferred the "maximum load." Whatever the case, enthusiasm for the mission ran high, and once permission was granted, the five pilots left the briefing room.

114 Ibid., 23.
115 Ibid., 23–24.

The nose of a Fiat showing the optical glass camera ports. The frontal port has a 15° down angle. The machinegun servicing access is just visible to the left. (Source: Força Aérea Portuguesa)

The pilots would be Costa Gomes, Vasconcelos e Sá, Nico, Captain Joaquim Costa Pereira, and Lieutenant José Firmino das Neves.[116]

Nico next visited the Photographic Reconnaissance Section and gave instructions to mount a camera in the frontal position of the nose, which had a 15° downward angle and would capture gun emplacements along his approach and the target just prior to his weapons release. In a 450-knot approach, photographs taken from this position would not be blurred. The other pilots each also gave instructions for his preferred bomb load.[117]

On the following day at the appointed hour the five pilots began their preflight and launched, climbing together to 10,000 feet over the Geba River. As they neared the target, Nico dropped to treetop level over the Cacine River, armed his bombs, and after identifying the prominent tree at Cassebeche, made a slight 2° correction. He set

116 Ibid., 24.
117 Ibid., 25.

ZPU-4 antiaircraft installation at Cassabeche in the border area of southern Guiné, as photographed from the forward oblique nose camera on José Nico's Fiat during his bomb run on 7 March 1968. Earthen ramparts in the form of an espaliered snail protect the gun crew from near-miss bomb blasts. The aircraft is approaching from a northeast direction, as the gun is expectantly trained to the south. (Source: Força Aérea Portuguesa)

the camera to run at eight frames per second, the maximum rate possible. The day was clear and bright, so this setting would yield excellent images with a sharp resolution. Suddenly the foliage disappeared, and he was passing over a network of antiaircraft positions with the "snail" in his bombsight dead ahead. Its sand earthworks were a brilliant white in the sunlight. In a fraction of a second Nico saw the multiple-barreled cannon was pointed away from him and the crewmen were looking elsewhere. As the target passed into the reticule on the sight, he pushed the release button.[118]

But no, there was no normal surge in airspeed or altitude and a quick glance to the left showed the bombs still on the rails. Nico had forgotten to arm the release system in the heat of the moment. Now the other four aircraft would arrive on top of a hornet's nest. A simple operation had now become complicated, and the enemy would be waiting with his hand on the trigger.[119]

Next Costa Gomes announced that he was attacking and dove at a 60° angle from 8,000 feet on a west-east heading. As he pulled out at 1,500 feet, he released his bomb load, and a rapid series of explosions blossomed around the target but did not destroy it. The smoke faded rapidly, and Vasconcelos e Sá began his attack. Costa Pereira was at 8,000 feet when he saw him approaching the target. There below him were

118 Ibid., 26.
119 Ibid.

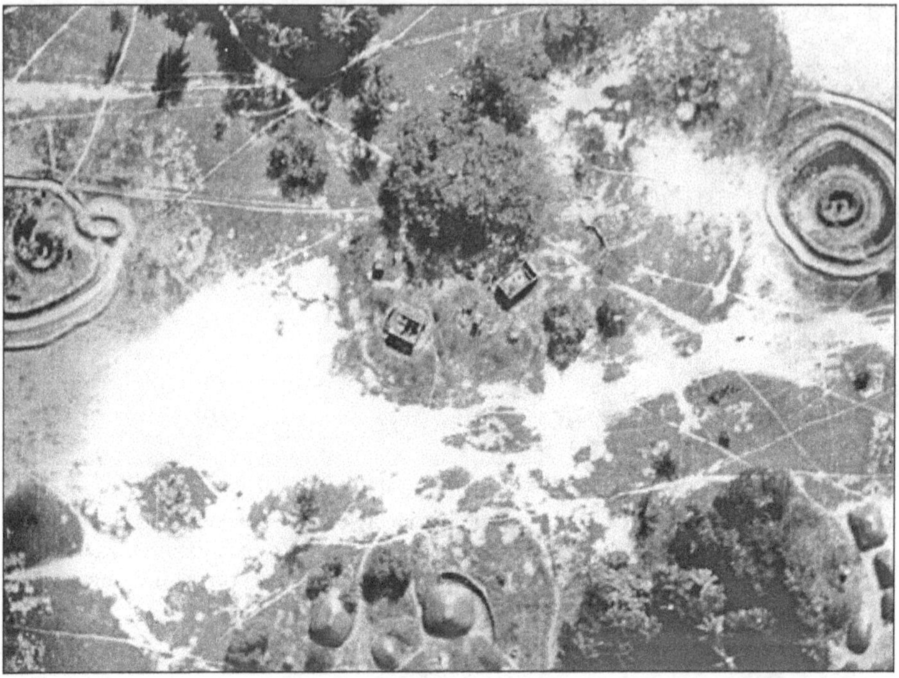

Overhead photoreconnaissance photograph showing the alignment of two of the many gun emplacements on the Quitafine Peninsula destroyed by the "Tigers" shortly thereafter. (Source: Força Aérea Portuguesa)

the gunners now melting the barrels of their machineguns, as they fired at the Fiats. Suddenly the target disappeared beneath the explosion of six bombs in a direct hit. Costa Pereira, who was rolling into the target after Vasconcelos e Sá, rapidly reached his bomb release point, but the target was completely obscured in an enormous cloud of smoke and sand. This meant that his delivery would be by guesswork, and indeed it fell slightly short. Despite some faults, the feeling of euphoria at success was enormous among the pilots.[120]

These attacks quieted the guns, as the surviving crews sought shelter in their holes against the inferno raging above. After recording the damage assessment photographs, the flight of five returned to Bissalanca. A second sortie of two Fiats found all quiet, and the gun emplacements strewn with wreckage. With no visible targets, the pair dropped their bombs in the surrounding tree cover. A sortie the following day revealed the wreckage removed from the destroyed sites. The post-flight analysis of Nico's photography confirmed that there had been a single ZPU-4 in the new site

120 Ibid., 27.

surrounded by several sites with 12.7mm tripod-mounted machineguns. Also visible in the photograph are the gun crewmen running to their positions in panic.[121] Even with this success and no ground follow up, the PAIGC would soon return.

In the meantime, FAP men and aircraft would be stretched to their limits. Aircraft flight records were not collated at the wing level until an air staff was created in 1965, but from then onwards it is apparent that the FAP was under enormous pressure. For example, in 1965 flight hours at Bissalanca totaled 6,351; in 1966, 13,776; and in 1967, 13,038.[122] In 1966 flight hours increased 117 percent over 1965, the peak period of the war in Guiné. As we have seen, maintenance struggled to maintain the air fleet and its reliability. General José Lemos Ferreira later observed that there were 50 or 60, 70 pilots at most during the war, and they risked their lives daily at all hours, because everything depended on them in one way or another.[123]

Arrival of Spínola

By late 1967 the overall campaign was clearly failing to address the insidious advance of the PAIGC, and towns, such as Cacine, Catió, Bedanda, and Gadamael, became besieged and isolated. As Lieutenant Alexandre Carvalho Neto, the military assistant to Brigadier António de Spínola would write, "The situation was absolutely catastrophic, on the brink of collapse. Bissau had become practically surrounded."[124] The President of the Republic, Admiral Américo Tomás, who visited Bissau in February 1968, returned to Lisbon to report that "the war was held by a thread."[125] Spínola confirmed this observation with an inspection visit of his own and described the situation as "desperate."[126] Spínola assumed command in May 1968 and saw his first task as one of drastically reducing PAIGC military capability and tilting the balance in favor of Portugal.[127] While these condemnations may be a bit hyperbolic, it was clear that Shultz's exclusively military approach was ineffective in the face of the civil organization and support by the PAIGC. Its strategy was unquestionably winning the population and thus gaining it the ability to maneuver at strength within Guiné. This void in Portuguese strategy needed to be filled, and Spínola had his answer. Airpower would again prove to be the critical element.

121 Ibid., 28.
122 Fraga, 123.
123 José Lemos Ferreira, "No Céu da Guiné" [In the Sky of Guiné] in *A Guerra de África, 1996–1974* [The African War, 1961–1974] by José Freire Antunes (Lisbon: Temas e Debates, 1996), 590–591.
124 José Pedro Castanheira, "Ao Serviço de Spínola e Marcelo" [In the service of Spínola and Marcelo], *Expresso* (20 September 1997): 56.
125 Ibid.
126 Willem S. van der Waals, *Portugal's War in Angola 1961–1974* (Rivonia: Ashanti, 1993), 181.
127 José Pedro Castanheira, "Memórias da Guerra e da Paz: Spínola" [Memories of War and Peace: Spínola], *Expresso Revista* (30 April 1994): 78.

Bissalanca in 1967 after ramp and runway improvements. Note the revetments around the Dakota hardstands as well as the now orderly appearance. (Source: Força Aérea Portuguesa)

9

The Last Frontier

Of the three theaters of operations, Mozambique was the final one and probably would not have developed as such but for the granting of independence to Tanzania in 1961. It was the emergence of Julius Nyerere, first as its premier and later in 1962 as its president, that was the most telling development, as he was an ardent nationalist who was passionately committed to supporting and providing a home for African liberation movements. These included the African National Congress (ANC) and the Pan African Congress (PAC) of South Africa, Zimbabwe African National Union (ZANU) and Robert Mugabe in their struggle to unseat the white regime in Southern Rhodesia (now Zimbabwe), and most importantly to our story, FRELIMO.

Nyerere was a generous host to FRELIMO and provided the ideal sanctuary from which to project power into Mozambique. Unfortunately, FRELIMO needed much more than a refuge to be successful, as both geography and internal conflict conspired to make its insurgency difficult. Mozambique is about 1,800 kilometers in length, and FRELIMO forces had to travel about half that distance through barren and hostile territory to reach significant numbers of people to proselytize. In the meantime, Portuguese airpower, which was originally headquartered at the far end of the colony in Lourenço Marques, relocated to Nampula in the north and subsequently made any insurgent incursion a costly and painful exercise. This is the story of how the FAP brought its airpower to bear on this enemy and made his life miserable.

Frelimo

Nyerere proved to be a skilled negotiator and peacemaker among his revolutionary guests, and as an avowed Marxist and advocate of the single-party state, encouraged a similar policy with these fellow travelers. Even in the beginning at independence in December 1961, when there was much optimism, needed foreign investment to fund Nyerere's new national plans was not forthcoming, as investors had noted the arrival in force of East German "advisors" in Zanzibar and were not prepared to risk any investment on his wooly idealism. Indeed, as Nyerere presided over an increasingly militant, xenophobic, and beggared state, he held divisiveness in line through the

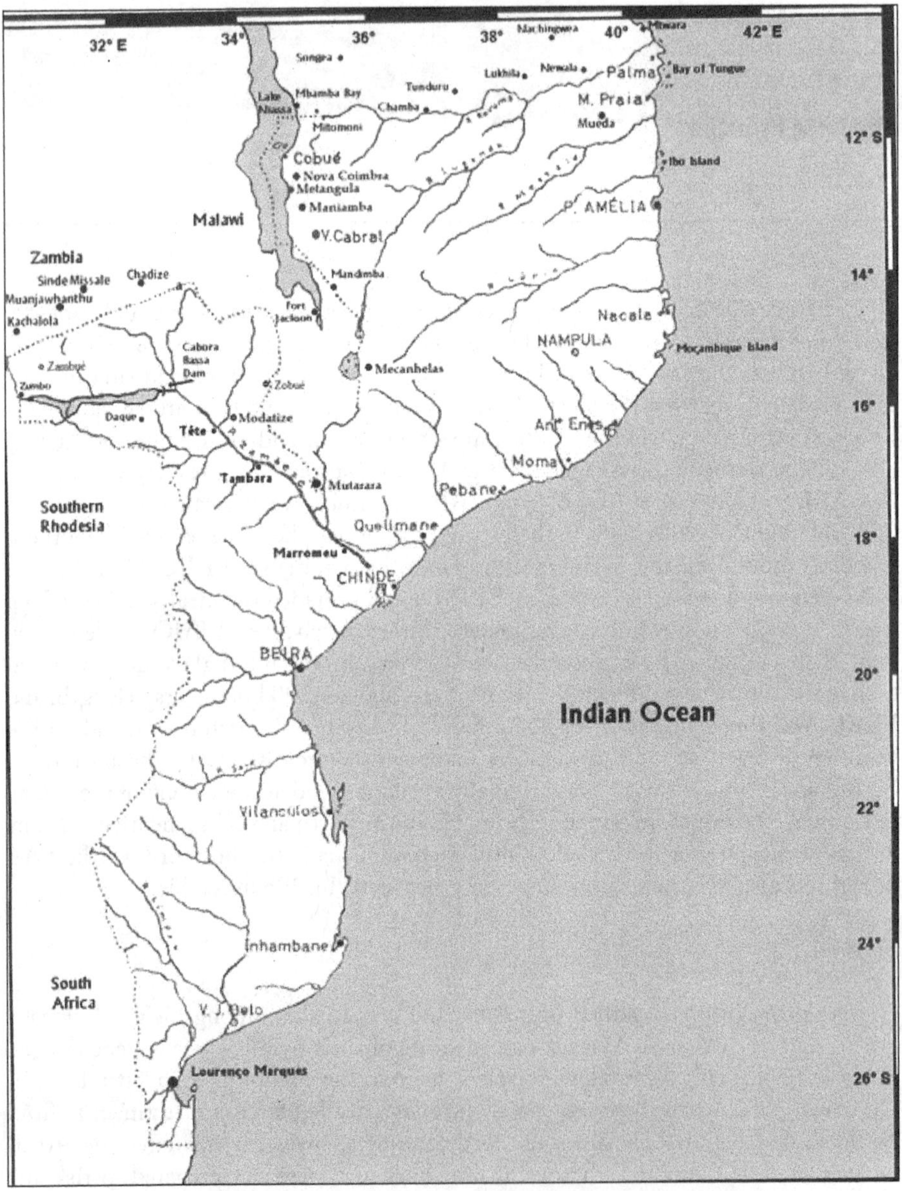

Map 9.1 Mozambique with Interior and Exterior FRELIMO Bases. (Modified by the author from a CNM briefing map)

liberal use of his Preventive Detention Act and East German-trained security police to imprison opponents.[1] His apparent commitment to human rights on the world stage was clearly at odds with his actions at home. The future FRELIMO government in 1975 would embrace similar policies and in so doing spawn its own insurgency.

Eduardo Mondlane, who with Nyerere's help had consolidated most of the disparate Mozambican nationalist movements into FRELIMO, was personally endorsed by him as its leader in a September 1962 inaugural national conference at Dar-es-Salaam.[2] Nyerere had selected him to lead FRELIMO while he was teaching anthropology at Syracuse University and brought him to Dar-es-Salaam, because he had not been involved in any of the earlier, bitter disputes between the several nationalist organizations resident in Tanzania. It was thought that this credential would make Mondlane a less polarizing and more capable peacemaker.

Mondlane was born in 1920, the son of a minor tribal chieftain, and was assisted in his education by protestant missionaries who helped him gain entrance in a secondary school in Lourenço Marques. Here he advanced and earned a primary school certificate in 1936. He continued his schooling in South Africa and ultimately moved to the University of Lisbon, where he met a number of fellow revolutionaries: Marcelino dos Santos, Agostinho Neto, Amílcar Cabral, and Marío de Andrade.[3] Association with these attracted PIDE surveillance, and he consequently transferred to Oberlin College, where he graduated with a BA in 1953. Oberlin, a small liberal arts college in Ohio, enjoyed a proud reputation for welcoming blacks and women at a time when few other institutions did so. He followed this with a MA and PhD in sociology from Northwestern University. In May 1957 following graduation, Mondlane was employed by the UN as a researcher on trust territories and through this job met Nyerere. Subsequently Mondlane joined the faculty at Syracuse University and, following Nyerere's conference, returned to the university to complete his teaching assignment. In his absence, turmoil erupted within FRELIMO, and when he returned to Dar-es-Salaam in mid-1963, he found the organization as divided as ever.

It was also soon after Mondlane's inauguration and in anticipation of the struggle that FRELIMO sent its first 250 young volunteers to the newly independent Algeria for military training.[4] Later the United Arab Republic provided similar support until 1971, when it was dissolved. It was a short-lived political union between Egypt and Syria that ran from 1958 to 1961, when Syria withdrew. Egypt continued offi-

1 Al J. Venter, *The Zambesi Salient: Conflict in Southern Africa* (Cape Town: Howard Timmins, 1974), 193.
2 John Marcum, *The Angolan Revolution: The Anatomy of an Explosion (1950–1962)*, Volume 1 (Cambridge: M.I.T. Press, 1969), 284.
3 Richard Gibson, *African Liberation Movements* (London: Oxford University Press, 1972), 277.
4 Ian F. W. Beckett, "The Portuguese Army: The Campaign in Mozambique, 1964–1974," in *Armed Forces and Modern Counterinsurgency* by I. F. W. Beckett and J. Pimlott, eds. (London: Croom Helm, 1985), 140.

cially as the "United Arab Republic" until 1971, when it adopted a new constitution, and Anwar Sadat became president and took the country on a different course. As training matured, and the conflict advanced, recruits were flown to more efficient, purpose-designed centers in the Soviet Union and China and given their initial instruction by seasoned insurgent professionals. This process was able to produce better trained personnel in volume not only to replace losses but also to expand the FRELIMO army. The graduates were then returned to the Tanzanian camps at Chingoe and Kongwa.[5] FRELIMO began with a disorganized force of uncertain strength and by the early 1970s had an active force of about 7,200 regulars and 2,400 popular militia.[6]

Mondlane and Nyerere first established a network of bases on the Tanzanian side of the border. The first encampment was near the settlement of Bagamoyo, but by May 1964 as operations expanded, so did training. The primary Tanzanian training and logistic camp became Nachingwea, which produced about 600 men per year under Chinese instructors. Other training centers and depots were built at Songea, Tunduru, and Lindi. There were also six bases next to the border for staging incursions: Mbamba Bay, Mbeya, Chamba, Lukwila, Newala, and Mtwara. Mondlane transferred his base to Mpitmbi or New Camp in 1966, and it was from this newly established network that he maintained his insurgent campaign into Mozambique.[7] As one would expect, Nyerere controlled the flow of arms, munitions, and supplies from China, the Soviet Union, and Eastern Europe on behalf of FRELIMO, and this matériel was always plentiful.[8]

Initial FRELIMO strength was derived from the overt support of the Makonde people, who lived in the north of the Cabo Delgado district and immediately across the Rovuma River on that portion of the Makonde Plateau in Tanzania. They formed a bellicose and itinerant nation that had been in place at least a thousand years.[9] René Pélissier describes the Makonde as a tribe of "intransigent warriors" who are "allergic to all forms of authority and outside influence."[10] Over the years they successfully rejected Islam, Christianity, and German and Portuguese attempts to administer them. They tended to remain animist and to worship their ancestors. By the early 1960s the Makonde in Tanzania had become thoroughly infiltrated by FRELIMO cadres and thus were the major source of insurgent recruits.[11]

5 Venter, 81.
6 Neil Bruce, "Portugal's African Wars," *Conflict Studies*, No 34 (March 1973): 22.
7 Venter, 48; Bruce, 8.
8 Aniceto Afonso and Carlos de Matos Gomes, *Guerra Colonial* [Colonial War] (Lisbon: Notícias, 2001), 168–173.
9 José Verdasca dos Santos, *Memórias de um Captão* [Memories of a Captain] (Lisbon: Universitária Editoria, 2004), 143.
10 René Pélissier, *Naissance du Mozambique: Résistance et Révoltes Anticoloniales (1854–1918)* (Orgeval: Pélissier, 1984), 19.
11 Venter, 48.

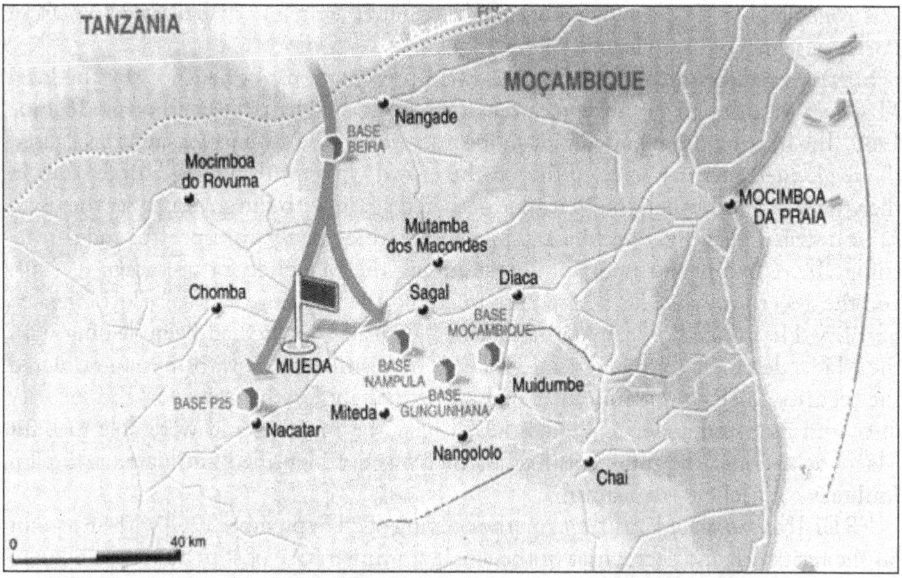

Map 9.2 FRELIMO Infiltration in Cabo Delgado.
(Source: Carlos de Matos Gomes, *Moçambique 1970: Operação Nó Górdio*)

The sudden welcoming of these outsiders was the result of a newly independent, black-ruled Tanzania and its hostility towards white-ruled Mozambique. Prior to 1961, the Makonde had been divided by a border with no fence or strict frontier enforcement. This nominal division did not affect their tribal unity, as they were free to move back and forth across the international border unhindered. The polarization initiated by Tanzanian independence brought almost complete restriction to intratribal communication, and the Makonde in Tanzania became concerned with fighting the Portuguese to reestablish their freedom of movement within their tribal region. Prior to this time the border had had little meaning to them. Now, however, insurgent activity had drawn Portuguese forces into the Makonde ancestral grounds of northern Mozambique in defense of the national border that was being threatened by a hostile Tanzania and its FRELIMO surrogate.

FRELIMO leadership was Bantu in origin and held in contempt by the Makonde; however, the Makonde in Tanzania, numbering about 340,000, were willing to align themselves with FRELIMO to expel the Portuguese and reestablish the link with their people in Mozambique, numbering about 190,000.[12] In turn, the Makonde held no interest in pushing further south outside of the Maconde Plateau, so FRELIMO

12 Afonso and Matos Gomes, 130.

was considerably stalemated in the north as much from intertribal rivalry as from Portuguese martial activity.[13]

Stirrings had begun with the orchestrated presentation of grievances by the local Maconde in an open forum before the Cabo Delgado district governor on 16 June 1960. The hearing degenerated into an intimidation of the authorities, and this caused the frightened security force to fire on the crowd with resulting casualties. Later in the winter of 1960, when the Maconde gathered again with the governor for the usual *banja* or tribal meeting, the chief and his retinue were openly disrespectful and threatening. There had been rumors of unrest among the Maconde for unknown reasons, and the governor, who was advised not to attend the meeting, had never experienced hostility. He insisted on conducting business as usual, for he had been dealing with the Maconde chiefs for a long time. Indeed, he dismissed the warnings and ordered the security force to remain out of sight. Fortunately for the governor, the security force and its commander remained poised near the meeting and were able to limit Maconde actions. The source of this hostility was not identified until later as foreign "military counsellors" in Tanzania.[14]

FRELIMO initiated hostilities on the evening of 24 September 1964 with an assault on the residence and secretariat of the small administrative post or *posto* of Cobué in the Niassa district. The following day the naval launch *Castor* (P 580) was fired upon while on patrol in Lake Niassa, and the *posto* in Chai, on the southern edge of the Makonde plateau in the Cabo Delgado district, was assaulted. Later in December, the *posto* in Olivença, just north and east of Cobué, was attacked. A FRELIMO prisoner taken there revealed that it was using N'Gombe, a remote settlement in northwestern Niassa, as a base. The navy struggled to organize a force to neutralize it.[15] The available airpower to support the operation, a Lockheed PV-2 Harpoon and two North American T-6G Harvards at Vila Cabral, remained grounded and unable to help because of the weather, as it was the rainy season. Eventually the Portuguese prevailed, and Radio Tanzania at Dar-es-Salaam confirmed two days after the operation that FRELIMO had suffered eight killed and many gravely wounded and that the insurgents had been trained in Algeria.

Initially, the FRELIMO military strategy called for assaults on the Cabo Delgado and Niassa Districts in the north of Mozambique opposite Tanzania and a second front in the Téte District in the northwest opposite Malawi and Zambia, when both received their independence in 1964. However, a lack of support from these countries,

13 Peter J. H. Petter-Bowyer, *Winds of Destruction* (Victoria: Trafford, 2003), 229–230.

14 Fernando E. Ribeiro dos Ramos, "Mozambique: Unfinished Campaign," in *Memories of Portugal's African Wars, 1961–1974*, proceedings of a conference held 10 June 1997, King's College London, ed. John P. Cann (Quantico, Marine Corps University Foundation, 1998), 107.

15 Abel Barroso Hipólito, *A Pacificação do Niassa: Um Caso Concreto de Contraguerrilha* [The Pacification of Niassa: A Concrete Case of Counterguerrilla War] (Lisbon: privately published, 1970), 38.

a difficult logistics problem for revolutionary power projection, and a scarcity of people in Téte and Niassa to proselytize forced FRELIMO to restrict itself initially to Cabo Delgado and the Maconde plateau. Its campaign there met with regular disappointment, but this did not stop it from establishing a shameful record of exaggerated war claims. FRELIMO spokesmen systematically lauded triumphant victories, padded the toll of enemy casualties, and made wild claims on the existence of extensive liberated zones within Mozambique. Their pompous, misleading, and inaccurate "war communiqués" were a constant embarrassment to FRELIMO's Cuban, Chinese, and Eastern European allies, and no journalist or international authority was ever able to visit any of the claimed zones, as they did not exist.[16]

FRELIMO had commenced operations with only a handful of trained combatants, and its first moves were tentative in that its leadership was divided over the nature of its campaign.[17] The internal divisions were in part generated by military failure on the single northern front five years into the conflict, and dissention grew so great that disaffected elements ultimately assassinated Mondlane on 3 February 1969. Portuguese observers pointed to the culprits as Maoist elements in FRELIMO and their growing opposition to Mondlane's pro-West position; however, these internal squabbles continued for years after Mondlane's death and resulted in stinging reversals militarily and a tendency to settle internal scores in blood.

FRELIMO had begun with a planned program of infiltration, subversion, and intimidation of the population, and raids against Portuguese infrastructure. There was the hope of mass insurrection against the Portuguese authorities, again in the vein of Holden Roberto's logic, but the population was too sparse, dispersed, and divided for this course to be a realistic expectation. The Makonde in Mozambique, for instance, did not necessarily share the revolutionary fervor of their kin across the border. Some Makonde were persuaded to join FRELIMO; however, true to their traditions, many rejected this intrusion and were subjected to the standard tools of terror, such as, murder, torture, and selective kidnapping, when they did not provide active help. In 1973 Neil Bruce made the observation that, according to defectors, more than 450 Makonde had been murdered by FRELIMO, 1,000 wounded, and 6,500 captured since 1964.[18] The accuracy of these figures is less important than the fact that atrocities were being committed on quite a large scale by African against African. In a country of 9 million people, the population was composed of 19 tribes and 17 different language groups. For FRELIMO, it was important to create a broad popular base across these myriad factions, and yet its practices destroyed any genuine hope of doing so.

Niassa, the northern border district immediately west of Cabo Delgado, was thinly populated because of its rough terrain and yet attractive because of its fertility and

16 Gibson, 281.
17 Malyn Newett, *A History of Mozambique* (London: Hurst & Company, 1995), 523.
18 Ibid.

Mondlane addressing the 2nd Party Congress, Matchedje, Niassa. (Source: FRELIMO)

ease of cultivation. It was the home of the Nianja, who numbered about 30,000 and were treated in the same rough fashion that FRELIMO had earlier delivered on the Makonde, the Ajauas, the Makua, and others who refused to revolt and exhibited a "negative" attitude. Indeed, few in the population were interested in independence or were persuaded by violence. It became obvious that the "liberation movement" would not easily progress either in the north or further south.

Until 1967, FRELIMO was not able to extend its activities beyond the border districts of Cabo Delgado and Niassa. Its insurgent incursions were limited to penetrating the territory short distances with small groups that made attacks. These were actually strikes in that they normally originated from bases in Tanzania just across the border, and after a brief hit-and-run action in Mozambique, the attackers quickly returned to their sanctuary.[19] There were some elements of FRELIMO who lived among the population in Cabo Delgado, and these provided intelligence and assistance.

When this strategy failure became evident in the late 1960s, FRELIMO reorganized itself at its 2nd Party Congress held in July 1968 at the settlement of Matchedje in a "liberated zone" in northwestern Niassa. The site was a few miles into Portuguese territory from the Tanzanian frontier in very remote and sparsely populated terrain and was only "liberated" in the sense that for the five days of the Congress the attendees went undisturbed. Its reorganization consisted of merging its civil and military components and establishing a military force within Mozambique organized

19 António de Jesus Bispo, "A Participação da Força Aérea na Guerra de África (1961–1975)" [Air Force Participation in the African War (1961–1975)], *Revista Militar* 2507 (December 2010): 1396.

by sectors, each with its own battalion formed with three companies of 150 men each. The regular forces were to be supported by a popular militia. This was the ideal; however, 450 men in a battalion at a FRELIMO base would quickly attract attention with devastating aerial bombardment, so in practice it was modified to fit the battlefield.[20]

The FRELIMO sectors within Mozambique were geographical divisions within the Portuguese districts, each with a central or provincial base for each sector, and under each of these there were sub-provincial bases. In Cabo Delgado, the center of the fighting, there was originally a single sector, and Base Moçambique was its center with 180 men. It had a sub-base Gungunhana and eleven operational bases: Angola, António Enes, Gaza, Inhambane, Limpopo, Lúrio, Manica, Maputo-Macomia, Marrupa, Nampula, and Beira. In most cases these names bore no relation to their actual location.[21]

In Niassa, the military was organized into three sectors: east (300 to 350 men), west (350 to 450 men), and south (150 to 200 men). There were also eight operational bases: Mepotxe, Meponda, Maniamba, Unango, Téte, Beira, Catur, and Liconnir. Later in Téte there were four military sectors: The first was north of the Zambeze River and west of the Capoche River with two encampments, one in Zambué and the other in Zumbo. The second was to the east of the Capoche River and northeast of the Téte-Zóbuè road. The third was south of the Zambeze River, and the fourth centered on the councils of Maotize and Mutarara.[22]

In 1968 FRELIMO acted to initiate a second front in the Téte District prompted by Zambian permission to do so.[23] In looking at a map of the area, one can readily see that it would have been far easier to use Malawi for that front; however, its president, Hastings Banda, would not permit it for fear both that his communication through Mozambique to the sea would be severed and that such activity would attract unwanted Portuguese, Rhodesian, and South African military incursions into his country. Eventually, however, Banda succumbed and gave tacit support in one of his tortuous changes in political direction.

Following Mondlane's death and Operation *Nó Górdio* (Gordian Knot) in June 1970, FRELIMO, under its new, aggressive leader Samora Machel, began to move its troops to Zambia and develop a series of bases along the Zambian border north of the Zambeze River and in the Téte District of Mozambique. These forces transited Malawi with Banda turning a blind eye much to Portuguese distress.[24] Indeed, Al

20 Ibid.
21 Ibid.
22 Ibid.
23 António José Telo, *História da Marinha Portuguesa: Homens, Doutrinas e Organização, 1824-1974 (Tomo I)* [History of the Portuguese Navy: Men, Doctrine and Organization, 1824–1974 (Volume I)] (Lisbon: Academia de Marinha, 1999), 616.
24 Adelino Serras Pires and Fiona Claire Capstick, *The Winds of Havoc* (New York: St. Martin's Press, 2001), 83.

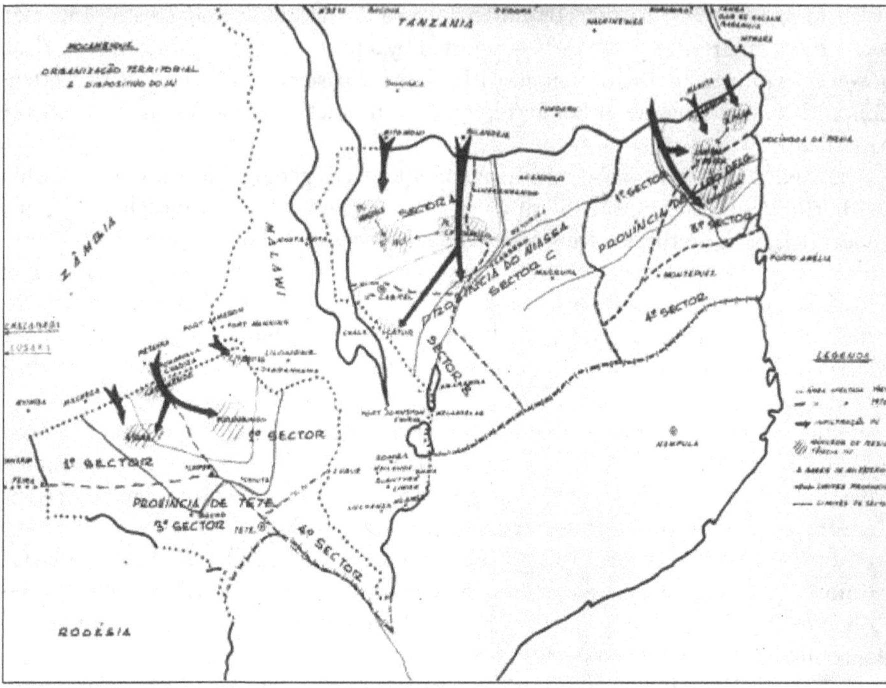

Map 9.3 FRELIMO Lines of Penetration and Territorial Organization. (Source: PIDE/DGS archives)

Aerial view of exposed huts at Base Moçambique in 1974. (Source: Personal archive of José Luís Silva)

Venter notes that FRELIMO insurgents were traveling openly by the busload from Tanzania and were supported by the Malawi police, who were making certain that they did not linger there.[25] This development forced General Kaúlza de Arriaga, who was now Commander-in-Chief of Portuguese Forces in Mozambique, to shift his attention to the northeast.

Téte was, like Niassa and Cabo Delgado, a sparsely populated area. Within the district, however, Portugal had embarked on an ambitious project in 1969 at Cahora Bassa (variously Cabora Bassa) to dam the upper reaches of the Zambeze River as a source of inexpensive electrical power for Mozambique and South Africa. The dam was also designed to create an enormous reservoir 280 kilometers in length and 38 kilometers at its widest point.[26] This source would irrigate here-to-fore dry but fertile lands in an effort to bring prosperity to Téte and Mozambique and harness this income for the *metrópole*. This project and its lake were seen potentially as an effective block to FRELIMO infiltration through Téte and into Rhodesia and southern Mozambique. The dam construction site was also a tempting target for FRELIMO with its Soviet 122mm Katyusha rockets. As it was, by 1971 FRELIMO was active in the area south of the Zambeze in Téte near the village of Daque and at the several neighboring villages of Deveteve, Nura, and Serra do Comboio.[27] These FRELIMO bases were both sources of attacks on Cahora Bassa and penetration southward. They also hosted Zimbabwean insurgents, a source of great concern to Rhodesia, which began conducting "hot pursuit" operations into Téte.

Still by the end of the war in 1974, FRELIMO penetration of Mozambique was limited largely to Téte, Niassa, and Cabo Delgado, and was stalemated militarily. Over the years of fighting, FRELIMO had created support only among a small minority of the population in the north, and so few further south, that it was exceedingly difficult for it to govern Mozambique for decades afterward. Indeed, by 1974 only perhaps 20 percent of Mozambicans wanted rule by FRELIMO.[28] Insurgencies, however, are primarily political wars, and Portugal was about to grant FRELIMO an unearned and unpopular political victory over Mozambique and its people, one that it could win neither militarily in the field nor politically at the ballot box on its own. FRELIMO was soon to discover that without political legitimacy it could not govern successfully, and the greatest suffering was yet to come for the Mozambican people.

25 Venter, 47–48.

26 Telo, 616.

27 Henrik Ellert, *The Rhodesian Front War, Counter-insurgency and Guerrilla Warfare 1962-1980* (Gweru, Zimbabwe: Mambo Press, 1993), 88.

28 Giancarlo Coccia, *The Scorpion Sting* (Johannesburg: Livraria Moderna, 1976), 141.

Portuguese Reaction

Anticipation of these problems began in late 1956, when Colonel Kaúlza de Arriaga, as Subsecretary of State for Aeronautics, authorized a military aviation presence in Mozambique, a move that was considered both bold and timely. The move had an enormously positive psychological impact locally in that it gave the people renewed confidence in the government and its commitment to protect them. It also sent an early message to the enemy that Portugal was prepared to defend Mozambique and its people with force.

The establishment of an air force in Mozambique required the accelerated construction of a network of bases from which airpower could be projected rapidly across the province. To be effective this infrastructure needed to be linked together by logistic transport and radio communications capabilities. The entire complex required time to put in place, and the FAP bent to this task with a will, as Portugal could not prevail without it.

This was not the first time that Portugal had attempted airpower projection in Mozambique, as in August 1917 the Expeditionary Squadron of Mozambique (*Esquadrilha Aérea a Moçambique*) under the command of cavalry Captain Teodorico Ferreira dos Santos (observer) arrived by steamship at Mocímboa da Praia, a seaport located about 50 miles from the northern frontier. The squadron consisted of three Farman F-40 aircraft, each powered by a Renault 8-cylinder, 130 horsepower engine, and manned by a pilot and an observer. The Farman was armed with a Lewis 7.7mm machinegun and diverse light bombs.[29] One of the three was promptly wrecked during its initial flight, killing the lone squadron pilot, cavalry Lieutenant Jorge de Sousa Gorgulho (pilot), and leaving the squadron with two operable aircraft and no pilot. Flight operations were resumed when cavalry Captain Francisco da Cunha Aragão (pilot) arrived in November. Other distinguished officers of the squadron were Lieutenant Pinheiro Corrêa (observer) and Lieutenant Francisco Higino Craveiro Lopes (pilot), the latter serving as President of the Republic between 1951 and 1958 after a distinguished career in the army air arm. The squadron moved to Lourenço Marques in December 1918 following the Armistice, had its name simplified to Aviation Squadron (*Esquadrilha de Aviação*), and was ultimately disbanded in 1920. The new 3rd Air Region was to prove a tribute to the audacity, combativeness, and idealism of these early aviators.

29 "*Esquadrilha Aérea a Moçambique*" [Air Squadron of Mozambique], *Momentos de História: Grande Guerra 1914-1918*, 10 October 2013, http://www.momentosdehistoria.com/MH_05_02_05_Exercito.htm (accessed 10 January 2014).

Farman F-40 preparing to launch on its disastrous initial flight at Mocímboa da Praia, Mozambique. (Source: Pinheiro Corrêa archive)

Infrastructure Development

The expansion of the 3rd Air Region may be divided into four phases: The first corresponds to the establishment of its headquarters in Lourenço Marques, which integrated the air combat elements that then existed in Mozambique. The FAP presence at the time was small, and much depended on the existing infrastructure. The first components of the airbase in Beira, for instance, were housed in the control tower and comprised a headquarters and ultimately a center for radio communications that linked FAP resources over the entire 1,800-kilometer length of Mozambique.[30] Simultaneous construction was underway on the basic infrastructure of key airfields as well, and these enabled aircraft to launch against any trouble spot in Mozambique. The airfields were located strategically in the north where the threat was developing and were positioned in a network that considered the radius of action of the FAP inventory of aircraft and initially covered Cabo Delgado and Niassa, where subversion was being immediately felt. Later the network would be extended to Téte when

30 Manuel Pinto Machado de Barros, "A 3.ª Região Aérea" [The 3rd Air Region], *Mais Alto* (May 1964): 39.

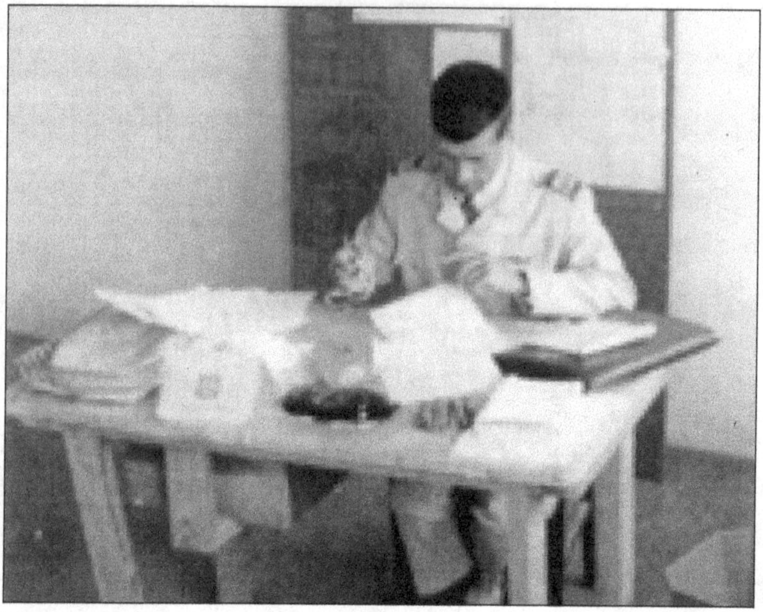

Lt Col Jorge M. Brochado de Miranda, commander of AB 5 in 1962 in his office. He would later become Chief of Staff of FAP. (Source: Força Aérea Portuguesa)

FRELIMO incursions developed there.[31] This network also brought economic benefit to remote areas and improved the well-being of the population, an important consideration in counterinsurgency.

The FAP acquired the land around Manga Loforte next to Beira and in December 1961 built what was at the time AB 5. The hangar of the local airline DETA was available in the first months of 1962, and the FAP aircraft were then able to benefit from the DETA maintenance services.[32] In this phase, which lasted about a year, FAP personnel exhibited great energy and enthusiasm in building the infrastructure that would both protect and enhance this important piece of Portugal. Air activity centered on transport missions for the army, pilot training, and familiarization flights throughout the province.[33]

The second phase concentrated on the expansion and full completion of the initial round of infrastructure projects, deployment of personnel, and the establishment of aviation squadrons. It also marked the beginning of a serious aviation presence in

31 Jesus Bispo, 1397.

32 Mozambique Airlines was founded in 1936 as DETA (*Divisão de Exploração dos Transportes Aéreos de Moçambique*), a division of the Ports and Railway Services.

33 Barros, 40.

Mozambique. Nacala was made AB 5 in late 1965, when all service headquarters were moved to Nampula, the capital of Cabo Delgado, where they would be nearer the vulnerable northern frontier. Beira then became BA 10, the only full-fledged Air Base in Mozambique. The next step down in capability was the Aerodrome Base, of which there were four: Nacala (AB 5), Nova Freixo (AB 6), Téte (AB 7), and Lourenço Marques (AB 8). The three of these in the north had maneuver airfields as dependencies. In all, there were 15 highly capable, major aviation bases built and over 200 small airfields scattered throughout the territory to support military and civil operations.[34]

Together with this construction, the FAVs were organized, and procedures established for operating in support of the land and naval forces. The FAV in Mozambique was similar to the same organization by that name in Angola, and the first ones appeared in June 1962, when the Air Club of Beira (ACB) was authorized to create a FAV and proceeded to construct a new headquarters and acquire training and observation aircraft. This was accompanied by the requalifying of old pilots to meet expanded needs.[35]

By October, FAV numbers 301 (Beira) and 302 (Lourenço Marques) were fully operational with 21 pilots and 53 pilots respectively. There was indeed a female pilot serving in FAV 301. As in Angola, the inventory consisted of a mix of light aircraft: Piper Tripacers, Colts, Apaches, and Super Cubs; De Havilland Chipmunks and Tiger Moths; Beechcraft Bonanzas, Cessna 140s, and Austers.[36]

Gradually during 1963, FAV 302 (Detachment João Belo), FAV 303 (AB 5), FAV 304 (Cabo Delgado) and FAV 305 (Inhambane) were established. In 1964, FAV 306 (Quelimane), FAV 307 (Vila Cabral), FAV 308 (Téte), FAV 309 (João Belo – old detachment FAV 302 JB), FAV 310 (Vila Pery – old detachment FAV 301 VP), FAV 311 (António Eanes – old detachment FAV 303 AE), and FAV 312 (Lumbo – old detachment FAV 303 L) were added. This expansion was reflective of a reorganization to cover the territory more effectively.

The role of the FAVs was similar to that of their counterparts in the north of Angola. They delivered mail and necessities to troops in the field and evacuated the sick and wounded in what often seemed like a relay race. The FAVs were supported by the FAP with fuel and lubricants and an operating budget. In early 1972 when Brigadier Manuel Diogo Neto became the commander of the 3rd Air Region, he reduced the FAV organization to activity south of the Save River, as operations in the north did not justify the use of FAP resources. The reduced force operated through the end of the conflict in April 1974.[37]

34 Jesus Bispo, 1398.

35 Pedro Manuel Ferreira, "FAV Formações Aéreas Voluntárias, Uma Singularidade Portuguesa no Teatro de Guerra Africano," II Parte [FAV Volunteer Air Formations, A Portuguese Peculiarity in the African War Theater, Part II], *Mais Alto* 349 (June 2004): 30–35.

36 Ibid.

37 Ibid.

De Havilland Chipmunk at the Air Club of Inhambane Hangar.
(Source: Força Aérea Portuguesa)

By the middle of 1963 the third phase had begun, and it focused on bringing the new airfields to the proper standards with functioning aids to navigation, such as radio beacons, new messing facilities and quarters, and hangars. A military mail system was established with the installation of robust air transport support. Frontier security was enhanced by the effective cooperation with ground forces and the addition of a photoreconnaissance capability. The expanding network of bases supported rapid reaction air detachments that could launch on short notice in support of local security needs.

The fourth and final phase saw the completion of the airfield network in which temporary construction was replaced with permanent structures. The 3rd Air Region continued to mature in its capabilities to support the ground and naval forces and the population. By the end of 1963 and without fanfare, aviation infrastructure had reached a wartime footing well ahead of hostilities.

During this building period, the FAP acquired a number of varied aircraft for its less orthodox missions. A Major Cardoso was quoted as observing that "ostensibly the registration of FAP 3406 (a Cessna 401 twin-engine light aircraft) was without any plausible reason…."[38] Yet, the FAP owned six Cessnas, four Piper Cherokees, and a number of light aircraft. This unmarked small transport fleet with its contract civilian

38 Tomás G. Conceição Silva, "A Verdadeira 3406 História" [The True 3406 Story], *Mais Alto* 332 (July–August 2001): 11.

FAV aircraft at Muidumbe delivering mail, beer, and fresh food during Operation *Gordian Knot*, July 1970. (Source: Personal archive of Manuel Serafim Matos Sousa)

Operation *Defoliant Dakota*, spraying system along wing trailing edge. (Source: Personal archive of Carlos Costa)

and military pilots flew sensitive cargo and personnel discretely to often controversial destinations. The euphemism for their flying, according to one pilot, was "Public Relations Work."[39]

There was also an important crop spraying mission that was performed on the large fields of crops necessary to feed the residents of the protected villages (*aldeamentos*) and other populations. In addition to crop management spraying there was an Operation *Defoliant*, which had as its objective the destruction of the maize crop planted by that part of the population supporting FRELIMO. The defoliant chemical was sprayed from a Dakota escorted by a Fiat G-91 in a tactic that proved extremely effective. The Fiat would fly at an altitude of 10,000 feet on a parallel course with that of the Dakota to avoid antiaircraft artillery or missiles and would provide direction to the Dakota flying at treetop level, guiding it to the targeted areas of enemy cultivation or *machambas*. When the Dakota was about to overfly the *machamba*, the Fiat would radio the distance to the release point for the defoliant. At a point two miles out (about one minute at the normal airspeed of a Dakota), the Fiat, which was armed with four 50 kilogram bombs, would drop one. The bomb explosion, which occurred 116 seconds after release at 10,000 feet, had two purposes: it distracted any anti-air reaction away from the Dakota and signaled the Dakota to begin releasing the defoliant.[40]

Building Offensive Capability

In February 1962 the first aircraft with significant attack capability arrived at the airfield in Beira (at the time AB 5) in the form of two Lockheed PV-2 Harpoons.[41] The Harpoon was a tired and obsolete aircraft slated for retirement with the advent of its replacement, the P2V-5 Neptune open ocean reconnaissance aircraft. However, as so many older "obsolete" aircraft find utility in the low technology of counterinsurgency, the Harpoon found a second life with the FAP in Africa in its adaptation for tactical bombing and close air support. It had first proved itself in Angola and was now to be tried in Mozambique. Prior to this development only transport operations had been conducted, and these centered on three C-47 Dakotas and six Noratlases at Lourenço Marques, then the headquarters of the 3rd Air Region.

In December 1961, Alverca had readied two PV-2s for their new mission, and these spent the next two months being nursed to Beira with the support of a Douglas SC-54D Skymaster from Lajes Field in the Azores.[42] On arrival the two Harpoons were assigned to Reconnaissance and Fire Support Squadron 103, which was

39 Conceição Silva, 12.
40 José Armando Vizela Cardoso, correspondence with the author, 2 July 2012, Lisbon.
41 Pedro Manuel Ferreira, "PV-2 em Moçambique, 1ª Parte" [PV-2 in Mozambique, 1st Part], *Mais Alto* 363 (September–October 2006): 38–42.
42 Ibid. The repair and refurbishing facility known as OGMA (Oficinas Gerais de Material Aeronáutico) is located in Alverca, Portugal, and is often simply called Alverca.

established on 3 February 1962 with Captain Marinho Falcão as its commanding officer.[43] The squadron was relatively small with five pilots (3 officers and 2 sergeants), a maintenance officer, and 35 enlisted specialists (radar and radio operators, mechanics, armorers, and electricians), and was housed in the local civil air hangar kindly loaned by the resident airline. Some weeks later the two aircraft were engaged in support of paratroops at Mueda in Cabo Delgado.[44]

By October of that year aviation in Beira had been reorganized into two squadrons and a flight under Operational Group 1001 (GO 1001). The squadrons were Squadron 101, formerly Squadron 103, with the Harpoon aircraft and Squadron 102 with the Noratlas and Dakota aircraft. The Liaison and Training Flight had the Auster and Harvard aircraft assigned. In November, the six Dakotas were transferred to Lourenço Marques (AB 8).

Squadron 101 suffered its first fatal accident on 5 February 1963, when a Harpoon flew a low-level reconnaissance mission about 30 miles south of Vila Pery in the region of the Mavuzi Dam and contacted a high-tension electrical pylon. An electrical wire became wrapped around the starboard propeller and caused the aircraft to crash, killing the pilot and five crewmen.[45] By a miracle two other crewmen were pulled alive from the burning wreckage. This aircraft had been modified to use the Williamson F24 aerial camera, the only one so equipped in the squadron.[46] Hence, its replacement needed to be modified similarly before its targeted delivery in September 1963.

When the war opened in October 1964, Squadron 101 had as its primary mission visual reconnaissance of the FRELIMO infiltration routes that crossed the frontier with Malawi. This meant a long round trip to the western border of Niassa. It also supported ground forces with close air support in cooperation with the 24 T-6G Harvards against internal FRELIMO bases. In May 1965, the last of the eight Harpoons arrived at Beira. In August, Squadron 101 was augmented with several Dorniers when the Liaison and Training Flight was disbanded. The fragile Austers went to Squadron 102. Shortly afterwards ten T-6Gs were also assigned to Squadron 101 and were sent as a detachment to Téte (AB 7). Towards the end of 1965, the Harpoon was clearly reaching the end of its useful life with a number of mishaps, but it remained an effective fire support platform with its explosive and incendiary bombs, rockets, and machineguns and was likewise employed extensively in low altitude

43 Ibid.
44 Ibid.
45 Ibid.
46 The F24 camera is a 1920s era British aerial camera used by Allied forces into the mid-1950s. Its conceptual design came from the Royal Aircraft Establishment at Farnborough in 1925, and its detailed design, and most production, was performed by the Williamson Manufacturing Company Ltd of Willesden Green, London. During World War II, the F24 was also manufactured by Vinten, and a lighter version was designed and manufactured by Eastman Kodak as the K-24 for use by U.S. forces.

reconnaissance and surveillance missions.[47] It was a noisy aircraft, so stealth was not part of its reconnaissance duties.

It had become clear that a replacement was needed for the Harpoons, which were plagued by maintenance shortfalls and a consequent low availability. The Fiat G-91 Gina was in the acquisition process and would soon appear in the skies of Mozambique.[48] Even as the G-91s were arriving, the Harpoons struggled throughout 1966 to maintain readiness in the face of engine failures and the loss of two additional aircraft in accidents.[49] The remaining four Harpoons limped on for four more years and were transferred eventually to the 2nd Air Region in Angola between 1970 and 1971.[50] Despite a struggle with the platform, it performed well and would be replaced with a combination of aircraft to perform its varied missions. The Dornier and Dakota would undertake surveillance, and the Fiats would assume a reconnaissance and attack role.

The arrival of the Fiats was important, as the FAP needed an attack aircraft that could carry substantial ordnance into the fight. The PV-2s were very useful with their five .50-caliber machineguns in the nose, eight 5-inch rockets beneath the wings, and a 4,000-pound bomb load in the internal bomb bay. The age of the aircraft and its consequent maintenance difficulties compounded by the remoteness of Mozambique continued to hamper its usefulness, and its numbers were eventually reduced to two through further accidents and lack of parts. The T-6G was a light attack aircraft known by its pilots as the "F-110," because it took off at 110 knots, cruised at 110 knots, and landed at 110 knots. The 24 Harvards assigned to Mozambique had served the FAP well, as they were rugged, easily maintained, and effective light attack aircraft with their twin pods housing 7.5mm machineguns and racks of fragmentation bombs, rockets, and napalm canisters. They stood up well in the face of ground fire, had good endurance, and were available in quantity when originally acquired by the FAP. By the time of their use in Mozambique, the once plentiful supply of spare parts and surplus engines had begun to dry up, and the airframes were wearing out. As an example, by this time, the rear cockpit instruments throughout the fleet had been cannibalized to replace front cockpit ones that had ceased to function.[51] So much depended on the six new Fiat G-91s, each with four Colt-Browning .50-caliber machine guns with 300 rounds each and external racks for up to 1,000 pounds (454 kilograms) of ordnance.[52] They did not have the bomb load of the PV-2, but their reliability, speed, and numbers would fill the gap.

47 Pedro Manuel Ferreira, "PV-2 em Moçambique, 2ª Parte" [PV-2 in Mozambique, 2nd Part], *Mais Alto* 364 (November–December 2006): 39.
48 José Manuel Correia, "Fiat G-91/R4 em Moçambique, 1ª Parte" [Fiat G-91/R4 in Mozambique, Part 1], *Mais Alto* 385 (May–June 2010): 26–30.
49 Ferreira, 40.
50 Ibid., 42.
51 Petter-Bowyer, 232.
52 Estado Maior da Força Aérea, 2ª Repartição, Memorandum No 6/61, Assunto: A Força Aérea em Moçambique, 9 June 1961.

A Difficult Fight

In the year following the September 1964 FRELIMO attacks, Mueda became the center of enemy activity and remained so throughout the war, as it was the point of convergence of every important line of communication in Cabo Delgado. Roads from Mocímboa da Praia on the Indian Ocean connected through Mueda to Nangade and further north to the Tanzanian border, to Mocímboa do Rovuma to the northwest, and to Nangololo to the south. FRELIMO worked to establish a substantial presence surrounding Mueda, making it somewhat of a besieged outpost, with bases in the areas of Nangade, Mocímboa do Rovuma, Mutamba dos Macondes, Mocímboa da Praia, and Nangololo.

The troops and aircrews assigned to Mueda were attacked often, and these attacks came at either dawn or dusk, periods of the day that became known in local parlance as the "Maconde Hour" (*Hora Maconde*). They were constantly awakened to alarms in the early morning hours to repel an attack, launch aircraft, and counterattack. They were subjected to insidious enemy propaganda that sapped their morale. They spent day after day under the tension of a combat zone and the uncertainty that such danger brought. For them it was also a frustrating war in that while they killed many of the enemy, he continued to flee to his sanctuary in Tanzania and return with his ranks refilled.[53] Mueda came to be the "land of war."

From time to time frustration with the enemy lodgment in the formidable terrain surrounding Mueda would reach a painful level and prompt the Portuguese commander-in-chief to address the situation. Just such an occasion occurred in mid-1965, and accordingly, a grand conventional operation styled *Águia* (Eagle) was put in place and designed to sweep the FRELIMO cadres from the vast area between the Rovuma and Messalo Rivers in Cabo Delgado and demonstrate the superiority of the Portuguese forces to the local population. It employed two battalions of infantry and a battery of 88mm artillery.[54] The FAP from its airfield in Mueda (AM 51) was able to support these ground operations with offensive fires as much as the evacuation of personnel and the transport of fresh food and mail.

This operation was a reflection of the personality of the first Commander-in-Chief of the Armed Forces in Mozambique, General João Caeiro Carrasco, who failed to gain control over the developing situation in the north with his conventional approach and its lack of subtlety. His actions encouraged the local population to flee to areas less accessible to military and civilian authorities, as the people were frightened by his heavy-handed military actions and consequently gave their support to the traditional

53 António de Almeida Tomé, "Mueda: A Mais Importante Base Operacional da Força Aérea no Norte de Moçambique" [Mueda: The Most Important Operational Base of the Air Force in the North of Mozambique], *Mais Alto* 236 (July–August 1985): 24.

54 Afonso and Matos Gomes, 192–193.

chiefs who were behind the emerging nationalist organizations.[55] Caeiro Carrasco directed the war from his headquarters in Lourenço Marques, about as far as one could get from the fighting and remain in Mozambique, and concentrated most of the troops and services in the south. *Águia* with all of its firepower was thought to be the answer to the situation in Cabo Delgado. Arriaga would repeat the prescription.

As in Guiné with Operation *Trident*, *Águia* had an initial phase in which the maximum available troops were deployed to occupy key points and conduct offensive operations from these sites. When troops began to deploy from Mueda on 2 July 1965 towards the enemy strongholds, the going became difficult because of enemy ambushes and the sweeping fire of automatic weapons. It should be noted that this modern form of subtle warfare was alien to the African character, which tends to be basic and direct. Africans have traditionally employed frontal attacks mustering hundreds of warriors with a disregard to the potential for casualties. The adaptation to this indirect method reflected the influence of seasoned foreign instructors.[56] These ambushes in the early days of the operation caused Portuguese casualties of four dead and 18 wounded.[57]

On 8 July, air and artillery bombardments were conducted within the Miteda-Nangololo-Muatide triangle and were accompanied by Portuguese ambushes established along the probable lines of insurgent retreat. Portuguese forces had their hands full, as FRELIMO proved very aggressive in countering their actions and even fired at the Auster airborne command post, as it did its work. Portuguese columns were constantly ambushed during the last two weeks of July and suffered seven more dead and 27 wounded.[58] The enemy proved numerous, motivated, well-armed, well supplied with munitions, and completely at ease in the area.

By the end of July, *Áquia* entered a new phase in which actions against the enemy were divided into nine color-coded subsets: *Águia Branca* (white), *Águia Amarela* (yellow), *Águia Vermelha* (red), etc. The operation continued under this new structure until 6 September. By the termination date the situation in the Maconde Plateau had reached a condition that would characterize the fighting in subsequent years: great violence through the use of mines and accompanying ambushes, especially over the roads and lines of communication; loss of control over a major portion of the population on the plateau with the exception of those who lived near military installations; and the presence of significant numbers of well-armed FRELIMO troops in relatively large bases hidden in the interior of the plateau and supported by logistic lines from Tanzania.[59] In the end, a modest amount of arms were captured, a great deal of FRELIMO infrastructure destroyed, and FRELIMO losses realized: 93 dead, 32

55 Carlos de Matos Gomes, *Moçambique 1970, Operação Nó Górdio* [Mozambique 1970, Operation Gordian Knot] (Lisbon: Prefácio, 2002), 13.

56 Ribeiro dos Ramos, 107–108.

57 Afonso and Matos Gomes, 111.

58 Ibid.

59 Ibid.

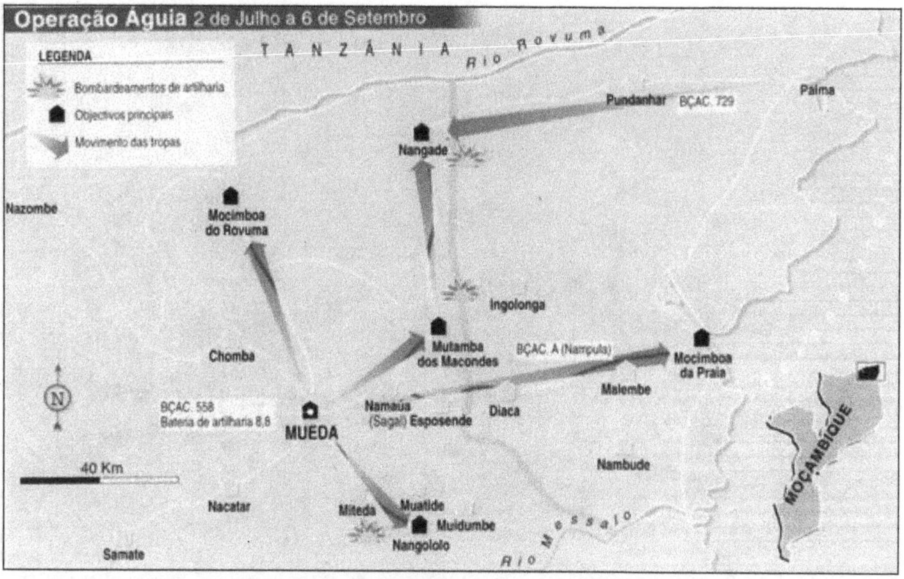

Map 9.4 Operation *Águia* (Eagle), 2 July to 6 September 1965.
(Source: *Guerra Colonial* by Afonso and Matos Gomes)

wounded, and 75 taken prisoner.[60] This was typical of grand operations in counterinsurgency, as invariably the enemy proves elusive, and the government is disappointed over the meagre results alongside the great effort expended.

In August towards the conclusion of *Águia*, a new ground force commander, Brigadier Francisco da Costa Gomes, was appointed, and the following month General António Augusto dos Santos relieved Caeiro Carrasco as overall commander. These two new generals transferred their commands to the north to be next to the primary area of operations and created a Northern Intervention Zone (*Zona de Intervenção Norte* or ZIN) with headquarters in Nampula. The districts of Niassa and Cabo Delgado were designated as Sectors A and B respectively and each was commanded by a general officer. Within these sectors the army established a checkerboard or *quadrícula* network of companies and battalions to blanket the terrain, particularly the area next to the Tanzanian frontier.[61]

This deployment of forces was the product of an operational concept that sought to interdict the FRELIMO insurgents as they crossed the Rovuma River and to bar their progress southward. A second barrier was established at the Messalo River and a third

60 Ibid.
61 Matos Gomes, *Moçambique 1970, Operação Nó Górdio*, 15.

Dornier during Operation *Zeta*. (Source: Força Aérea Portuguesa)

at the Lúrio River.[62] The effect of these three barriers was to stalemate FRELIMO in the north and limit its advance to the border area and the Maconde Plateau. This strategy was supported primarily by the airbases of Nacala (AB 5), Mueda (AM 51), and Nampula (AM 52). Nevertheless, within this Sector B it was difficult to prevent the establishment of bases and arms and munitions depots, for the terrain was difficult, as in the Dembos of Angola, and as soon as one enemy installation was discovered and eliminated, another would spring up. Hence, it was constant work to counter the persistent FRELIMO activity.

Also by 1965, the army had established a similar *quadrícula* network of units centered around Nova Frexio in Sector A and supported by the airfield AB 6 there. While these forces were sufficient to counter the advance of FRELIMO into that area, it was positioned well south of the northern border and defaulted to the enemy much of Niassa. Accordingly, by mid-1965 FRELIMO had penetrated southward by way of Lake Niassa and established a presence along the shore of the lake at five villages: Nova Coimbra, Metangula, Maniamba, Mandimba, and Mecanhelas.[63]

62 Ibid.

63 Luís Alves de Fraga, *A Força Aérea na Guerra em África: Angola, Guiné e Moçambique, 1961–1974* [The Air Force in the War in Africa: Angola, Guiné and Mozambique, 1961–1974] (Lisbon: Préfacio, 2004), 132.

On 8 January 1969, two Fiats departed BA 10 in Beira and landed in Nampula. The next day two more Fiats were deployed to AB 5 in Nacala. These four aircraft began their familiarization flights, and by April there were a total of eight Fiats in northern Mozambique, comprising Squadron 502, the *Jaguares*. The initial close air support mission for the new arrivals was flown on 9 April. In May a two-plane detachment was sent to AB 6 in Nova Freixo to execute planned attacks on enemy installations. In June the squadron became committed to supporting the various phases of Operation *Pesquisa* (Search), in which its aircraft conducted independent attacks utilizing general purpose (GP) bombs of 50 and 100 kilograms and 350-liter napalm canisters along the routes of infiltration that paralleled the Capoche, Manga, and Chissiboe Rivers from Zambia into northern Téte.[64] These were not apparently followed by ground force sweeps but were an attempt to reduce the enemy by airpower alone, always an inexact proposition. Meanwhile during this early period, the small squadron was dealt a difficult blow with the tragic loss of two pilots while they were flying a T-6G. The Fiat pilots were also qualified in the T-6G and often flew missions in these as well as their Fiats. Later in August, the Fiats flew similar missions in the area along the Camassanzo, Cassolo, Luangue, and Chissiboe Rivers and the infiltration routes from Zambia into northern Téte.

While the Fiats were busy, other FRELIMO infiltration routes were receiving attention in Cabo Delgado, and an important, specifically targeted operation on the Maconde Plateau styled Operation *Zeta* began on 7 June 1969. It featured the insertion of two companies of paratroops, one from BCP 31 (Beira) and the other from BCP 32 (Nacala), into the zone of Malambuage just south of the Rovuma River, where there were FRELIMO depots that supported infiltration from bases in Tanzania, particularly Base Limpopo. It was an area that lent itself to an air envelopment by paratroops, but they had to act promptly before the FRELIMO infiltration moved to its objective, the heart of the Maconde Plateau and the triangle bounded by Mueda, Nangololo, and Sagal.[65] The operation was based on strong intelligence collection, and its planning was done under the tightest security and on very short notice, as it was in reaction to a window of opportunity that would shut quickly.[66]

The operation represented perhaps the most significant paratroop drop in Mozambique, as it was the first and largest operational drop in that theater and one of the rare ones during the entire war. The aim of *Zeta* was to overcome by surprise a series of enemy supply bases whose defenses were completely oriented to a ground force assault and thus vulnerable from the air. In fact surprise was total, and FRELIMO forces fled in typical insurgent fashion. The bombardment inflicted by the PV-2s and

64 José Manuel Correia, "Fiat G-91/R4 em Moçambique, 2ª Parte" [Fiat G-91/R4 in Mozambique, Part 2], *Mais Alto* 386 (July–August 2010): 32–33.

65 Afonso and Matos Gomes, 192.

66 Victor Elias, "A Operação Zeta," http://cuamba.blogspot.com/2010/05/operacao-zeta.html (accessed 6 May 2014).

T-6Gs on the fugitives softened their enthusiasm for combat and cut off their lines of flight. Indeed the operation was so well executed and met with such weak resistance that over its three days it became in truth an exercise in teaching tactical maneuvers that would be useful in future operational deployments. At the abandoned enemy Base Limpopo and in the settlement of Balade the paratroops raised mines, disarmed booby traps, and linked with army units that cooperated in a sweep of the plateau. The relatively new Alouette III helicopters were present in full and moved forces rapidly around the battlefield to outmaneuver the escaping FRELIMO cadres.[67] In addition to the helicopter transports, other air support was critical and consisted of:

1 Alouette III gunship armed with a 20mm canon
8 T-6G Harvards that attacked the escaping insurgents
2 PV-2 Harpoons for close air support
4 Noratlas transports to insert the paratroops
3 Dakota transports to insert the paratroops

Despite its apparent success, the insurgents acted predictably and disappeared when a vastly superior force descended from the sky, so there were few killed or captured. They abandoned extraordinarily large dumps of weapons and munitions and made no attempt to hold ground. Normally their bases were in terrain that was inhospitable to a parachute assault and safe from Portuguese forces with their shortage of air drop capability, so they were rarely disturbed.[68] *Zeta* was the exception.

Alouette IIIs Arrive

Prior to the arrival of the Alouette III, there was but a single Alouette II in Nampula that had been transferred there some months before from Squadron 94 in Luanda. The obvious need for helicopter support in ground operations prompted augmentation of the lone vehicle with first one and later a second squadron of the new Alouette III.

The Alouette III helicopters that distinguished the success of Operation *Zeta* arrived in Mozambique between July and October 1966 and began their first operational flights in April of the following year. The initial six delivered were assigned in November 1966 to Squadron 503, the *Indios* (Indians), at Nacala (AB 5).[69] Presumably the name was adopted from the *Indios* being "on the warpath." Immediately the helicopters were deployed to bases next to the combat zones where insurgent activity and fighting was the most intense. The typical deployment sites were Nampula (AM 52), Mueda (AM 51), and Vila Cabral (AM 61).

67 Ibid., 193.
68 Ibid.
69 Mário Diniz and Luís Proença, "AL III História breve de uma vida longa, 1963–2013" [AL III brief history of a long life, 1963–2013] *Mais Alto* 404 (July–August 2013): 22.

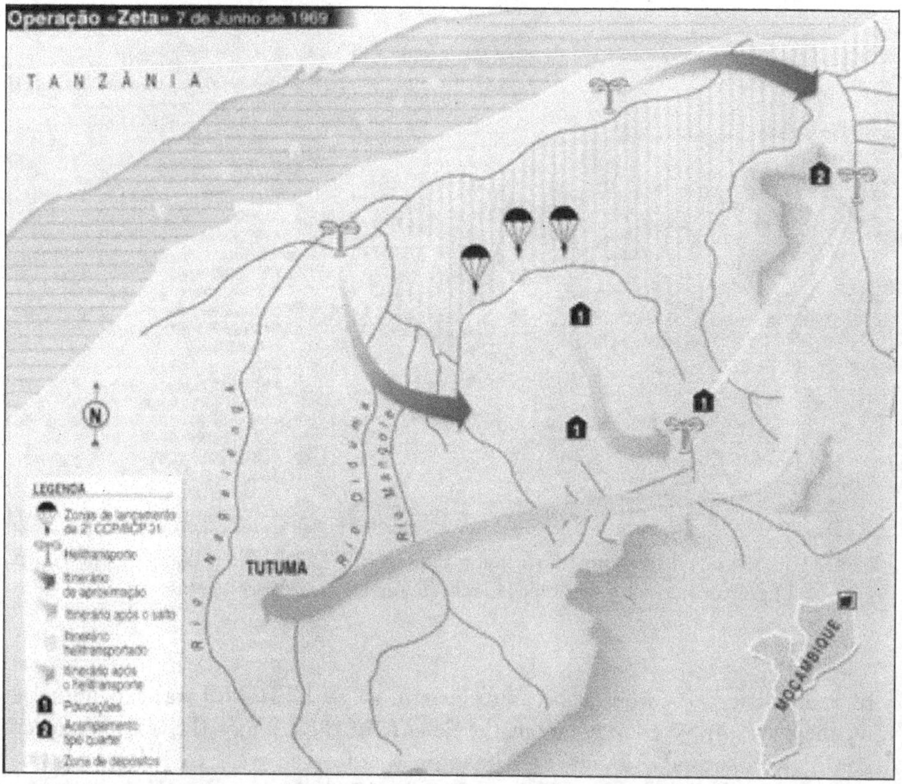

Map 9.5 Operation *Zeta* – Paratroop Insertion into the Maconde Plateau.
(Source: *Guerra Colonial* by Afonso and Matos Gomes)

The helicopter proved to be an aircraft of immense utility in the area of Mueda and in Niassa, which was described as the "State of Widespread Mines" (*Estado de Minas Gerais*).[70] The *Indians* acquired an additional helicopter during Gordian Knot and afterward maintained two detachments: a two-plane one in Vila Cabral and a five-plane one in Mueda. During these deployments the helicopter ground crews developed a system for hot refueling an Alouette during intense, high-tempo combat operations and in eight minutes or less were able to fill the tanks of the helicopter with the specially filtered jet fuel.[71] This minimized the ground time of the helicopter and rapidly returned it to flight, often a critical matter with only a very limited number available and urgent ground support needed.

70 Ibid., 23.
71 Ibid., 24.

Alouette IIIs parked on the ramp at Mueda in the early 1970s with their doors propped open in preparation for a scramble launch. (Source: Força Aérea Portuguesa)

In September 1970, when it became apparent that FRELIMO was expanding in Téte, there was grave concern about the defense of the Cahora Bassa dam project. Accordingly, a second Alouette III squadron was formed with six helicopters and based at Téte (AB 7). The new squadron, Squadron 703, the *Vampiros* (Vampires), also maintained detachments at Furancungo (AM 71), Chicoa (AM 72), and Mutarara (AM 73). By the end of 1970 the number of Alouette IIIs in the two squadrons with their many detachments had risen to 25.[72] By 1974 the number was 31 and reflected the increased use of helicopter airlift of troops to take advantage of intelligence on enemy columns penetrating the northern border area. With the rapid mobility offered by the helicopter, Portuguese forces could be quickly positioned against a vulnerable enemy force to isolate and destroy it. The Alouette III carried five armed men, and a flight of five could put a combat group of 25 on the ground quickly. In 1971, AM 72 was moved southwest to Estima, and in 1973 the squadron received three of the SA-330 Puma medium helicopters. This number would be expanded to five in 1974.[73] The Puma carried eight armed men and made for a more efficient operation.

The Puma was an all-weather, medium-lift helicopter designed and built by Sud-Aviation for the French army (*Armée de Terre*). Its twin turboshaft Turmo IIIC-4 engines delivered 1,320 shaft horsepower each and gave the helicopter great versatility.

72 Jesus Bispo, 1399.
73 Ibid.

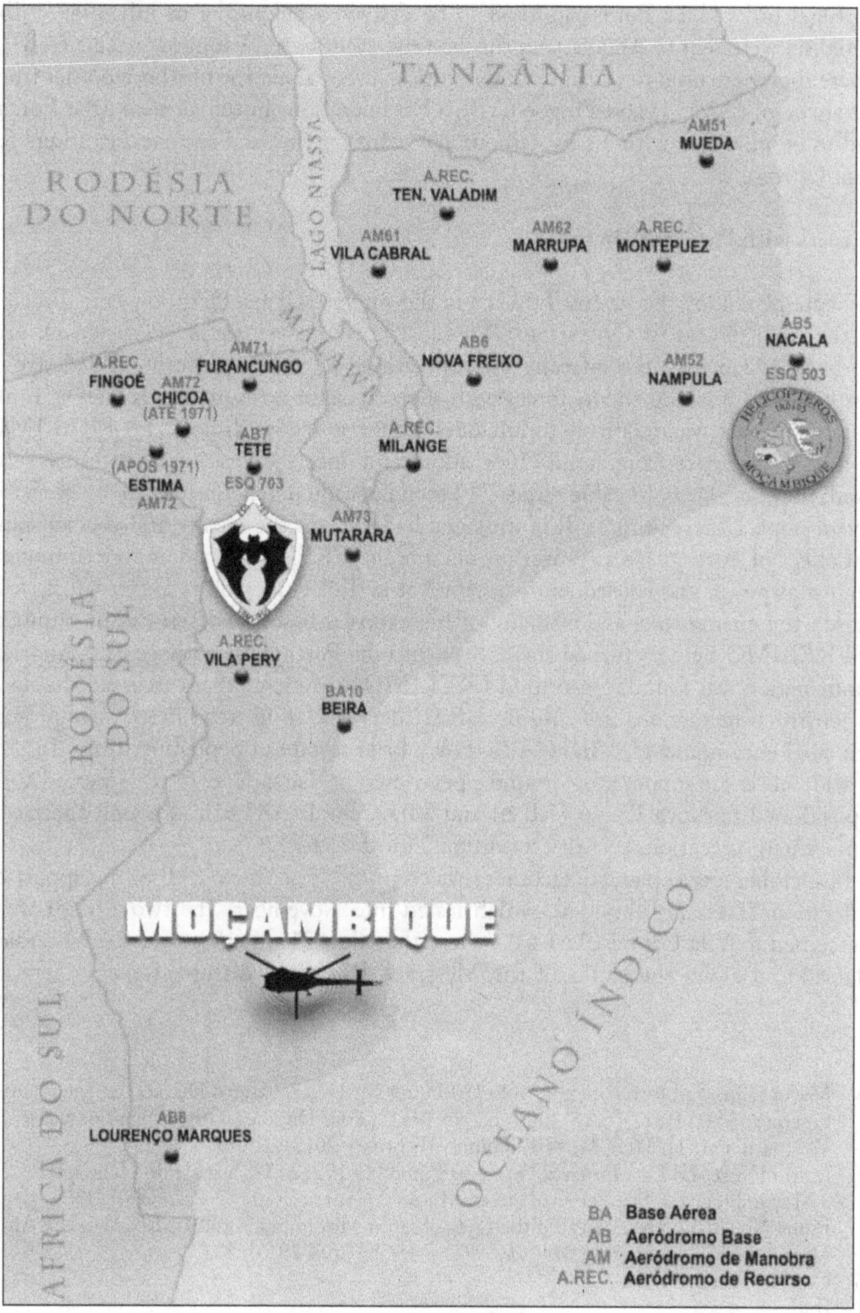

Map 9.6 Airbases in the North of Mozambique Showing Helicopter Basing. (Source: Força Aérea Portuguesa)

Portugal ordered 12 Pumas in 1968 to be delivered beginning in July 1969.[74] The initial six were sent to Angola, and the next five went to the Vampires at Téte (AB 7), where they were used to transport intervention forces along the northern border from Furancungo (AM 71) and Fingoé (AR).[75] For operations in the *ultramar*, the Pumas had to be modified with a new camouflage paint scheme and special dust filters for their jet engines.

Success with Patient Policing

In September 1969 the several T-6Gs and the single available Harpoon were directed to the final phases of Operations *Pesquisa* (Search), *Persistência* (Persistence), and *Cacimbo*. Meanwhile the ongoing Fiat operations in Téte apparently yielded strong results in diminishing enemy forces with their constant pressure on his lines of infiltration in what was relatively uninhabited country. It was difficult for FRELIMO to hide its presence from aerial observation, and once the location of its cadres was fixed, they became vulnerable targets.[76] Much later intelligence reports pointed to the severe impact these regular Téte missions had had on the enemy and the substantial effect of Portuguese airpower on his morale. FRELIMO cadres were impotent against airpower and consequently terrified of it. This vulnerability and resulting fear caused the enemy to abandon many of his exposed bases, and significant numbers of FRELIMO fighters turned themselves into the Portuguese authorities.[77] Another confirmation was found in captured FRELIMO documents that discussed the need to construct shelters, acquire anti-aircraft artillery, and train crews in its use to protect principal encampments.[78] The G-91s took a brief respite in September from Téte to provide close air support for ground operations in Niassa, and for this period they repositioned to Nova Freixo (AB 6) and Vila Cabral (AM 61), although the latter represented "precarious security conditions" for the jets.[79]

In October a two-plane detachment was sent to Nova Freixo (AB 6) in support of Operation *Trovoada* (Thunder), which lasted into November. The two aircraft were reassigned to Vila Cabral (AM 61), where they flew 21 missions over the Miandica Plateau and along the banks of the Messinge River and disrupted enemy activity

74 Mario Diniz and Luís Proença, "SA-330 Puma No 19512, Quatro Décadas na Vanguarda Operacional, 1° Parte" [SA-330 Puma No 19512, Four Decades in the Operational Vanguard, Part 1], *Mais Alto* 401 (January–February 2013): 19–24.

75 Mario Diniz and Luís Proença, "SA-330 Puma No 19512, A Chegada do Helicóptero a Moçombique e o Seu Percurso até ao Museu do Ar em Sintra, 2ª Parte" [SA-330 Puma No 19512, The Arrival of the Helicopter in Mozambique and Its Route to the Air Museum in Sintra, Part 2], *Mais Alto* 402 (March–April 2013): 43.

76 Correia, Part 2, 33–34.

77 Ibid.

78 Ibid.

79 Ibid.

Squadron 703, The Vampires, Alouette IIIs in the hangar at Téte.
(Source: Força Aérea Portuguesa)

A Puma helicopter based at Nampula. Note the special engine dust filters and the camouflage paint scheme for African operations. (Source: Força Aérea Portuguesa)

with their machineguns and rockets. In the second phase of *Trovoada*, which was to the east, the G-91s flew eleven attack missions in conjunction with the T-6Gs from Mueda to render FRELIMO Base Beira completely unusable. Base Beira was the key FRELIMO logistic depot in Cabo Delgado. These actions softened the enemy for the final phase in December, in which the G-91s flew five additional support missions with diverse weapon loads. The *Jaguares* now had seven G-91s assigned, and the additional aircraft allowed more flexibility in deployment and mission assignment. As 1969 ended, Squadron 502 operations were expanding substantially in that fully one-third of the annual flight hours were flown in the final month.[80] Of course, while the addition of the seventh Fiat helped these calculations, still the workload had accelerated substantially.

The year 1970 in the 3rd Air Region was an eventful one in that the FAP began planning for a second Fiat squadron, and General Kaúlza de Arriaga assumed the duties of Commander-in-Chief of the Armed Forces there. Arriaga had for some time been anxious to return to the army from his duties with the Caetano cabinet, and he did so in the late 1960s. His initial operational command as a general was ground force commander in Mozambique, which he assumed in July 1969, replacing Costa Gomes. Here he chafed to command the entire theater and pressed Caetano to remove Augusto dos Santos and install him. Caetano knew well that Arriaga had no experience in operational command, but as the last of the Salazarists, he was staunchly loyal to the regime. So in March 1970, Augusto dos Santos was reassigned as Chief of Staff of the Army, and Arriaga replaced him.

Arriaga's notions of how to fight the war differed from those of his predecessor in that rather than the constant pressure of army patrols with aviation support within the *quadrícula*, he envisioned a large conventional sweep of the Cabo Delgado area infected by FRELIMO in an echo of *Águia*. Augusto dos Santos and Costa Gomes believed that FRELIMO could be contained and victory achieved by the constant, indeed relentless, pressure of small actions gradually grinding down the enemy at a sustainable cost. From the time that Arriaga was a professor of military tactics during the academic year 1964–1965 at the IAEM, the premier graduate school for military officers in Lisbon, he advocated the attack in force supported by artillery and aviation fires as the preferred method of winning war, any war. This was a counterinsurgency theory that had been abandoned by most Portuguese generals and certainly was not held by either Rhodesia or South Africa, two allies in the insurgency wars of austral Africa.[81] Now with his first operational command, he had the opportunity to put this theory into practice and prove its validity.[82] He believed that such a grand operation in which the enemy was swept from the contested area was the path to victory over

80 Ibid.

81 Aniceto Afonso and Carlos de Matos Gomes, *ALCORA: O Acordo Secreto do Colonialismo* [ALCORA: The Secret Accord of Colonialism] (Lisbon: Divina Comédia, 2013), 155.

82 Matos Gomes, 6.

an insurgency despite the cost. Arriaga had visited the United States and witnessed in demonstrations the type of helicopter envelopments being used in Vietnam and believed that such tactics would work in Cabo Delgado. Exercise *Himba*, which was discussed earlier, was indicative of his thinking. Such large-scale operations were not counterinsurgency. They took much time, talent, and manpower to organize, and when launched, were no secret. Fully alerted, the insurgent could avoid contact or choose the battlefield. As a consequence, results for government forces were invariably disappointing. In fact, the constant pressure of small operations based on sound intelligence, as in the thinking of Augusto dos Santos and Costa Gomes, was far more effective in its low-key approach, and in a subsequent chapter we will see Costa Gomes effect this policy with great success in the east of Angola.

Operation *Gordian Knot*

In June 1970, the small village of Mueda again became the center of the grand operation that Arriaga mounted to conquer FRELIMO, Operation *Gordian Knot*. Its airbase was the only infrastructure of any consequence and even that was wanting. Its buildings appeared old and dirty. The tarred streets connecting them were covered in a fine powder, as the nearby army encampment continually stirred up dust coming and going on unpaved roads. Indicative of the situation was a crude sign nailed to a pine tree in the middle of the base that read: "Welcome to Mueda. Here we work, we fight, and we die. A 'nugget' is worse than a terrorist"[83]

Arriaga believed that with this victory he would dispel the doubts held by a group of skeptical generals and politicians and, for one who had been loyal to Salazar and awarded a major operational command, secure with it the future leadership of the country.[84] Arriaga established the Operational Command of the Intervention Forces (*Comando Operacional das Forçes de Intervenção* or COFI) next to his headquarters in Nampula and in April moved it to Mueda. The COFI commanded the detachments or companies of the elite troops: paratroops, commandos, and *fuzileiros* (marines), which in Arriaga's concept would be dropped by helicopter on top of the enemy to destroy him.

Arriaga's plan was relatively simple, but many adjustments had to be made in its execution because of the difficult terrain. For instance, the altitude of Mueda itself was 2,789 feet, and this required lengthening the runway by 1,200 feet to accommodate fully fuelled and armed Fiats on their takeoff roll. A further factor was the nature of the terrain. The principal rivers in the operating area were the Mueta, Litembo, and

83 Alfredo Cruz, *O Voo do Falcão: Piloto de Combate, Moçambique* [Flight of the Falcon: Combat Pilot, Mozambique] (Lisbon: Fronteira do Caos, 2014), 58–59. The word *checa* was a local military term for a newcomer or "nugget" in equivalent U.S. military terminology.

84 Mato Gomes, 6.

"Welcome to Mueda".
(Source: Força Aérea Portuguesa)

Muatide, all of which held water throughout the year and became veritable cauldrons during the rainy season. These would provide formidable barriers to maneuvering ground troops and would require a significant number of helicopters. The valley of the Muera River, situated to the east of Mueda, constituted a critical area for Portuguese forces. Here the terrain consisted of mountainous areas with steep valleys whose slopes held tall impressive trees surrounded by grass. Along the riverbanks of these forested valleys the enemy had established its key bases and its strongest presence in all of Mozambique. From these bases attacks were launched on the surrounding Portuguese forces during the Maconde Hour.[85]

The remainder of the terrain, particularly north of Mueda, consisted of shrub-like plants three to four meters in height that formed a dense, nearly impenetrable matted growth. It was a formidable barrier to ground forces attempting to maneuver, as this aggressive vegetation contained nasty briars, prompting the troops to nickname it the "zone of whips" (*Zona dos Paus*).[86] This difficult, enemy-infested topography was ringed with army encampments: Sagal and Diaca in the north, Muidumbe in the east, Nangololo in the south, and Mueda to the west. However, successfully flushing FRELIMO from its sanctuary was considered an unlikely occurrence with the normal contingent of assigned forces. Arriaga intended to fix this problem.

85 Cruz, 212.
86 Ibid., 65.

Troop insertion in the "zone of whips" during *Gordian Knot*.
(Source: Personal archive of Manuel Serafim Matos Sousa)

Helicopter insertions in this terrain were difficult and fraught with unexpected danger. Generally they were preceded by a well-coordinated bomb attack by the *Jaguares* with the troop transport helicopters arriving on top of a series of "maximum load" (1,000-pound or 454-kilogram) GP bomb explosions. The bomb blasts would clear a landing zone in the brambles, and the troops would be deposited on the target while the enemy was hopefully caught unprepared and still stunned by the concussions. During the months of the dry season when the natives burned the grass to clear their fields for cultivation, a helicopter troop insertion might unwittingly deposit the troops onto still smoldering and hot ground.[87]

The vast territory also lacked aerial navigation aids, and visual navigation could be confusing. There are many stories of pilots being lost in this immense territory, and some did not turn out well. Probably the most famous one is recounted by General Alfredo Cruz from his time as a junior officer in Cabo Delgado. The story begins with a mission assignment given to Lieutenant Blasco to transit in an Alouette III from Nampula to Mueda and his flawed flight planning. A direct route tested the extreme

87 Ibid., 64–65.

range of the helicopter, so normally pilots landed in Porto Amélia to refuel. Blasco, however, was new to Cabo Delgado and inexperienced in operations there. He filed a flight plan to go directly to Mueda and launched despite the fog that day and reduced visibility. Altogether, this was now an accident waiting to happen.[88]

The route to Mueda involves three large rivers easily visible from the air. The first was the Lúrio, the second the Messalo, about 25 miles south of Mueda, and the final one was the Rovuma, which was the frontier with Tanzania about 50 miles north of Mueda. Blasco crossed the Lúrio without seeing it. A bit later he crossed the Messalo thinking that it was the Lúrio. The wind from the southeast and the poor visibility caused him to pass east of Mueda without realizing it. When he reached the Rovuma, he believed it to be the Messalo and thought he would be in Mucaba in 10 to 12 minutes. He contacted the tower and gave the controller his particulars. As time passed Blasco could not find the runway and became disoriented. Just at this moment he saw a village and resolved to land and establish a fix. On landing, he kept the rotors turning and sent his crewman to discover where they were. The Africans in the village responded in an unfamiliar language and made it clear that Blasco was in Tanzania. The crewman returned to the helicopter in a panic, and Blasco quickly lifted off and turned south to the river. With a mere 10 minutes of fuel remaining, Blasco landed on one of the islands in the middle of the Rovuma and informed the tower of the situation. Quickly two helicopters were launched, one with two cans of petrol and a pump and the other a helicopter gunship (*heli-canão*), call sign *Tubario* (Shark). Finally, at the end of two hours, Blasco was rescued, and the three helicopters returned to Mueda. The great fear of every pilot was being lost in this wilderness and not being rescued, as was Blasco.[89]

It should also be noted that virtually all of the new pilots arriving in the *ultramar* and destined for mixed (T-6G, Dornier, or Auster) and Alouette III squadrons were weakly prepared for their duties. It took an extensive adaptation period for them to become comfortable in both their new aircraft with their combat loads and the difficult flying environment, and during this phase of their development they were at an increased risk of getting into difficulty. It simply took time to gain adequate experience in combat and to acquire tempered judgment.[90]

This wilderness terrain favored the insurgent, as it was difficult to gather intelligence on the enemy within it, and even if intelligence was accurate and timely, it was very difficult to act quickly on it with ground troops. Aerial observation was often impossible. Searching for a downed pilot could thus be like looking for the proverbial needle in a haystack.

88 Ibid., 52–53.
89 Ibid.
90 Aleixo Corbal, Oliveira Simões, Fidalgo Ferreira, and Heitor Almendra, "2° Região Aérea, Lições Aprendidas entre 1961 e 1975" [2nd Air Region, Lessons Learned between 1961 and 1975], proceedings of a conference "Almoços da AFAP," January 2011, 13.

The concept of *Gordian Knot* called for defining the battlefield as the area encompassed by the villages of Mueda, Miteda, Nangololo, Muidumbe, Sagal, and Namaua. This was known as "Nucleus Central" and was surrounded by a circle of forces. It was to be penetrated from the south and west from Water Posts 14 and 15 and the settlement of Capoca, as depicted on the battlefield map. The ground forces making the penetrations would be supported by aviation resources and combat engineers. In the three phases, numbered on the map, there were planned advances on the Bases Gungunhana, Moçambique, and finally Nampula (designated G, M, and N respectively). It was expected that the insurgents would flee, and to counter this, ambush sites were established on the ground and aircraft patrols initiated to kill or capture those in flight.

It was estimated that there were 2,500 insurgents in the targeted area and about 20,000 local inhabitants who were largely controlled by FRELIMO.[91] The insurgents had infiltrated through the logistics Base Beria and established themselves in Nucleus Central. To sweep them from this wilderness, Arriaga assembled a force of 8,000 men and another 2,000 for the logistic support necessary to sustain them for the month-long assault, for it was on the scale of a conventional battle in the vein of World War II.[92] Indeed, the necessary resources came from every corner of Mozambique, and ultimately over the months-long operation an estimated 30,000 participated in one form or another.[93]

Alongside Arriaga's *Gordian Knot*, the 3rd Air Region prepared an autonomous operation with the code name *Santuário* (Sanctuary) to support the ground forces. The primary components of *Sanctuary* were independent operations against enemy targets, troop envelopment and support with Alouette IIIs, reconnaissance, airborne command, liaison, and medical evacuation.[94] Mueda (AM 51) was filled to capacity with:

4 Fiat G-91s
4 T-6G Harvards
7 Alouette IIIs
1 C-47 Dakota
2 Noratlas
3 Do-27s[95]

Beginning on 25 June, the FAP launched successive attacks on FRELIMO Base Lúrio, located south of Nucleus Central, as a diversion from the true targets of G, M,

91 Matos Gomes, 10.
92 Ibid., 28.
93 Ibid.
94 Ibid., 10.
95 Correia, Part 2, 36.

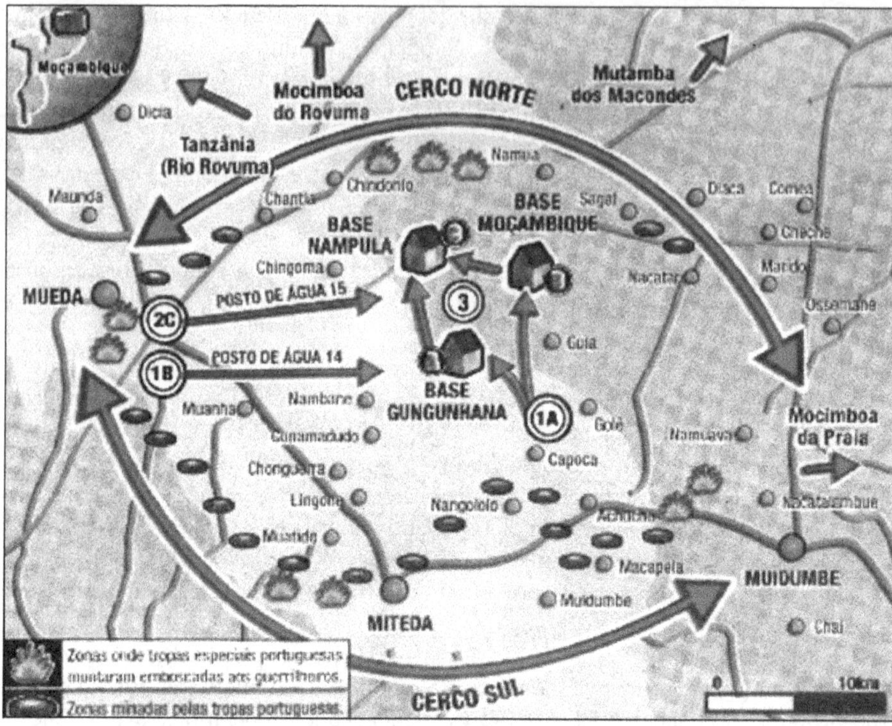

Map 9.7 Operation *Gordian Knot* Battlefield.
(Source: "Operação Nó Gordio" in *Jornal 24 Horas*)

and N.[96] Throughout *Gordian Knot*, air operations were hindered by weather conditions and especially the morning fog. Nevertheless, the Fiats and T-6Gs were airborne during critical operations on the ground over the five-week period and in the closing week neutralized the fleeing enemy groups. Throughout they supported the land and heliborne assaults. As the operation opened, ground troops attacking G were unable to locate it initially and after a day found it abandoned.[97] The column attacking M encountered stiff resistance and was unable to neutralize a small force of 58 insurgents.[98] The *fuzileiros* attacked N and reduced an enemy force of 60 insurgents.[99] The large number of insurgents thought to be in Nucleus Central seemed to have vanished. Indeed at the first scent of *Gordian Knot*, they had used the local population as cover and melted into the flow of people fleeing the area and the coming fight.

96 Ibid.
97 Matos Gomes, 61.
98 Ibid., 66.
99 Ibid.

On 22 July, the Fiats and T-6Gs took off from Mueda (AM 51) to furnish fire support along the northern route running between Mueda and Sagal in a series of attacks using their maximum bomb load to destroy targets. During these operations, the aircraft managed to identify and destroy the enemy locality of Chipungo, a concentration of insurgents who had escaped detection by the ground forces, as it was outside of Nucleus Central.[100]

One of the most important helicopter operations during *Gordian Knot* was casualty evacuation. Integral to these operations were the para-nurses who assumed the same risk as the pilots in entering a combat zone and proved a providential addition in the care of the sick and wounded. They performed selflessly and showed enormous courage in discharging their mission. Many of these evacuations were "zero hours" (*zero horas*), that is, situations in which time was of the essence and the patient required immediate medical attention to survive. Sometimes these occurred end-on-end, and the nurses' spirit of sacrifice in the face of the fatigue associated with such operations moved all who saw them. In these cases particularly, the skills of the para-nurses as much as those of the pilots made a life or death difference.

Following *Gordian Knot*, FRELIMO largely reduced operations in Cabo Delgado and Niassa and shifted its focus to the west and Téte, where there was a thin and unfriendly population. The opportunity to penetrate Mozambique and reach the south through Téte, however, held more promise. While FRELIMO had been surprised by the scale of *Gordian Knot*, the largest operation undertaken during the war in any theater, it was hardly defeated. Arriaga now moved his troops to Téte in an attempt to foreclose FRELIMO incursions in this district; however, Portuguese forces now were overextended and faced great difficulties in stopping infiltration completely. Additional planned airpower would help bridge this gap. Neighboring Rhodesia would now be drawn into Téte to help cover the Portuguese shortfall, as FRELIMO began to share its bases with the Rhodesian insurgency movements. In contrast, Niassa and Cabo Delgado became for the moment temporarily quiet.

Arriaga was delighted with the success of his operation and spent Christmas of 1970 with his wife in a former FRELIMO base in the north. Caetano, however, was less delighted at the rising financial and military cost, and further such offensives were forbidden.[101] Militarily, *Gordian Knot* was well executed but was a misuse of limited resources. These most certainly could have been used to better effect by devoting them to the classic counterinsurgency elements of policing the population and gathering intelligence on the enemy. The toll that it took on the troops and equipment was unaffordable in the long run from the strategic perspective.

100 Correia, Part 2, 36.
101 Malyn Newitt, 531–532.

Map 9.8
Air Targets of Base Lúrio and Locale Chipungo in *Gordian Knot*.
(Source: Força Aérea Portuguesa)

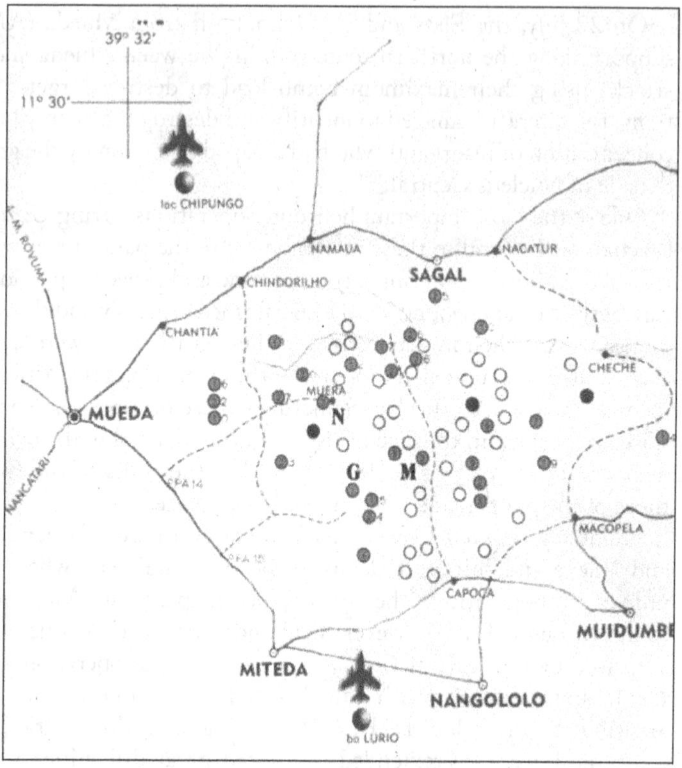

Alouette III evacuating wounded at Miteda during *Gordian Knot*.
(Source: Personal archive of Manuel Serafim Matos Sousa)

A Dornier on its takeoff roll at Miteda during *Gordian Knot*. Note the unimproved packed-earth airstrip. (Source: Personal archive of Manuel Serafim Matos Sousa)

2nd Fiat Squadron

The second Fiat squadron, designated Squadron 702, the *Escorpiões* (Scorpions), began its life in July 1970, when the first of its G-91s arrived in Beira. For the remainder of the year the squadron built its inventory of eight aircraft and conducted training at BA 10, as the facilities at Téte, its ultimate destination, were unprepared to host the squadron. As late as November, the Téte base commander responded to Arriaga's demands for Fiat deployment with a letter through his chain of command noting that his airfield remained unequipped to receive the Fiat squadron. His key complaint was that there was but a single hangar for the eight Fiats, and this did not contain a specialized, climate-controlled storage space for the delicate electronics of the aircraft. Further, in the Téte countryside that was tantamount to a wilderness, there were inadequate or nonexistent navigation aids or radios to cover the operating area of AB 7 and its maneuver airfields of Furancungo (AM 71), Mutarara (AM 73), and Chicoa (AM 72). By December Fiat basing was set:

- Nacala (AB 5), Squadron 502 (*Jaguares*), 8 aircraft assigned, 7 available, 1 in repair.
- Téte (AB 7), Squadron 702 (*Scorpions*), 8 aircraft assigned, 5 available, 3 in training.[102]

102 Correia, Part 2, 38.

By January 1971, the G-91s began to arrive in Téte and were assigned reconnaissance missions. Meanwhile in Mueda, high usage had caused deterioration of the main runway, and this meant that it could no longer accommodate the full range of G-91 operations. Accordingly, deployments were shifted to the civil airport at Porto Amélia despite the fact that its runway did not have the length to accommodate a fully loaded G-91 on takeoff. The airfield was set on a long, high promontory with the Indian Ocean on its eastern side and the deep blue water of the natural harbor of Porto Amélia on the western side. At the northern end of the runway lay the narrow entrance to the largest natural deep-water harbor in the world.[103] A three-plane detachment was initiated with pilot rotations from Nampula every two weeks. Mueda was now used only as an alternate for Fiats.

The primary missions in Cabo Delgado became fire support, armed reconnaissance, and destruction of enemy concentrations. This latter mission was performed in FAP-designated "free fire" zones (*Áreas de Intervenção Livre* or AIL), which were established through coordination with the sector commanders and in cooperation with the local ground force commanders. The AIL itself was carefully controlled and strictly limited in its days and hours of activity. Despite the now limited utility of Mueda for the Fiats, many pilots found themselves landing there to refuel after searching for their targets in the AIL and performing their other missions because of its strategic location next to the battlefield. The tempo of operations was such that between April and September 1971 there were over 650 sorties flown by the Fiats, and these consumed a vast amount of ordnance. This heightened activity resulted in very low magazine stock levels, particularly of bombs and fuzes, and this deficiency limited Fiat attack missions for the remainder of the year.[104] Clearly FRELIMO had returned unfazed by *Gordian Knot*; however, it was being pounded from the air by the new Fiats, an unexpected and unwelcomed development for which it had no defense.

In October, the Fiats targeted numerous insurgent supply columns originating in Tanzania. In one incident a T-6G pilot detected a group hiding in the grass bordering the Nacalenga River and confirmed the sighting with numerous low passes. He directed the airborne patrolling Fiats from Porto Amélia to the target, and after its bombardment, they landed in Mueda to refuel. Subsequently they returned to machinegun the suspected positions.[105] As the year concluded, the Fiat squadrons were very satisfied with their work in bringing security to the north and reducing enemy activity. Indeed, the enemy was preoccupied with survival in the face of airpower and continued to live in great fear of it.

103 Petter-Bowyer, 229.
104 Ibid., 39.
105 Ibid.

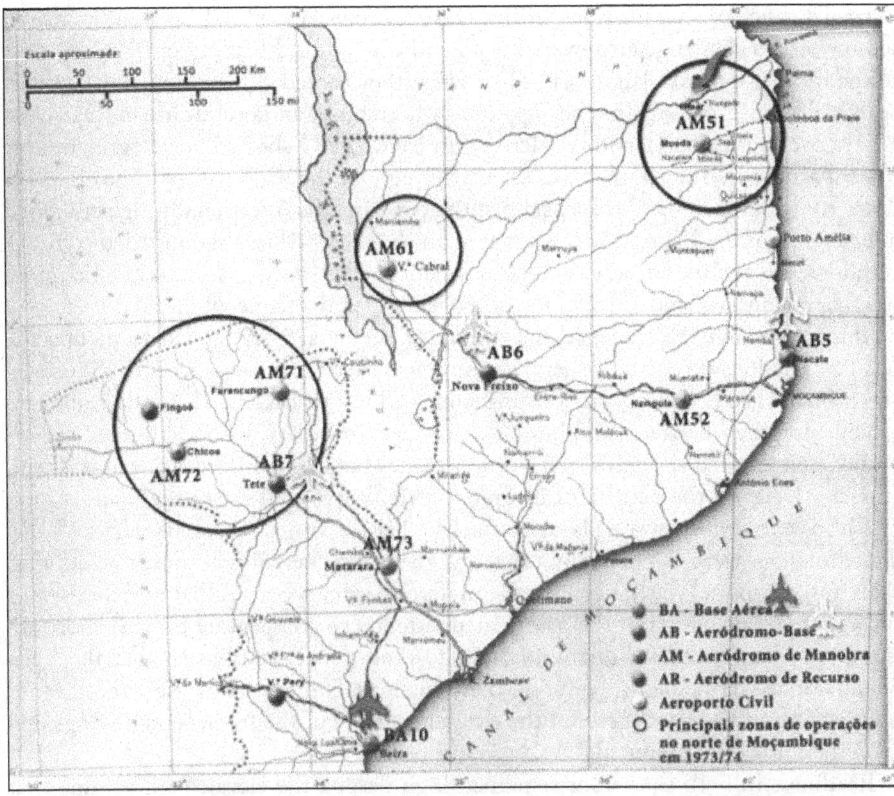

Map 9.9 Mozambique High Threat Areas, 1971. (Source: Força Aérea Portuguesa)

Primacy of Airpower

Because of the munitions shortage, the Fiats in the first three months of 1972 flew 67 missions from Nacala and only 18 from Téte.[106] Subsequently the Republic of South Africa stepped into the breach and supplied the 3rd Air Region with the needed bombs and fuzes. The enemy remained comfortable with the environment on the Maconde Plateau and returned to the Nucleus Central area, where he easily established himself in the fastness of the region with its luxuriant foliage and difficult terrain. Typical operating problems at the time were reflected in two incidents with aircraft from Squadron 502. The first occurred 60 miles southeast of Mueda when a Fiat was hit with small arms fire. While this was hardly surprising in light of the

106 José Manuel Correia, "Fiat G-91/R4 em Moçambique, 3ª Parte" [Fiat G-91/R4 in Mozambique, Part 3], *Mais Alto* 387 (September–October 2010): 33.

normally assigned Fiat missions, still it was very unusual and indicated an increased willingness to challenge airpower. The second was a forced landing in Nangade next to the frontier by two Fiats short of fuel when they exceeded their time on station in search of a missing T-6G that had apparently been hit by antiaircraft fire in the area of Kitaya over Tanzanian territory. Here again FRELIMO showed new aggressiveness toward airpower. The weather was severe throughout the region at the time, and this likely contributed to the low fuel state of the two aircraft. Additionally, lightning had affected one of the Fiats, as it had suffered an electrical failure that impeded communications and the use of radio aids to navigation. The landing occurred in marginal weather conditions, and indeed, one of the jets suffered a flameout on final.[107]

The terrain and the presence and nature of enemy activity tended to dictate the primary missions executed by the Fiats. In Téte, visual reconnaissance (RVIS) proved the primary need, that is, the locating and tracking of FRELIMO incursions and establishing their behavior and intent. Once their trajectory was determined, appropriate ground force action or heli-borne assault could be initiated to destroy them. In Cabo Delgado the pattern of insurgent activity was well established, and so the missions were almost always planned attack (ATIP) or armed reconnaissance (ATIR). These missions were hampered by a lack of 2.75-inch rocket rails, as these rockets with their anti-personnel warheads were the preferred ordnance for ATIR and destroying insurgent groups hiding in the grass. In an attempt to compensate for this shortfall, the *Jaguares* attached 50-kilogram bombs to these rocket stations so that the Fiats would not be sacrificing firepower on their missions. Unfortunately the arming vanes on the fuzes of the bombs scored the external surface of the fuselage and wings and soon their use was prohibited.[108]

Readiness in both squadrons remained high despite the elevated flight time that averaged about 200 hours per month per squadron. This achievement occurred in spite of Squadron 502 being short two pilots and 702 being short four.[109] This shortage of pilots persisted, and as the workload now fell on the contracted numbers, aviators literally lived in the aircraft. Their work was vital to keeping the enemy at bay and the conflict limited to the north, for it was largely through the Fiat missions that the vast and difficult area could be policed, the infiltrating FRELIMO cadres identified, and their bases destroyed. The ground forces likewise depended on attack aviation being available in a timely fashion. These were big demands, and the FAP delivered.

By 1973, FAP operations had become a severe and serious threat to FRELIMO and its attempt to move the conflict forward. Stalemated in the north and bleeding troops, it desperately sought a solution to its primary vulnerability—Portuguese airpower.[110]

107 Ibid., 33–34.
108 Ibid.
109 Ibid., 35.
110 Estimates for FRELIMO deaths during the almost 10-year conflict range from a high of 85 thousand to a low of 60 thousand from multiple sources.

In February, the *Jaguares* were targeted unsuccessfully by antiaircraft fire, as they attacked Base Gungunhana, evidence that FRELIMO was still seeking an anti-air solution. It had not been able to protect its bases with the same effectiveness as had the PAIGC, because the circumstances were considerably different. Difficult, well policed terrain made moving heavy weapons nearly impossible and precluded such artillery as the dangerous 14.5mm quad-barreled ZPU appearing in Nucleus Central. By April, however, there was concern that FRELIMO could acquire the Soviet SA-7 Strela surface-to-air missile (SAM), which was a hand-held, man-portable, heat-seeking antiaircraft missile. These had appeared in Guiné in March with devastating results, a situation that we will examine shortly, and there was every reason to expect them in Mozambique. In the meantime, the pilots experienced increasingly aggressive antiaircraft fire.

In March, Squadron 502 lost a pilot and his Fiat when a defective fuze exploded a 50-kilogram bomb while it was still attached to the aircraft in flight. This was the single Fiat loss in the entire war in the 3rd Air Region and tragically accentuated the shortage of both pilots and aircraft. The FAP now had only 35 Fiats from an original buy of 40: 10 at Bissalanca (BA 12) in Guiné, 15 in Nacala and Téte, and 10 in the *metrópole* used for pilot training or under maintenance rework. While there was an urgent need to reinforce the squadrons in Guiné and Mozambique with additional Fiats, the shortage of Fiat-qualified pilots to man them pointed to a decision not to send additional aircraft to the theaters. The required number of Fiat pilots for Guiné was eight, one per available aircraft; however, in light of the irregular demands for air support, there were twelve assigned. The Chief of Staff of the FAP argued that because there were only 25 G-91s in the *ultramar*, there should be only 35 Fiat-qualified pilots assigned. But in fact there had never been more than 21 assigned, so it was decided to reduce the number of pilots allocated.[111] The decision not to augment either the Fiat inventory or Fiat pilots assigned in the *ultramar* was bad news for the campaign, as the existing Fiat resources would now be overworked to meet commitments. Indeed, the campaigns in Guiné and Mozambique now depended on a handful of Fiats and their pilots to deliver aviation fires. This important and pivotal resource was too thin by any measure, and the aviators and their machines remained under heavy stress. The pilot shortage was alleviated somewhat by requesting that the Fiat-qualified pilots serving on the staff in Nampula join Squadron 502 to help with its missions. Even this addition of three or so pilots was an enormous help.

In June a Fiat was unable to deploy its drogue chute on landing and overran the end of the runway in Beira. Somewhat later structural cracks associated with the firing of the Fiat .50-caliber machineguns were discovered in another of the aircraft. This development necessitated inspecting all of the Fiats in the 3rd Air Region, which

111 Correia, Part 3, 37.

affected availability.[112] The high usage rates of the aircraft were clearly taking a toll in stress on the airframe and maintenance.

As Squadron 502 continued to conduct its ground support mission along the northern frontier, the pilots found little and encountered weak antiaircraft fire. Yet attacks on the principal insurgent bases drew concentrated fire, and with each attack the protective fire became more intense. The Fiats began to adjust and vary their attack profiles to reduce their risk. For instance, in planning a mission against a well-protected base, a Fiat flight of two would depart Porto Amélia in time to arrive at their target in the early hours and half-light of dawn when the semi-darkness would reveal the flashes of the antiaircraft artillery firing at them. These conditions in the Maconde Hour helped to refine the target, and the Fiats attacked the gun emplacements with their maximum total load, the rising sun behind them, and the eyes of the FRELIMO gunners blinded. Some of the pilots with the typical ironic humor of aviators quickly named this tactic the "hot breakfast." For the pilots themselves, the "heat" of the mission occurred on their own takeoff roll with a full load of fuel and munitions. The jets needed the full 6,000 feet of runway at Porto Amélia and only reached rotation speed in the last 100 feet of the paved surface. Mueda remained open to the Fiats, although it was unsuitable for heavy use, and pilots continued to land, refuel, and rearm in Mueda, as circumstances permitted.[113] In May, a Fiat flying from Mueda suffered three hits from a 14.5mm antiaircraft piece, a certain sign that FRELIMO air defenses were being strengthened.

Night alerts for support of ground forces were frequently executed at Porto Amélia, and the normal procedure of launching flights in pairs was quickly abandoned, as the risk of an accident on the narrow runway was high. Runway lighting was also poor and other important safety technologies were not available.[114] The rails for the rockets remained unavailable, and this shortfall denied an important weapon to the pilots. In the case of attacking small groups of insurgents hiding in the grass, the perfect selection was a rocket with an anti-personnel warhead.[115]

A new and effective weapon appeared in the hands of insurgents around mid-1973 in the form of 122mm rockets. Mueda became a prime target, and the rockets were fired at its airfield from the cover of dense grass at a range of up to 10 kilometers. The accuracy was not good, but this required Mueda to be reinforced. Eventually, as we shall see, the FRELIMO aim would improve.[116]

The lack of pilots persisted. In the final months of 1973 there were eleven pilots between Squadrons 502 (six) and 702 (five). Five additional aviators were needed to complete the allocation originally set. Despite the dedication of those remaining,

112 Ibid.
113 Ibid., 37–38.
114 Ibid.
115 Ibid.
116 Ibid.

there was a limit, and flight hours flown had now descended to the 1969 level. This had an effect on the enemy, as the FAP could no longer apply the same pressure consistently. Maintenance likewise became a problem, as the Fiats were wearing out, and the demands on maintenance personnel escalated. In the short term, FAP solved the maintenance problem by sending a team from Alverca to help.[117] It should be noted that the selfless sacrifice of the maintenance personnel in keeping the aircraft flying safely was exemplary. These shortcomings in personnel and aircraft, however, reduced the rhythm of operations and resulted in a rise in intensity of enemy activity across Cabo Delgado and Téte. Clearly the enemy could be held at bay only with the proper augmentation and replacement of aircraft, pilots, and maintenance personnel. FRELIMO cadres were now able to penetrate further south from Téte, and it became easy for the insurgents to appear at inopportune times in very small numbers quite a distance to the south and thus to create anxiety in the population. This threat was modest and containable, but there remained an unsettling low level of incidents.[118]

Operation Rescue

One of the most notable of these incidents is described by Colonel António de Almedia Tomé in which a group of FRELIMO insurgents managed to penetrate the District of Cabo Delgado and travel south some 350 kilometers from the Rovuma River frontier to the small settlement of Boma, where there was no military presence. Here they attacked the resisting population with its "negative" attitude and left more than fifteen dead or wounded. In order to ensure their safe return to Tanzania, they took twelve women and children hostage as a shield to protect them from being hunted and destroyed by the FAP, which sought to intercept the party before it reached its sanctuary.[119] Military intelligence also detected the presence of a heavily armed group of about 50 insurgents crossing the Rovuma to act as a covering force for the fleeing insurgents and their hostages and ensure a safe return to Tanzania. This force had crossed in boats from Tanzania and had brought antiaircraft artillery from its base in Nachingwea.[120]

A two-phase rescue operation was proposed, first, to surround the covering force and neutralize it, and then to attack the returning insurgent group and separate the hostages from their masters. This last action would be the most difficult, as the FAP would have to separate the hostages from the insurgents while simultaneously avoiding injury to the hostages and preventing the party from joining the covering force. If the two FRELIMO forces were seen to join, then the operation could not proceed, as it

117 Ibid., 39.
118 Ribeiro dos Ramos, 107.
119 António de Almedia Tomé, "Mueda, Norte de Moçambique" [Mueda, North of Mozambique], *Mais Alto* 237 (September–October): 21–24.
120 Ibid.

would endanger the hostages and be beyond the capability of the small, select force allocated. If the hostages were taken into Tanzania successfully, then they would be publically exhibited in Dar-es-Salaam as prisoners, an undesirable outcome, so there was great incentive to rescue them. To the PIDE/DGS agent in Mueda, Inspector Casimiro Monteiro, who knew the local area and its people well and held the confidence of those population segments loyal to Portugal, the operation was certainly feasible, and he encouraged it.[121] In 1969 Caetano had changed the name of PIDE to the *Direcção-Geral Segurança* or General Security Directorate, now the DGS.

The rescue force was to consist of a reinforced section of 15 paratroops from BCP 32 in Mueda. These men were very experienced in this type of operation and were all to be volunteers. They were to be transported to the Rovuma River bank in three helicopters supported by two gunships, four Fiats and T-6Gs to neutralize the covering force. As the entire rescue operation depended on total surprise, execution of the various components had to be carefully timed and coordinated. This meant that the helicopter assault force had to arrive at the bank of the Rovuma just after the four Fiats made their bombing runs. The paratroops were to be deposited in the wake of the bombing runs to address the remains of the covering force, which was to be allowed to flee across the river as an expedient tactic to remove it from the equation. Meanwhile, the antiaircraft artillery had to be identified and destroyed, and the fleeing insurgents with their hostages had to be located and isolated. Once the transport helicopters had dropped the paratroops, they would standby to pick up the hostages when the gunships had dealt successfully with the situation. Radio silence had to be maintained until the enemy was engaged, as FRELIMO had sophisticated HF and VHF listening capability and knew the frequencies most used.[122]

The helicopters launched from Mueda in the semidarkness of early dawn and flew a nap-of-the-earth profile to shield the noise of their blades. It took great concentration by the pilots, as there were many obstacles in the irregular, semi-dark terrain, and from the Mueda plateau there was a steep downward slope to the Rovuma. The helicopters were minutes from "H" hour for the attack, when suddenly a series of flares lit up the river bank and the first target. The Fiats began their bombing runs to neutralize as much of the enemy covering force as possible before the paratroops arrived. Any inequality on the riverbank would be overcome by air support. After the paratroops were on the ground, the gunships searched for the 12.7mm antiaircraft artillery and destroyed it. As daylight increased, one gunship sighted the insurgent group with the hostages. The enemy force by the river had not connected with the fleeing group, so the operation proceeded. It appeared the enemy force in flight numbered about 20 and was pushing the hostages ahead of it. The sudden, unheard, and unanticipated appearance of the gunship paralyzed the insurgents, who reacted to the new threat by attempting to disappear into the forest. The gunner fired into the forest areas nearest

121 Ibid.
122 Ibid.

the now-abandoned hostages and protected them so that the transport helicopters could land and retrieve them. As the helicopters touched down, the hostages raced to the machines and, after all were safely aboard, were flown to a nearby army post for first aid. Later they would be examined at the hospital in Mueda. The Fiats and T-6Gs remained on station until the helicopters returned to collect the paratroops. Two combat groups of paratroops then swept the area to capture the insurgents who had sabotaged Boma. The aviators counted some 15 bullet holes in their aircraft, and yet all had returned safely. That evening there was much celebration in Mueda over the success of the operation.[123]

Téte in 1973

Earlier adaptation of the Fairchild K-20 aerial camera to a napalm canister that could be carried on a wing station of the Fiat for overhead reconnaissance photography now came into play in Téte. A series of photographic reconnaissance (*Reconhecimento Fotográfico* or RFOT) missions were initiated the length of the long northern border between January and May 1973. These missions combed both sides of the frontier with Zambia looking for FRELIMO bases and activity at such sites. They were flown at 15,000 feet particularly to avoid the heat-seeking, ground-fired Strela anti-air missile envelope or that of any other ground antiaircraft fire, and were successful in revealing in detail the depots used by FRELIMO to supply its infiltrated fighters in Téte with their munitions and supplies. At the beginning of March, overhead photography revealed troops massing for a large attack aimed at the remote outpost of Gago Coutinho situated just south of the frontier with Zambia and somewhat opposite the important FRELIMO staging base of Chadize. At Gago Coutinho there was an infantry company and a small unimproved airfield to support it. The enemy was identified as hiding in an area of tall grass en route to Gago Coutinho. This was duly targeted, and on the morning of 3 March, two Fiats were launched with their maximum bomb load and set a course for the FRELIMO assembly point in the grass. The mission was to destroy the force in place before it had a chance to disperse, as often insurgents would split into small groups and travel independently to the target before reassembling for the attack. In the early afternoon, the Fiats found the target and released their bombs, causing heavy losses to the enemy and forcing his remnants to retreat into Zambia.[124]

Six months later in September and again in Téte there was another notable success. An Alouette III flying a RVIS mission identified Base Damião, an important FRELIMO depot hidden in exceptionally difficult terrain. A helicopter RVIS was an unusual selection, as helicopters are noisy, and while the enemy knew that he was

123 Ibid.
124 José Manuel Correia, "Fiat G-91/R4 em Moçambique, 4ª Parte" [Fiat G-91/R4 in Mozambique, Part 4], *Mais Alto* 388 (November–December 2010): 34–40.

detected, he did not abandon the base, as he felt secure in its fastness. Accordingly an attack sortie was planned for the Fiats from Tête (AB 7), which could easily address the inhospitable terrain and negate the enemy's ground defenses. The Fiats were completely successful in destroying the base, and such success was welcomed.[125]

Targeted enemy supply routes from bases in Zambia were becoming increasingly difficult, as they were now defended by 14.5mm antiaircraft artillery. These weapons proved a lethal threat to Dorniers that constantly connected the remote outposts in liaison duties and the FAV light aircraft that brought the mail and supplies. This would lead the way for more dangerous anti-air defenses. The insurgents lived in fear of airpower and were determined to protect themselves from it.[126] Certainly experience over time showed improved FRELIMO results. The use of this weapon was a step up from the earlier, widespread use of the 12.7mm artillery, which was concentrated in the Maconde Plateau. The accelerating danger from antiaircraft fire can be seen in the numbers. For instance, in 1965 there were 14 recorded hits: five Dorniers, eight T-6Gs, and one Auster—all light aircraft. While there were far more FAP aircraft in 1973 to serve as targets, the FRELIMO gunners were doing more damage with 40 recorded hits: seven G-91s, four C-47s (one of which resulted in the death of a radioman), five Noratlases, eleven Dorniers, three T-6Gs, and ten Alouette IIIs (which resulted in the death of a pilot and two crewmen, and two pilots wounded).[127] Certainly FRELIMO was learning through experience, and the threat for FAP airmen continued to deepen.

In the closing months of 1973, FAP acquired airborne infrared technology mounted in a Dakota that enabled it to detect and record the body heat of insurgents otherwise hidden in the foliage-covered terrain below. This capability permitted the identification of FRELIMO bases and their quick elimination by passing the coordinates of the insurgents to loitering Fiats that bombed them. This effective localization and destruction of FRELIMO cadres created a recruitment crisis with the enemy, as hundreds of its men were being lost at a rapid pace, a fact confirmed after the war to General José Vizela Cardoso by Joaquim Chissano, a member of the FRELIMO Central Committee.[128]

1974—The Long Year

Given the shortage of pilots and maintenance personnel and the wear and tear on the men and aircraft, Diogo Neto, who was now a brigadier and the chief of the 3rd Air Region, wrote to the Chief of Staff of the Air Force in a secret note about conditions with the FAP in Mozambique. He proposed expanding the FAP presence with additional squadrons and new equipment embodying advanced capabilities. Further

125 Ibid.
126 Ibid.
127 Jesus Bispo, 1401.
128 José Armando Vizela Cardoso, correspondence with the author, 2 July 2012, Lisbon.

there was concern about the Strela, as intelligence reports cited its likely presence in Cabo Delgado, Niassa, and Téte. He also wished to construct a series of new runways to give the FAP additional flexibility in its operations and install new runway lighting at Nacala (AB 5), Mueda (AM 51), and Nova Freixo (AB 6). Likewise, he requested mobile radar and anti-air missile batteries to defend the dam construction at Cahora Bassa and several key airfields despite the lack of a coherent threat from the air.[129] While the Soviet Union and China did give a number of emerging African nations a token air force of obsolete fighter aircraft, the pilots could barely fly them much less conduct an extraterritorial air strike. These air forces were exclusively symbolic. Nor realistically would such a strike ever be initiated against a colonial power, as the retaliation would be such that the essentially defenseless perpetrator would suffer a major destructive blow to its infrastructure and economy. Finally, Neto went to great lengths to defend his proposal to locate all Fiat training in Mozambique, an idea that had gone nowhere in Lisbon when previously aired.[130]

The initial response to Neto's note was a procedural one directed at Fiat operations and based on the experiences of pilots in Guiné who had been targeted by the Strela. From now on the Fiats would only be assigned ATIP missions or support of selected objectives, avoiding protection missions in which the aircraft was relatively slow and within the altitude envelope of the Strela. Attacks against the enemy must now be made with the maximum bomb load and delivered in a single pass from a steep angle of attack (AOA) of 45° beginning at 8,000 feet with the sun at the pilot's back.[131] This attack profile would severely limit the Strela threat, as the missile had a tendency to be confused by the sun and had no head-on acquisition capability. Its acquisition envelope reached only to 8,000 feet, so the Fiat attack was initiated from a safe altitude. The threat of the shooter being destroyed by the release of a maximum bomb load would also likely disturb his aim. Finally, the AOA put the heat source of the Fiat skyward and away from any possible missile acquisition. These procedures contained the Strela threat against the Fiat in Mozambique and did not hinder FAP offensive missions.

Mueda—The Preferred Target

Mueda was likely the most attacked airbase in the FAP, for as we have seen, it was situated relatively close to the frontier.[132] The surrounding terrain offered excellent cover for insurgents to stage an attack, and it was a logical and tantalizing target for the enemy. It was also sited in the midst of the heaviest fighting and relatively close to the primary interior bases and depots of FRELIMO—the crossroads of the war. The

129 Correia, Part 4, 34–40.
130 Ibid.
131 Ibid.
132 Ibid.

most devastating attack occurred 20 January 1974, in which a barrage of Katyusha 122mm rockets set the aviation fuel farm ablaze. Earlier attacks, such as the one at first light, the Maconde Hour, in the closing months of 1966, used Chinese 60mm mortars sited at the approach end of runway 16 and damaged a helicopter and a T-6G with its shrapnel. Afterwards the vegetation at each end of the runway was cleared, and a free fire zone created, first, for a 40mm cannon and, later, for a 20mm antiaircraft artillery that could fire a high volume horizontally.[133]

The attacks usually came from the north and from increasingly greater distances, as the range of the new weapons used by FRELIMO also increased. Later attacks used the 82mm mortar and then the 122mm Katyusha rocket, both of Soviet origin. The 82mm mortar weighed 56 kilograms (123 pounds) and had a maximum range of 3,000 meters. It could be disassembled into several parts to make it portable; nevertheless, even in its component parts it was a heavy load to be carried any distance by insurgents. Its ammunition weighed 3 kilograms per projectile and was transported in a weapon cache crate of ten rounds, or about 30 kilograms (66 pounds). This too was quite a load for the insurgents, so an attack on Mueda required perhaps a dozen men simply to put the weapon and its projectiles in place. A barrage of more than ten rounds would require a significantly expanded team, as the mortar could fire at a rate of 15 to 25 rounds per minute.

The Katyusha was similarly demanding, although a single rocket and its launcher could be handled by perhaps half as many men. A rocket was three meters in length and weighed 66 kilograms (147 pounds). Each was launched from a simple, relatively lightweight, steel frame that was quickly deployed and easily abandoned, as a swift evacuation was standard wisdom. The extreme range of a Katyusha was some 18 kilometers (11 miles), so the insurgents could conceivably be evacuating the launch site while the weapon was in flight, unless they were conducting a lengthy barrage. By the time of impact, counterbattery fire and punishment of the launching crew in its flight could become complicated, unless the barrage endured and provided a target fix. In this latter case counterbattery fire in any form was generally successful in making the enemy pay a heavy price, as a barrage of any magnitude would require many porters to transport the rockets and force them to remain in place during the action. This generally produced a large and vulnerable target. Counterbattery fire was also an important check on attacks, as the hit and run tactics had the potential to change to a more aggressive action with follow-on infantry attacks.

After a number of diverse incidents at Mueda, the base commander installed concrete revetments to protect the flight line and dug trenches or fuel pits for the approximately 1,500 200-liter drums of JP-4 jet fuel. The protective trench system

133 Mueda (attaque)/Guerra do Ultramar (Mueda) [Mueda (attack)/War in the Overseas Provinces (Mueda)], Livro de Acts 2002, item 29, pages 69–70, Archivo Histórico de Força Aérea, Alfragide.

was also adjacent to a preparation area for napalm canisters.[134] In short, the base with its valuable aircraft and their fuel supply was an overwhelmingly attractive target for FRELIMO and represented an opportunity to destroy the aviation threat that it could not otherwise neutralize.

During a conversation in the bar at the officers' mess in Mueda on Christmas Eve, Captain José Augusto Queiroga remembered the PIDE/DGS agent, Inspector Casimiro Monteiro, cautioning that a great attack had been planned for that evening but would not happen, as logistic difficulties had delayed it. In fact, it was postponed for nearly a month, as it took that long for FRELIMO to assemble the needed porters to transport the 122mm rockets from its base of Nachingwea in Tanzania to Base Beira and from there to the plateau of Mueda. There were also other developments delaying the attack. The free fire zone around Mueda had been extended outward and now represented a circle of cleared terrain five kilometers in radius. Further, during periods of stand-down and consequent vulnerability, such as mealtimes or dawn or dusk (the Maconde Hours), a *heli-canhão* conducted visual reconnaissance of the area to prevent mortar attacks. Enemy mortar firebases now must be mounted within the cleared circle to reach AM 51.[135] The Katyusha with its extended range could be launched successfully outside of the cleared perimeter, but such a firebase would be out of visual range and require spotters to correct the rocket aim. Pilots searching for the firebase could identify the spotters and destroy them, thereby blinding the rocketeers. The position of the spotters also revealed the position of the firebase.

Monteiro revised his predictions in early January, as intelligence revised the progress of the porters and the attack party. His new estimate held that the attack would occur between 16 and 24 January. Captain Queiroga and his copilot Lieutenant Carreira, both "Indian" pilots, made their daily RVIS along the perimeter of the cleared approaches to Mueda in a *heli-canhão*, and on the evening of 19 January, all seemed normal. Later that evening an enormous storm moved through the area, a veritable wall of cumulonimbus clouds that dumped torrential rain and was accompanied by violent thunder and lightning. The electrical disturbance disrupted FAP radio communications, potentially isolating the base. The storm itself was perfect cover for the FRELIMO insurgents who were busy through the night establishing their firebase.[136]

134 Ibid.

135 José Augusto Queiroga, "Histórias Vividas: A Prenda de Natal" [Tales Lived: The Christmas Present], undated deposition, Caixa 181, Archivo Histórico de Força Aérea, Alfragide.

136 António de Almeida Tomé, "Mueda: A Mais Importante Base Operacional da Força Aérea no Norte de Moçambique," 22–29.

Attack on Mueda, January 1974. Note Revetments for Aircraft.
(Source: Força Aérea Portuguesa)

At 0430 the next morning Aviation Specialist Zé Luís was awoken with a great shock and tremor that reminded him of the 1969 earthquake in the *metrópole.* His friend Marques yelled, "It is an attack! It is an attack."[137]

This attack broke the normal pattern in several ways and certainly reflected the hand of foreign advisors in its execution. The most convenient site for FRELIMO to launch an attack on Mueda was from the north, as it was the closest to the Tanzanian border. This was logical, as the weapons for bombardment were heavy, and the insurgents were always anxious to shed their weight as soon as possible. Sites to the north were the first obvious opportunity to do this. While this attack also used 82mm mortars to the north, its primary assault was launched from a depression to the east-southeast of Mueda about 10 kilometers from the airfield. Such a site would be counterintuitive to the normal FRELIMO fighter, as it would involve picking his way around the base with a heavy rocket and its launching frame and travelling well beyond the airfield. There were other factors too, for the rocket salvos were to be followed by a ground assault from this unexpected direction to overrun and capture the base and hopefully

137 Zé Luís, "Mueda Terra da Guerra: 19/20 Janeiro 1974" [Mueda Land of War: 19/20 January 1974], undated deposition, Archivo Histórico de Força Aérea, Alfragide.

eliminate Mueda as an air threat. This was to be a great FRELIMO victory. Much was riding on its success, as much had been committed to the assault.

The airfield defenses were very vulnerable from this sector, as attacks from the south were considered improbable, or at least less probable, in light of the rugged character of the terrain. Clearly on the advice of foreign technicians, the enemy installed his fire base in the bottom of a ravine well below the level of the airfield and attacked with salvos of four 122mm rockets for about 20 minutes. Although many of the rockets overshot the airfield and landed on the far side in the valley below, by chance one hit the fuel pits and ignited several fuel drums. The blaze spread from drum to drum and generated an enormous fire that consumed all 300,000 liters of aviation fuel on the base. Columns of flame and smoke shot hundreds of feet into the air, and the heat from the pits was so intense that it lasted well after the blaze was extinguished.[138] Afterwards fuel had to be airlifted, until a truck convoy could be organized.[139] Fortunately, the wind at the time was from 090° at 28 knots and took the smoke away from the airfield and towards the Valley of Miteda.[140]

Alerted by the preliminary mortar rounds, Major Vaz Afonso, commander of AM 51, manned his Fiat, taxied to the runway amidst the rocket attack, and began his takeoff roll with four 50-kilogram bombs loaded beneath his wings. Seconds later a rocket impacted the exact position on the runway where he had begun his roll.[141] With the alarm sounded, the entire base moved to action in its defense. Every man knew his job and bent to it with a will. The control tower became the command post in time of attack, as it offered a vantage over the battle zone and provided communication with all units on the ground or in the air.[142]

Queiroga was starting his Alouette III at the same time and rose vertically to search for the enemy firebase position. Immediately he saw an insurgent spotter in a treetop directing the rocket fire.[143] The Fiat was informed that the firebase was 160° at eight kilometers from the field and that the 82mm mortars to the northeast and southeast were three to four kilometers more or less.[144] It turned south to follow the trail of rockets and bomb their source. The attack was over in twenty minutes, and the Fiats, all of which had scrambled after Afonso, landed at Porto Amélia and later in the day returned to Mueda. Fortunately there were no injured or killed, and indeed two

138 José Augusto Queiroga, "Histórias Vividas: O 20 de Janeiro 1974" [Tales Lived: 20 January 1974], undated deposition, Caixa 181, Archivo Histórico de Força Aérea, Alfragide.
139 Livro de Acts 2002, item 29, pages 69–70.
140 Zé Luís.
141 Correia, Part 4, 36.
142 Tomé, "Mueda: A Mais Importante Base Operacional da Força Aérea no Norte de Moçambique," 24.
143 Zé Luís.
144 Tomé, "Mueda: A Mais Importante Base Operacional da Força Aérea no Norte de Moçambique," 24.

rockets hit and failed to explode when they buried their warheads in the tarred surface of the runway.

It was believed that the enemy had planned an infantry assault on the airbase to be coordinated with the rocket attack. At the conclusion of the rocket launches, the insurgents were to spring from their hidden positions and overrun the base. However, in the thick fog and intermittent rain there was much confusion. Perhaps FRELIMO was satisfied with setting the fuel on fire and was unable to do more. Paratroop Captain José António Jerónimo Gonçalves, who was responsible for the defense of AM 51, at first ordered the artillery to shell preselected sites to the north, the direction of the 82mm mortar fire, but then when the source of the rocket threat was identified, his antiaircraft artillery poured its fire in that direction.[145] When the enemy positions were inspected with an RVIS later in the day, there was widespread evidence of blood on the ground. Apparently FRELIMO had removed all of its dead and wounded and had not been able to stage its attack in the face of withering defensive fire. Much later it was learned that the FRELIMO commander who had directed the unsuccessful attack was condemned to death for his failure to take the base and subsequently executed.[146]

Strela Incident

There was a preoccupation among aviators with the inevitable appearance of Strela in northern Mozambique, as in November 1973 intelligence gathered by the PIDE/DGS in Tanzania indicated that a group of 40 FRELIMO insurgents had returned from their training in the Soviet Union and were planning attacks on Mueda under the protection of the new weapon. The Soviet shipment of Strelas would arrive in Tanzania and be delivered to FRELIMO depots at the end of 1973 or the beginning of 1974. Before the arrival of the missiles, the commander of the 3rd Air Region sent one of his experienced aviators, Captain Rui Balacó Moreira, who had previously served in Guiné (1967–1969), to Bissalanca (BA 12) to be briefed on the experience with the missile there and the methods used to evade it successfully. As a result, a directive was issued by the commander of the Fiat squadron in Téte describing the Strela and giving evasion instructions.[147]

The first encounter occurred in Cabo Delgado on 10 April, when Lieutenant Carlos Mocário performed a high AOA bombing of Base Gungunhana in his Fiat and saw a long plume of smoke from the Strela engine. He followed the recommended evasive maneuvers and lost the missile. Neto sent a secret note to Lisbon the next day asking for anti-reflective paint for the Fiats to aid them in defeating the Strela. The French had developed this green-tinted, anti-SAM paint, and a team from Alverca

145 Queiroga, "Histórias Vividas: O 20 de Janeiro 1974."
146 Zé Luís.
147 Correia, Part 4, 37.

was dispatched with a small quantity to repaint the Fiats. Meanwhile in Guiné there appeared to be a new version of the missile that left no telltale smoke plume, and this development raised further concern.[148]

In Téte the first appearance of the Strela also occurred on 10 April during an operation without consequence. Later on 14 April and again on 20 April, when the Rhodesians sent several small reconnaissance aircraft to locate a FRELIMO base, two were shot down by Strelas.[149]

In May, with the rainy season nearing its end and the Carnation Revolution having occurred in Lisbon on 25 April, the war continued unabated on the Maconde Plateau. FRELIMO remained stalemated in the north and felt that it could now exploit the ceasefire with the new Portuguese government to expand its presence by force and gain as much control of the country as possible. The nationalist movement had achieved little traction with the people of Mozambique and would be defeated in a proposed referendum, so the need to expand its presence forcefully was seen as crucial, notwithstanding the ceasefire.

For an unexplained reason the new military chiefs following the Revolution believed that this was an opportune moment for the military attachés accredited to Portugal to travel from Lisbon to Mozambique for a visit to the theater combat zone. Seven senior officers accordingly arrived at Beira and were put in the hands of the FAP for their tour.[150] They boarded a Dakota named "Silver Arrow," which had been specially prepared and provided with luxurious accommodation for the illustrious visitors.[151]

The crew was very experienced, and the trip went well with a visit to Nampula and then to Mueda. Here the base commander briefed the visitors on air operations and the advent of the Strela, paying particular attention to the danger zones and the means of evading the missile. The commander's final observation was that the Fiats always flew at 10,000 feet above the terrain for safety.[152]

The visitors' next stop was Nangade, a vulnerable base next to the Tanzanian border, where the visitors would hear from a battalion commander. Nangade was a Portuguese island in the center of FRELIMO-held territory. The base was a fortress. The airfield was ringed with revetments and nests of quad-barreled antiaircraft artillery aimed outward horizontally to fend off ground attacks using their high volume of fire. Still, it had been attacked many times. The journalist Giancarlo Coccia described an April attack occurring just prior to the arrival of the foreign dignitaries:

148 Ibid., 38.

149 José Armando Vizela Cardoso, correspondence with the author, 2 July 2012, Lisbon.

150 The seven attachés were Brigadier Alfredo Berenguer César (Brazil), Colonel Roberto França Domingues (Brazil), Colonel Peter Brackley (USA), Commander Robert Thompson (USA), Commander Thomas Binney (UK) Lieutenant-Colonel Edward Grubbs (West Germany), and Major W.A. Kempeu (South Africa).

151 Ricardo Cubas, "Maio de 1974: Uma Missão em Moçambique [May of 1974: A Mission in Mozambique], *Mais Alto* 381 (September–October 2009): 39.

152 Ibid., 40.

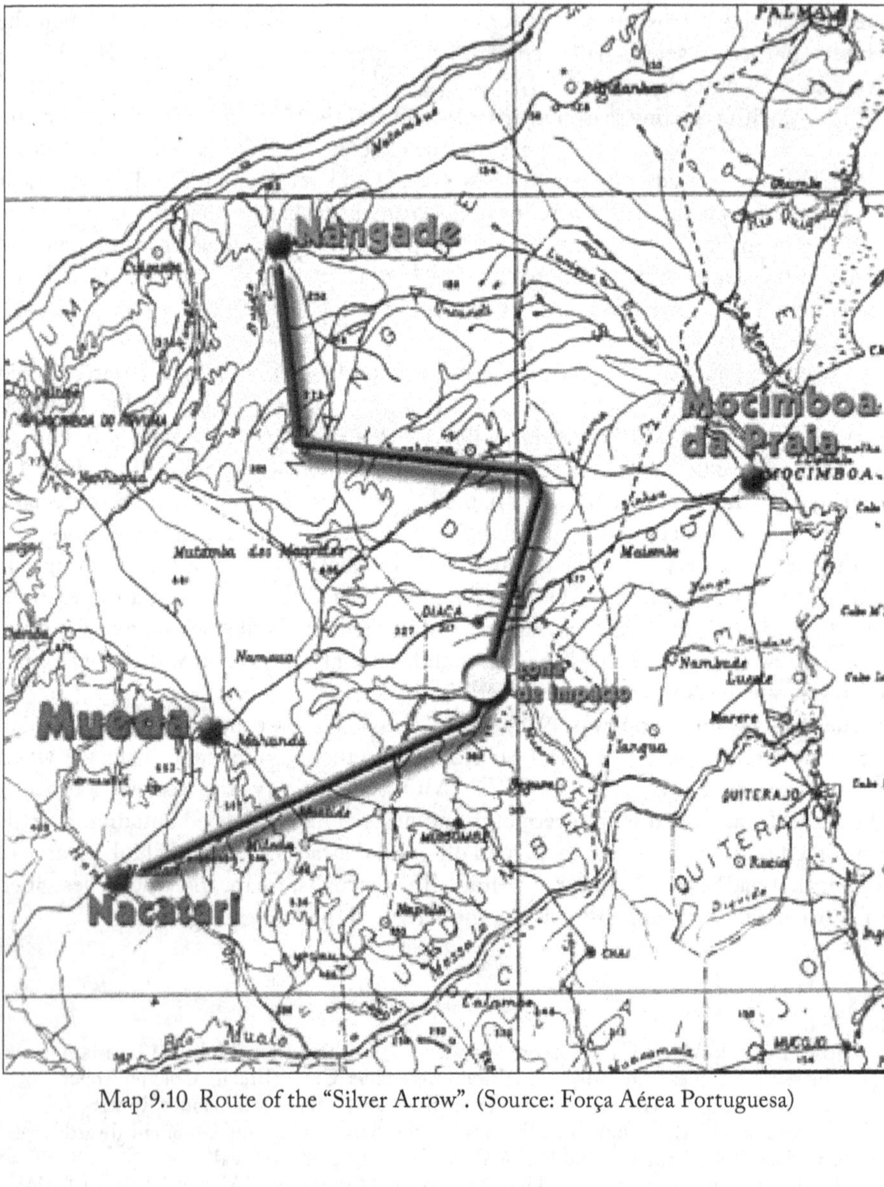

Map 9.10 Route of the "Silver Arrow". (Source: Força Aérea Portuguesa)

> Only a week before we arrived a FRELIMO group had installed itself on a ridge between two nearby hills. From there they had fired more than a hundred and fifty 122mm rockets at the base. Nothing important was damaged in the action which lasted thirty minutes, and ended when a Fiat G-91 fighter dispatched by the Portuguese scored a direct hit.
>
> We passed through the gap—named "The Bosom" in honor of Gina Lollobrigida—some days later. Apart from the crushed arms and legs, the insufferable smell, and a single hand still holding onto a piece of electric wire, there was little left of the FRELIMO men.[153]

Following the briefing at Nangade, all boarded the aircraft for the return flight to Nampula. It was raining at the time with multiple cumulonimbus clouds predicted along the route that passed near Mueda and over the most hostile parts of the Maconde Plateau. The safe altitude was 10,000 feet above the terrain, which was itself 1,500 feet, so the Dakota would have to climb to 11,500 feet. Once airborne, the weather conditions proved extremely difficult with driving rain and lightning. The command pilot, Captain José Luís Henriques Assunção, turned eastward toward Mocímboa da Praia in an attempt to find clear weather, but conditions in this direction proved worse. It was impossible to climb above the clouds or even to the anticipated 11,500-foot altitude, and the Dakota returned to its original route. Unfortunately, it was now flying over the principal FRELIMO base, Base Moçambique, where there were Strelas. The opportunity of a large target within Strela range was simply too much of a temptation for the enemy.[154]

When fired, the missile sought the heat of the engine exhaust and found the right engine of the "Silver Arrow." The explosion rocked the Dakota violently and shattered the Perspex in the nearby window next to the German attaché, who went into shock. At the same instant the mechanic yelled, "My captain, we have been hit by a missile! The engine is stopped and burning."[155] The engine, however, continued to run and the propeller continued to turn. Black smoke poured from its exhaust, and there was enormous vibration throughout the aircraft, as the explosion had blown a hole in the wing. The pilot with great aplomb verified the state of the right engine and gave the order to secure it. Its fuel and ignition were cut, and it stopped. With the left engine working well, the pilot turned toward Mueda and contacted the tower to request an emergency landing because of missile damage. After reducing speed to 100 knots, extinguishing the engine fire, and stabilizing the aircraft, the pilot verified that there were no injuries. Mueda tower responded that the weather was extremely bad with torrential rain, fog, lightning, and thunder and that conditions there were below minimums. The pilot was directed to land at Nacatari, a small packed-earth airfield about five minutes

153 Coccia, 10.
154 Cubas, 40.
155 Ibid.

"Silver Arrow" after landing. (Source: Força Aérea Portuguesa)

southwest of Mueda. Accordingly, the pilot turned southwest and maneuvered around the weather to maintain visual contact with the ground. As Nacatari had no navigation aids, he intercepted the radio beacon at Mueda and followed a line of bearing to locate the small field with its 400-meter (1,300-foot) runway—manifestly insufficient for a Dakota. When the field came into sight, the pilot judged that he would have to turn 270° to align with the runway and keep the damaged wing with its questionable lift on the outside of the turn. If he turned into the damaged wing with its weakened lift, he could easily lose control of the aircraft. With each maneuver, the vibrations heightened, and the passengers believed that death was certain. They remained silent, waiting for the right wing to separate from the fuselage with catastrophic consequences. The pilot established the aircraft on final and called for 10° flaps to verify that the system was working properly. He aimed for the runway threshold and now called for landing flaps. Airspeed bled to 65 knots, and the Dakota seemed to hang on its remaining engine. At touchdown, the engine was immediately secured to prevent a fire. To all on board it was instantaneously obvious that the aircraft would overrun the small runway. On the pilot's right the land was cleared, and on his left there was grass. He chose the grass, as it would provide some braking traction for the aircraft. He hit the left brake with all of his strength, and the Dakota spun left, skidding to a halt with an almost complete reversal of direction. Immediately all abandoned the aircraft.[156]

156 Ibid., 41–42.

Captain Assunção inspecting the damage to "Silver Arrow".
(Source: Força Aérea Portuguesa)

A flight of Alouettes was in the area supporting paratroop operations and landed to evacuate the dignitaries. Soon they were in Mueda, as the weather had lifted temporarily, and transferred to a Noratlas bound for Nampula, the end of their dramatic and unvarnished battlefield tour.[157]

While the Strela was no longer effective against the Fiats with their defensive procedures and anti-reflective paint, it still represented a lethal threat to the slow and low and hence vulnerable helicopters, transports, and light aircraft. The FRELIMO leaders in another overstated war communiqué crowed that they had eliminated airpower, "the one obstacle to effective control of Cabo Delgado by the guerrilla forces," and yet the Fiats continued to fly with complete freedom of action throughout the district.[158]

Criticality of Airpower

Initially in Mozambique the most important aviation mission was to detect the FRELIMO incursions into the northern districts of Cabo Delgado, Niassa, and Téte and the massing of insurgents for a subsequent attack on either a military post or the population. This was performed by armed reconnaissance missions flown initially by

157 Ibid., 43.
158 Correia, Part 4, 39.

the T-6G and the few Harpoons available. The distances were vast, and even though the flights were long, they were the only means to police the lengthy frontier with any hope of success. Visual reconnaissance was performed by the Dornier, and photoreconnaissance by the Dakota. These aircraft worked well, as they were proven technology with a consequent ease of maintenance. Their only fault was their slow speed, and thus covering their assignments was clearly time consuming. On the other hand, after a certain point, speed is a liability, as the pilot or observer could easily miss a significant clue in the blur of passing terrain.

From 1968 onward the Fiat improved the equation with its increased speed, firepower, and flexibility. It could respond quickly to deliver close air support to a ground unit and search a far greater area than the Harpoon or T-6G on an armed reconnaissance mission. Yet none of these aircraft were state-of-the-art, leading-edge technology, as no such requirement existed in counterinsurgency.

The T-6G was an armed trainer, and the Harpoon an outdated open-ocean reconnaissance aircraft adapted to bombing and fire support. The Fiat could no longer survive in any projected NATO encounter with Soviet aviation and became surplus. Yet these aircraft had a bright future in counterinsurgency, a type of war in which the enemy possessed no airpower. With this array of aircraft and their proper use, the FAP was able to create uncertainty and insecurity in the enemy with a projection of force that he could not counter successfully. With airpower Portugal maintained a constant presence and kept the insurgents off balance. It disrupted their incursions and caused debilitating casualties. This presence competed successfully with the FRELIMO attempt to do the same, particularly with the population. Portugal was able with its outdated transports, the Dakota and the Noratlas, to support the local population and offer a better way of life or punish the disloyal.

In summary, airpower created an enormous disequilibrium in the north of Mozambique in favor of Portuguese forces. The enemy was constantly on the defensive—hiding from aerial observation, attempting ineffectively to camouflage his encampments, and living in constant fear of aerial bombardment. The advent of the Alouette III helicopter and its gunship version transported and protected ground troops engaging the enemy directly. Consequently, FRELIMO suffered a constant stream of losses, as its columns more likely than not found themselves isolated far from help and mortally threatened, if not destroyed, by devastating force. Its morale suffered deeply. Airpower in this type of conflict, even with technologically obsolete aircraft, proved the great determinant in the hands of skilled FAP strategists and pilots and is testimony to its vital importance in counterinsurgency.

10

The Road to Alcora

Portugal had two neighbors in Africa that were key supportive allies in its fight to defend its possessions: Rhodesia and South Africa. Their relationship with each other and with Portugal was initially mixed but matured in the face of mutual threats to their national interests. Here we will examine the events leading to national cooperation and the signing of a tripartite treaty in October 1970 known as Exercise ALCORA. This led to mutual military support of the three against the incursions of nationalist movements aiming to overthrow the governments of each, and coordinated military operations and air support were features of the treaty important to our story.

Initial Indifference

Originally in 1961, all three of the countries comprising ALCORA held fundamental differences. Certainly, their political, economic and military links were essential, but a cultural difference held them apart. For instance, few in Rhodesia remembered that in 1922 it had rejected the South African offer of integration into the Union of South Africa. During World War II, Rhodesians proudly fought for the British against the Germans, while many South African leaders were proud of their years spent in detention at the Koffiefontein internment camp with 800 of their fellow South Africans on suspicion of being pro-Nazi. John Vorster, who would become prime minister of South Africa in 1966, was the most prominent. The preceding government of Dr. H. F. Verwoerd, prime minister from 1958 until his assassination in 1966, was friendly but cool to Rhodesia.[1] South African officials were unenthusiastic about the Rhodesian push for independence from Britain and behind the scenes tried hard to dissuade its politicians from such a rash course. South African leaders remained correct in all of

1 Dr. Verwoerd was assassinated on 6 September 1966 as he entered the House of Assembly in Cape Town, when a parliamentary messenger stabbed him in the neck and chest four times. He died immediately, and the blood-stained carpet remained in Parliament until its removal in 2004.

their dealings with Rhodesia, and any support offered would come reluctantly and only because of the accident of geography.

In contrast to the South African links with Rhodesia, a relationship between Rhodesia and Portugal hardly existed at this time despite the fact that the bulk of its imports of oil and goods came either by pipeline or rail directly through the Beira corridor. In September 1965, Salazar with his Council of Ministers established a doctrine to address potential flashpoints on the horizon, of which the brewing Rhodesian Unilateral Declaration of Independence (UDI) was the foremost. Regarding Rhodesia there were two points: First, Portugal would uphold the right of free passage and access to the ocean ports under its sovereignty. This would be granted irrespective of the political ideology or position held by any country, and these included the Congo, Zambia, Swaziland, Rhodesia, and South Africa. Second, Rhodesia was a problem of Great Britain. Portugal would not take sides in this quarrel, just as other nations had no right to intervene in Angola and Mozambique.[2] Portugal had not argued with the British path in Tanzania, Malawi, Zambia, or Swaziland, despite the fact that all four shared borders with Mozambique. As Rhodesia moved closer to UDI, Portugal came into sharp focus.

On the other hand, South Africa had enjoyed a long and positive relationship with Portugal and especially Mozambique. Mozambique, or at least the southern half of it below the Save River, was in many ways an extension of South Africa. Lourenço Marques was a major port and railhead serving South Africa, and the beaches south of it had been a favorite destination of vacationing South Africans for many years. The Mozambican economy was substantially integrated with that of South Africa, and indeed, motorists drove on the left-hand side of the road, as in South Africa, Rhodesia, Tanzania, Zambia, and other former British colonies. There were perhaps 500,000 or about 10 percent of Mozambicans working in South Africa at any one time, primarily in the mining industry. Both Portuguese and South African investments in Mozambique during the war years produced a 9 percent growth rate, and by 1971 South African investment in the colony had surpassed that of Portugal, as Portugal began to look more toward Europe. Politically both Portugal and South Africa were semi-pariah states because of their national policies. In summary, there was a close tie between Portugal and South Africa on economic and political issues, and security would quickly come to the forefront.

In August 1963, Salazar wrote a long letter to Verwoerd that established a basis for future cooperation and ultimately led to ALCORA. The letter arrived at an opportune moment, one in which South Africa was suffering international condemnation for the Sharpeville massacre and felt isolated with its forced departure from the British Commonwealth and the formation of a Republic. The Sharpeville massacre occurred in March 1960 at the police station in that township, where a crowd of 5,000 to 7,000 black protesters were demonstrating over the Pass Laws. While accounts differ,

2 Jorge Jardim, *Sanctions Double-Cross* (Lisbon: Intervenção, 1978), 12.

in essence the crowd intimidated the besieged police station by hurling stones and advancing on it. The police then fired into the crowd, killing 69 people. In May 1961, South Africa was expelled from the British Commonwealth because of its apartheid policies. Its expulsion was driven by the threat of former British African colonies to leave the Commonwealth, if South Africa remained a member with its apartheid. The British sided with their newly-independent black-ruled nations. On top of these two setbacks, the UN in August 1963 had imposed an arms embargo. This action had a long history that began with its founding when in June 1946 the Indian government requested that the discriminatory treatment of Indians in South Africa be included on the agenda of the very first session of the UN General Assembly. In April 1960, the UN Security Council, in its first action on South Africa, adopted Resolution 134 deploring the Sharpeville massacre and called upon the government to abandon its policies of apartheid and racial discrimination. In April 1963, the newly established UN Special Committee against Apartheid held its first meeting to monitor South African race relations, and in August 1963, the UN Security Council adopted Resolution 181 calling upon all states to cease the sale and shipment of arms, ammunition and military vehicles to South Africa. Salazar's letter was thus a bit of welcome support in an otherwise dark period.

The letter was long and reflected Salazar's thinking on the type of relationship he wished to establish with South Africa. He noted that the two nations were alone in defending Africa and Western civilization from communism and that every form of cooperation with each other would be most advantageous to their resistance and defense.[3] Portugal, in fact, needed South Africa very much, as the *metrópole* was a long way from the African battlefields, and indeed, when help was required in the 1961 Angolan emergency, South Africa had stepped forward to supply military matériel.

In September, Verwoerd responded that South Africa had been honored to furnish "the material assistance in the scope of military equipment during the rebellion in Angola over the past year (1962) to the extent that our resources permitted."[4] Verwoerd indicated that South Africa was prepared to continue with defense cooperation "if such assistance is required," as "it is always our desire to maintain and strengthen our policy of being good neighbors with the overseas provinces of Portugal."[5] In November, South Africa assigned senior military officers as its vice-consuls to Luanda and Lourenço Marques to strengthen the relationship.[6]

Angola was of particular strategic importance to South Africa, as its League of Nations mandate, South West Africa, lacked water and energy, two key ingredients for its development. It was a mineral-rich area, and South Africa wished to proceed in

3 Aniceto Afonso and Carlos de Matos Gomes, *ALCORA: O Acordo Secreto do Colonialismo* [ALCORA: The Secret Accord of Colonialism] (Lisbon: Divina Comédia, 2013), 73.

4 Ibid.

5 Ibid.

6 Ibid. The vice-consul assigned to Luanda was Brigadier Willem van der Waals, who wrote *Portugal's War in Angola 1961–1974* (Rivonia: Ashanti Publishing, 1993).

opening a mining industry there. In October 1964, South Africa signed several accords with Portugal, the most prominent of which was the one relating to the Cunene River and its hydroelectric project to supply power to South West Africa. Others related to the supply of Mozambican mine workers for the Transvaal mines and the guarantee of their treatment and protection, and the establishment of cold storage facilities at Lourenço Marques to facilitate the South African export of its citrus fruit through that port.[7] Later TAP and South African Airways (SAA) entered an accord that made accommodation for two flights weekly between Lisbon and Johannesburg and addressed the overflight difficulties with certain African countries and the mutual need to support the air routes of each.[8] These accords were the beginning of a mutually fruitful relationship that would cover the gambit of political, economic, and military issues and ultimately be capped by the Cahora Bassa dam, one of the most ambitious infrastructure projects in Africa.

UDI and Sanctions

A few weeks after Prime Minister Ian Smith led Rhodesia to its UDI in November 1965, the UN Security Council passed its resolution 217, which called on all member states to withhold recognition of Rhodesia, refuse assistance to its government, sever economic relations with it, and embargo petroleum shipments to the former colony. This resolution would serve as the basis, however flimsy, for the subsequent British maritime blockade and the expectation that other member states would cooperate. In December, Britain banned selected imports from Rhodesia and prohibited the export of British oil to it.[9]

This set the stage for the initiation of the British naval blockade as part of a maritime embargo. Already, however, there were holes in such a plan, as Rhodesia received oil via three routes: by road across Beit Bridge from South Africa; by rail through South Africa or Mozambique from either Durban, Port Elizabeth, or Lourenço Marques; and by pipeline from Beira to the Rhodesian refinery at Umtali. The British did not have the resources to block all of these routes, and thus there was grave concern over "seepage" and "serious leaking in the oil sanctions machinery" that would enable Rhodesia to circumvent the oil embargo with increasing effectiveness.[10]

By February 1966, there were indications that two Greek-registered tankers were headed for Beira with oil for Rhodesia: the *Joanna V* and the *Manuela*. Britain swung into action and committed an aircraft carrier, two frigates, and a support ship to close Beira to Rhodesian-bound oil. After successful negotiations with the French, a

7 Ibid., 74.
8 Ibid.
9 Richard Mobley, "The Beira Patrol: Britain's Broken Blockade against Rhodesia," *Naval War College Review* (Winter 2002): 65.
10 Ibid., 66.

three-plane detachment of maritime patrol Royal Air Force (RAF) Avro Shackleton Mk2C reconnaissance aircraft from 204 Squadron were added to the force and based at Majunga, Madagascar, to support the effort. These began their single-aircraft daylight patrols on 19 March, flying 12-hour missions daily between 0600 and 1800 hours local time along the shipping lanes. They went north as far as the end of the Mozambique Channel and south to a point 50 miles beyond Beira. Their primary sensor was radar, and it reached between 40 and 60 miles each side of aircraft track. Contacts were identified visually once acquired by radar. If the contact was designated a "suspect tanker," it would be reported for intercept by the awaiting warships.[11]

The blockade crisis began on 1 April, when a Shackleton initially sighted the *Joanna V* steaming northward toward Beira, and the Royal Navy warned it not to enter port. The Greek government refused permission for the British to board the merchant vessel, and the *Joanna V* continued toward Beira. The British warned that force would be used to prevent the discharge of oil at Beira, saying that the *Joanna V* "will be the first and last to enter Beira."[12] In defiance of the British, the *Joanna V* anchored two miles offshore in Beira roads. Britain, determined to prevent the discharging of oil, pressured Portugal. Portugal claimed neutrality and reiterated that it would not supply oil to Rhodesia, or acquire oil in the name of Rhodesia, or transport it in Portuguese ships.[13] Portugal went on to say that it was not prepared "to assume the role of culprit in a situation for which it formally and categorically rejects any responsibility."[14] Whereupon Britain obliquely threatened to use extreme measures to prevent the oil from being transferred ashore and pumped through the pipeline to the refinery at Umtali. Salazar was unwilling to run the risk of seeing English troops ashore in Beira and face another Goa. At the time that Goa was taken by India, the anti-colonial government of Great Britain had approved of the action, so Salazar took British colonial hostility seriously. This threat of force prompted the Portuguese to reinforce Beira with aircraft and crack troops.

In anticipation of the crisis, six of the F-84G Thunderjets from Squadron 93 in Luanda had been flown to Beira in November 1965 and operated from BA 10 until May 1966.[15] These were reinforced by two P2V-5 Neptunes from Squadron 61 at Montijo (BA 6) in the *metrópole*, which were also flown to Beira and operated there monitoring the Royal Navy from November 1965 to mid-1966. Likewise the four remaining Harpoons of Squadron 101 spent most of their flight time on patrol between April and November 1966, again monitoring the Royal Navy ships off Beira.

On 14 April, the Portuguese moved paratroops into Beira and seized the *Joanna V*. These moves denied the British the option of a simple "police" strike ashore to

11 Ibid., 69.
12 Afonso and Matos Gomes, 80.
13 J.T.R. Wood, *A matter of weeks rather than months: The Impasse between Harold Wilson and Ian Smith, Sanctions, Aborted Settlements and War, 1965–1969* (Victoria: Tafford, 2008), 71.
14 Ibid.
15 The six Thunderjets were serial numbers 5141, 5144, 5164, 5199, 5203, and 5245.

seize the ship. Even so, Portugal realized it would be foolish to deepen the crisis, and in a series of meetings between Ian Smith and several Portuguese diplomats, it was agreed that Rhodesia did not need oil from the *Joanna V*. This resolved the flashpoint; however, both South Africa and Portugal continued to facilitate the delivery of oil to Rhodesia by train from Port Elizabeth and Durban in South Africa and Lourenço Marques in Mozambique. As to the *Joanna V*, Shell Oil unloaded half of its cargo, and a month later the balance was taken off by the Royal Navy auxiliary tanker *Appleleaf* attached to the blockading force. Afterwards, the *Joanna V* quietly left Beira.

Britain appeared to have won the "Battle of Beira" and in the process had destroyed Beira as a port. It never really recovered, and the refinery at Umtali became another casualty, as it lay dormant for 14 years. On balance, it was an empty victory for Britain, as it had earned the enmity of the Portuguese, and they would now happily assist Rhodesia to evade sanctions without conscience.[16]

The blockade continued until Mozambique independence in 1975 and appears to have accomplished remarkably little. Britain did not have the resources for an extended blockade along the Mozambican and South African coastlines to isolate Rhodesia completely, and simply maintaining a force off Beira proved an enormous drain on the Royal Navy at a time when British commitments were contracting. Some 76 ships supported the ten-year effort, and one estimate placed the cost of the unilateral blockade at £100 million.[17] While Britain made numerous overtures to other countries for help, none did.

Concern for Téte

Northern Rhodesia was granted its independence and became the Republic of Zambia in October 1964, following Tanzania by some three years. Kenneth David Kaunda, a former schoolteacher turned politician, became its first president. He had become an ardent nationalist in the 1950s, had led the drive for Zambian independence, and like Julius Nyerere of Tanzania, was passionately committed to supporting and providing a home for African liberation movements. His landlocked country shared a frontier with no less than six countries, four of which were fighting active insurgencies, Angola, Mozambique, Rhodesia, and South Africa, and this gave him ample opportunity to make trouble for his neighbors. Within two years of coming to power, Kaunda offered sanctuary to the ANC, FRELIMO, MPLA, South West African People's Organization (SWAPO), União Nacional para a Independência Total de Angola (National Union for the Total Independence of Angola or UNITA), ZANU, and Zimbabwe African People's Union (ZAPU). FRELIMO would find a new front in Téte and host ZANU and ZAPU. The MPLA and UNITA would open an eastern front in Angola. SWAPO would open a front in the South African border areas of

16 Wood, 76.
17 Mobley, 79.

Caprivi and Ovamboland. All of these threats would come from Zambia and would cause the three threatened countries of Portugal, Rhodesia, and South Africa to combine against them.

Experience in the "Battle of Beira" reinforced a common interest between Portugal and Rhodesia in Mozambique, and cooperation expanded rapidly, particularly in the military sphere. In July 1965, Rhodesian troops were airlifted by helicopter in

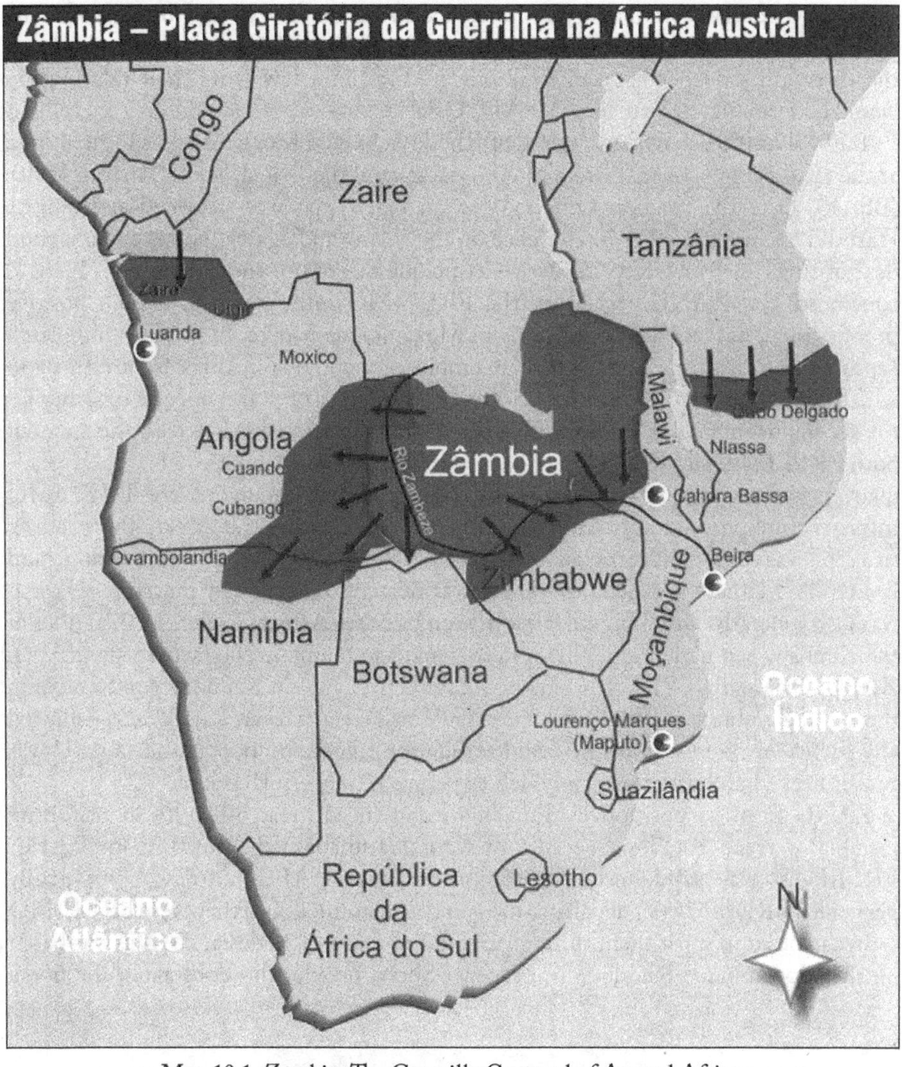

Map 10.1 Zambia: The Guerrilla Carousel of Austral Africa.
(Source: António Pires Nunes, *Angola* (Lisbon: Prefácio, 2002))

pursuit of ZAPU insurgents in Operation *Pagoda II*. These were members of a liberation movement espousing majority rule in Rhodesia, sponsored by the Soviet Union, and afforded sanctuary in Zambia. The enemy had entered Téte by accident when its elements lost control of the canoe they were using to cross the Zambeze River into Rhodesia and were swept downstream from the border post of Kanyemba, which was located in the northeastern most part of Rhodesia at the intersection of the Rhodesian, Mozambican, and Zambian borders. Ten of the 11 enemy were captured in this operation, which involved close coordination between the Portuguese police and the Rhodesian forces and was concluded in September. It was the first of many such coordinated operations that would occur until April 1974. Cooperation between the three allies deepened from 1967 onward, and in a few years their relationship would be formalized in an accord—ALCORA.

The vast northeastern border area of Rhodesia, where Kanyemba was located, had at the time a single Special Branch representative of the British South African Police (BSAP), Detective Section Officer Winston Hart, who was stationed at Bindura. Hart did his best to gather intelligence on the enemy and against orders spent a good deal of time relatively deep inside Mozambique talking to the Portuguese. He was convinced that ZANU was using the FRELIMO infrastructure, but no positive proof existed, as combined operations in Mozambique had yet to produce important captures or confirming documents.[18] Combined operations in this case are those in which Portuguese and Rhodesian or other armed forces or police operate together for the accomplishment of a common purpose. Hart's suspicion was based on the fact that both FRELIMO and ZANU were backed by Communist China and had their headquarters and training sites in Tanzania. Hart's work in cataloging the FRELIMO infrastructure proved important, even if he was anticipating the threat. The strategic situation was somewhat complicated in that ZAPU was headed by Joshua Nkomo and favored by Kenneth Kaunda, as Nkomo was related to Kuanda by marriage. Nkomo was also generally, although mistakenly, seen by other African leaders as the father of the Zimbabwean nation and relied on this myth to maintain the favored standing of ZAPU. Originally FRELIMO favored ZAPU to maintain Kaunda's good graces in the use of Zambia as a sanctuary in its assault on Téte; however, ZAPU never showed any particular interest in using Téte to reach the Rhodesian population. ZAPU was Soviet-backed and based on the N'debele people in western Rhodesia.

ZANU from its position in Tanzania experienced great difficulty in transiting Zambia to reach Rhodesia because of Kaunda's unabashed favoritism for ZAPU. FRELIMO was assiduously courting a reluctant ZAPU, until ZANU finally persuaded FRELIMO leadership to dress and arm its forces as if they were FRELIMO cadres in order to spirit them through Zambia to assault Rhodesia. ZANU was based on the Mozambique-Rhodesia transborder Shona people who dominated the north

18 Ron Reid-Daly, *Pamwe Chete: The Legend of the Selous Scouts* (Johannesburg: Covos Day, 2001), 4–5.

and east of Rhodesia. Even as late as 1970, ZANU remained committed to a direct assault on Rhodesia across the Zambeze from Zambia, a situation that left the insurgents with wide stretches of difficult, open, and unpopulated terrain to cross to reach the populated areas. This terrain likewise extended along the Zambeze into Téte and provided a similarly difficult obstacle for the insurgents there, providing it was policed properly. Despite Hart's following these developments and developing intelligence, he was considered an alarmist at the time.[19] His instincts, however, were correct, and his initiative helped to lay the groundwork for what was to become close cooperation with the Portuguese in stalling both FRELIMO and ZANU.

The next covert coordinated effort was a combined operation in Mozambique that began on 16 April 1968, when a group of 17 Lusaka-based nationalists infiltrated northwestern Téte. Their presence was reported to Portuguese authorities on 20 May by local tribesmen from Catumbula, a small village just north of the Mazóe River, which in that area formed the border. A Portuguese patrol made contact with the insurgents, who had scattered by the next day. The dispersed insurgents split into two groups, each moving toward Rhodesia. One headed southwest and the other south. Acting on intelligence provided by the Portuguese, the Rhodesians on 23 May formally initiated their intercept as Operation *Flotilla* and established its command post at the border town of Nyamapanda, about 15 miles south of the Mazóe River and on the main road to Malawi. Rhodesian Light Infantry (RLI) patrols located the tracks of six men on the northern (Mozambican) side of the river, and a counterinsurgency effort by the Portuguese military and police reduced their number to three. The remaining crossed into Rhodesia in early June and were captured by the Rhodesian authorities.[20]

ZANU continued to fail in spectacular form, as it attempted a direct assault on Rhodesia from Zambia and came to realize that the shortest route to the Rhodesian African population that it needed to reach was through the lightly-defended western extension of Téte. This flanking move on Rhodesia began after September 1968, when Rhodesia concluded Operation *Excess*, an intercept of ZANU insurgents conducted in the tri-border area that captured 36 of the 38 infiltrators, and thereafter, ZANU began to explore the possibilities of Téte.[21] FRELIMO had opened a second front there but had not been able to exploit it, as it had neither adequate personnel nor a proper logistic system. At the time FRELIMO was concentrating its resources in the eastern border districts of Cabo Delgado and Niassa. Following Operation *Excess*, when there was a lull in the fighting in Rhodesia, border patrols were conducted by the Rhodesian Air Force (RhAF) in a number of aircraft. When it used its English Electric Canberra bombers for this mission, its pilots shutdown one of the two jet

19 Peter Petter-Bowyer, *Winds of Destruction* (Victoria: Tafford, 2000), 252.
20 Wood, 460.
21 Ibid., 475.

engines to conserve fuel and extend engine life.[22] The Canberra was a British first-generation jet-powered light bomber that proved highly adaptable to tactical bombing and photographic reconnaissance in counterinsurgency. During this slack period, the Rhodesians responded eagerly on 7 December 1968 to a Portuguese request for military assistance in protecting the construction site of the formidable Cahora Bassa gorge hydroelectric dam with its anticipated 2,075 megawatt output. It was an impressive project costing at the time $135 million and slated to supply power to both Mozambique and South Africa.[23] The RhAF committed four Alouette III helicopters the following day under Wing Commander Ken Edwards in great secrecy—an important feature of this and all Portuguese-Rhodesian inter-service cooperative deployments. It would also be extended to all aspects of ALCORA. Other participants included the accomplished reconnaissance pilot Peter Petter-Bowyer and two future commanders of the RhAF, Norman Walsh and Hugh Slatter. With the helicopters were Rhodesian Army advisors and tracker units, including Captain Ron Reid-Daly, who would later command the Selous Scouts. The detachment flew to Bene, a village located about halfway between the dam site and the northern border of Téte. While the operation lasted only ten days, there would be many more. This and future such combined operations were broadly styled by the Rhodesians as Operation *Natal*, while the classified service codename was Operation *Tripper*.[24]

This particular deployment to Bene was a first in combined inter-service cooperation, and it sought to assist the Portuguese in reducing FRELIMO and rendering it incapable of providing either ZANU or ZAPU safe passage to Rhodesia through its infrastructure in Téte.[25] Indeed, the FRELIMO offensive in Téte might have begun sooner had an agreement been reached with Zambia. When this was eventually achieved in 1968, FRELIMO easily and rapidly advanced to the Zambeze, as Portuguese force levels in Téte were very low and focused on the defense of the Cahora Bassa site. Then in 1970, FRELIMO conducted a large offensive and crossed the Zambeze near Chicoa. This was a decisive development, as it opened direct access for ZANU through FRELIMO graces to the northeast of Rhodesia. Portuguese forces were quite thin south of the Zambeze, and this enabled the advance of FRELIMO and ultimately its dominance of western Téte.[26] To the Rhodesians it was worrisome that FRELIMO could move so rapidly and potentially facilitate ZANU and ZAPU penetration of eastern Rhodesia. The combined operations continued under *Tripper* with varying deployments to Chicoa, Tembue, Nova Mague, and similar sites until 1971, when the Portuguese had acquired sufficient helicopters to conduct their

22 Ibid., 618.
23 Ibid.
24 Ibid., 618–619.
25 Petter-Bowyer, 202.
26 Barbara Cole, *The Elite: The Story of the Rhodesian Special Air Service* (Amanzimtoty, South Africa: Three Knights, 1984), 38.

Portuguese and Rhodesian briefing at Cancombe Tactical Headquarters, Téte, during Operation *Tripper* with General Augusto dos Santos in the center. (Source: Arquivo Histórico Militar)

own joint operations.[27] Joint operations are those conducted by an integrated and complementary force composed of the various branches of a nation's armed forces. Nevertheless, according to Rhodesians and judging by the results of the combined effort prior to 1971, a continuous presence of Portuguese forces operating with their Rhodesian counterparts after that date would have made a substantial difference in curbing FRELIMO and denying ZAPU and ZANU the use of Téte. Combined inter-force cooperation was fundamental, and the cross-pollination reinforced both Rhodesian and Portuguese effectiveness. Combined reconnaissance flights in Aermacchi AL-60 Trojans, Dorniers, and Alouette IIIs yielded a wealth of information to support follow-on activity. From the Rhodesian point of view, if access had been granted to the southern bank of the Zambeze in Téte on a continuous basis, its strategic needs would have been met.[28] Even so, this early coordination set the stage for ALCORA.

27 Wood, 619.
28 Petter-Bowyer, 208.

The Aermacchi-Lockheed AL-60F-5 Trojan was a light utility aircraft with a tail-wheel, fixed main landing gear, and a single Lycoming 400-horsepower engine driving a three-bladed propeller. The engine was underpowered for the bulky airframe, and the aircraft was very noisy, particularly in the higher RPMs, as the propeller wingtip speed went supersonic. Aviators joked about its performance, saying that it was the only aircraft that received bird strikes from the rear.[29]

Téte Operations

A review of the major operations in Téte after 1971, sometimes by the Rhodesians, sometimes by the Portuguese, and sometimes together, illustrates the usefulness of combining against the common threat emanating from Zambia. The first such attempt was Operation *Lobster*, which occurred between 31 August and 3 September 1971, and in which Rhodesian forces contacted FRELIMO cadres in the triborder area. By early 1972 ZANU was poised to make its first probes into Rhodesia aided by FRELIMO, which was now thoroughly entrenched along the border on the Téte side. By March ZANU had established itself at the FRELIMO base of Matimbe, just east of Mukumbura. The Rhodesian Special Air Service (SAS) mounted a raid there with Portuguese permission and killed several enemy who were wearing FRELIMO uniforms. Captured documents, however, revealed that they were ZANU and that they had aligned with FRELIMO to gain a toehold in eastern Rhodesia.

This situation in Téte was monitored by the Rhodesians primarily through SAS operations; however, when the small, covert "stick" of four men located either FRELIMO or ZANU cadres, there was a Rhodesian inability to follow through with a sizable troop insertion to destroy them. At the time this was due to a lack of helicopter airlift capability, as these were always in short supply.[30] The alternative was a parachute drop, but only the SAS was parachute qualified, and in 1971 there were neither enough SAS troops nor parachutes to effect a sizable insertion. In summary, Rhodesia was unable to capitalize on the intelligence generated by the SAS in Téte. The Portuguese were similarly thin on the ground.

One of the more fundamental difficulties in Portuguese-Rhodesian combined operations was the lack of accurate maps to aid in their planning and execution. Consequently, in mid-1972 Operation *Junction* was initiated in coordination with the Mozambican military authorities. RhAF No. 5 Squadron detailed a Canberra to conduct low altitude (2,000-5,000 feet) reconnaissance and mapping missions along the Téte-Rhodesia border from the Mázoe River, a tributary of the Luenha River, to the border post of Zumbo on the Zambeze River.[31] A detachment of South

29 Ibid, 245.
30 Ibid., 292–293.
31 José Armando Vizela Cardoso, "A Guerra Untramarinaem Moçambique, Operações Militares" [War in the *Ultramar*: Military Operations], presentation given at the Instituto

African Air Force (SAAF) Canberras joined the Rhodesians in this project. From these photographic reconnaissance missions the FRELIMO base of Damião was identified along with extensive activity around the Rhodesia-Mozambique border post of Mukumbura, and these both were marked for attack in a combined Operation *Sable* during September and October 1972.

Operation *Sable* for the RhAF began with the deployment of four Alouette IIIs, a Trojan, and four Provosts to support the Rhodesian Light Infantry (RLI) and SAS troops at the small, remote, and spartan border airfield of Nyamasoto near Rushinga, about 75 miles southeast of Mukumbura. The Percival P-56 Provost has the distinction of being the last piston-engine basic trainer operated by the RAF. It was a two-seat, side-by-side, low-wing monoplane with fixed landing gear and a tail wheel. It was powered by an Alvis Leonides nine-cylinder, air-cooled radial engine developing 550 horsepower at 2,750 RPM and driving a three-bladed propeller. Its fixed landing gear and pneumatic flight control and brake assist took some skill to master, as did the cordite starter cartridge. The Provost Mk52, like the Harvard, was an armed trainer effective in counterinsurgency.[32]

Cross-border reconnaissance began immediately and soon trails were observed and followed into heavily treed and broken rock cover. The reconnaissance Trojan judged it to be a base without structures but was unable to determine if it was occupied. Four Alouette IIIs with troops were assigned to the Trojan for direction to the camp. The assault force landed south of the base and began its sweep northward, uncovering some ZANU insurgents who broke northward. Out of 24 observed, seven were killed, two escaped wounded, and three were captured.[33]

This was one of a number of logistics bases that served as waypoints between Zambia and Rhodesia and was only lightly occupied by ZANU porters returning to Zambia. Further reconnaissance revealed a largely empty Téte with many villages abandoned. FRELIMO had stolen everything of value from the local inhabitants in an intensive foraging campaign, and as a consequence the locals were hiding from both FRELIMO and the Portuguese, who wanted to put them into protected villages or *aldeamentos*. In confirmation of this, the Trojan picked up the trail of a large herd of cattle that had obviously been stolen from the Mozambican tribesmen and was being driven north to Zambia and the FRELIMO bases there. These conditions pointed to the fact that FRELIMO had been in control of Téte for some time.[34]

In the meantime on 3 September the Fiats attacked the base of Damião in support of a helicopter-borne combat group inserted there. This destroyed base would be reoccupied and enlarged over the next year.[35]

de Estudos Superiores Militares, Lisbon, 20 June 2012.

32 Petter-Bowyer, 22–23.

33 Ibid., 264.

34 Ibid., 265.

35 José Armando Vizela Cardoso, correspondence with the author, 2 July 2012, Lisbon.

Operation *Sable* was expanded into a more permanent Operation *Hurricane*, which began with its initial action on 19 January 1973 and was the first sizable military engagement outside the borders of Rhodesia. It would last until the Rhodesian ceasefire in January 1980. Such operations became known as "externals," and this first one was in coordination with the Portuguese authorities who agreed to allow the Rhodesians to operate in depth south of the Zambeze River and between it and the Rhodesian border with Téte.[36] Accordingly, Rhodesian authorities decided to deploy two groups of SAS for an extended period in this targeted area.

Rehearsals were held for the operation, and finally when all was ready, a Dakota took off from the New Sarum airfield to drop two pathfinder teams at last light. The two four-man SAS teams were to free-fall into the selected area west of the Musengezi River about 22 miles into Téte, where there was a known enemy rendezvous point at two crossing trails. Following the pathfinder jump, the rest of the SAS team of 20 jumped into the designated area. There were two casualties who were evacuated while the remaining 26 stayed in Téte for a month. The rainy season was at its height, and the men were soaked constantly. The insurgents were regularly surprised by the SAS in areas where they thought they were safe. As a result, a great deal of equipment was captured when the insurgents abandoned it to escape. The success of the SAS team provided a valid reason for it to remain in Téte, and from this point forward western Téte became a major battleground for Rhodesia.[37]

A month later on 16 February 1973, a Trojan on a reconnaissance mission in Téte located a large camp just north of the Dague River close to the Rhodesian border. This discovery was a new development, as the camp had not been in existence ten days earlier and was clearly a combined FRELIMO-ZANU base. The target area appeared to be approximately 700 by 600 meters, and its destruction would be assured by a force of four Canberras and four Hunters after having been marked by a Vampire T11.[38] The Hawker Siddeley Hunter was an obsolete subsonic British jet aircraft developed in the 1950s as a single-seat, single engine, highly maneuverable fighter-bomber and reconnaissance aircraft. It used the same engine as the Canberra and was employed by Rhodesia in counterinsurgency. The De Havilland T-11 Vampire was an obsolete British single-engine, tandem two-seat, jet fighter-trainer with a new life in Rhodesian counterinsurgency.

Several small insurgent bases were also identified in the border area of Téte, and one was selected for attack. The target was set in heavy bush on the east bank of a dry river where a distinctive bend with visible water made identification certain. On 23 February a flight of jets levelled it.[39]

36 Cole, 45.
37 Petter-Bowyer, 53.
38 Ibid., 312–313.
39 Ibid., 314.

During March two additional bases near Mukumbura were identified and marked for destruction. One had 15 head of cattle stolen from the local tribesmen and was very close to the border. This would be attacked by Hunters and quickly followed by a RLI-Alouette III insertion. When the RLI troops were on the ground, they observed fresh tracks of fleeing insurgents headed southeast to another known base. Cattle tracks were observed headed north and were followed for three kilometers before finding 55 head, which were taken to Rhodesia.[40]

Among the dozens of combined or joint operations effected in Mozambique, Operation *Auréola* (Glory) launched on 15 September 1973 by the Portuguese against Mount Calambagupa (16°34' South – 33°07' East), about midway between Chioco and the Mazoé River, is a classic example of a joint operation. The target was Base Damião, destroyed a year earlier, which was now reoccupied and very much in use. Attention first came to this target when a FRELIMO insurgent surrendered and revealed its location and the details of its operation to his army interrogators. By 7 September the FAP liaison was talking to the prisoner to pinpoint the location of "Base of Commander Damião of FRELIMO." The next day as a deceptive reconnaissance mission an Alouette III was assigned a medical evacuation in the area of Chioco during which it was timed to pass over the Base Damião infrastructure at noon on the way to its destination. The Alouette III pinpointed the coordinates, and after landing at Chioco, pretended to receive a sick patient and continued to the Téte Hospital in order not to raise suspicion. Immediately preparations began for an operation to reduce the base. The ground force assigned was a combat group of 30 troops transported in three Alouette IIIs and supported by two gunships. Just prior to the troop insertion, the base would be bombed by the Fiats with their maximum load. The attack was launched on 15 September and completely destroyed Base Damião. It also resulted in the capture of extensive supplies of arms, munitions, and matérial.[41]

The combined Operation *Marble* began on 6 April 1974 with a small team of Rhodesian reconnaissance pilots and their Trojans prepositioned at Estima (AM 72), where there was a Portuguese brigade headquarters. The task of the Trojans was to locate FRELIMO targets for a helicopter assault. The operational area extended westward along the northern bank of the Zambeze from the Cahora Bassa dam, which was filling and thus changing the shoreline. The next day the Rhodesian reconnaissance flight happened upon a large FRELIMO logistic base and immediately called for the assault force to rendezvous at one of the river junctions selected for such a purpose.[42] The plan was to lead the assault force in the new Pumas flown by the *Vampiros* directly to the target. The task of the noisy Trojan was to mask the sound of the approaching Pumas and mark the target with rockets. When sighted by the Trojan, the helicopters

40 Ibid., 315.
41 José Armando Vizela Cardoso, correspondence with the author, 2 July 2012, Lisbon.
42 Ibid.

were in a loose formation. The Trojan marked the target and its pilot, Peter Petter-Bowyer, observed the assault:

> Coming around in a long line astern for a left-hand orbit around the base, the two gunships went into action immediately because there were FRELIMO people running in all directions.... All four Pumas dropped their troops south of the target but so close together I could not believe it. They then lifted, turned about in a hover and disappeared the way they had come.[43]

The conditions were perfect for Strela attacks, as the sky was a cloudless blue. The Trojan moved to 6,000 feet and followed a number of converging paths leading to four additional large bases, all in use. There were many small dependencies associated with each camp, and these were linked to fields of maize, rapoko (a small reddish grain also known as finger millet), and cassava and tended by a coopted population.[44]

The bases provided work for the combined forces, and on 11 April it was agreed that the Portuguese assault force would attack Base 2, the RhAF would strike Bases 3 and 5 with attack aircraft consisting of two Canberras and four Hunters, and the *Scorpion* Fiats would hit Base 4. Major José Armando Vizela Cardoso, a FAP pilot, was assigned to the Trojan with Petter-Bowyer to help coordinate the FAP aircraft, as he spoke English.

The strikes went well with mixed results, as FRELIMO had become jittery and abandoned Base 4. In the case of the remaining three, there were numerous dead and evidence of many having been wounded. Great amounts of enemy arms were recovered.[45] During the mission spotting and marking assignments, a Strela was fired at Vizela Cardoso and Petter-Bowyer's Trojan without consequence, as the missile was unable to acquire the small aircraft at altitude.[46]

On 14 April, the Vizela Cardoso and Petter-Bowyer Trojan moved east of the destroyed targets in coordination with another Trojan, which radioed that it was onto something big but then lost contact. A search was unable to locate the aircraft. Meanwhile, additional aircraft were engaged in the search. On 20 April another Trojan was lost, and its wreck was spotted by the smoke coming from its ordnance. On investigation there appeared to be a large FRELIMO base hidden by the density of the trees from which Strela attacks had been launched on the two Trojans. Rhodesia had lost two Trojans and 4 aircrew in a week and as a consequence withdrew from Operation *Marble*.

For both the Portuguese and the Rhodesians these missions typify the pattern of those that were executed during this period in Téte and until the conclusion of the war. The enemy was initially located through a reconnaissance mission or missions,

43 Petter-Bowyer, 326.
44 Ibid.
45 Ibid., 330.
46 José Armando Vizela Cardoso, correspondence with the author, 2 July 2012, Lisbon.

and once his location and strength were appraised, an appropriate force was tailored to address the threat. This depended on the aircraft and troops available and their location and readiness. The preferred profile was an initial bombing by a flight of Fiats after the target was marked by a Dornier or Harvard. This bombing was followed by forces inserted by Alouette IIIs or Pumas and protected by one or more gunships. For the Rhodesians it was the same with an attack by Hunters and Canberras after the target was marked by a Vampire or Trojan. The bombing was followed by an Alouette III troop insertion. It was a tried and tested method and, despite its predictable profile, was successful. It should be noted that invariably the troops inserted in such helicopter envelopments were specially trained forces. For the Portuguese, these were the paratroops, commandos, and *fuzileiros*. For the Rhodesians they were the SAS, RLI, and on occasion the Selous Scouts.

South West Africa

If Téte were the strategic concern of Rhodesia, South West Africa represented a similar problem for South Africa, one in which Angola was the key. SWAPO insurgents maintained bases in their sanctuary of Zambia and transited the Angolan district of Cuando Cubango with the increasing help of the MPLA after 1966, as its cadres gained experience in the area. The SWAPO objective was to infiltrate the Caprivi Strip and Ovamboland, border areas of South West Africa, and from Zambia the most direct route was across southern Angola.

The question of South West Africa had its origins in the partitioning of German possessions following World War I. Known then as German South West Africa, it was given to South Africa as a League of Nations mandate, and South African ownership survived a number of rulings by the World Court, in which the UN asserted that League of Nations trusteeship had passed to it. The territory was vast at 823,000 square kilometers and had a population of about one million, or a density of 1.2 people per square kilometer. SWAPO had its origins as a labor organization, the Ovambo People's Congress (OPC), which was formed in 1957 to contest the migrant labor system in which thousands were hired under a two-year contract to work in other parts of South West Africa. It was a primary grievance, as it was administered without much compassion or interest in the workers' welfare and was the target for condemnation by many organizations.[47] The shift of its focus from labor abuse to independence came on 10 December 1959, when the authorities decided to relocate the residents of a long-standing black-populated area of Windhoek known as "Old Location" to the new township of Katatura further from the city center. A large group of residents gathered to confront the authorities, and while the police tried to disperse the crowd, they were unsuccessful and in the end opened fire. Eleven were killed in the incident

47 Willem Steenkamp, *South Africa's Border War, 1966–1989* (Gibraltar: Ashanti Publishing, 1989), 18.

and 54 were wounded.[48] The next year the OPO became SWAPO. This was seen as an attempt at a broader appeal; however, over the course of the war about 95 percent of the captured or killed insurgents were Ovambos, and the few non-Ovambos captured complained of racial discrimination. In the early years of the insurgency there was some activity in Caprivi, but over the term of the war the insurgency never moved beyond Ovamboland.[49] In 1962, SWAPO located its headquarters in Lusaka and founded the People's Liberation Army of Namibia (PLAN).[50] This meant that there would be a shooting war.

The first PLAN infiltration from Zambia occurred in September 1965 with six trained insurgents slipping over the unguarded Angola-Caprivi border. Except for surveillance little action was taken. In February 1966 a second small group managed to murder two shopkeepers and an itinerant Ovambo in Angola, believing that they had crossed the border. The group dispersed, and three were eventually arrested on tips from local inhabitants. In July 1966, a third group crossed and attacked tribal chiefs, white farms, and a border post. The South African reaction was swift and decisive, as on 26 August helicopter-borne troops in Operation *Blue Wildebeest* knocked out the nascent SWAPO base at Ongulumbashe in South West Africa and killed two, while rounding up 54. It was the only SWAPO base ever to be established on South African soil. In early 1968, PLAN began new infiltrations from Zambia into east Caprivi. By March there were a total of 160 insurgents behind bars in South Africa. In October two large groups slipped in from Angola and restarted the insurgency. Reaction was again swift, and 56 insurgents were captured, while the rest fled over the border.[51] The nearly deserted Cuando Cubango district of Angola was now clearly a transit area for PLAN, and this situation set the stage for further cooperation between Portugal and South Africa.

Support to Rhodesia

South Africa not only helped Rhodesia avoid sanctions but gave it support in its counterinsurgency struggle. In 1967 South Africa sent a contingent of South African Police (SAP) to Rhodesia, as the ANC was operating in conjunction with ZAPU in the northeast of Rhodesia. Long-range insurgent strategy could involve alliances of convenience to reinforce the campaigns of each other, at least in theory; however, these liaisons rarely worked for long, as each was in competition for the spoils that come to a single-party state. The SAP was a paramilitary force, and its detachment was supported by Dakotas, Cessna 185s, and Alouette IIIs. The moniker of "police" attempted to obscure the fact that there were many military personnel in

48 Ibid.
49 Ibid.
50 Ibid.
51 Ibid., 22.

A SAP Cessna 185 at a forward airfield in Rhodesia in 1967 before receiving a camouflage paint scheme. (Source: Personal archive of William Marshall)

the detachment as well as 50 SAAF helicopter pilots, maintenance personnel, and communications technicians.[52] From their deployment until their withdrawal in 1975 there were on average about 2,000 SAP in Rhodesia.[53] The number of SAAF Alouette III helicopters deployed to Rhodesia can only be guessed, although the range varies between 50-plus to 24. The fact that the incoming Zimbabwean government in 1980 took possession of a mere eight functioning Alouette IIIs suggests that a great many more than this range were deployed.[54]

South Africa not only began to provide military aid, eventually to the extent of funding about half of the Rhodesian defense budget, but joined Portugal in circumventing sanctions against Rhodesia. South Africa began to see a relationship with its two neighbors that covered not only the military but the economic and political spheres as well.[55] While military support for both Portugal and Rhodesia was relatively

52 Paul L. Moorecraft and Peter McLaughlin, *Chimurenga! The War in Rhodesia 1965–1980* (Marshalltown: Sygma Collins, 1982), 31.

53 Ibid., 167.

54 Peter Baxter, *SAAF's Border War: The South African Air Force in Combat, 1966–1989* (Solihull: Helion, 2012), 21.

55 Afonso and Matos Gomes, 88.

straightforward, the remaining two spheres were somewhat different. Rhodesia required imports that would move through South Africa and Mozambique to keep its economy afloat and exports to leave the same way to earn foreign exchange. Portugal, however, was different in several ways, as its colonial economies were underdeveloped, and it needed investment. While the SAP was moving into Rhodesia, Portugal and South Africa were completing negotiations on the Cahora Bassa dam project. This would be by far the largest infrastructure project in the *ultramar*.[56] Between the Cahora Bassa and the Cunene River dam, South Africa would be taking 80 percent of the electrical output of these two massive facilities on opposite sides of Africa. The relationship between Portugal and South Africa was further strengthened with the visit of P.W. Botha, the South African defense minister, to Lisbon that same year.[57] By February 1968, South Africa had agreed to support Portugal with five Alouette IIIs and later increased this to eight Alouette IIIs and four Cessna 185s.[58] These would be assigned to an Air Detachment based at Rundu on the border between South West Africa and Angola. [59] At the same time Portugal and South Africa established two aviation command centers at the recourse aerodromes (*Aeródromo de Recurso*) of Cuíto Cuanavale (AR) and Gago Coutinho (AR), which was built in 1967. They were termed a Joint Air Support Centre (JASC) in English and a *Centro Conjunto de Apoio Aéreo* (CCAA) in Portuguese.[60] At each of these sites South African aircraft, pilots, maintenance personnel, and communication specialists supported Portuguese operations.

By 1969 it was evident that Salazar's stroke had permanently incapacitated him, and Marcelo Caetano became prime minister. He represented a new, more engaging personality with his television presentations *Conversas em Família* (Family Conversations) and made a number of important visits abroad.[61] These overtures were, however, cramped by national policy, for if the status quo were disturbed, then the traditional regime would crumble. It was thus impossible to grant concessions to the nationalist movements and find a solution to the eight-year, three-front war without compromising the regime. For Portugal, an African solution must be found in Africa, as it had few friends elsewhere, and there developed a heightened profile to its conversations with South Africa and Rhodesia. Despite the number of visits between important figures in these three governments, Caetano denied that there was anything afoot. Yet in the east of Angola during the 1968 dry season there were successful combined operations, and in 1969 Operation *Siroco* obtained excellent results. Priorities in Angola shifted to the east, and this suited South Africa, as its threat was from the east.

56 Ibid., 90.
57 Ibid., 96.
58 Ibid., 104.
59 Ibid., 104 and 125.
60 Ibid., 121.
61 Ibid., 140.

On 4 March 1970, South Africa held a briefing or review of the current and future combined operations for the military commanders of Portugal and South Africa. Exercise ALCORA was the outgrowth of this meeting. There were many issues that worried the South Africans, such as the cost of operations and the differing counter-insurgency theories. The South Africans were very conscious of the cost of supporting the Portuguese and shared their concerns. On theory, the South African one held the political aspect in the forefront, as all insurgencies are foremost political wars, and this guided the operational strategy. Because of the nature of the Portuguese government as virtually a single-party state, the political element had to be finessed.[62] Notwithstanding these differences, the accord known as Exercise ALCORA was formalized in October 1970 between Portugal, Rhodesia and South Africa. The security services of the three signatory countries kept the pact absolutely secret despite the fact that the three were actively assisting each other with matériel, troops, and intelligence.[63] Sir John Nicholls, British ambassador to South Africa at the time, reported that South Africa was not giving military aid to Portugal and holding Rhodesia at arm's length. How little he knew.[64]

Angola

Frustrated with Léopoldville's restrictive position on operations in northern Angola, the MPLA resolved at its January 1964 Conference of Cadres in Brazzaville to begin a calculated move to Zambia. From there it could re-establish its offensive. While Agostinho Neto elected initially to remain in Brazzaville, he would put his trusted associate Daniel Chipenda in charge of the eastern front, and from this point, things developed rapidly for the MPLA.

The supply chain in support of this offensive began with Julius Nyerere in Dar-es-Salaam, through which arms flowed from such supplier-states as the Soviet Union, Communist China, Czechoslovakia, Bulgaria, and Cuba. Dar-es-Salaam was not only a port of entry but also the center for training the armed insurgent elements. From this center of mischief, arms supplies and trained insurgents flowed to eastern Angola over a lengthy trail that wound through Tanzania and Zambia for nearly 1,000 kilometres to Lusaka. From here there was a northern route of another 400 kilometres that served forces entering the Cazumbo Salient of Angola and a southern one of about 300 kilometres that supplied thrusts both across the Cuando River and along the Zambeze into Angola, and ultimately for SWAPO into South Africa. As we shall see in the next chapter, Portugal and South Africa moved swiftly to address this looming threat.

62 Ibid., 176–177.
63 Wood, 343.
64 Ibid.

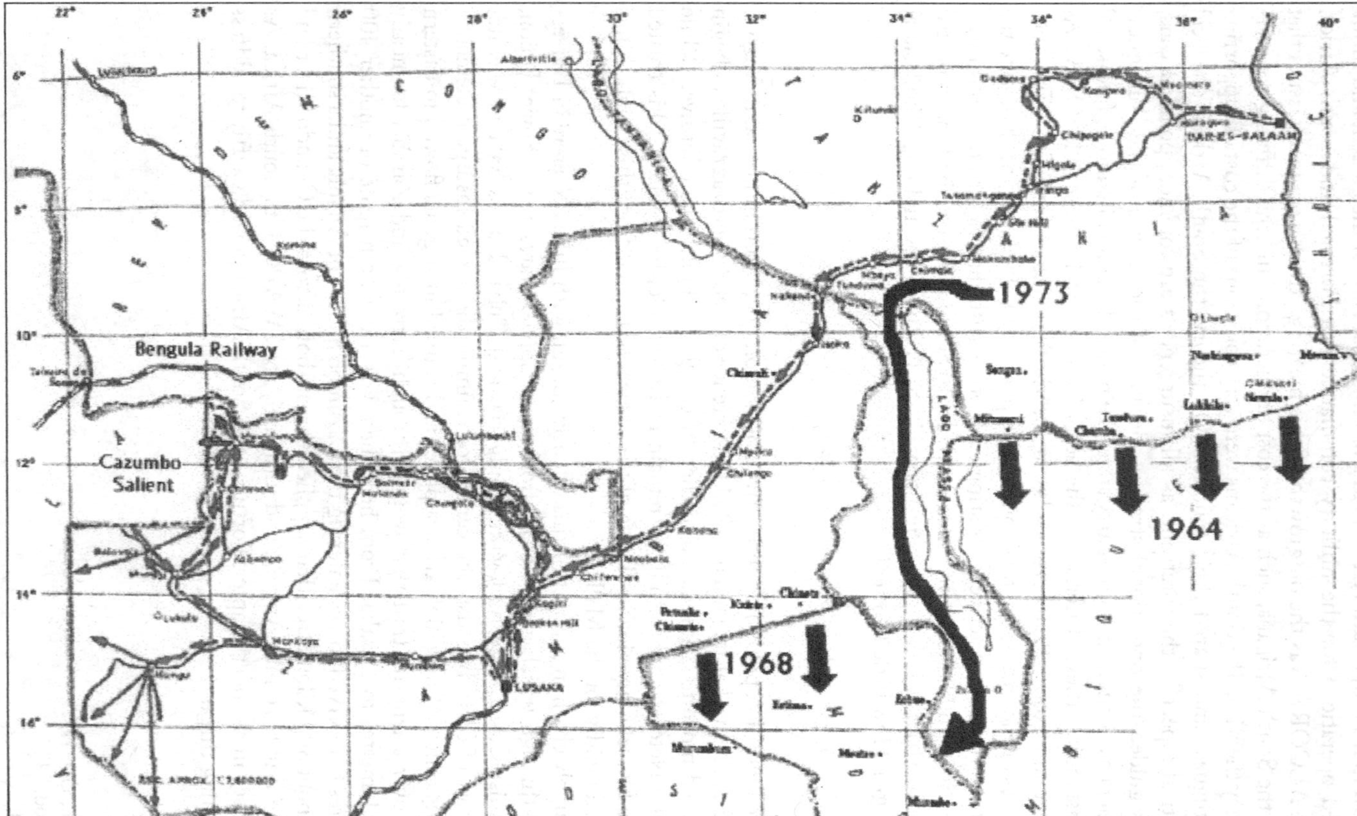

Map 10.2 Infiltration routes from Tanzania into Mozambique, Zambia, Angola and South Africa.
(Source: PIDE/DGC Archives, as modified by the author)

11

Angola's "Eastern Front"

The Eastern Military Zone (*Zona Militar Leste* or ZML) of Angola, previously the Eastern Intervention Zone (*Zona de Intervenção Leste* or ZIL) between September 1961 and March 1968, was a vast plateau of some 700,000 square kilometers (270,272 square miles) that comprised the districts of Lunda, Moxico, Bié, Cuando Cubango, and portions of Malange and was about 7½ times the size of continental Portugal.[1] It was a wasteland for insurgent proselytizing, as there were only 1.3 million inhabitants in this eastern theater or about five people per square mile. This latter figure was misleading, as the bulk of the population lived along the Benguela Railway (CFB) or in the primary towns, so in the bush there was actually less than a person per square mile. The CFB connected the Atlantic port of Lobito through the eastern border town of Teixeira de Sousa and to the rail networks of south-eastern Congo and Zambia, which led to vast and rich copper mines. The goal of the insurgents, of course, was to penetrate far enough westward to where there was a denser population that would respond hopefully to the insurgent message and significant commercial infrastructure to attack. The problem with this approach was that it was a long and forbidding route to the developed area toward the Atlantic Ocean, and much could go wrong.

This eastern area is a seemingly endless plain with an altitude of about 3,000 feet. Within it there are two elevations, one in Bié and the other in the Cazombo Salient, where the land rises irregularly to about 5,000 feet.[2] The Salient is a square protrusion into Zambia that is about 240 kilometers (144 miles) per side and is divided by the Zambeze River running diagonally through it. On the east side of the river it is mountainous and rocky. To the west of the river it is forested, sometimes densely. It had about 2,000 inhabitants, of which three or four dozen were Europeans.[3]

1 António Lopes Pires Nunes, *Angola 1966–74, Vitória Militar no Leste* [Angola 1966–74, Military Victory in the East] (Lisbon: Prefácio, 2002), 10.

2 Ibid.

3 Ricardo Cubas, Um Destacamento Atribulado" [A Painful Detachment], *Mais Alto* 323 (January–February 2000): 36.

Overall the ZML was a sprawling savanna sparsely dotted with trees. From the point of view of the insurgent, there were ideal concealment areas scattered through the more elevated land characterized by ravines and dense forest.[4] Both the vast savannas and mountainous terrain posed significant policing problems for Portugal, for it meant that finding an insurgent column or small "squadron" in either setting was extremely complicated, and gathering the vital intelligence about its activities and intentions was an extraordinary challenge. When insurgent infiltration began in 1966, there were four army battalions in the ZIL.[5] At the beginning of 1968 there were six battalions, and by the end of the year there were twelve.[6] Nevertheless, the ZML was a big place and moving even the increased numbers of troops to contact with the enemy and supporting them was difficult. The key was the mobility and flexibility provided by airpower. In one example, during the ten-day period 16–26 February 1971, two Puma helicopters, each able to carry 15–18 men, moved a total of 2,400 men to intercept the enemy in difficult and otherwise inaccessible terrain.[7] It was airpower capability such as this that was indispensable in locating the enemy, moving troops to engage him, defeating him, and in the end helping to deliver a victory to Portugal and its people in the east of Angola.

Insurgent Strategy

The strategy of the nationalist movements was to seek independence for Angola through armed conflict rather than negotiation, although the UNITA strategy was less clear. Accordingly the MPLA from its Zambian bases planned a two-pronged assault. The southern one would be mounted from Mongu, Shangombo, and Sikongo and known as the Route of the Cuando, as shown on the nearby maps that depict the UPA/FNLA, MPLA, and UNITA external bases and internal military zones. It would follow this river valley westward with the plan of reaching the populated and wealthy district of Bié and the central plain of Huambo, the heart of Angola. From this point the insurgents hoped to control the entire country and to reach all the way to Malange through an axis of advance along the Cuanza River valley.[8]

The northern one was called the Route of the Luena, or the "Agostinho Neto Route" by the MPLA, and was to be launched from Chipango and Cassamba and aimed along this river directly at Luso and from there to the highland plain of Malange.

4 António Lopes Pires Nunes, *Siroco: Os Comandos no Leste de Angola* [Siroco: The Commandos in the East of Angola] (Lisbon: Associação de Comandos, 2013), 45.

5 Comissão para o Estudo das Campanhas de África (1961–1974), *Resenha Histórico-Militar das Campanhas de África (1961–1974), 2° Volume, Dispositivo das Nossas Forças, Angola* [Historical-Military Report of the African Campaigns (1961–1974), 2° Volume, Disposition of Our Forces, Angola] (Lisbon, Estado-Maior do Exército, 1989), 127.

6 Pires Nunes, *Angola 1966–74, Vitória Militar no Leste*, 26.

7 Ibid., 30.

8 Pires Nunes, *Angola 1966–74, Vitória Militar no Leste*, 40.

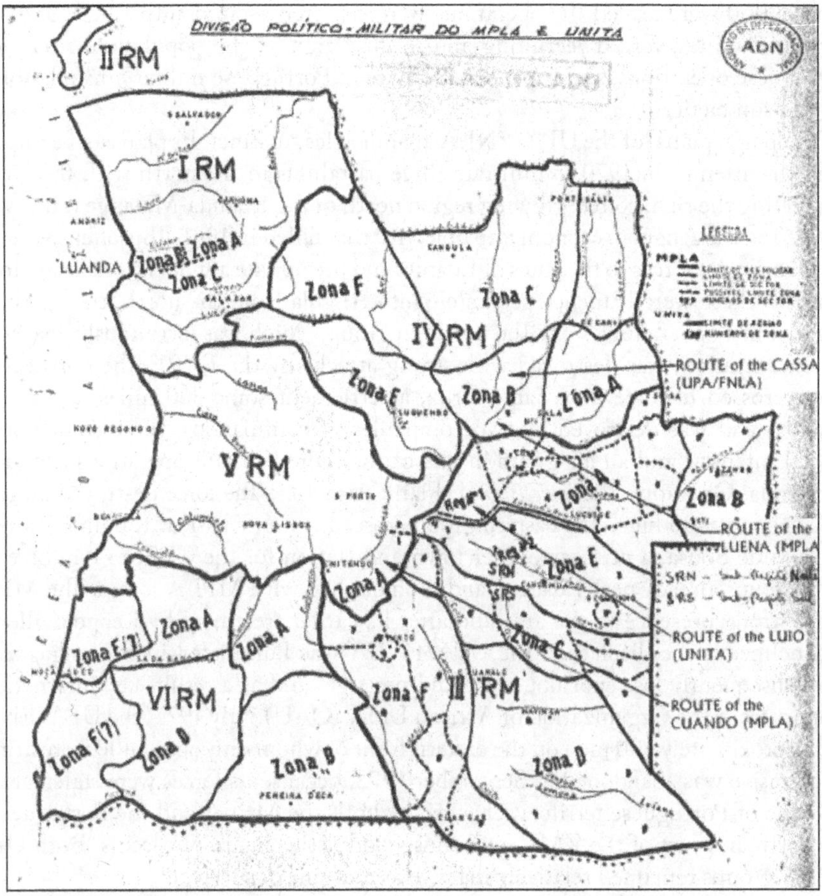

Map 11.1 Insurgent Military Zones and Incursion Routes into Eastern Angola. (Source: Arquivo da Defesa Nacional)

The hope was to gain control of the Luanda-Malange railway, reach Luanda, and link with forces coming from the north.[9] These routes pushing out of the Salient and following the Luena River never successfully penetrated a front defined by an interior line connecting the population centers of Cazoa, Cazage, Cassai-Gare, and Lucusse with Luso as its hub.[10] The MPLA could muster about 4,000 men, of whom

9 Ibid.
10 Comissão para o Estudo das Campanhas de África, *Resenha Histórico-Militar das Campanhas de África 1961–1974, 6 Volume, Aspectos da Actividade Operacional, Tomo I, Angola–Livro 2* [Historical Military Report on the African Campaigns 1961–1974, 6

about 3,000 were armed. Its operations were centered around anti-vehicle and anti-personnel mines, forced recruiting and intimidation of the population, reaction to Portuguese operations, and some bursts of fire at Portuguese military installations in hit and run tactics.[11]

Campaign plans for the UPA/FNLA are much less distinct. Its plan can be imputed from the attempt in 1970 to infiltrate three battalions in the north with the goal of controlling the rich, coffee-growing region north of the Luanda-Malange railway and paralyzing the Angolan economy, an objective that failed in 1961. This offensive action would bring its forces to the gates of Luanda and precipitate a sudden conclusion to the conflict. These were launched separately into Angola from the north, the northeast, and east in a general invasion. The northern group, which was fortuitously discovered in concentration, was destroyed with strong attacks by the FAP.[12] The northeastern group crossed in force near Santa Cruz, a settlement some 360 miles northeast of Luanda, and was destroyed by two companies of commandos, three companies of special infantry, and an air group in one of the more brilliant operations conducted in Angola, Operation *Golpe de Flanco* (Flank Attack).[13] The force destroyed an enormous base of 600 huts. The eastern group crossed from its base at Kahundu through Teixeira de Sousa, a primary border town and station for the CFB, to reinforce one of its groups already near Luacano and soon clashed with MPLA forces. The MPLA had a strong presence in the area and after its earlier treatment by Léopoldville was not inclined to be displaced. The UPA/FNLA was humiliated in this engagement and consequently lost enormous face and prestige, and as a result, key international support from the Organization of African Unity (OAU).[14] By 1971, the UPA/FNLA had approximately 600 men on the eastern front of whom only about 400 were armed. Its offensive was abandoned in September 1972, because its forces were stalemated at the edge of Portuguese territory, checked by both the MPLA and the Portuguese in the north and east of the ZML, and thus stood little chance of success. Both UPA/FNLA groups remained relatively inactive as a means of survival.[15]

UNITA simply intended to establish a base deep in the interior of Angola and, using some notion of the oil-spot theory of expanding control, mobilize an ever increasing number of the population to gain the "total independence of Angola." Indeed, Savimbi was forced by circumstances into this improbable posture, claiming that a true nationalist movement should operate from bases within Angola, as in July 1967 he was

Volume, Aspects of Operational Activity, Tome I, Angola, Book 2] (Lisbon: Estado-Maior de Exército, 2006), 353.

11 Ibid.

12 Pires Nunes, *Angola 1966–74, Vitória Militar no Leste*, 25

13 Ibid.

14 Comissão para o Estudo das Campanhas de África, *Resenha Histórico-Militar das Campanhas de África 1961–1974, 6 Volume, Aspectos da Actividade Operacional, Tomo I, Angola–Livro 2*, 223–228.

15 Ibid.

finally and completely expelled from Zambia. Certainly the implementation of such a policy ensured his isolation. By 1971 UNITA was surrounded, and Savimbi counted fewer than 1,300 followers. In order to survive in defeat, he and his men came to an accommodation with the Portuguese. Between 1971 and 1973, UNITA activities were restricted to a prescribed zone at the headwaters of the Lungué-Bungo, and it agreed to cease operations against the Portuguese.[16] As part of this understanding, UNITA would receive arms and medical support and would be free to engage the MPLA.[17] By this time Savimbi had about 500 armed men.[18]

There were substantial weaknesses in the campaign design of all three in that their lines of communication would be long and vulnerable, or in the case of UNITA, non-existent. The goals of each would rely increasingly on the weak strategy of hope, as each moved further and further from its external support and progressively isolated itself deep in the hostile interior of Angola. If these organizations were going to rely on classic proselytizing of the population for support, then it would not come willingly. The Angolan people were experiencing new political and economic freedoms and financial prosperity and would hardly welcome what would be tantamount to Napoleonic-type foraging by foreign armies. Then too, the basic premise of all three insurgencies had been removed with successful responses to the people's grievances. Nationalism for the sake of nationalism and the political goal of a single party state running a country for the benefit of its oligarchy or chief-of-state held little popular appeal. Lastly the greatest enemies of the insurgents were each other, as all three were in competition for the spoils to be mined from governing Angola. There was no thought of a coalition government for the greater good of the diverse population.

In contrast, the army in the ZML by December 1971 had 55 infantry companies, two companies of commandos, two squadrons of cavalry, an artillery battery of 88mm field guns, two engineering companies, and a construction company. There were also Auxiliary Forces consisting of three groups of *Fiéis* (Faithful) (2,318 men), 71 groups of *Grupos Especiais* (Special Groups) (2,279 men), one group of *Leais* (Loyals) (128 men), 31 groups of *Flechas* (Arrows) (1,665 men), and various armed militias (5,331 men), a total of 11,721 additional men over the approximately 7,500 regular forces.[19] Further there were approximately 1,200 air force personnel and perhaps 300 naval

16 Adelino Gomes, "Exército e UNITA Colaboraram antes de 74" [Army and UNITA Collaborate before 74], *Público* (19 December 1995), 2-4; and Óscar Cardoso, "Criador dos Flechas" [Creator of the Flechas] in José Freire Antunes, *A Guerra de África, 1961–1974, Volume 1* [The War in Africa, 1961–1974, Volume 1] (Lisbon: Temas e Debates, 1996), 409–410.

17 Dalila Cabrita Mateus, *A PIDE/DGS na Guerra Colonial 1961–1974* [The PIDE/DGS in the Colonial War 1961–1974] (Lisbon: Terramar, 2004), 199–207.

18 Pires Nunes, *Angola 1966–74, Vitória Militar no Leste,* 19.

19 Ibid., inside front cover; and João Paulo Borges Coelho, "African Troops in the Portuguese Colonial Army, 1961–1974: Angola, Guinea-Bissau and Mozambique," *Portuguese Studies Review* (Spring-Summer 2002), 129–150.

personnel, bringing the total number of Portuguese forces to about 20,700.[20] This number yields a ratio of 3.2 to 1 over the insurgent aggregate of 6,350 men. While comparable ratios of other counterinsurgencies were higher, such as, Malaya at 37.5 to 1, Cyprus at 25 to 1, Algeria at 16.7 to 1, Kenya at 4.6 to 1, Vietnam in 1964 at 4.1 to 1, and Vietnam in 1968 at 8.75 to 1, isolation of the insurgents in a sparsely inhabited and barren region belied the apparent weakness of this figure.[21]

Portuguese Campaign Strategy

In countering this multi-axis assault, the Portuguese were quick to see that UNITA and the UPA/FNLA would be easily contained and that the MPLA represented the real threat. They thus developed a three-phase theater strategy of first checking the expansion of the insurgent penetration. This would mean defining the population battlefield and bringing security to those threatened. Next, the security forces would surround the insurgents and limit them to a geographic area bounded by the Cuito, Cuanza, Munhango, and Cassai Rivers, as depicted on the nearby map by a line running from Dirico in the south to Teixeira de Sousa in the northeast. This was admittedly a substantial tract, but it was thinly populated by any standard and would provide little sustenance. The insurgents could thus do little damage in this isolation and would indeed slowly starve. Within it Portuguese security forces would relentlessly pursue the enemy, who would become increasingly harried and besieged. Finally in 1973 and 1974, as the cordon drew progressively tight, the enemy would be completely destroyed.[22]

The need for this lengthy approach was prompted by a number of factors. First, Portugal had limited manpower to police the vast region of the east. This limitation dictated an approach that enlisted the sparse terrain, geographic remoteness, and harsh climate as allies in isolating the insurgents in an attritional war. Second, these three regional characteristics appeared to favor the insurgent in that he could easily hide or move largely unhindered by Portuguese forces, so time was needed to track and destroy him.

The problems of sparse terrain and geographic remoteness were nowhere more evident than in the Chana da Cameia, an extensive plain of some hundreds of square miles completely covered in tall grass. Its extension, vastness, and lack of topographical references made it difficult to navigate and search for the enemy successfully. It was bounded in the north by the CFB and in the south by the Luena River, the Route of Agostinho Neto. In the rainy season it was largely impassable because of flooding, but in the dry season it could be easily crossed on foot, in vehicles, or on horseback.

20 Ibid., 67.

21 John P. Cann, *Contra-Subversão em África* [Counterinsurgency in Africa] (Lisbon: Prefácio, 2005), 30.

22 Pires Nunes, *Angola 1966–74, Vitória Militar no Leste,* 42.

Alouette IIIs refueling in the Chana da Cameia. (Source: Força Aérea Portuguesa)

Very few people lived in the Cameia, but the small numbers who did were forced to aid the MPLA. MPLA troops in transit through this *chana* alongside the Luena were difficult to find hiding in the grass and required tailored tactics to weed them out. The proximity of the trackless Cameia to the eastern frontier made it a magnet for insurgent columns trying to elude security forces and move westward.

The weather was another important factor and was indicative of the harsh climate. The rainy season in the east ran from October to March and left enemy columns wet, miserable, and mired. Conversely, it could severely affect flight operations with its electrical activity and towering cumulonimbus clouds. Captain Carlos Acabado described flying in the east on his way to Sete in the rainy season:

> I saw the airfield in the distance as a trace on the ground, then became preoccupied with the possibility of not having sufficient fuel to circumnavigate the storm that covered the horizon in front of me. Black clouds, so characteristic of the rainy season, rose like high walls from the ground, pregnant with electricity, striking frighteningly, with long and impressionable discharges that appeared immobilized against the dark background, in a spectacularly terrifying but also beautiful force.[23]

23 Carlos Acabado, *Kinda e outras histórias de uma guerra esquecida* [Kinda and Other Stories of a Forgotten War] (Linda-a-Velha: DG Edições, 2011), 9.

Alouette IIIs resupplying a cavalry squadron in the Chana da Cameia, east of Angola. (Source: Força Aérea Portuguesa)

This was followed by a dry but cool season characterized by heavy morning fog, the *cacimbo*, which was caused by the long, cool African nights. Indeed, the transformation from daylight to darkness in the eastern savanna during this period came with such velocity that pilots had the sensation, according to Lieutenant José Queiroga, that the rotation of the earth accelerated.[24] While this fog hid the insurgents, it invariably hampered flight operations and required great flying skill. Again Captain Acabado describes one such helicopter insertion next to the Quembo River northeast of Cuito Cuanavale during the *cacimbo*:

> The first section of commandos was launched onto the humid banks of the Quembo, where an intense ground fog persisted. The helicopters touched down and rose in a way that put the men on the ground as rapidly as possible without scattering them. Everything unfolded in a strange atmosphere, blending into a single color in which sunlight seemed absent and in which the riverbank rose out of the fog...."[25]

24 José Augusto Queiroga, "Histórias Vividas: Por Terras do Fim do Mundo, Uma viagem entre os 18 e os 9 graus de Latitude Sul" [Tales Lived: For the Lands at the End of the Earth, A trip between 18 and 9 degrees South Latitude], page 6, undated deposition, Caixa 181, Archivo Histórico de Força Aérea, Alfragide.

25 Acabado, 51.

Map 11.2 Eastern Military Zone, 1968, with Insurgent bases in Congo and Zambia. (Source: Modified from a map by António Pires Nunes)

Another difficulty navigating eastern Angola lay in the hundreds of water courses that permeated the Angolan landscape and were deeply affected by the seasons. The extensive riverine network had its origins in the elevations of Bié and Cazombo and made certain that any ground movement by either Portuguese forces or the enemy would involve one or more river crossings. Some were easily fordable and others extremely difficult, depending on the season and what the rains had done. Portuguese forces learned to excel in negotiating this extensive river system. They developed the skills to cross any river, day or night, and to use it as a natural obstacle to enemy maneuver and as ideal defensive terrain. The rivers were also natural divides for visual reconnaissance zones and important for orientation in the immense area of the east where flying in marginal weather conditions was a frequent occurrence. They likewise proved to be avenues of enemy approach for combat and lines of communication for enemy infiltrations. These latter characteristics were to help in predicting the location and routes of enemy incursions. Rivers also served as natural boundaries to the battlefield and provided navigation references and orientation both on the ground and in the air.

Air navigation relied on the visual use of prominent terrain features, such as counting the rivers crossed. However, the great stretches of plain in the east, the *chanas* that were characterized by a monotonous and unbroken horizon, offered few references.[26] The air charts available were scale 1:1,000,000 and were insufficient in detail to be useful in visual navigation. Consequently, there were numerous hand-drawn charts that provided the necessary detail for air operations. Communications were also a problem in navigating these vast spaces, as neither the Alouette III nor the T-6G had any HF radio capability, and their VHF capability reached only about 30 miles. As a precaution, these aircraft always flew in pairs when transiting long distances. It should be noted that when the Alouette III began operations in 1963, it had no radio compass (radio direction finder or RDF) installed. Five years into the conflict this void remained. Admittedly there were few radio beacons available, particularly in the east, and often these were unreliable, for their support came from Lisbon, where because of indifference or incompetence (or both) this urgent need was never given proper merit.[27]

Because of the nature of the battlefield and the enemy, a containment and attrition strategy was well suited and logical. It was one in which Portuguese forces would define and then gradually reduce the area of insurgent contamination, and the enemy would be forced to fight from a progressively circumscribed and isolated position and

26 Ricardo dos Santos Cubas, Geraldo Estevens, Guilherme Costa Santos, and José Augusto Barrigas Queiroga, "Ensinamentos da Guerra em Angola" [Teachings of the War in Angola], proceedings of a conference "Almoços da AFAP," January 2011, 22.

27 Aleixo Corbal, Oliveira Simões, Fidalgo Ferreira, and Heitor Almendra, "2° Região Aérea, Lições Aprendidas entre 1961 e 1975" [2nd Air Region, Lessons Learned between 1961 and 1975], proceedings of a conference "Almoços da AFAP," January 2011, 27. There were three radio beacons: Henrique de Carvalho, Luso, and N'Riquinha, a shameful few for such a large area.

be ultimately eliminated. The few local people living in the east and southeast could hardly subsist themselves, much less support the insurgents with food, recruits, intelligence, and shelter. In summary, Portuguese theater strategy was based on sound reasoning that followed from a considered appraisal of opponent strengths, battlefield terrain, resident population, and available Portuguese resources.[28]

While the military dimension was important in this strategy, it had to be integrated with and built around the notion that the war in the east was a fight to preserve and protect the population from insurgent intimidation. This meant organizing it to defend itself and supporting it in every way with such skills as medical care, animal husbandry, education, etc. The job of the military was to suppress violence and provide a security umbrella under which Portugal could earn the loyalty of the population. This way of fighting required extensive coordination across civil-military tasks as much as military ones. The FAP was in the forefront in supporting the population and fighting for its loyalty with its airlift and medical capabilities. It was believed that in time as a result of civil efforts all-important intelligence on the enemy would flow from a loyal and confident population. This was an important development and indicator in counterinsurgency, as it reflected a population that believed in its government. This view did not diminish the military aspect, but it did bring into focus the fact that all insurgencies are primarily political wars, and this one was no exception.

Captain Acabado recounts an event illustrative of the Portuguese effort to secure this loyalty during one of the many actions in Operation *Siroco*, which we will examine shortly. A commando company was launched in a surprise vertical envelopment at dawn on a zone where it was known that the enemy was present and could only survive with the help of the population or with the knowledge of the local tribal chief. The raid was considered a great success and produced insurgent losses, captured arms, and valuable documents. The village of the tribal chief and other local settlements were located adjacent to the insurgent encampments, so obviously the people knew of the enemy, yet neither the chief nor his people were molested by the commandos by order of their commander.

The next day at dawn four jeeps transported the commander and six of his commandos, all unarmed, to the primary village to pay a courtesy call on the chief. He was surprised at the visit and wondered about the motives of the commander, who explained what his commandos were doing and that they would be in the area for some additional days. In turn the chief told of the insurgents coming to his village, where the people were unarmed and defenseless, and of his inability to inform the authorities. Later the chief came to the commando encampment to bring some goats and chickens which he laid at the feet of the commander in a very African ritual. The chief was given a chair next to the commander, smoked a cigarette, drank some beer, and finished with a glass of brandy. Finally he thanked the commander emotionally

28 Comissão para o Estudo das Campanhas de África, *Resenha Histórico-Militar das Campanhas de África 1961–1974, 6 Volume, Aspectos da Actividade Operacional, Tomo I, Angola–Livro 2*, 356.

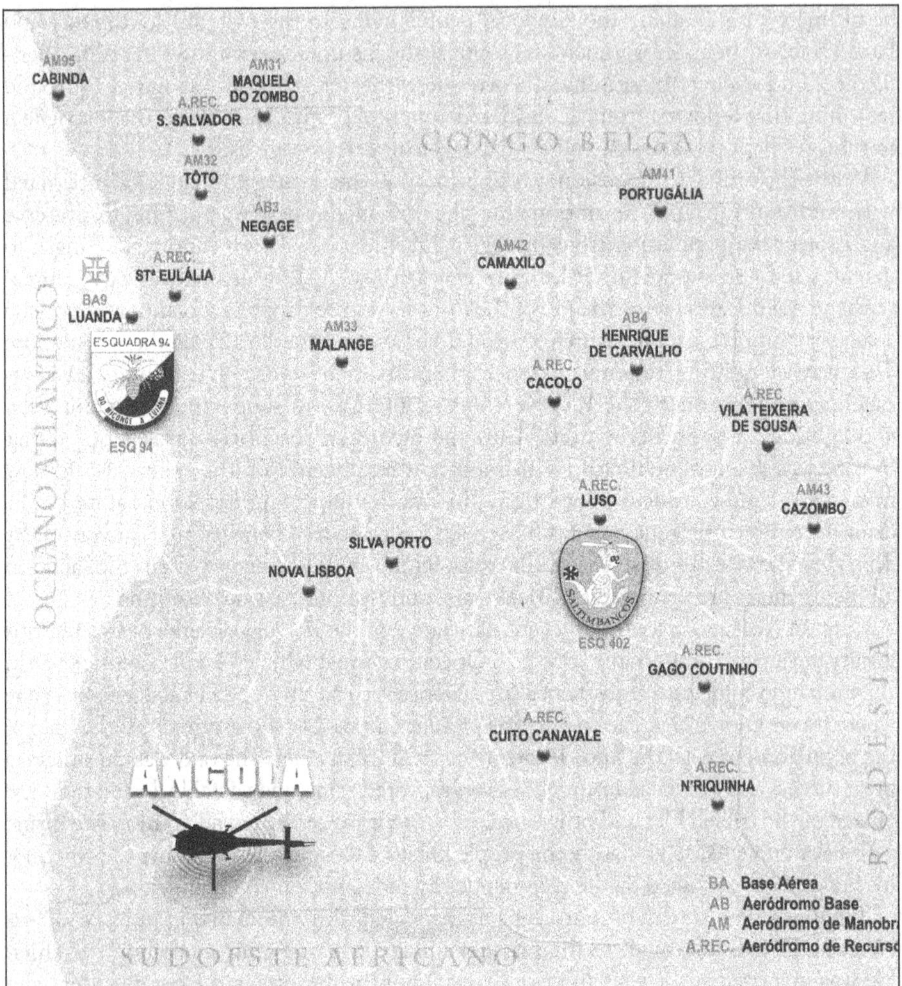

Map 11.3 2nd Air Region Airfields. (Source: Força Aérea Portuguesa)

for taking care to avoid his men, women, and children in the assault. Much later the commander explained to his men how important this meeting with the chief had been. The maintenance of trust in relations with the population was important both for the people as well as the commandos. In the conduct of war, it was important to be generous with the weak, who cannot always be protected, as they should be. The duty was upon the strong to practice good manners and not intimidate the population.[29]

29 Acabado, 136–137.

Opening Moves

Subversion of the population in the east of Angola was estimated at 6 percent in 1965 and had steadily risen from then to 42 percent in 1968, when the MPLA offensive reached its height.[30] In 1966, General Amadeu Soares Pereira, CCFAA, was heavily occupied with the northern front, as while the MPLA and UNITA in the east were making their initial forays, there remained a substantial threat from the north that required attention. In November 1967, the new CCFAA, General João Anacoreta de Almeida Viana, relieved General Soares Pereira and as late as August 1968 was overseeing Operation *Nova Luz* (New Light) against the MPLA and UPA/FNLA columns in the north that required both troops and helicopters to execute—resources that he could not yet release in any numbers to the east. His priority remained the north throughout his tenure despite developments in the east. Nevertheless, in 1966 four Harpoons had been redeployed to airdrome base Henrique de Carvalho (AB 4) and two additional Harpoons to recourse airdrome Luso (AR). It should be remembered that until 1966 the ZML was for all practical purposes a peaceful area where the population showed little interest in the ideas offered by strangers, so there was little reason for more than a token military presence.[31] During 1968, a Beechcraft C-45 Expeditor (often called a "Twin Beech"), 11 Harvards, and 11 Dorniers were added to AB 4.[32] Also in February 1968, South Africa agreed to support Portugal with five Alouette IIIs and later increased this to eight Alouette IIIs and four Cessna 185s, all which would be assigned to an Air Detachment based at Rundu on the border between South West Africa and Angola. This much needed help bridged an important gap. Coincidently, Portugal and South Africa established two aviation command centers at the recourse aerodromes of Cuito Cuanavale (AR) and Gago Coutinho (AR), which were freshly built facilities. These were termed Joint Air Support Centers (JASC) in English and a *Centros Conjuntos de Apoio Aéreo* (CCAA) in Portuguese. At each of these Portuguese airfields South African aircraft, pilots, maintenance personnel, and communication specialists supported Portuguese operations.

This mixture of aged and light aircraft may seem inadequate even in a counterinsurgency; however, these were important resources in the east. The Dorniers, for instance, with their special characteristics of endurance, unimproved field capability, and good visibility, provided an indispensable freedom of action for troops in establishing links and contacts with local people, in studying the terrain, its trails, and traffic, and depositing troops in remote areas to conduct reconnaissance patrols.[33] In the several *chanas* of the ZML, the Dorniers were able to orient and guide troops in

30 Ibid., 6.

31 Pires Nunes, *Siroco*, 347.

32 António de Jesus Bispo, "A Participação da Força Aérea na Guerra de África (1961–1974)" [Participation of the Air Force in the African War (1961–1974), *Revista Militar* 2507 (December 2010): 1390.

33 Pires Nunes, *Siroco*, 195.

South African aircrew in the east of Angola refueling their Alouette III during an operation. Helicopter has no national markings and carries only a tactical mark. (Source: Personal archive of Melo Vidal)

their searches.[34] Likewise they resupplied helicopters in the field with fuel to make their operation more efficient and eliminate a long, slow return trip to a formal airfield to refuel.[35] Indeed, the troops became very adept in making improvised airstrips for the versatile and welcomed Dornier, whose skilled pilots grew accustomed to landing on these rough *pistas* that barely passed for an airfield. Landing areas frequently had inclines, declines, or curves on the intended landing stretch, and these irregularities posed a special challenge.[36] For instance, during Operation *Luena Grande* (Great Luena) in October 1966 in the east, one of the early engagements in which an Air Group was formed at AB 4 with two Alouette IIIs, two T-6Gs armed with 37mm rockets, a PV-2, and a Dornier. The helicopter and Dornier were deployed first to Cazombo and then to Lumbala, where the tactical headquarters was located. The MPLA had crossed the border and headed west across the Salient. The force available to engage the enemy column was inadequate; nevertheless, it did so, and a soldier was gravely wounded.[37] The Dornier evacuated him to Luso for care and throughout

34 Ibid.
35 Ibid.
36 Cubas, 37.
37 Ibid., 39.

Luena Grande performed all of the tasks normally expected. It certainly proved its worth in augmenting the capabilities of the ground forces, and this opened the way ultimately to a very close air-ground relationship.[38]

Invariably helicopters were the big deficiency. There was wide demand for the limited numbers, as there was no other way to move troops rapidly in the Angolan terrain to surprise, overwhelm, outflank, and outgun the enemy. Every significant operation ultimately followed the pattern of transporting elite troops in a helicopter assault, protecting the force with one or two gunships, and evacuating the wounded. Hence there were never enough of them, and the accelerating numbers over the period of the conflict are testimony to this need.

The first 12 Alouette IIIs arrived in Luanda in July 1963 and were assigned to Squadron 94, the *Moscas* (Flies). The squadron maintained Alouette III detachments at Negage (AB 3) and Henrique de Carvalho (AB 4), and variously at Ambrizete, São Salvador, Santa Eulália, Quibaxe, Cabinda, Santo António do Zaire, and Toto, depending on operational demands. The first truly large engagement using helicopters occurred in the north during the six-day period 14–21 March 1965 with Operation *Aniversário* (Anniversary). It involved about 2,000 troops and the eight available Alouette IIIs, which staged from Santa Eulália.[39]

In August 1966, two Alouette IIIs, one a transport and the other a gunship, flew to Henrique de Carvalho (AB 4) to support Group Command No 1974 (*Comando de Agrupamento* No 1974) in its operations. The helicopters engaged UNITA at Lucusse in their first armed action and afterwards operated near Luvuei, an extremely isolated outpost in the "Lands at the End of the Earth." During the month-long operation, the helicopters carried commandos and paratroops and proved invaluable in giving the troops mobility and thus surprise. In the final sortie, the Alouette IIIs headed for Cazombo (AM 43), where they remained, and their crews were rotated. This might be called the genesis of Squadron 402, the *Saltimbancos* (Travelling Entertainers), which was commissioned in October 1969 at Henrique de Carvalho (AB 4) and maintained a detachment at Luso (AR).[40] Its helicopters were the gift of South Africa, the first five of which arrived in June 1968 at Cuito Cuanavale in C-130 Hercules transports in Portuguese livery.[41] These were quickly followed by two more in July and another two in August.[42] Because of their origin, they carried no national identification. It was thought that the casual observer could not differentiate between unmarked

38 Pires Nunes, *Siroco,* 195.

39 Mário Diniz and Luís Proença, "AL III, História Breve de uma Vida Longa, 1963–2013" [AL III, A Brief History of a Long Life, 1963–2013], *Mais Alto* 404 (July–August 2013): 18–19 and 22. There were also two Dorniers armed with rockets.

40 Ibid.

41 Queiroga, "Histórias Vividas: Por Terras do Fim do Mundo, Uma viagem entre os 18 e os 9 graus de Latitude Sul," 4. Serial numbers of the helicopters were 9317, 9318, 9319, 9320, and 9321.

42 Diniz and Proença, 24–25.

Alouette IIIs of Squadron 402, *Saltimbancos*, refueling in a *chana* in the east of Angola. Note the lack of a national markings. (Source: Força Aérea Portuguesa)

Portuguese and unmarked South African Alouette IIIs, both of which operated in the east from this period to the end of the war. Ultimately there were 17 new Alouette III helicopters delivered to Angola by South Africa that carried no national markings or serial numbers to protect this relationship and the secrecy of Exercise ALCORA. This series became known as the "Black Block" of helicopter serial numbers, because they "did not exist."[43]

By 1972 Alouette III numbers in Angola had reached their maximum at 29 units, which altogether flew about 4,500 hours annually.[44] In early 1970, six Puma helicopters arrived in Squadron 94, and these flew about 1,200 hours annually.[45]

War in the east began in earnest in November and December 1966, when some MPLA insurgents penetrated the northeast corner of the Cazombo Salient near the border settlements of Jimbe and Caianda. UNITA also entered the fray, and Savimbi's first attack occurred in December 1966 on the post of Cassamba. He made another two weeks later on Christmas Eve, when 500 UNITA insurgents attacked the Portuguese garrison of Teixeira de Sousa and suffered a disastrous defeat with 234 dead. UNITA

43 Ibid.
44 Jesus Bispo, 1390.
45 Ibid.

General Francisco da Costa Gomes at his desk in Luanda. (Source: Força Aérea Portuguesa)

and the MPLA were soon alienated from their vital patron, as Portugal struck at the fragile Zambian treasury by closing the CFB to copper exports for a week. Again in April 1967, there were two UNITA acts of sabotage on the CFB infrastructure, and again Portugal closed the railway and threatened lengthier closures unless attacks ceased. Thereafter UNITA was unwelcome in Zambia, and the MPLA likewise was put on notice not to interfere with the CFB.

Portuguese leverage had its roots in Zambian dependence on the CFB to export its copper ore across Angola. The railway always represented a tempting target for insurgents, and thus Zambia found itself caught between the practical need for foreign sales and the ideological desire to support nationalist movements. CFB closures carried a strong message for both Zambia and the Congo. As a consequence, during early 1967 the MPLA was ostensibly pushed out of Zambia and infiltrated the east in two primary areas, Lake Dilolo on the western edge of the Salient, and a larger area located about 160 miles due south of Luso and bounded by the settlements of Sessa, Muié, Cangombe, and Cangamba. By 1968 this surge had penetrated as far west as Bié district, and as a result of such overextension, various MPLA columns found themselves quite vulnerable. Considerable numbers were destroyed or captured, but by 1970 there were still 1,300 men in Zambian bases supporting regular columns into Angola, a constant threat.[46] Clearly there were limits to diplomatic and economic pressure in the face of Kaunda's strong feelings about African nationalist movements.

46 Diniz and Proença, 24.

General Francisco da Costa Gomes arrived in Luanda in May 1970 as CCFAA in the midst of this enemy offensive. He began by reorganizing his headquarters staff and establishing an Office of Irregular Troops, which coordinated the specialized African troops, various militias, and other irregular forces. These troops with their unique tracking skills and familiarity with the local terrain did much to locate the enemy columns in the vastness of eastern Angola. Costa Gomes also shifted focus to the east and selected Brigadier José Manuel Bethencourt Rodrigues as the ZML commander in March 1971. He was granted wide latitude in executing the containment and attrition strategy. Bethencourt Rodrigues was an infantry officer who had stepped into the vacuum left by the disastrous 1962 Dakota aircraft accident at Chitado, which had killed most of the general staff in Angola. Since then he had enjoyed a wide portfolio of important and highly visible assignments in which he had built his reputation as a cohesive and competent leader and gained the confidence of Costa Gomes. His new headquarters was in Luso, where the FAP had only a recourse airdrome.

The aircraft permanently assigned to the ZML were dependent logistically and administratively on the Commander, 2nd Air Region, in Luanda. This dependency and its failure to respond had the effect of immobilizing numbers of these aircraft routinely.[47] The situation continued until the end of 1971 and significantly affected air support capabilities. This ineffectiveness also encouraged increased MPLA adventure in the east. There were as a consequence a number of failed missions and many recriminations. The situation became a painful error, and to correct it General Manuel Simão Portugal, Commander of the 2nd Air Region at the time, Costa Gomes, and Bethencourt Rodrigues created a new structure at Luso, the Eastern Air Sector Command (*Comando do Sector Aéreo Leste* or COMSECAR). COMSECAR, however, still required fine tuning. Luso was a dependency of AB 4, where the aviation logistic and administration functions for the entire eastern sector were located. Immediately the dependency status of Luso, the demands of COMSECAR, and those of Bethencourt Rodrigues clashed.[48] The solution was to consolidate the functions of these several entities into a single person: the commander of AB 4 was made COMSECAR.[49] This finally created a unified command that rapidly reestablished the needed air support capacity to generate effective operations in the ZML.[50]

COMSECAR was part of a wider organization of air operations in the 2nd Air Region. Command of the Region was organized into a General Staff and organs of personnel and logistics. The General Staff assumed a particular importance in operations and intelligence and there were arms to oversee these vital functions. Military air

47 Corbal, Simões, Ferreira, and Almendra, 17.

48 António Silva Cardoso, *Angola, Anatomia de uma Tragédia* [Angola, Anatomy of a Tragedy] (Lisbon: Oficina do Livro, 2000), 286.

49 Ibid., 287.

50 Corbal, Simões, Ferreira, and Almendra, 17.

operations were controlled by the Military Flight Information and Control Authority (*Orgão de Controlo e Informação de Voo Militar* or CCIVM), and it was responsible for all aviation activity in the three Air Sectors of the Region, as shown on the nearby map: Air Sector AB 3 (Negage), Air Sector BA 9 (Luanda), and Air Sector AB 4 (Henrique de Carvalho), which included COMSECAR.[51] The map also shows the reorientation of the air infrastructure. Originally the air infrastructure was oriented toward the Congo and not Zambia. This orientation placed AB 4 north of the active insurgent penetrations in Moxico and Cuando Cubango. Its dependency airfields to the south opposite the insurgent bases in Zambia were better positioned to address the new threat and were thus expanded to accommodate the full range of air support required in the southeast. There were nine dependencies of AB 4, six of which were located in the southeast: Luso, Vila Teixeira de Sousa, Cazombo, Gago Coutinho, Cuito Cuanavale, and N'Riquinha. These were the most active in supporting ground forces and the population, and those designated as recourse airfields were not always as spare in their facilities, as one might imagine from the name. Reflective of this, Colonel Silva Cardoso remarked in 1972 on visiting an AR, "One day I landed and stayed for two days in Cuito Cuanavale at its enormous airfield with the capacity for every type of aircraft."[52] The building of this airfield infrastructure, which began in the 1950s, continued unabated during the entire 13-year war.[53].

Every FAP mission had an objective or purpose that was detailed in its Mission Order (*Ordem de Missão* or ORMIS), which described the type of aircraft, its ordnance, its assignment, its operational area, its period of execution, and instructions for coordinating with other air or ground units.[54] Each mission command pilot was required to submit a report on completion and provide copies to the operations department at the base to which he was attached and the General Staff. The report was considered with other information by the General Staff in developing its picture of the enemy, appraising coordination with ground forces, and addressing other relevant aspects so that it could fine tune its activities. Daily there was a briefing for the Commander, 2nd Air Region, that featured his intelligence and operations chiefs and included logistics and other pertinent developments. In this briefing all relevant activity for the period was raised and discussed. Of particular interest and concern was the refinement of air-ground support and thus was born the concept of "associated command" and the Air Group.[55]

Aviation command and control in the ZML was required to adjust to the increased enemy activity and to redeploy aircraft rapidly to maintain the FAP capability in

51 Ricardo dos Santos Cubas, Geraldo Estevens, Guilherme Costa Santos, and José Augusto Barrigas Queiroga, "Ensinamentos da Guerra em Angola" [Teachings of the War in Angola], proceedings of a conference "Almoços da AFAP," January 2011, 4.

52 Cardoso, 297.

53 Corbal, Simões, Ferreira, and Almendra, 4.

54 Santos Cubas, Estevens, Costa Santos, and Queiroga, 4.

55 Ibid., 5.

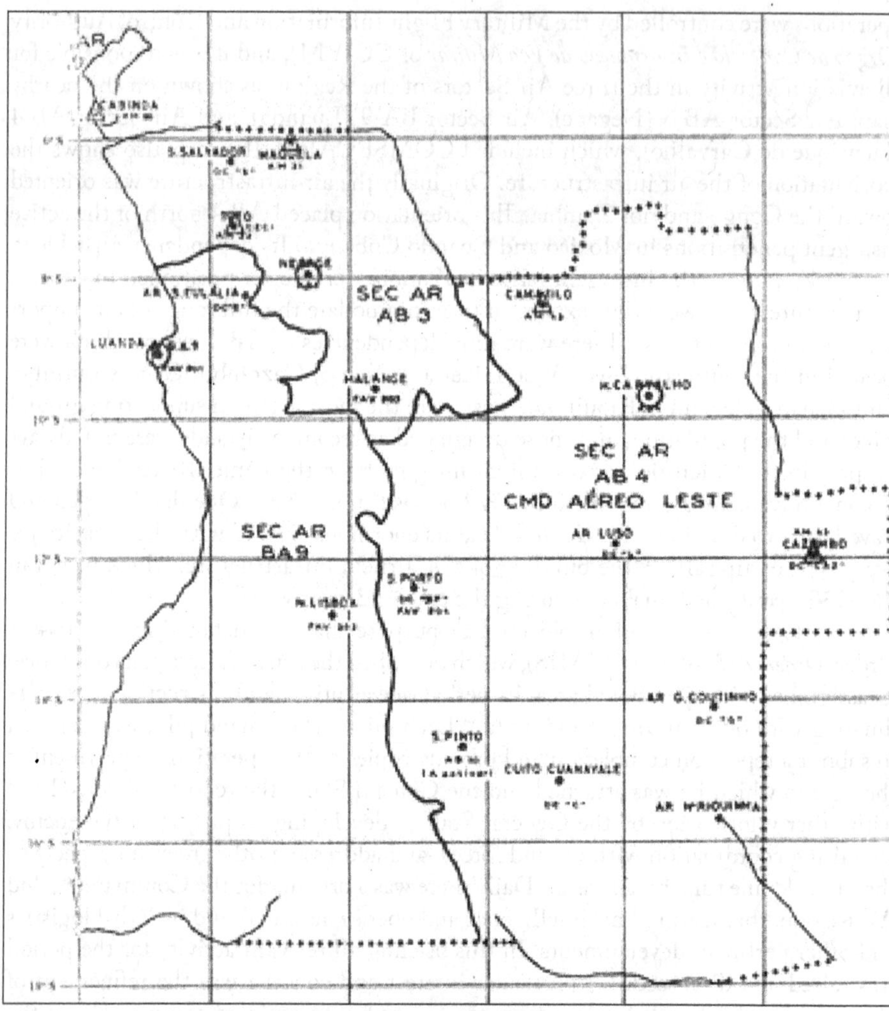

Map 11.4 Air Sectors of the 2nd Air Region.

support of ground forces chasing the invading enemy. This called for decentralization, and the solution for this ground support was seen at the time to be the creation of air detachments. Known as Air Groups (*Agrupamento Aéreo* of AgrAr), they were commanded by a junior officer and attached to an army unit, as needed.[56] The

56 Comando da 2° RA, "Destacamentos de Cooperação e Agrupamentos Aéreos" [Detachments of Cooperation and Air Groups], Luanda, 10 March 1968, AHFA, 335, B1, Cx 177.

composition of each was fixed by the commander of the 2nd Air Region in light of the need for support, the local situation, the distances involved, and obviously the availability of aircraft, pilots, and mechanics.[57] Many took on an air of permanence, if the unit was in prolonged contact with the enemy, such as during an entire five-month dry season campaign.

The FAP also established detachments of light aircraft, such as Austers, Dorniers, and sometimes T-6Gs, at various airfields, again under the command of a junior officer. These were known as Detachments of Cooperation (*Destacamentos de Cooperação* or DCs), and they performed the everyday tasks of general transport, evacuation of wounded, visual reconnaissance, and sometimes armed reconnaissance in the frontier areas. Their operations supported troops operating within a radius of the DC consistent with the operating capabilities of the aircraft available.[58] In some cases, such as DC "L" (Luso), heightened reinforcement was needed to meet a higher tempo of operations in the area.[59] At DC "C" and others in the extreme south where the South Africans operated, it was necessary that commanders speak English or the South Africans speak Portuguese, a condition not always met. Indeed, as Lieutenant Queiroga noted, the South Africans had very different dietary and cultural habits from the Portuguese that included speaking Afrikaans—the language of the Boer with its Dutch origins. Briefings were in English, but otherwise each national spoke his own language.[60]

The six DCs, all in the southeast, were:

DC "H" at Henrique de Carvalho (AB 4)
DC "L" at Luso (AR)
DC "CAZ" at Cazombo (AM 43)
DC "G" at Gago Coutinho (AR)
DC "C" at Cuito Cuanavale (AR)
DC "SP" at Silva Porto (AR)[61]

It should be noted as well that operations extended to the South African airbase at Rundu opposite the small Portuguese airfield at Calai, an outpost about 164 kilometers (about 98 miles) due west of Dirico.[62]

57 Corbal, Simões, Ferreira, and Almendra, 18.
58 Ibid.
59 Ibid., 17.
60 Queiroga, "Histórias Vividas: Por Terras do Fim do Mundo, Uma viagem entre os 18 e os 9 graus de Latitude Sul," 5–7.
61 Força Aérea no Leste de Angola, Comando Aéreo Leste, Luso, 30 November 1972, Directiva 2/72, Annex A, AHFA, 355, A3, Y3.
62 Queiroga, "Histórias Vividas: Por Terras do Fim do Mundo," 4.

Refueling an Alouette III at a DC in the east of Angola.
(Source: Personal archive of Melo Vidal)

In the case of both the Air Groups and DCs, the army commander being supported was invariably senior to the junior aviation officer commanding the Air Group or DC, and this potentially created tension. When air support was ordered by the army, but the order went against such guidelines as the type of missions authorized for the Air Group or DC to perform, maintenance problems with the assigned aircraft, and weather limitations, there was the possibility of conflict.[63] In an attempt to minimize this friction, the commander of the 2nd Air Region created the concept of an "associated command" (*associação de commando*) under which the flight directives governing the conditions for operating aircraft stood behind the Air Group or DC commander, and under them he had exclusive authority. This arrangement made responsibility clear; however, for a senior army officer, being an "associate" with a junior officer in the command structure could at times cause awkwardness.[64]

Siroco 1969

Operation *Siroco* 1969, an early example of air-ground coordination, began in the east in September and was a counterinsurgency action which, like the hot dessert wind for which it was named, lasted the entire dry or fighting season for each of three years and swept through the endless savannas in the ZML. It likewise served as a vehicle for development and perfecting air assault and close air support of ground forces that

63 Corbal, Simões, Ferreira, and Almendra, 18.
64 Ibid.

would later be so devastating to the enemy. Group *Siroco* (*Agrupamento Siroco*) was formed at Luso, and its Land Group (*Agrupamento Terrestre*) consisted of four companies of commandos (CCMDS), a brigade of PIDE/DGS, and about 300 auxiliary troops.[65] The commando teams were the centerpiece of *Siroco* and were expected to address enemy columns uncovered by the PIDE/DGS and the auxiliaries. They were commanded by Lieutenant Colonel Gilberto Manuel Santos e Castro. Their purpose was to execute the "hunt" (*caça*), which was a type of action based on the Portuguese commando concept of always being the hunter rather than the hunted. The hunt aimed at locating and reducing the enemy "squadrons" penetrating the ZML.[66] For *Siroco* there was to be a minimum of three CCMDS, and at least two of these were formed into two or three hunting teams, depending on the character of the terrain and the disposition of the enemy. The third was held in reserve or on rest from earlier operations. The teams were linked by radio to each other and to a command post and were ready to move by air transport at the first generation of reliable intelligence. The Air Group composition was flexible in light of the type of operation envisioned and the required transport. The limitations of the Alouette III meant that a number of trips might be made by the limited aircraft allocation to move the CCMDS. This would change with the arrival of the Pumas, but that was a year into the future. The initial *Siroco* Air Group consisted of a Dornier and four Alouette IIIs, and was commanded by Captain Carlos Acabado.[67]

The idea behind this new hunting organization was that its elite troops were able to chase down the insurgents in the difficult countryside where they hid and to maintain contact with them by using the helicopters to move faster than their quarry. They were put on the ground as a group even though they may have arrived at the landing zone by a separate helicopter in the formation. The helicopters would approach a landing zone and one by one hover two or three meters above the ground while the troops jumped from the helicopter. Once emptied, the helicopter would move ahead, thus opening a spot for the one behind it and its men. This evolution was repeated by each helicopter until the entire force was deposited on the ground. Once on the ground, the assembled team went to work. The teams had the ability to call for reinforcements, if needed, and these came by helicopter. Ultimately they were very effective in isolating and destroying any enemy formation.

The Air Group and its aviation resources were collocated with the ground forces, meaning that the aircrews spent some months living in the bush with the commandos, a situation that brought its own adventures. The Dornier was constantly in need of improvised landing strips, and these were carved from the *chana* by troops advancing on foot and wielding blades against the tall grass. They were followed by Mercedes-Benz

65 Pires Nunes, *Siroco*, 160–161.
66 Ibid., 140.
67 Ibid., 163.

A South African Alouette III in the east of Angola with troops making the two-meter jump to the ground. Note the lack of national insignia and only the tactical marking "4" visible. (Source: Personal archive of Pierre Fargeas)

Unimog trucks to pack the earth and complete the task.[68] During night landings, the airstrips were lit by the headlamps of the jeeps or *Unimogs.* Captain Acabado remembers such operations clearly, living side by side with his army comrades and sharing the dangers and hardships that form such bonds. His favorite time was in the evening in the field when he was able to share the taste of an old whiskey with his commando friends, drink it from an enameled cup, and always soften it with water from the *chana.*[69]

The Air Group attached to *Siroco* drew the usual tasks of performing:

- Helicopter maneuver or tactical transport of the hunting teams to their areas or zones of operation and then recovering them;

68 Acabado, 133.
69 Ibid.

- Limited aerial visual reconnaissance by helicopter aimed at orienting the teams on the ground to physical obstacles, fleeing insurgents, or other dangers;
- Liaison with PIDE/DGS, command posts, commanders, etc.;
- Fire support when anticipated or required; however, this would be light in view of the fact that no gunship was initially assigned;
- Resupply missions by the Dornier to transport food, fuel, ammunition, etc.;
- Bombardment missions, which again were light, unless a heavier aircraft was assigned specifically for the purpose; and
- Airborne command post (*Posto de Comando Volante* or PCV) duties, which were normally assigned to the Dornier.[70]

The Air Group was normally limited to a three-month cycle of support because of aircraft maintenance limitations; however, maintenance requirements could be extended somewhat by reducing flight hours. This would also reduce the effectiveness of Group *Siroco*, as it would pose operational limits overall. The flexibility and effectiveness of *Siroco*, which depended heavily on air maneuver, would be reduced. On the other hand, keeping the teams together, including the aviation element, was important, as the individual members became increasingly familiar with each other's duties and procedures and over time were molded into a cohesive team.[71]

The first series of operations for *Siroco* in the ZML occurred between August and November 1969 and were aimed at the Salient. The concept of operations was to explore the areas little frequented, make contact with the population, collect intelligence, and capture or destroy enemy elements identified.[72] The teams were deployed from a command post located at an airfield for the Air Group and subsequently relocated, as contact with the enemy or intelligence dictated. This particular operation began at Mucussuege, a station on the CFB in the northwest corner of the Salient. As the reconnaissance developed, the post was moved progressively to:

- Caianda – a small town on the northern frontier of the Salient
- Jimbe – a small town in the northeast corner of the Salient
- Luacano – a station on the CFB just northwest of the Salient
- Cavungo – a village in the center of the Salient and on the eastern edge of the Chana da Cameia
- Camafuáfua – a station on the CFB on the northern frontier 42 kilometers west of Caianda
- Lumege – a station on the CFB at the western edge of the Chana da Cameia and about 100 miles west of the Salient.

70 Pires Nunes, *Siroco*, 163–164.
71 Ibid.
72 Ibid., 166.

The insurgents had left tracks along the northern edge of the Salient, had penetrated the frontier around Mucussuege, and had followed a southwest route toward Lake Dilolo and the Chana da Cameia. The Air Group moved the hunting teams along the route, shifting airfields progressively westward, as they made contact with UPA/FNLA elements. As the teams progressed, they visited the population to brief the village chiefs and gather intelligence. The PIDE/DGS agents during and following *Siroco* interviewed or questioned some 660 persons captured in suspicious circumstances.[73]

Siroco had begun on 1 September, a mere month before the rainy season. It thus became increasingly difficult to operate the aviation support, as the weather deteriorated, and the suggestion on the conclusion of *Siroco* was to begin the operation much earlier in the year. Certainly the insurgents were taking advantage of the good weather beginning in March. The hunting teams should be likewise engaged.

As the three-month period drew to a close, the helicopters were flown less frequently because of approaching maintenance requirements and a lack of spare parts. As the operation progressed, declining helicopter availability affected ground mobility, and the teams became less effective, as the balance between walking and flying shifted.[74] The lesson here was that increased helicopter availability, or perhaps more helicopters, were needed to maintain pressure on the enemy during an extended operation. The deficit was corrected with the delivery of six Puma helicopters between June and December 1969 and their appearance in Angola in 1970.

Detachment *Benedito*

Following the 1969 *Siroco* campaign in December, the MPLA coincidently sought to resupply its isolated force in the north in the Dembos region on the Dange River, which was near Quitexe in the District of Uíge, and decided to make an attempt from the east to effect this. This was the fourth attempt after three columns, two of which failed, had been sent to reinforce the MPLA outpost there. The first, *Cienfuegos*, departed the Congo at Songololo in August 1966 with 100 Cuban-trained men and negotiated the 190 miles to the Dembos, successfully crossing the M'Bridge and Loge Rivers, as it was the dry season.[75] Fired by this success, the MPLA leadership launched a second column, *Kamy*, again trained and armed by Cuban advisors, which made its way to Songololo in mid-January 1967 and entered Angola. It became lost, despite having two guides, and faced wide, flooded rivers with raging currents. At the Loge River, 47 turned back, and about 70 continued. Of the original numbers, only 15 successfully regained the Congo border, and only 21 arrived in the Dembos.[76]

73 Ibid., 194.
74 Ibid., 195.
75 Piero Gleijeses, *Conflicting Missions: Havana, Washington, and Africa, 1959–1976* (Chapel Hill: University of North Carolina Press, 2002), 180–181.
76 Ibid.

MPLA leadership attempted a third column, *Bomboko*, in June 1967 and sent about 180 Cuban-trained men to Léopoldville, where they were to find their way to Songololo and enter Angola. The large group did not break into smaller, less obvious numbers and attracted the attention of Mobutu, as they travelled en masse. Rounded up and imprisoned, the detainees were released eventually, but the MPLA decided to make no more attempts to reinforce this Dembos base from the north. Neto now decided to send a reconstituted *Bomboko* to the Dange River from its eastern front in Zambia.[77] It took virtually a year to put the column in place, for there was a wait for Soviet air transport to Dar-es-Salaam, another delay in Tanzania, and a long trek through Zambia itself. By July 1968, the force had finally reached the base at Cassamba. The column was to be named in honor of João Gonçalves Benedito, a former member of the MPLA central committee who had led its clandestine wing in the Congo. He was captured in November 1966 by Mobutu's army and subsequently delivered to the UPA/FNLA, which detained him for a year before executing him.[78]

The resupply of Dange, which lay 1,000 kilometers (about 600 miles) from Cassamba had become an obsession with MPLA leadership after the three attempts from the north, but *Benedito* had to wait another 18 months before it was able to begin its journey in December 1969 in the midst of the rainy season. There are various estimates of its strength; however, it probably had about 80 porters and perhaps 50 fighters. It crossed into the Salient from Cassamba next to the border post of Caripande, where the Zambeze River crosses the frontier into Zambia.[79] From there *Benedito* followed a northwest route across the Chana da Cameia, the CFB, and then the Luso-Henrique de Carvalho road just north of Dala (100 kilometers north of Luso).[80]

It was the intent of the MPLA planners of this "long march" to have *Benedito* create along its route small military detachments with depots of arms and supplies to support future incursions. Likewise it was to extend its influence in the east to the districts of Moxico, Lunda, and Malange and win the population to its side.[81] However, launching an operation of this type in the rainy season when the rivers were at their flood stages and the endless *chanas* were inundated was begging for problems. Further, as in most cases with insurgent movements, there was dissention. Almost immediately there were some cases of indiscipline among its fighters and porters, and this manifested itself in the region of Cazage just before Dala, where several insurgents fired their weapons to intimidate some villagers and then burnt their village. The inhabitants fled to the closest Portuguese authority and revealed *Benedito*.[82]

77 Ibid., 182.
78 Pires Nunes, *Siroco*, 214–215.
79 José Augusto Queiroga, "Histórias Vividas: Uma Operação pouco ortodoxa" [Tales Lived: A slightly orthodox operation], page 4, undated deposition, Caixa 181, Archivo Histórico de Força Aérea, Alfragide.
80 Ibid., 12.
81 Ibid., 10.
82 Pires Nunes, *Siroco*, 219.

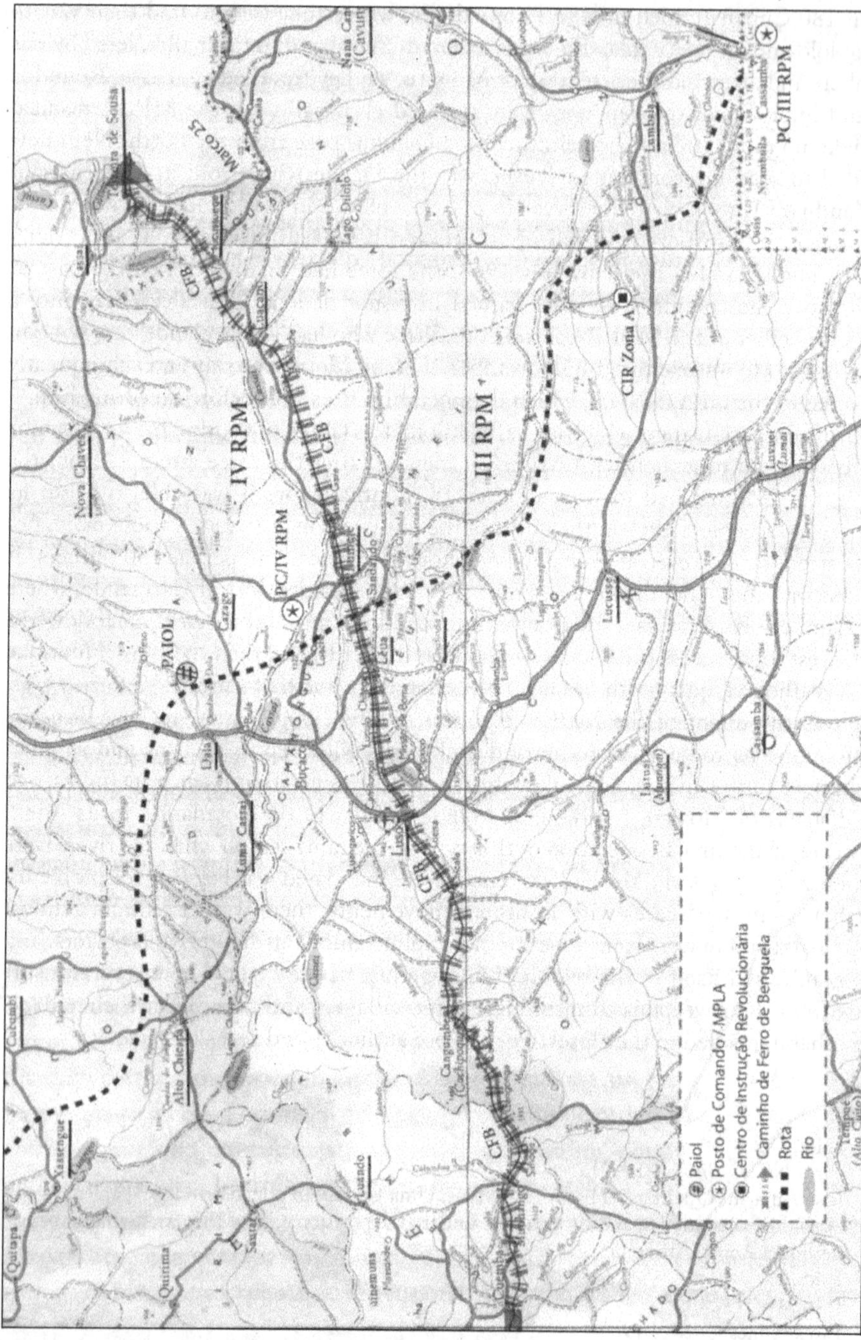

Map 11.5 Route of *Benedito* from Cassamba to the Cuango River. (Source: Modified from a map by António Pires Nunes)

PIDE/DGS had been tracking the development of *Benedito* since its inception in Zambia through its agents both in Zambia and within the MPLA. As it struggled through the ZML, the tracking continued. Portuguese strategy was to track its progress towards the known destination, allow it to struggle forward against the weather and terrain obstacles, and wait for an opportune time to attack. A likely point was the Cuango River, which it had to cross during flood stage. This significant barrier would likely throw the enemy into disarray, force him to openly expose himself, and make him an extremely vulnerable and ripe target.[83] Portugal was following its strategy of letting the great distances and appalling weather of Angola exhaust the enemy.

As *Benedito* progressed along the route to Alto Chicapa on its way to the Cuango, the terrain changed dramatically with irregular elevations, aggressive mountains, and abrupt ravines, some with dense vegetation, before opening onto the Malange plateau. While this was difficult going for the column, it was able to hide, making Portuguese attack less likely. However, time was on the Portuguese side. Another hindrance was the guide, as he was taken sick and slowed progress. Even so, he knew the route only as far as the Xassengue-Cucumbi road and did not know the Cuango River crossings.[84]

In an attempt to solve the crossing problem, a number of insurgents posed as Portuguese troops and approached a village to talk with its chief and ask directions for passage across the Cuango. The chief explained that it was impassable in the rainy season. The river had a mighty current some 60 meters across, and its margins were flooded, conditions that made it impossible to use the traditional canoes. After attempting a crossing with air mattresses and failing, the detachment took refuge in the grass next to a village and was shortly overflown by an aircraft at about 1000 hours on 4 February 1970.[85] This was followed by two additional passes an hour or so apart, and three hours later three Alouette IIIs arrived. Panic ensued in *Benedito*.[86]

Operation *Carnaval* (Carnival) began with the launching of three Alouette IIIs from Squadron 94 at Luanda (BA 9) at 0600 on 4 February, a significant date for the MPLA, and followed the highway leading eastward to Malange, where they landed to refuel.[87] Two of the Alouette IIIs were transports flown by Lieutenant Queiroga and Sergeant Armindo Rocha, and the other a gunship flown by Major Luís Almada de Oliveira, the squadron commanding officer.

The next landing was at Quitapa, where the fuel was adjusted to 250 liters each for the transports so that they could maintain their maneuverability with the troops aboard. The transports again launched with five commandos each and followed Major Almada in the gunship to locate *Benedito*.[88] The last intelligence on the enemy column

83 Queiroga, "Histórias Vividas: Uma Operação pouco ortodoxa," 12.
84 Pires Nunes, *Siroco,* 219.
85 Ibid., 220.
86 Ibid.
87 Queiroga, "Histórias Vividas: Uma Operação pouco ortodoxa," 15.
88 Pires Nunes, *Siroco,* 222.

placed it northeast of Quitapa on the eastern bank of the Cuango. The helicopter pilots saw below vast reaches of water where the Cuango had exceeded its banks and flooded the adjacent land. There were tufts of trees here and there, but all was completely inundated. After 20 minutes of flying, Major Almada radioed that he had located *Benedito* and that the transports should rendezvous with him. Immediately the gunship could be heard firing into the column and causing numerous losses among the insurgents and porters. Meanwhile, the two transports searched for a dry landing zone in the flooded *chana*. The commandos were finally put on the ground and pursued the now broken *Benedito*. In the melee there were 37 insurgents and 20 porters killed, and the rest were taken prisoner or escaped. The commander of *Benedito*, Rafael Zembo Faty, committed suicide rather than suffer the humiliation of prison. Major Almada, who had led the way in destroying the dream of Agostinho Neto, was awarded the Cruz de Guerra.[89]

This type of operation with heli-borne elite troops supported by one or more gunships became known as "pirate" operations and developed as the standard air assault and ground engagement format against insurgent columns.[90] Their perfection in the east resulted from the close coordination between Brigadier Manuel Maria Delgado e Silva, commander of the ZIL from late 1970 until March 1971, and Santos e Castro, the commando commander.[91] The remnants of *Benedito* were pursued in two subsequent operations, *Teste/H1* and *Teste/H2* on 6 and 7 February respectively. These concluding actions yielded a handful of prisoners, a few dead insurgents, enormous amounts of arms and munitions, and complete destruction of the column.[92]

Siroco 1970

Following *Siroco* 1969, a new hunting group was assembled in June 1970 and began operations in July. Its initial assignment was an area containing the Chana da Cameia, and later it moved southward and west of Gago Coutinho to pursue the enemy along the Cuango River. Its overall command was again given to Santos e Castro and the ground elements consisted of three CCMDs and two companies of special infantry. The supporting Air Group commander was Captain Acabado.[93] At its conclusion in October there remained the same frustrations that had plagued operations a year earlier. Despite substantial losses in men and matériel by the insurgents, they generally understood the terrain better than *Siroco* and thus proved highly elusive.[94]

There were many lost opportunities because of poor radio contact between ground units and their supporting aircraft. The onset of the rainy season in the midst of both

89 Ibid., 224.
90 Ibid., 222.
91 Ibid., 348.
92 Ibid., 226–227.
93 Ibid., 295.
94 Ibid., 332.

Siroco 1969 and 1970 contributed to poor air-ground coordination, and operations needed to begin well before June to take advantage of good weather.[95]

Siroco 1971

A third and final Group *Siroco* was authorized in April 1971 and would operate for the three months of mid-June through mid-September along the Cuito River north of Cuito Cuanvale.[96] It was commanded by Lieutenant Colonel Fernando Manuel Jasmim de Freitas, and its ground force consisted of five CCMDs and a combat group of *flechas*, the PIDE/DGS auxiliary troops and trackers. The aircraft assigned to the Air Group were one Alouette III gunship, four Alouette III troop transports, and a Dornier.[97] As operations progressed, the teams moved eastward toward the Zambian frontier. This sweep involved four major operations from west to east the length of the Route of the Cuando and uncovered a number of large enemy groups that were able to return fire with strength when challenged. Given the direction of the sweep, the MPLA columns moved eastward across the frontier to the safety of their Zambian bases to avoid direct confrontation.[98] Bethencourt Rodrigues's chief of staff, Lieutenant Colonel Ramires de Oliveira, noted that *Siroco* day after day "felt the guerrilla" but had little contact. The *Siroco* commandos believed, "It is necessary to capture a live terrorist."[99] But no one knew when it was likely to happen, as the enemy was well led and experienced. The MPLA forces truly played a "cat and rat" game, hiding their main "body" of troops during the *cacimbo*, as they were at a disadvantage.[100] Ramires de Oliveira noted in his "Notes on the ZML" that, "That very evening we captured a prisoner, and he confirmed that the enemy columns had passed through our area."[101]

Group *RAIO*

Like its name, thunderbolt, *Raio* shocked the MPLA columns. The CCFAA authorized its formation and deployment in June 1972, and its operations ran from July to October, three months. It was the largest such operation in the east yet with a ground force of three CCMDs, two companies of infantry, one of cavalry, six of auxiliary troops, and six groups of PIDE/DGS *flechas*.[102] *Raio* was tasked with sweeping clean two primary enemy routes from Zambia into Angola: Route of Agostinho Neto

95 Ibid.
96 Ibid., 407–408.
97 Ibid., 411.
98 Ibid., 503.
99 Ibid., 505.
100 Ibid.
101 Ibid.
102 Ibid., 506.

(Luena River), and Route of the Lungué-Bungo. The results were impressive by any standard. Several major columns were destroyed, and the MPLA suffered 121 deaths, with many more wounded, and 138 captured.[103] This broke the back of the MPLA.

Group *Raio* operations were extended into Cuango Cubango in January 1973 to neutralize MPLA squadrons "Big Man" and "Sandalo." The ground force remained of similar size; however, it was augmented by five FAP Alouette IIIs and seven South African Alouette IIIs, all coordinated through the CCAA/JASC at Cuito Cuanavale.[104] The entire force was organized into ten hunting groups, by its conclusion on 17 October, the MPLA had suffered 121 dead, of whom seven were leaders and captured about 150.[105]

The final and decisive operation by *Raio* was *Bizarra*, which ran between mid-September and mid-October 1973. This offensive in the southeast was aimed at the area bounded by Gago Coutinho, Muié, Chiume, and the frontier with Zambia. It aimed to sweep the zone of MPLA columns and eliminate the MPLA encampment *Capoche* next to the Luate River and about 10 miles into Angola from the Zambian frontier.[106] *Capoche* held about 30 insurgents.[107] The ground force contained two CCMDs and about 1,200 auxiliary troops.[108] The Air Group was a combined South African-FAP force with the following aircraft:

Period 29 September-7 October:
- 1 Alouette III gunship (FAP)
- 5 Alouette III troop transports (FAP)

Period 7-17 October additions:
- 1 Alouette III gunship (FAP)
- 1 Alouette III gunship (South Africa)
- 5 Alouette III troop transports (South Africa)
- 1 T-6G Harvard (FAP – based at Ninda)
- 1 Dornier DO-27 (FAP – based at DC "G")
- 1 Cesna 185 (South Africa)

The FAP aircraft operated alone between 29 September and 7 October, and were joined by the South Africans and an additional FAP gunship between 7 and

103 Ibid., 519.
104 Ibid., 581.
105 Comissão para o Estudo das Campanhas de África, *Resenha Histórico-Militar das Campanhas de África 1961–1974, 6 Volume, Aspectos da Actividade Operacional, Tomo I, Angola–Livro 2*, 373–376.
106 Pires Nunes, *Siroco*, 598.
107 Ibid., 597.
108 Ibid., 599.

17 October.[109] In the first period the aircraft flew 143 hours of ground support, and in the second, 321 hours.[110]

Group *Raio* identified a number of heavily armed enemy columns in encampments scattered through the targeted area. These were capable of sustained resistance, that is, fighting for between one and two hours.[111] The insurgents would then abandon their camps and retreat across the border into Zambia. The abandoned camps contained substantial numbers of weapons and munitions.[112] For the Portuguese, the auxiliary forces proved the weak link in the operation, as they required additional training to give them confidence in combat. None understood the operation, their assigned patrol areas, or the use of helicopters.[113] This weakness is not surprising, as the most effective troops against the enemy were the specially trained ones. These were the commandos, paratroops, and *fuzileiros*. It was they who performed the "pirate" operations on a routine basis. Other units with advanced training were the specialized infantry companies, which were very competent forces. Cavalry squadrons also fell into this category, although these were specialized in their own way. Outside of these types of units, the Portuguese soldier was little different from his counterpart in any other force, as he was a reflection of the investment that had been made in his training and development.

Bizarra was the last significant operation in the east. From this point onward, MPLA activity was confined to the border. Groups of insurgents came across from time to time to plant the odd mine and perhaps set an ambush, but the war became very quiet on the Zambian border.

Looking Back

War in the east had benefitted from the earlier learning period between 1961 and 1969 in addressing the enemy in the north. During the opening enemy moves in the east, specific task forces were assembled to address a targeted incursion and threat. The overall strategy remained directed to the north, until proselytization of the eastern population arose alarmingly in 1968. When the east required attention and resources, it received them. Bethencourt Rodrigues proceeded by defining the battlefield and developing a strategy of containment and attrition to address the vast open spaces of the east and the limited Portuguese forces. Integral to his solution was the use of highly mobile groups of elite troops as hunter-killer forces. Airpower was the key to this approach and its success.

109 Ibid.
110 Ibid.
111 Ibid., 601.
112 Ibid., 601–603.
113 Ibid., 602.

Puma helicopter collecting troops following an operation in a *chana* in the east of Angola. (Source: Força Aérea Portuguesa)

Bethencourt Rodrigues understood precisely the intentions of the enemy and easily predicted his behavior. Costa Gomes was able to release the needed resources to the east with the arrival of the SA-330 Puma helicopters in 1970 and their increased carrying capacity. This new air resource enabled wider helicopter airlift, and when it came to the east, it was accompanied by the companies of commandos under Santos e Castro.

There developed a perfect resonance between the helicopter pilots and the ground forces, particularly the elite forces or "intervention troops." The logistics of supplying aviation and ground support was strengthened to the point that operations against insurgent columns were extended from the normal three or four days to 22 days.[114] Lastly, the melding of intelligence and the aviation-ground force team enabled the very effective pursuit and destruction of the enemy columns. The assiduous practice of this concept through the *Siroco* series of seasonal campaigns produced tactical teams of devastating capability in a difficult physical environment.

By the conclusion of *Bizarra*, the UPA/FNLA and the MPLA were destroyed. For the UPA/FNLA the crisis was complete. For the MPLA, it was reduced to sneaking across the border from its bases in Zambia and planting mines along the frontier roads. Its actions along the border were violent and short in both duration and penetration.

114 Corbal, Simões, Ferreira, and Almendra, 24.

The notion here was to strike a blow against the Portuguese that would open the way for a column to penetrate inland. This never worked.

In the observation of Major Carlos Acabado, all of the military superiority that Portugal ultimately enjoyed in the ZML was the result of the arrival of Brigadier Bethencourt Rodrigues and the staff that accompanied him. His warfighting methods transformed the theater of operations. There was a profound revolution in the strategic and tactical concepts, and for the first time in the history of subversive war, the guerrilla was conquered. More importantly, however, was the mobilization of the population, with funding support, to effect an economic recovery that led to indispensable social stabilization.[115]

In an appraisal of the situation by Brigadier Hélio Felgas, Bethencourt Rodrigues's relief, he reported that by the beginning of April 1974 there was no contact between Portuguese forces and the insurgent groups of any of the nationalist organizations. By the end of April the southeast of Angola was calm, and insurgent activity was practically nonexistent.[116]

In an interview with the journalist José Freire Antunes following the war, Daniel Chipenda of the MPLA had this to say about the effectiveness of Portuguese strategy:

> From 1972 we faced a crisis in expanding, because we had lost the capability to control a major part of the population that had retreated from the frontier. Conversely the Portuguese army, with its psychological action, succeeded in controlling the strategic villages where the population was administered by the army. This was a difficult phase between 1969 and 1972. We felt that this Portuguese offensive was a very real one. It was a phase of great difficulty. We could not deliver on the promises of a better way of life for the population, as an incentive in the struggle against the Portuguese administration. And the population that we were able to affect was united, more or less, in understanding this.

The population sought to escape control of the MPLA and the other nationalist movements and nowhere was this more evident than in the explosion of education during the war in Angola.[117] By the 1970s under the protection of the security umbrella one saw daily thousands of children in their white smocks walking to school. The undisputed statistics prove this learning explosion. A decade previously such a phenomena would have been hard to imagine.[118] Bethencourt Rodrigues, an army general who understood airpower, was indeed the author of a victory that permitted the people to envision a future more worthy and different from the tragedy that was to follow.

115 Acabado, 154–155.
116 Hélio Felgas, "Opinião" [Opinion], as quoted in *Resenha Histórico-Militar das Campanhas de África, 1961–1974*, 6th Volume, Tome I, Book 2, Angola by Comissião para o Estudo das Campanhas de África (1961–1974) (Lisbon: Estado-Maior do Exército, 2006), 450.
117 Silva Cardoso, 302.
118 Ibid., 292.

General Bethencourt Rodrigues; the smile of victory.
(Source: Força Aérea Portuguesa)

12

Securing the North of Angola

The north of Angola had always presented a tempting target for insurgents, as it was the most direct route to Luanda, the perceived center of gravity for the Portuguese. Despite its appeal, it was a difficult course, as the terrain was forbidding, making the short distance an illusion. This situation worked also against them in another way, as once they had reached the mountainous jungle of the Dembos, they could not so easily mount offensives from their redoubts. On the other hand, the Portuguese could limit the enemy to his fastness. The answer to routing him from his sanctuary was the prevention of his reinforcement. This meant intercepting the columns from the north bringing supplies and fighters. Again, in an attrition strategy, if the redoubts were isolated, they would become increasingly a besieged and starving lot. This isolation was achieved through the use of specialized airmobile teams that were more agile, powerful, and persistent than the insurgent reinforcement columns and could locate and destroy them, even as they hid in difficult, indeed forbidding, terrain. These Portuguese units were conceived and born in mid-1972 and from this point onward militarily the days of the enemy in the north were numbered.

UTCI

Infiltration of enemy units into the north of Angola from what by 1972 was known as the Republic of Zaire seemed to be an endless security problem that demanded a solution. For instance, in one of the few effective reappearances of UPA/FNLA insurgents in the Dembos area, a new column appeared in March 1971 and established a base camp named José Dias in honor of one of its luminaries. This camp was well defended and quickly became the source of much mischief. *Flechas,* the PIDE/DGS tracking and intelligence gathering troops, operating in the area on 7 March, identified the encampment at coordinates 08°49' South – 013°58' East and laid plans to assault it the next day. At about 0430 hours just prior to dawn, when the *flechas* were encircling the headquarters, they were taken under fire by the insurgents. In the ensuing action, the *flechas* rescued 2 men, 6 women, and 11 children and took a *canhangulo,* a homemade firearm. In debriefing the rescued group, the *flechas* learned that the insurgents maintained a rigorous lookout, as every evening the area was subject to aerial reconnaissance.[1] Now located, its further prosecution was given over to heavier forces.

In July of 1972, after a year of increasing insurgent presence, the Commander of the 2nd Air Region, General Rui Bras de Oliveira, and his chief of staff, Major Aurélio Benito Abeixo Corbal, assembled what they hoped would become a solution to this insidious infiltration.[2] The two officers sought to utilize the capabilities of Squadron 94, the *Moscas*, with its Alouette IIIs and Pumas and the BCP 21 to form heli-borne teams based on a select group of paratroops with an as yet unperfected specialization in tracking. These were to be embarked in helicopters for mobility and use their skills to identify suspicious trails from the air. If the situation required a closer look, they could then be inserted onto the trail in question. The tactical intelligence so developed was to be followed by a powerful heli-borne team to destroy the insurgents. This team with its methodology was named the Tactical Counter-Infiltration Unit (*Unidade Táctica de Contra Infiltração* or UTCI).[3]

The UCTI initially was composed of two echelons. The first had two Alouette IIIs, one a transport with two trackers and the other a gunship for their protection. The second echelon was composed of a Puma, which transported 15 paratroops, a gunship, and a Noratlas with three additional combat groups. Once the first echelon located the column, the second swooped down on the unsuspecting insurgents and destroyed them. The entire force acted only on intelligence that confirmed the enemy had crossed the border with Zaire in force. The UCTI was based at the maneuver airfield of Toto (AM 32), which was about 80 miles from the northern border in the District of Uíge.[4]

After about a year of these "reconnaissance-in-force" actions and their refinement, Bras de Oliveira issued a formal document defining and regulating these missions and providing guidelines for those who performed them. He also established a permanent command at Toto to run them: the Special Command for Counter-Infiltration (*Comando Especial de Contra Infiltração* or CECI). This command operated in a zone defined by the CCFAA and had the following boundaries as depicted by the dashed line on the nearby map:

- To the north, the frontier with Zaire,
- To the east, the Zadi and the Zadi-Luquiche Rivers to their source just north of Carmona,
- To the south, the Loge River from the Vale do Loge at Carmona to the Atlantic Ocean, and
- To the west, the Atlantic Ocean.

1 Direcção Geral de Segurança, Acções das Nossas Forces [Actions of Our Forces], DGS-Angola, 22 June 1973, SDGN 1802, Intelligence Report 677, Arquivo da Defesa Nacional, Paço do Arcos, Portugal.

2 Ricardo dos Santos Cubas, Geraldo Estevens, Guilherme Costa Santos, and José Augusto Barrigas Queiroga, "Ensinamentos da Guerra em Angola" [Teachings of the War in Angola], proceedings of a conference "Almoços da AFAP," January 2011, 7.

3 Ibid.

4 Ibid.

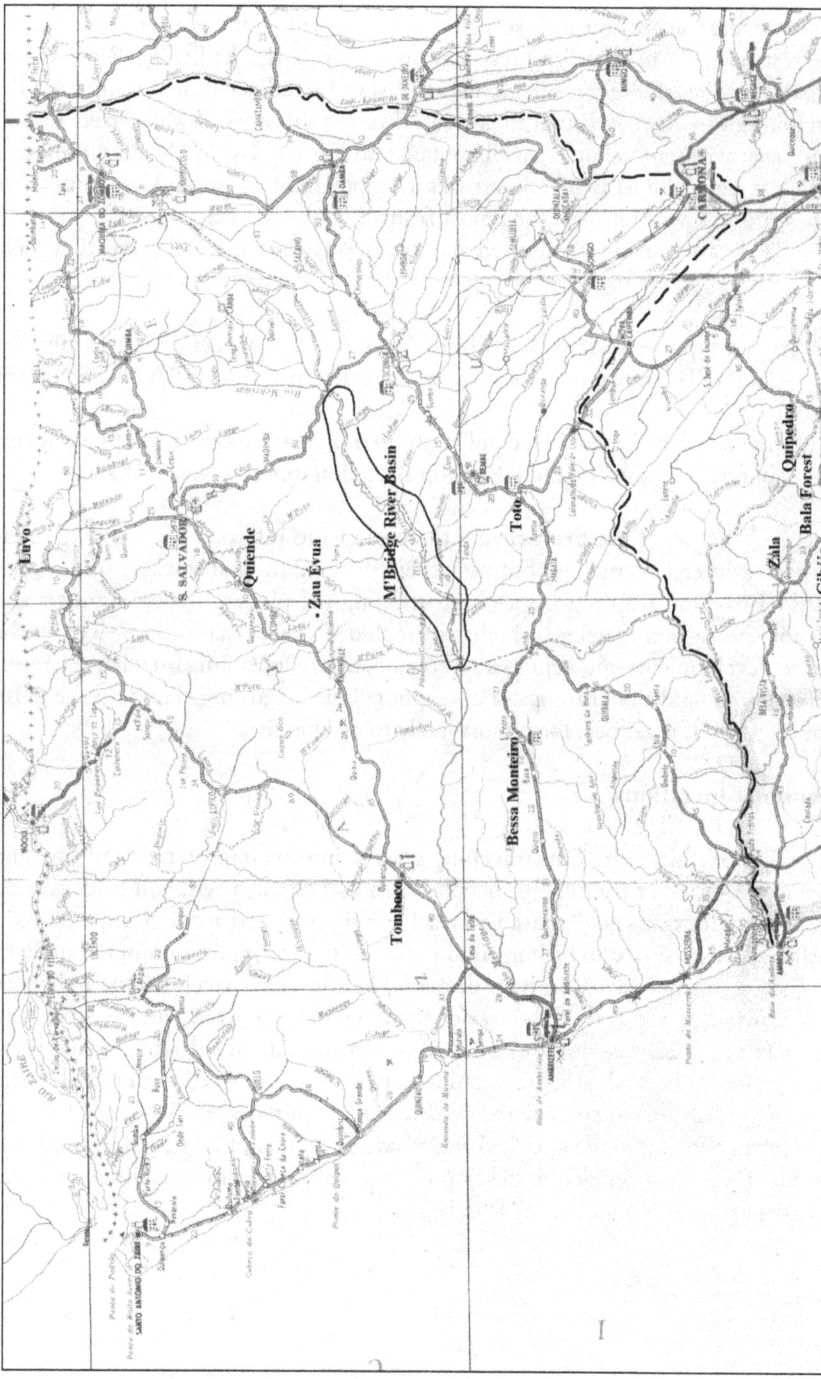

Map 12.1 North of Angola Showing the UCTI Area and the M'Bridge River Basin. (Source: the author)

The area where the infiltrating insurgents were most vulnerable—the one presenting the greatest obstacle to their progress southward—was the M'Bridge River Basin. Their goal was to reach the forest region northeast of Luanda, and their route from the Zairian border directly south to it took them through this area. It was perfect terrain for them, as it was remote and sparsely populated. For the UCTI this basin midway along the course of the M'Bridge River was a "happy hunting ground," as it was here that interception operations were the most fruitful.[5]

This formalization also restructured the forces available to CECI, and they were now:

- An Air Group, designated the "Royal Falcon," of an Alouette III gunship and two Alouette III transports. These were supported by Luanda (BA 9) and Negage (AB 3).
- A permanent detachment of a combat group of 20 paratroops, including trackers and sappers, furnished by the Paratroop Battalion in Belas, south of Luanda.

It was still the job of the first echelon to develop and interpret the tracking intelligence for an intercept mission, but now there were additional trackers in a second Alouette III to allow leapfrog tactics in developing the picture. Troop numbers and aircraft for the second echelon, which still included the 20 paratroops, were to be tailored to developments and kept as a reserve on alert. Depending on what the interception required, additional troops or air support must be arranged and, in the judgment of the CECI, must be a force most likely to be effective.

Interception Operations

The detection of trails created by travelling groups and the interpretation of tracking intelligence in terms of the type of people, their direction, speed, and loads carried tended to be an inexact science based on tracker judgment and acquired experience in the bush. One almost had to be born into it. Moreover, its application in the modern military environment had been largely displaced by advances in technology, and hence it was somewhat of a lost art.[6] The war of the guerrilla, however, with its "hide and seek" nature required that this ancient skill be reacquired, and Portugal made great investments to do so. Both Rhodesia and South Africa were faced with the identical problem of tracking insurgent incursions and had acquired specially trained troops with experience in this skill. Portugal reviewed the results of its neighbors and was impressed. It was thus decided to develop this capability within the elite forces, and accordingly, arrangements were made to begin.[7]

5 Ibid., 8.
6 Ibid., 9.
7 Ibid., 10.

The first course for trackers was held at the headquarters of the commandos in Luanda, and the paratroops attended. It was conducted by veteran Rhodesian officer and sergeant trackers over a three-month period. During the course the Rhodesians issued ten invitations to the most promising students, four to paratroopers and six to commandos, to attend the practical tracking course in Rhodesia. Upon the return of these students, they constituted the nucleus of tracking capability in Angola and were responsible for preparing a course at Belas for officers and sergeants that began in July 1973.[8] The course had both a theoretical and practical side. The first taught the techniques of tracking and methodology of interpreting the signs of the trails, such as the number of people passing on it, the age of the signs, the weight of the travelers, and the difference between the fighters and the porters. The second was a practical application. The course ran continuously, and its students were immediately integrated into front line units as both aerial and terrain trackers. The immediate results were very positive and clearly reinforced the decision to invest in this capability. As the course matured, other army units were able to send personnel for training. The third course was held in Luso and aimed at the locally recruited forces in the east.[9]

The first FAP intercept operations were prompted by intelligence from tracking reconnaissance gathered in late 1971 and early 1972 during the height of the rainy season, when the rivers were swollen and incursion columns were experiencing difficult progress southward from Zaire. Their slow pace and extended presence in areas of vulnerability, such as river crossings, provided an opportunity for discovery. The FAP tracking effort gradually intensified, and as it did so, the technique of both the Alouette III pilots and the paratroop trackers improved measurably. In fact, the coordination between the two, as each came to appreciate the role of the other, became very close. The most fruitful area for interception and tracking—and thus employing the teams—was the intersection between a north-south line from São Salvador to Toto and the M'Bridge River, which ran from east to west. It was a natural obstacle to insurgent columns and forced them to pause and regroup for a crossing. Further, the river could only be crossed at certain points, particularly in the rainy season, so the regroupings predictably occurred at these locations and forced the enemy to mass and reveal himself. Notwithstanding this, even when the enemy did arrange to cross the M'Bridge, and the FAP knew his intention, the intercept was never certain.[10]

Within this operational context, one of the first truly successful intercepts occurred over the last days of 1972 and the first ones of 1973, again during the rainy season, to the north of the M'Bridge and east of the village of Zau Évua, headquarters of an army battalion that was part of the quadrille defense system. The enemy force was a supply column consisting of between 70 and 80 elements. Its progress was detected and followed by tracker teams operating from two Alouette IIIs working north to

8 Ibid.
9 Ibid.
10 Ibid., 11.

south between São Salvador and Quiende. As the reconnaissance progressed, the trackers came closer and closer to the column and just before the intercept spent two days undetected overlooking the locale where the column bivouacked for two nights.[11]

When the commander of the Air Group had fixed the column, he activated his second echelon composed of two Pumas, each carrying a combat group of 15 paratroops and each escorted by a gunship. Further, the reserve of three additional "para" combat groups were loaded onto a Noratlas and departed Luanda for the area.[12]

The Pumas flew a nap-of-the-earth profile directly from Toto to the target area and deposited the paras on the terrain around the enemy's trail, a move that allowed them to comb the area as efficiently as possible in an attempt to gain contact. Three of the paras entered an uneven bit of terrain made so by the presence of a stream covered by very dense vegetation, and one of them suddenly found himself in the midst of the enemy in hiding. He quietly and quickly retreated and signaled for the gunship to rake the tuft of foliage. Several bursts of fire followed. The paras were fortunate in not being hit, but the fire of the gunship caused the enemy to flee into the open. The gunship, now exhausted of ammunition, departed for Zau Évua to rearm, while another gunship was inbound. As soon as the two gunships were on station, they raked the areas where the enemy had sought refuge. The paras rounded up the survivors and accounted for the entire column. Aside from the killed and captured enemy, the paras found great quantities of light arms, anti-personnel and anti-vehicle mines, and supplies.[13]

The Key: Intelligence

The key to an insurgent intercept was a timely knowledge of the time and place that he intended to cross the frontier. Intelligence on his intentions was gathered from informers who had infiltrated the insurgent centers in the Zairian sanctuary or from the interrogation of prisoners. Once a planned incursion was anticipated, the FAP pilots and trackers were launched and commenced searching for signs of the column. There was, however, often a great lag in time between the decision to form a column by the insurgent leadership and its actual appearance in the north of Angola. This delay meant that intelligence, while accurate, suffered from a certain operational friction within the insurgent movement in simply getting things done. There generally was little sense of urgency, or even if there was, organizing and launching a column generally proved time consuming. Eventually though, as the column was launched and progressed, it generated clues that advertised its presence when it finally crossed the border and proceeded along its route. As the column left signs, FAP aerial reconnaissance and the tracking teams began to pinpoint its location and define its movement.

11 Ibid.
12 Ibid.
13 Ibid., 12.

This ultimately led to the destruction of the large majority of enemy columns entering Angola from the north, and from the time that the UTCI tracking units were initiated in mid-1972, insurgent columns were destroyed with disturbing regularity for the enemy. His bases began dying, and his ability to maintain them at any level of readiness was severely weakened. This translated into an inability to conduct operations, and within the north insurgent bases struggled simply to survive.[14]

It is worth describing two further cases to illustrate these intercept operations, and in the first we begin with Paulo Panda, who was a UPA/FNLA insurgent commander with a great deal of experience leading groups infiltrating from Zaire. In early 1973 he commanded a column destined to resupply the UPA/FNLA base of *Gibóia* (Boa Constrictor), which lay to the northeast of Luanda and was responsible for the subversion of the coastal villages of Ambriz, Capulo, Barra do Dande, and Tabi. Several days after entering Angolan territory, he was overflown by a number of Dornier reconnaissance flights originating from Toto, as they monitored the traditional trails of likely columns. Panda was forced to proceed cautiously, and this slowed his progress. Further, Panda became sick, and this necessitated his remaining behind. Without its leader, the column began to make mistakes and revealed itself. It took only a day before it was attacked by three combat groups of paras (about 60 troops) supported by two gunships. The first attack was made by a single combat group that was dropped at the rear of the column, and this forced the insurgents and porters to disperse and abandon great quantities of matériel. Meanwhile, the second combat group in a Puma was dropped in the front of the column blocking the direction of flight. Surrounded and confronted, the insurgents and porters became disoriented and attempted to flee in all directions. Three were killed, three captured, and an undetermined number wounded. The third combat group landed and helped to "fine comb" the area. This proved very difficult, as the terrain was highly irregular with escarpments and ravines, all with dense vegetation. It was likely that the insurgents dispersed in groups of two or three and hid until the paras left.[15]

Panda was ultimately captured and in captivity furnished intelligence on a column that had entered Angola shortly after his in the area of Noqui in the spring of 1973 and was composed of insurgents escorting a large group of heavily laden porters. It represented an excellent target for the UTCI, which quickly identified the trail used by the large group.[16] The trail was followed across the Tomboco-São Salvador road, and the trackers estimated that they were four to five days behind the column. From this point the FAP tracking teams began their work using the usual two Alouette IIIs as transport and a gunship for protection. The trackers followed the trail some kilometers to the south along the Quiende River, which ran south to north through Quiende, and entered an area of dense foliage where the track became difficult to

14 Ibid.
15 Ibid., 13.
16 Ibid., 13–14.

follow. By mid-afternoon they were joined by a combat group of paras who arrived in two Pumas. The trackers followed the trail until the end of the afternoon. At about 1900 hours, when the light began to fade, they saw that the track skirted a small thicket, entered a larger one, and disappeared into its interior where there were footprints in all directions. Because the light was fading quickly and quite dim, it was difficult to know which direction the column had taken. After a short time another trail was detected that entered a thicket opposite the one under investigation. This reversal was a normal precautionary procedure designed to make detection difficult just as the column was preparing to bivouac overnight. A quick check of the immediate area confirmed that the column had remained overnight there recently, as the paras found a spade and a canteen next to extinguished campfires. There were three of these campsites about 30 meters apart and within each the grass was pressed flat from sleeping porters and insurgents. The remains of animal skins were also evident, but there was little else of note. Interestingly there was also a lookout post, a small detached area where sentries had stood watch to protect the principal group and make certain that its leaders were not surprised while sleeping. In consideration of the lateness of the day and the impossibility of making further progress, the para commander decided to bivouac nearby.[17]

The next day at 0500 the tracking was begun anew and followed to a point where the trail left the grass and headed west. It then inflected north to cross the Quiende River and wound along its left (east) bank where another overnight campsite was discovered not far from the previous one. This one was different, however, in that it extended along the riverbank with numerous hot fire sites, some still burning. After a quick reconnaissance of the campsite, the paras continued their advance to the south between two trails that ran parallel and close to each other. By 0730 the trackers had closed to within 300 meters of the enemy, whose trail exited a small thicket, proceeded across a small clearing, and entered major vegetation. This was certainly the present location of the enemy.[18]

After the paras informed their command post, they began a cautious approach with two sections moving in parallel, one taking up a position on the left edge of the thicket without detection and the other moving around the thicket to the right about 80 meters from the target, a position that permitted a vantage for either defense or attack. The approach of the paras and the nearness of the gunships signaled the beginning of the assault. As the paras on the left began to fire and advance in a "hammer and anvil" tactic, the enemy reacted with wild fire from automatic weapons and sought to abandon his position as quickly as possible. The gunships overflew the enemy position and made his movement very difficult. In the disorderly flight, porters and insurgents alike abandoned their loads and attempted to disappear into the high grass. Simultaneously the section on the right, which was some distance away, unleashed its

17 Ibid., 14.
18 Ibid.

fire. The heli-transported reserve now arrived and assumed the just-vacated position on the left. The gunships raked the thicket in front of the paras as they advanced. As the sections of paras worked their way through the thicket, the enemy fled in all directions. By 1130 the reserve was combing the battleground and found signs of the explosive loads from the gunship cannons, the remains of bloody clothes, and many abandoned loads. These largely consisted of anti-tank and automatic weapons, hand grenades, anti-personnel and anti-vehicle mines, munitions and fuzes.[19] No figures were offered on the number of enemy killed, wounded, or captured.

The second example occurred on 3 July 1974, three months after the April revolution, which had little immediate influence on the fighting in Africa. On this late date the UTCI led by Captain Fidalgo was composed of two Alouette III transports with trackers and a gunship and deployed to an area northeast of Quiende and west of São Salvador. The UTCI was acting on intelligence that had identified a border crossing by insurgents to the west of Luvo.[20] This small search force looked for several days without success and eventually moved further west where earlier infiltrators had passed. Finally the trackers located a recently used trail along which a group of more than a hundred infiltrators had passed two to three days earlier. Pursuit of the column proceeded rapidly, as the trackers could now follow the trail from the air and needed to land only occasionally to confirm signs on the ground and refine their information. These quick glances allowed the two Alouette IIIs to leapfrog one another and move quickly to contact with the enemy. By the end of the day there was sufficient knowledge to move an intercept force into position, and the Alouette IIIs returned to Toto to prepare for pursuit the following day.[21] Meanwhile, notice was given to the 2nd Air Region and the paratroopers to be prepared for an intercept.

On 4 July before dawn the tracker force of three Alouette IIIs launched from Toto and flew to the point where the search had been broken the preceding evening. Here there were the remains of fires and other signs of a recently abandoned bivouac. This new information was relayed to Luanda and reinforcements were prepared. There were two Pumas, a company of paras, and two B-26 Invaders armed with fragmentation bombs, an aircraft we will examine shortly. This ordnance was aimed at softening the column before the paras were inserted.[22]

The two Pumas flew to Toto each with a combat group of paras aboard to preposition for the operation. Meanwhile, Fidalgo was flying over the trail scouting for an appropriate landing zone for the paras when, as he skirted a small elevation in the terrain next to a thicket, he saw in the middle of the trail an insurgent holding a rifle and immobilized, indeed hypnotized, by the sight of the helicopter. Fidalgo reacted immediately and flew low over the enemy, who jumped aside to avoid being "hit" by

19 Ibid., 15–16.
20 Ibid., 16.
21 Ibid.
22 Ibid.

the helicopter and then quickly disappeared into the bush. Clearly the column was nearby, and it was important to act rapidly. While the gunship continued to circle the thicket, Fidalgo requested B-26 fire support and the Pumas with their combat groups. The B-26s raked the site with their bombs and machineguns and were followed by the Pumas. The landing zone was on the north rim of the dense vegetation, and the landed paras quickly moved south into the area still smoldering from the passes by the B-26s. The paras encountered no resistance and observed a number of bodies left by the staffing.

The paras were commanded by Lieutenant Colonel Almendra, who on reconnoitering the area from the air decided to call for a second insertion of two combat groups on the far side of the thicket with the intention of capturing the remaining elements of the column who sought to escape in that direction. However, as his plan was being executed, the Alouette III gunship departed for Bessa Monteiro to refuel and rearm, and this left the ground troops without close air support.[23]

Meanwhile, on the ground the paras returned to the south rim and reentered the thicket, where they found no insurgents. As they emerged on the other side, they found that the fleeing insurgents had set fire to the grass. It now burned with great ferocity and forced the paras to detour around the flames. This move put them in an unfavorable position, and the enemy, taking advantage, opened fire with automatic weapons and grenade launchers and wounded two paras. Fidalgo flying his Alouette III had no arms and could only guide the para attack. Suddenly the enemy fire practically ceased.[24]

During this moment a vulnerable Alouette III landed to evacuate the two wounded without the protection of the gunship that was still refueling. An urgent and unprotected evacuation in combat was always worrisome, and now as the wounded were being loaded as quickly as possible, a blast of machinegun fire hit the two wounded and aggravated their condition. The helicopter was also hit but executed a liftoff away from the source of the fire and, skimming the treetops, headed for Toto.[25]

Sometime later after an intense search for the enemy, the paras agreed that as night approached, they would maneuver toward an elevation about 300 meters from the thicket and establish a defensive perimeter for the night bivouac. Before doing so, however, the paras asked the Alouette IIIs to inspect the site from the air after again refueling in Bessa Monteiro. While there, Fidalgo's helicopter refused to start following its refueling, and he changed places with the gunship pilot to return to the scene of the action. On doing so, he was told that the enemy was now located along a small stream that wound through the thicket. Positioning the gunship to rake the banks of the stream, the gunner fired into what was a largely obscured or only partially visible target. After a few long bursts from the cannon, the enemy returned fire with

23 Ibid., 17.
24 Ibid., 18.
25 Ibid.

his light arms from the small openings in the dense vegetation.[26] The return fire hit the gunship in several places and wounded the gunner in his right leg. Fidalgo immediately broke contact and turned towards Toto, which was a 20- to 30-minute flight. The damage to the helicopter was minor, but the gunner had lost much blood and made a tourniquet from an old, oily rag that he had found in the helicopter to stem the flow. Fidalgo landed next to the Noratlas that was evacuating the two wounded paras and loaded his gunner on board with them.[27]

The following morning, the paras now with the support of two gunships made a final attack on the enemy position. Enemy elements fled in part, while some porters were captured or simply surrendered exhausted. The paras found a great deal of supplies, some automatic weapons, a large quantity of ammunition, and many anti-personnel and anti-vehicle mines.[28]

It is certainly possible to describe many more UCTI operations that illustrate the effectiveness of its mixed air-land components. The fine tuning of UCTI allowed it to prevail over the enemy in the north and to choke his supply lines to the point that he became virtually impotent.[29]

The Doctor in Zala

The insurgent temptation to fire on the wounded and the helicopter evacuating them was generally overwhelming, if given the chance. It is worth mentioning another such case that occurred during the middle of 1973 in the region of Zala, a small town about 75 miles northeast of Luanda and in the midst of the celebrated Bala Forest, where the enemy was always very active, and the terrain difficult. The FAP operation followed the normal UTCI profile with the now classic bombardment from F-84s, one of their last operations, followed by a helicopter envelopment to deposit troops in three different landing zones designed to encircle the column. The attack was not held until all of the troops were in place, as there was a casualty at the initial landing zone.[30]

The evacuation procedure was routine. The Alouette III equipped with litters and its gunship escort were directed to the site by the PCA. Meanwhile, the wounded para was prepared for evacuation, while the gunship scouted the area and gave a "green light" for the *Mosca* to land. At the moment that the Alouette III was about to touch down, the pilot yelled that he was being targeted "like a lamb being led to the slaughter." He waived off, as intense fire erupted around the landing zone. There were various mortar rounds, black smoke from hand grenades, and excited commands on

26 Ibid., 19.
27 Ibid.
28 Ibid.
29 Ibid., 20.
30 Ricards dos Santos Cubas, "Episódios Áfricanos: O Médico de Zala" [African Episodes: The Doctor of Zala], *Mais Alto* 306 (April–May 1997): 38–39.

the radio. As the exchange died, the gunship again checked the area, and the evacuation proceeded. Despite the shooting no one had been wounded and no helicopter hit—a testimony to the poor marksmanship of the insurgents. The pilot was informed that in fact there were two casualties on the ground, and one was a 12-year-old black girl who was gravely wounded. The Alouette III rushed both to the small medical post at Zala, and the doctor who met the helicopter immediately understood the situation. The girl was semi-conscious and had sustained a massive wound on the inside of her right thigh. A tourniquet next to her groin showed that she had lost a great deal of blood and that her femoral artery must be sutured immediately.[31]

The doctor was a young, recently commissioned lieutenant, who was physically unimposing—short and very thin with blondish hair. He wore a stained and yellowed operating smock many times too large that he had secured with a sash around his waist to prevent it from dragging on the floor. He hurried to begin the surgery. Two nurses, one white and the other black, assisted him, and as he gave the orders, they reacted with some puzzlement, for they had little experience in such a situation. A long five minutes passed, when the nurses called attention to the accelerated respiration of the girl. She would not survive much longer because of her dangerous blood loss. A transfusion must be made, but there was no time to arrange for the blood. As if by a miracle the donor appeared. The doctor placed his left arm next to the nurse on that side and commanded her to roll up his sleeve to make the transfusion. The nurse hesitated, uncertain of the doctor's instructions. He insisted, knowing that his blood was "O-negative"—the universal donor type. Soon the blood ran directly into the vein of the girl through a transfer system run by the nurse. Meanwhile, the doctor proceeded with his surgery indifferent to the fact that he was the donor.[32]

The hemorrhaging was finally stopped. It was not known for certain how much blood the doctor had lost, but it was substantial. When the girl's wound was finally closed and the last stitch made, the nurses had to support him, as his legs trembled. While he was weak, he was smiling, for he had saved a young girl's life that had almost slipped through his fingers.[33] He was transferred immediately by a Puma to the hospital in Luanda, and Brigadier Ricardo dos Santos Cubas, its pilot, remembers that he was covered in blood and his eyes peacefully closed. His chest rose and fell regularly, so he was clearly alive. He arrived at the hospital safely and with his sacrifice had earned the respect and admiration of all.[34] It was also a powerful message to the population that all lives were priceless regardless of political viewpoint, age, or sex.

31 Ibid.
32 Ibid.
33 Ibid.
34 Ibid.

B-26 *Invader*

By the end of 1961, it was apparent that the conflict in Africa was going to be a long one that required a powerful aviation ground support capability. The PV-2 Harpoon was obsolete and considered simply a makeshift bridge to the acquisition of more capable and modern aircraft. Briefly the FAP considered a quantity buy of the De Havilland Canberra; however, this alternative was a relatively expensive route in both acquisition and continuing operating costs.[35] FAP attention was drawn to a more affordable and proven counterinsurgency platform, the Douglas B-26 Invader, a World War II light attack bomber highly modified by the United States for counterinsurgency.

In the early 1960s the U.S. Air Force had drawn 40 old B-26s from storage and converted them to the potent B-26K with new engines of an additional 500 horsepower each and the ability to carry an incredible 8,000 pounds of external ordnance with a strengthened wing. The B-26K also carried eight .50-caliber machineguns in its nose. The up-engined and re-winged bomber was purpose-designed to interdict the Ho Chi Minh Trail in Southeast Asia. If Portugal could acquire a number of these old but proven airframes and modify them for its purposes in Africa, so the reasoning went, then this could be the perfect answer to the Harpoon replacement.

The B-26 had first appeared in August 1943 as the A-26 in the U.S. Far East Air Forces and proved deeply flawed in its rushed production. Douglas modified the aircraft with a high-visibility cockpit and other adjustments and refinements and reintroduced it in September 1944 with the U.S. 9th Air Force in the European theater. Here it proved a complete success.

The Invader was originally a single-pilot aircraft in the vein of the De Havilland Mosquito and was manufactured in two variants. The A-26B had a solid nose that mounted six .50-caliber machineguns, and the A-26C had a "greenhouse" or bombardier nose. With a few manhours in maintenance one nose could be replaced with the other. Both aircraft had a three-man crew. In the B model the navigator/gun-loader sat to the right of the pilot. In the C model the navigator/bombardier also sat to the right of the pilot and crawled into the glass nose to operate the bombsight on horizontal bombing missions. The third crewmember sat in the tail and operated the dorsal and ventral turrets from a slaved gun sight. The aircraft proved fast and capable because of its highly efficient laminar flow wing design. This soundness of design had lent the aircraft to successful modification. At the end of the war the aircraft was retired from service, some 2,452 airframes having been built.

When the U.S. Army Air Force became the U.S. Air Force in 1947, it changed the designation of the aircraft to B-26 the next year and returned it to service in 1949. The

35 José Manuel Correia, "B-26 Invader: Um Bombardeiro Misterioso na Força Aéria Portuguesa, 1° Parte, Uma Chegada Sinuosa e Atribulada" [B-26 Invader: A Mysterious Bomber in the Portuguese Air Force, Part 1, A Sinuous and Troubled Arrival], *Mais Alto* 375 (September–October 2008): 20.

aircraft was used in Korea between 1950 and 1953 to destroy North Korean supply lines and proved itself again and again in this role. It was once more largely removed from service in 1958.

In January and February 1951 the United States lent 113 B-26s to France to fight in Indochina. At the end of the conflict 88 were returned, as 25 had been lost in combat or accidents.[36] Later during its war in Algeria, France purchased 76 B-26s, and these proved highly successful. The B-26 was also operated by the Biafran, Cuban, and Indonesian governments and by the CIA in its various assignments, most notably in 1964 against the Simba rebels during the Congo Crisis.

The B-26 under consideration by Portugal was powered by two Pratt & Whitney R-2800-27 Double Wasp 18-cylinder engines producing 2,000 horsepower each. The normal weapons load was six .50-caliber machineguns in the nose, 4,000 pounds of ordnance in the bomb bay, and another 2,000 pounds on wing stations. This 6,000-pound load was six times that of a T-6G!

Portugal sought to purchase 20 B-26s and have them rebuilt to return them to "zero hour" airframes. The nose would retain the six .50-caliber machineguns, and there would be an additional three in each wing. The bomb bay and wing stations must have the flexibility to accommodate the various munitions available to the FAP, including drop tanks for extended loitering capability.[37]

As the aircraft were made airworthy by an aircraft rework company in the United States, a contract pilot was hired to ferry the aircraft to Portugal. His misadventures, including a violation of the restricted zone around the White House, led to a review of the export licenses, a halt in deliveries, and the ultimate acquisition of only seven B-26s (serial numbers 7101–7107). After arriving in Portugal in mid-1965, the Invaders were sent to Alverca in January 1966 to begin a general program of revision. The bomb bays and wings were reconfigured to accommodate the ordnance available to FAP. The radios were upgraded, and the navigator's seat in each was converted with copilot controls. The anti-icing system was removed, as the Invaders would be flying only at low altitudes. The dorsal and ventral turrets were also removed to lighten the aircraft. The French were able to furnish the operational and technical manuals and spare parts for the aircraft from their experience in Indochina and Algeria, and Brazil, which had a squadron of 18 Invaders at the time, was able to help with replacement parts and the important "Rock Island" wing reinforcing kits.[38] This latter support from France and Brazil circumvented the reluctance of suppliers to sell war matériel to colonial powers. All of these relationships took time to develop, and progress at Alverca was consequently delayed.

36 Philippe Gras, *L'Armée de L'Air en Indochine (1945–1954): L'Impossible Mission* (Paris: Harmattan, 2001), 335–358.
37 Corriea, 1° Parte, 20–21.
38 Ibid., 24.

By early 1971 the seven aircraft were ready for deployment, and two (7104 and 7106) were sent to Guiné and operated there for five months from Bissalanca and Nova Lamego. At the end of this period, 7106 was replaced with 7107, and the two continued to operate in varied missions. The aircraft performed well but was not completely integrated into the FAP operations and remained a specialized assignment for pilots. Pilots who flew the B-26 came to appreciate the armor around the cockpit and the thick Plexiglas windscreens that offered superior protection against ground fire, for unless a round hit the Plexiglas squarely at 90°, the material would deflect it.[39]

In October 1972, Invaders 7101 and 7106 departed the *metrópole* for Luanda. These were followed by 7107 in December and 7105 in January 1973. The first operational missions began in February, and these were flown in support of Puma and Alouette III insertions of ground troops, although there were in addition the usual assigned bombing targets and armed reconnaissance missions.[40]

In March, 7102 with its bombardier nose was added to the Angola fleet, and this brought the total to five. In the middle of October, the Invaders underwent anticorrosion maintenance and were given their anti-radiation green as protection against the Strela. By December they were beginning to need major inspections and repair attention, and the schedule appeared to reduce the available numbers over the next 18 months.[41]

During 1973, the B-26s in Angola together flew 653 hours, a fairly low average number of 130 hours annually or about 11 hours monthly per aircraft, and in the opinion of many who flew the aircraft it was underutilized. This remains puzzling, as the small number of Pumas was well integrated into FAP operations and made a major contribution. Certainly the Invaders could have made a similar contribution; however, they remained stepchildren.[42]

Success of Constant Pressure

In June 1973, *flechas* detected and followed the tracks of a group of 25 insurgents who when surprised fled, abandoning some salt and bananas and parts of Claymore mines.[43] This was a group attempting to resupply *Gibóia*, a UPA/FNLA base that was subsisting exclusively on meat from hunting, a situation similar to that of *Tigre* (Tiger) in the area of Tabi, a small coastal village about 40 miles north of Luanda and bordering the game park Coutada-Ambriz. Intelligence reported symptoms of

39 Ibid., 26–28.

40 José Manuel Correia, "B-26 Invader: Um Bombardeiro Misterioso na Força Aéria Portuguesa, 2° Parte, As Missões no Teatro de Guerra Angolano" [B-26 Invader: A Mysterious Bomber in the Portuguese Air Force, Part 2, Missions in the Angolan Theater of War], *Mais Alto* 376 (November–December 2008): 20–21.

41 Ibid., 23.

42 Ibid., 28.

43 Cubas, Estevens, Costa Santos, and Queiroga, 20.

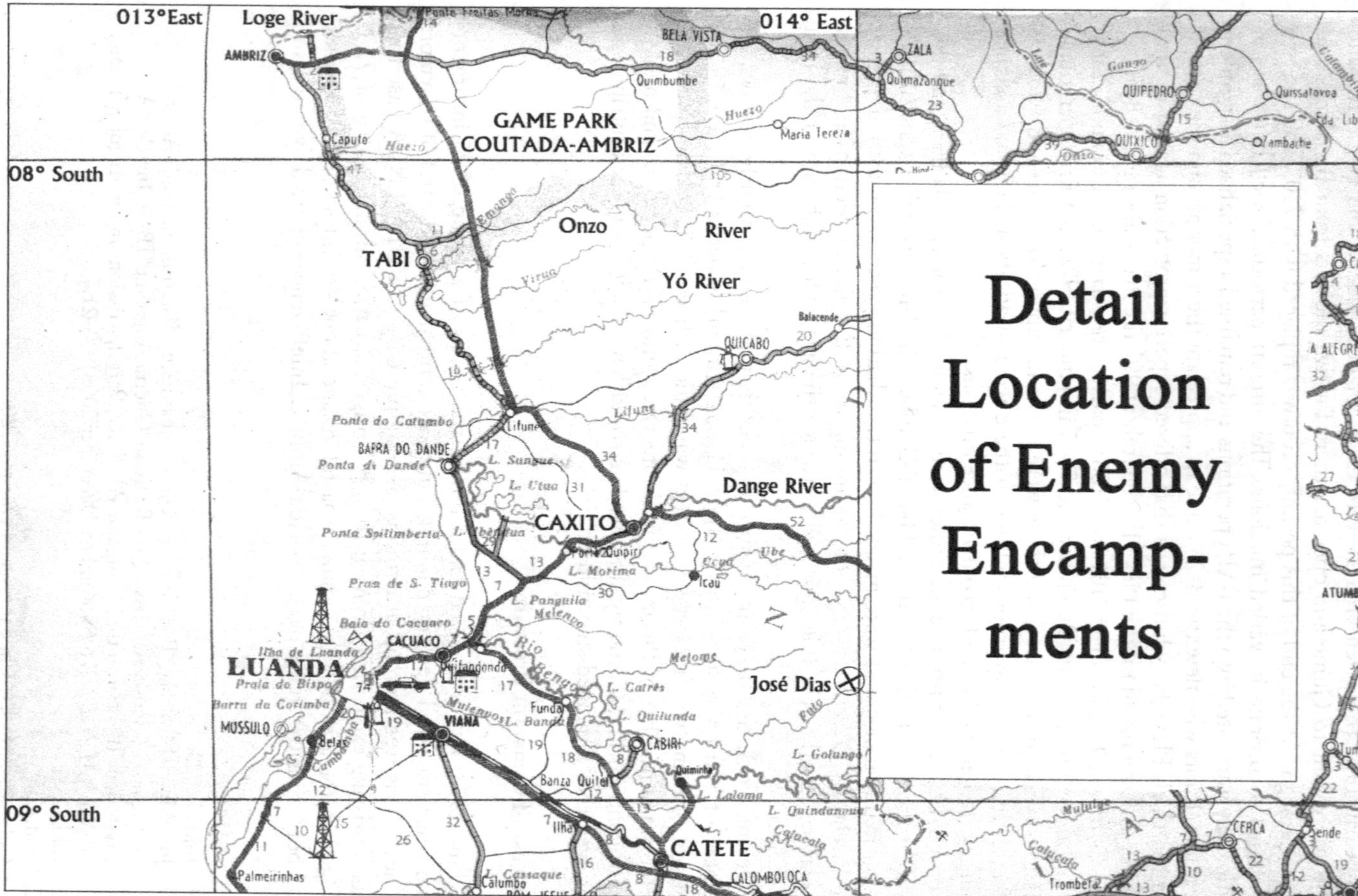

Map 12.2 Enemy encampments – Dembos forest region, 1970–1973. (Source: the author)

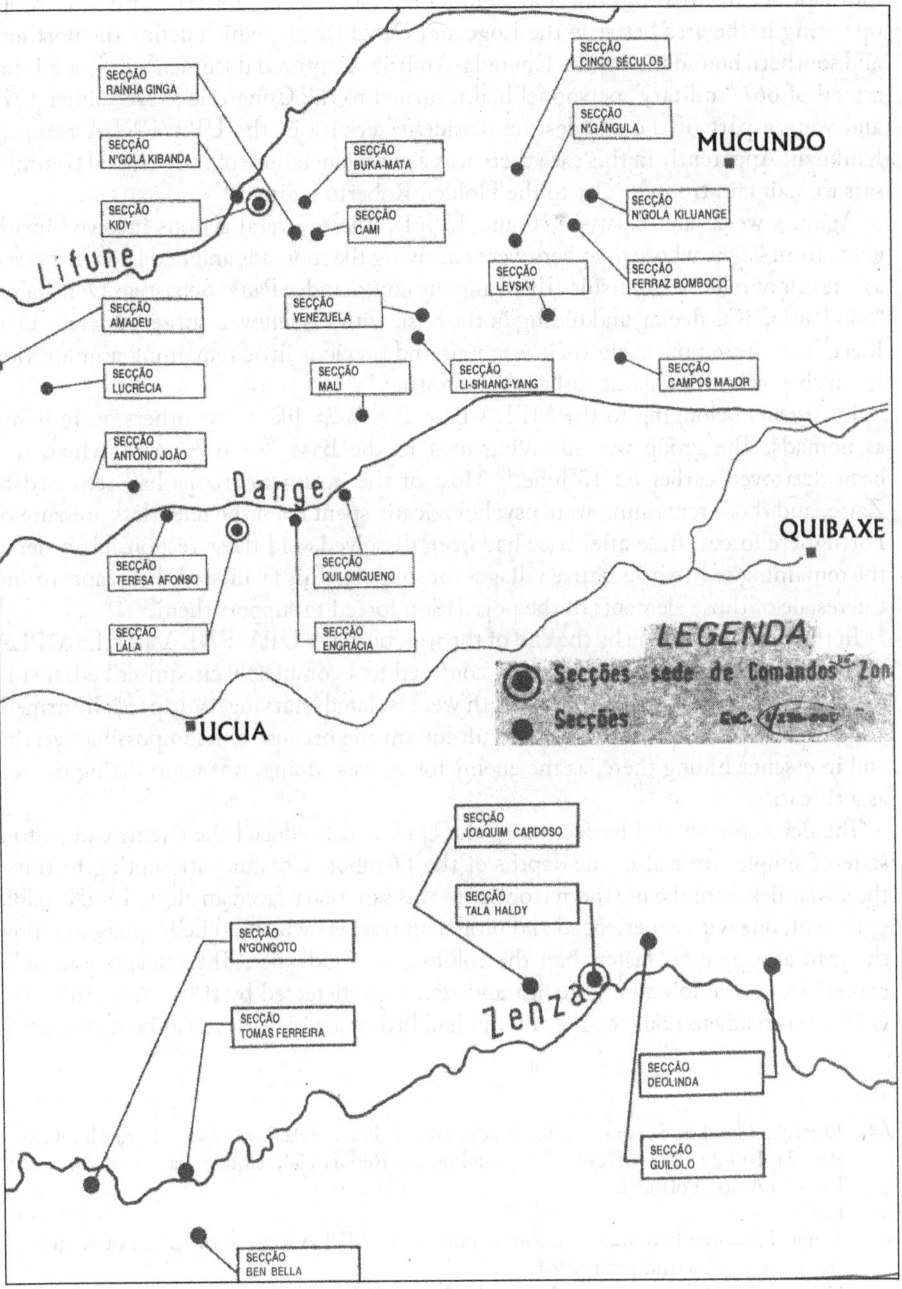

Map 12.3 Enemy Base Locations, Dembos Region. (Source: Estado-Maior do Exército, Comissão para Estudo das Campanhas de África)

indiscipline, disorganization, and demoralization among the UPA/FNLA troops operating in the area between the Loge and Onzo Rivers, which define the northern and southern boundaries of the Coutada-Ambriz. Captured documents indicated that a total of 667 "military" personnel had returned to the Congo since December 1971 and were a part of the clandestine dissident activity in the UPA/FNLA camp at Kinkuzu. Apparently in this case there was a complete failure of the political commissars to maintain troop loyalty to the Holden Roberto regime.[44]

Again a week later between 10 and 13 July, These several actions involved insurgents from *Tigre*, whose members were surviving like nomads and had been dispersed as a result of relentless pursuit. The company commander Paulo Sebastião Domingos, "São Paulo," was fleeing and hiding in the bush with his fellow insurgent Mateus Dias Ideia. They were now using their weapons and precious little remaining ammunition to survive by hunting game rather than fighting.[45]

Insurgents belonging to the MPLA base *Ben Bella,* like many others, were living as nomads. The group was surviving next to the Base *Tomás Ferreira*, which had been destroyed earlier on 17 June.[46] Most of the insurgent troops had returned to Zaire, and those remaining were psychologically spent from the relentless pressure of Portuguese forces. Base after base had been dissolved, and those responsible ordered the remaining to live near native villages for support. This facilitated their capture and the rescue of those elements of the population forced to support them.[47]

In the north of Angola by the end of the war, both the UPA/FNLA and the MPLA were forces strangled, exhausted, and confined to a completely circumscribed area in the fastness of the Dembos region. Both were isolated, starving, only partially armed, and desperate. The Dembos was difficult for anyone because of its impossible terrain, and in essence hiding there, as the enemy forces were doing, was neutralizing oneself as a threat.

The development and perfection of the UTCI had reduced the enemy cadres to a state of simple survival in the depths of the Dembos. Columns attempting to travel the 150 miles from the northern frontier to this sanctuary faced an alerted and capable opponent, one with experienced and proficient trackers who used helicopters to follow the trail at a pace far faster than the column could advance. The trackers and helicopters were able to employ stealth and remain undetected by the column until the UTCI was ready to pounce. The column had little warning that it was being targeted,

44 Direcção Geral de Segurança, Acções das Nossas Forces [Actions of Our Forces], DGS-Angola, 10 July 1973, SDGN 1802, Intelligence Report 754, Arquivo da Defesa Nacional, Paço do Arcos, Portugal.

45 Ibid.

46 Tomás Francisco Ferreira was a commander in the MPLA areas in the north of Angola along the Congo frontier in 1961.

47 Direcção Geral de Segurança, Acções das Nossas Forces [Actions of Our Forces], DGS-Angola, 18 July 1973, SDGN 1802, Intelligence Report 766, Arquivo da Defesa Nacional, Paço do Arcos, Portugal.

and when it was surrounded and cornered, there was little choice but to abandon the supply loads and weapons to permit unencumbered flight. Members of the broken column, if not killed or captured, became refugees in an unfamiliar land, easily identifiable and ultimately captured. Some, of course, gained their destination base or returned to Zaire, but those captured talked readily and easily and revealed what they knew. The intelligence gathered from the prisoners insured the destruction of subsequent columns and surviving bases. New columns were regularly broken and surviving bases reduced to ineffectiveness as they withered. The UTCI was indeed a powerful counterinsurgency tool.

13

Spínola and the New Air War in Guiné

By March 1968, when Brigadier António de Spínola arrived in Guiné, all was certainly not lost. Under the aggressive and determined air campaign from mid-1967 onward, the PAIGC had been put on the defensive and truly weakened. It was forced to swiftly dismantle its large internal bases that were attractive and vulnerable to air attack and replace them with semi-permanent smaller ones that allowed more flexibility, mobility, and agility. These smaller bases were supported by 26 exterior bases surrounding Guiné—16 in Senegal and 10 in Guinée-Conakry.[1] This air offensive had nearly paralyzed the PAIGC, as entire villages had fled to neighboring states and abandoned their crops. Without these, PAIGC troops were forced to forage for edible plants, and many began to question the war effort. The PAIGC now had to explain to fleeing villagers why they should endure death and destruction to feed starving troops.[2]

It was also at this time that Amílicar Cabral formed the first antiaircraft units. These were as much a practical defense against the feared FAP as they were a morale builder, and at least now his troops were able to challenge the free hand that the FAP enjoyed. He had been intensely negotiating for aid over the last five years within the communist bloc, and after he had secured agreement for the Chinese to train his insurgent force at the Nanking Military Academy, he turned to the Soviets, who then agreed to supply heavy weapons. By 1965 war matériel was abundant, and the FAP began to see the increased threat from this. Additional heavy weapons and technical help came from Cuba, and this relationship was strengthened when Cabral was invited to the initial meeting of the Tricontinental Conference in January 1966 in Havana. From this point onward PAIGC leaders made frequent visits to the Soviet Union (10),

1 Mustafah Dhada, "The Liberation War in Guinea-Bissau Reconsidered," Journal of Military History 62 (July 1998): 581.
2 Ibid., 583.

Cuba (3), and Eastern Europe, North Korea, and China (9).[3] Initially Cabral saw his new antiaircraft weapons as simply defensive and felt that something more needed to be done to provide relief. He decided to try an offensive move against the "metal birds" by attacking them on the ground in Bissalanca, and a small team of thirteen saboteurs attacked BA 12 on the night of 19 February 1968. For both the PAIGC and the Portuguese this was the first such attack, and the PAIGC troops used about a dozen mortar rounds against the base. There was no damage to the aircraft, although there were some casualties in the local militia.[4] Now, however, Cabral had opened a hornet's nest, and the G-91 attacks on the new PAIGC antiaircraft positions and bases, particularly in the south, accelerated and developed into an unabated duel between these two forces. Continued FAP pressure coupled with Spínola's social and political outreach to the population fostered an opportunity to conclude the fighting in Guiné with honor. Here we will explore the role of airpower in achieving this end.

Weakening the PAIGC

In the opening days of January 1968, Operation *Faísca* (Spark) was launched against antiaircraft positions in the region of Cassequelã, again on the Quitafine Peninsula. The Fiats destroyed one ZPU-4 at Banir and a DShK 12.7mm gun at Dameol. Four additional gun emplacements were hit and abandoned. On 5 January two flights of Fiats were taken under fire with heavy weapons hidden in the bush around Cassebeche, and in the afternoon of the following day the scenario repeated itself, prompting the Fiats to attack the position and destroy the gun with direct hits by 200 kilogram bombs. Two days later an attack on Cassebeche encountered fire from a newly installed battery, and the Fiats returned to destroy it with a number of high-angle dive bombing attacks. On 9 January two missions, one in the morning and another in the afternoon, attacked the settlement of Canefaque with seven Fiats equipped with napalm. The object was to destroy the base, but the PAIGC, if nothing, was persistent, and follow-on reconnaissance discovered a new gun emplacement nearby with extensive protective revetments. Again on the morning of 11 January and on the following day two Fiats with their maximum bomb loads destroyed it. Further missions over the area on 13 January drew no antiaircraft fire.[5] Ten days later on 23 January in bombing Cantanhez and Quitafine, the Fiats encountered a revetment at Ponta Cacine and together with a T-6G engaged it immediately with rockets and machineguns to silence it.[6]

3 Mustafah Dhada, "Guinea-Bissau's Diplomacy and Liberation Struggle," *Portuguese Studies Review* 4, No 1 (Spring–Summer 1995): 31.

4 José Manuel Correia, "Fiat G-91 na Guiné 1966–68: Os Dois Primeiros Anos de Combate" [The Fiat G-91 in Guiné 1966–68: The Two First Years of Combat], *Mais Alto* 399 (September–October 2012): 34.

5 Ibid.

6 Ibid.

As the air planners were preparing for Operation *Hércules* (Hercules), reconnaissance sorties unearthed a DShK 12.7mm gun at Cafal Balanta, which fired at the Fiats and attracted several barrages of their 2.75" rockets. Reconnaissance continued along the southern frontier of Quitafine. Operation *Crepúsculo* (Twilight) was opened on 4 February with an attack on a PAIGC camp at Caiar Island. As one of the Fiats made its bomb release over its target, immediately a hidden magazine was hit and exploded violently. The blast startled the follow-on Fiats coming onto the target and initiating their bomb runs.[7]

On 8 February in pressing attacks over the Quitafine, the Fiats encountered ferocious anti-air resistance in the area of Calaque Manjaco and attacked it subsequently with two missions of four aircraft each, all eight carrying their maximum bomb load. Two positions were destroyed and the remainder abandoned by the gun crews.[8]

The next day, 9 February, Operation *Martelada* (Hammered) opened with the goal of silencing antiaircraft emplacements in the region of Cassebeche and Calaque Manjaco. During the second phase of the operation on the morning of 14 February, a pair of Fiats was attacked from three positions in the area of Cassebeche. That afternoon seven Fiats descended on the positions, destroying one and forcing abandonment of the other two.[9]

After almost a month of quiet in which some of Cabral's abundant heavy weapons were moved to the peninsula to replace those destroyed, there was a resurgence of PAIGC anti-air activity. In the second week of March there appeared a ZPU-4 and four DShK 12.7mm guns around Cassebeche, and this prompted Operation *Martelada III* (Hammered III) described at the end of Chapter 8 by Lieutenant José Nico. This was followed by a para heli-borne assault from Catió that destroyed the enemy camp near Cassebeche and captured much matériel. These successes relieved some of the pressure on the FAP pilots, and close air support of ground operations occupied their time following *Martelada III*.[10]

The rainy season set in, and this limited sorties to low level flight and attack because of the cloud formations. This restriction gave the PAIGC gunners a better opportunity, if they could anticipate aircraft appearance, and they were aided in this by a network of lookouts arrayed around each emplacement. The Fiat pilots employed varied deceptions to achieve surprise, such as arriving on top of the target in sequence from different directions, and became very adept at these maneuvers to avoid being targeted and hit. This precaution of extensive lookouts applied not only to the antiaircraft artillery emplacements but to enemy camps as well. As the PAIGC gained experience, it improved its methods of defense and detection for both of these installations and made it increasingly difficult to be surprised by either air or land forces.

7 Ibid.
8 Ibid.
9 Ibid.
10 Ibid., 35.

Surprise attacks on the ground against PAIGC internal bases had been very successful between 1963 and 1966. However, by 1968 it became nearly impossible to surprise a PAIGC encampment because of the sentries and mined approaches to it. Often these defensive measures were situated several kilometers from the base. Only a heli-borne assault after 1968 stood any chance of success against PAIGC redoubts.[11]

In the second week of June a photoreconnaissance mission was flown over Quitafine to identify new antiaircraft emplacements, and several weeks afterward on 28 July, Captain Vasquez, the commanding officer of the "Tigers," and Lieutenant Colonel Francisco Costa Gomes, the air group commander, were flying a photoreconnaissance mission along the southwestern border with Guinée-Conakry to collect images of new gun positions situated across the frontier. These had fired on Vasquez's Fiat the previous day, and he guided Costa Gomes to the target. The two flew parallel to the border and employed their lateral cameras to look south across the frontier. As they were making their pass, a PAIGC DShK 12.7mm hit the Fiat of Costa Gomes in the combustion chamber of its engine and caused a catastrophic fire that forced his ejection in the vicinity of Gandembel, where there was a paratroop caserne. This was the second aircraft lost by the "Fiat squadron," Squadron 121, and this one by its air group commander no less. The paras watched the ejection and parachute deployment from the parapet of their fortification but lost sight of the pilot as he floated into the tall grass on landing. Immediately they left the caserne en masse to search for the downed aviator, and only the duty sentries remained to man the base. Costa Gomes drew his pistol and, wielding it, walked toward the caserne and entered its parade without being challenged or even noticed. Suddenly one of the young paras spied a gray-haired man in a blue flight suit approaching and wanted to know why he was there. Costa Gomes was about to answer, when the young para on looking at Costa Gomes's flight suit added that he looked too old to be a pilot![12] Analysis of the mission photographs revealed that the guns were DShK 12.7mm and situated in Guinée-Conakry with the capability of firing across the border to protect the PAIGC encampments and particularly PAIGC use of its infiltration and logistics corridor around Guileje.

On 20 August Operation *Júpiter* (Jupiter) was launched in the region of Gandembel as a paratroop envelopment with close air support from BA 12. The assault resulted in the capture of light weapons and caused dozens of PAIGC casualties. The antiaircraft artillery was silent. The next month a flight of Fiats conducting armed reconnaissance along the border came under fire from heavy weapons around Guileje. The pair avoided the fire and attacked immediately with rockets, silencing the gun emplacement.[13]

Taking advantage of the temporary reduction in PAIGC activity, a series of air missions were conducted in October in support of the army and *fuzileiros*. These

11 Hélio Felgas, "Papel do Helicópterio na Guerra na Guiné" [Role of the Helicopter in the War in Guiné], Jornal do Exército (August 1968): 33.

12 Correia, 35; and José Correia, correspondence with the author, 31 July 2012, Lisbon.

13 Ibid., 36.

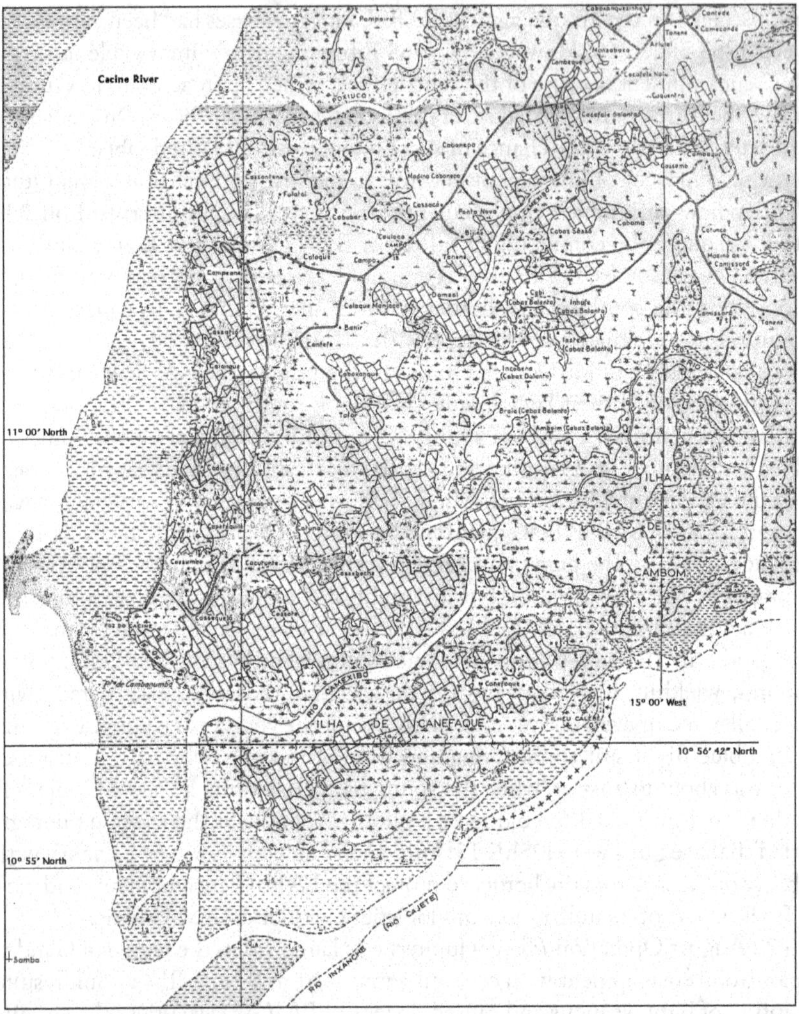

Map 13.1 Quitafine Peninsula and the southwest of Guiné. (Source: the author)

operations (codenamed "Flying Fish" and "Welcome") were primarily air mobile maneuvers centered on the fleet of Alouette IIIs, which executed heli-borne assaults on the areas around Concolim and Bricama, both in the Oió Forest. These were supported by gunships, Fiats, and T-6Gs as close air support.[14]

14 Ibid.

On 3 November Operation *Tubarão* (Shark) was launched in the area of Buba Tombo. During the same period two T-6Gs were targeted by antiaircraft artillery in Guinée-Conakry, but the absence of antiaircraft reaction to the heli-borne assaults of Operations *Baco* (Bacchus) and *Apolo* (Apollo) seemed to indicate, according to intelligence, that authorities in Conakry had ordered the PAIGC to withdraw its fighters from the frontier zone. Presumably Conakry did not want the FAP making forays into its territory to silence PAIGC anti-air weapons. The PAIGC regrouped even as the PIDE/DGS warned the FAP of new antiaircraft artillery being moved into Quitafine next to the frontier between Cacoca and Sangonhá.[15]

In December heli-assault operations and their close air support were limited by a lack of 200-kilogram bombs; however, the numbers available from the magazine were used effectively to destroy a PAIGC DShK 12.7mm gun. Fresh reconnaissance of Quitafine revealed a new ZPU-1 next to Cassebeche. The photoreconnaissance Fiat was taken under fire but escaped undamaged. This anti-air duel with the PAIGC in southern Guiné, it appeared, was a "contest without rest."[16]

Spínola

When Brigadier António de Spínola, who habitually wore the traditional monocle of a cavalry officer, arrived in Guiné, he had a broad portfolio of experience on which to base his observations that the war there was a disaster. His résumé included leading a volunteer force in the Spanish Civil War in 1938 and acting as an observer on the German eastern front opposite Leningrad in November 1941. In 1955 he was made a member of the Administrative Council of the steel concern Siderurgia Nacional in addition to his normal military duties and thus became a beneficiary of Salazar's strategy for managing his military through such appointments.[17] Spínola thus had Salazar's ear and confidence.

At age 51 as a lieutenant colonel, he had volunteered to command Cavalry Battalion 345 and later in November 1961 departed for Luanda as part of the army reinforcement following the outbreak of violence in the north. In this assignment he proved himself brave and courageous and gained the respect and loyalty of his officers and men by leading from the front and enduring the dangers and hardships that they suffered. Likewise he won the admiration and affection of the local population and its leaders through his pacification efforts. He traveled widely in Angola during his assignment and departed in 1964, when his tour of duty was completed, to return to Lisbon and to senior staff duty. It was while in this staff assignment that he undertook

15 Ibid.

16 Ibid., 37.

17 Douglas L. Wheeler, "The Military and Portuguese Dictatorship, 1926–1974: 'The Honor of the Army,'" in *Contemporary Portugal*, ed. Lawrence S. Graham and Harry M. Makler (Austin: University of Texas Press, 1979), 199.

his survey of Guiné and, on his making his report, remembered Salazar saying, "It is urgent that you leave for Guiné."[18]

A few months following Spínola's appointment to Guiné, Salazar suffered a stroke and was replaced by Marcello Caetano. Spínola viewed Cartano's appointment with optimism, as he believed that there would now be a new colonial policy that would resolve the war and the international ostracism. Caetano in the early 1960s had confronted Salazar and argued forcefully for a federal system in which the overseas territories would have their own governments and participate equally with the *metrópole*.[19] Spínola openly identified with this concept and had great optimism that such a solution would be put in place.[20] Indeed, he was a strong supporter of the new prime minister.

Spínola assumed command in May 1968 as a newly-promoted brigadier and saw his first task as one of drastically reducing PAIGC military capability and thus tilting the balance of war in favor of Portugal.[21] He proceeded to implement an aggressive and imaginative offensive against the PAIGC. He sought both a political and military victory, and as part of the political dimension initiated a plan to win the population through a carefully orchestrated program of social and economic measures that lifted its standard of living. Education, health, infrastructure, animal husbandry, and agricultural cultivation were improved markedly under his comprehensive blueprint, "A Better Guiné."[22] These improvements were publicized and supported in a well-designed psychological operation to explain the changes and secure the backing of the people. The military dimension of this plan was the protection of the population from insurgent intimidation and the defeat and destruction of PAIGC armed capability. To this purpose the armed forces assumed an aggressive posture in its new modified organization that emphasized the recruiting of Africans.[23] Spínola's sense of urgency translated into accelerated activity on the ground and in the air.

Operation *Vulcano*

There remained a festering anti-air sore in Quitafine, and in the closing days of 1968, a pair of Fiats patrolling along the southern border were fired upon by a newly-installed

18 José Pedro Castanheira, "Memórias da Guerra e da Paz: Spínola" [Memories of War and Peace: Spínola], *Expresso Revista* (30 April 1994), 26.

19 Luís Nuno Rodrigues, "*For a Better Guinea!* Winning hearts and minds in Portuguese Guinea," presentation at the seminar "Legacies of Conflict, Decolonization and the Cold War," Lisbon University Institute, Instituto Superior das Ciências do Trabalho e da Empresa, May 2010, 2.

20 Ibid.

21 Ibid., 28.

22 António de Spínola, *Por Uma Guiné Melhor* [A Better Guiné], (Lisbon: Agência-Geral do Ultramar, 1970), 24.

23 Castanheira, 26..

ZPU-4 provocatively, indeed tauntingly, positioned in the center of the ancient settlement of Cassebeche. Earlier these quadruple 14.5mm artillery pieces and the smaller single barrel DShK 12.7mm had been destroyed or captured by attacks, some combined and others independent, of Fiats, gunships, and heli-borne ground forces. The problem was so insidious that Spínola had the air commander, Colonel Manuel Diogo Neto, report to him daily at his headquarters in the Amma Fort in Bissau. The southern border was difficult to police and thus easy to cross, so replacement of destroyed anti-air artillery pieces was a fairly simple task, particularly as the supply was described as "abundant."

In earlier instances gun emplacements of this type had been routinely eliminated by a Fiat releasing its maximum bomb load from a 45° dive angle with the sun at its back. This was repeated until the gun was destroyed. Frequently this was followed by a vertical envelopment of ground troops to clear the area of surviving munitions, hidden weapons, and lingering PAIGC insurgents.[24]

The antiaircraft weapons had been taken out individually as they appeared since Operation *Resgate*. The thinking of Spínola's staff now was to wait a bit longer until the PAIGC had committed a large number of weapons to the Quitafine Peninsula and then destroy the installations in a single large operation. This worried those pilots who would have to fly the helicopters with the paras. For them it was not difficult to imagine what would happen, if the enemy armed himself too much.[25]

As the days slid by, antiaircraft artillery flourished in the Quitafine, and its proliferation became very visible to the reconnaissance pilots. Increasingly it appeared that a simple problem was now becoming ever more complicated. The proliferating installations were protected by double parapets, and this meant that bomb delivery must impact precisely in the center of the fortifications to achieve destruction. This requirement significantly increased the danger to the pilots who were anxious to begin the operation before the threat escalated further.[26]

The concept of operations was divided into three phases:

1. The initial strike would be made by seven Fiats in three formations of three, two, and two, each with the maximum bomb load.
2. Ten Alouette IIIs would transport two companies of paras, the first landing to the south of the targets and the other to the north, both supported by gunships.
3. Close air support by the Fiats with 2.75" rockets and .50-caliber machineguns.

24 Manuel Diogo Neto, "Operação Vulcano, Cassebeche, 7 Março de 1969" [Operation Volcano, Cassebeche, 7 March 1969], *Mais Alto* 257 (January–February 1989):15.
25 Ibid.
26 Ibid.

This operation would stretch the limits of air capability in Guiné:

- 7 Fiats out of 8 available
- 10 Alouette IIIs out of 11 available
- 10 T-6Gs out of 14 available
- 5 Dorniers out of 7 available
- 2 Dakotas out of 2 available
- 34 pilots out of 35 available
- 2 companies of paras out of 2 available[27]

The airfield at Catió would provide refueling facilities for the helicopters. It was entirely a FAP operation, as there was little interest in helping from the army, although there were a number of army installations in the area of the targets.[28] Spínola's staff intervened and suggested that the ground force be reduced to a single company of paras, Company 121 under Captain Silva Pinto. The paras were organized into twelve combat groups of ten men each and were heavily armed in anticipation of strong resistance, as ground operations had not been conducted on Quitafine in over a year. Intelligence indicated the presence of strongly armed groups in the targeted area, so much depended on air support. In the opinion of the more experienced pilots, if it was possible to destroy the ZPU-4, then the air defense would collapse and its demise would facilitate para operations, which was no small effort with 20 troops in two waves (5 in each Alouette III) and a reserve of 40 in Catió. In preparation for the operation, the Dakotas moved the paras to Catió the preceding day, and on the day of the operation it was planned to have four of the eight Alouette IIIs lift the first wave to a position 1,500 meters north of the ZPU-4 objective at Cassebeche.[29]

In the early hours of 7 March 1969, a Dornier flew a respectable standoff distance from the landing zone near the target and found the sky clear, the wind calm, and visibility excellent. As the sun rose resplendent on the horizon, it foretold a scorching day.

The eight transport helicopters and two gunships launched to begin their work, and the first flight of three Fiats accelerated down runway 03 at Bissalanca, rotated, and climbed to 8,000 feet. Flight time to Cassebeche was eleven minutes, and the three aircraft on passing Cacine reduced power to 70 percent and speed to 250 knots. They aligned with the sun at their backs and passed along the southern frontier in preparation for their attack. Suddenly the enemy commenced firing, and tracers from the guns revealed seven active positions: six DShKs and a ZPU-4. The ZPU was the primary target.[30]

27 Ibid.
28 Ibid., 16.
29 Ibid.
30 Ibid.

On the initial pass the Fiat unloaded in the center of the "snail," and the ZPU-4, target ALFA, disappeared. Position FOXTROT, a DShK, was also destroyed. The helicopters arrived as the second flight of Fiats began their attacks and took out positions DELTA and ECHO, while the paras advanced through the bush and were met with light arms fire. As the paras progressed, they were met by stronger resistance and heavy machinegun and rocket fire. The orbiting Dornier reported three active antiaircraft sites.[31]

The Fiats landed to be rearmed for close air support. This was the plan; however, the antiaircraft artillery still needed to be destroyed. In attempting to support the paras, a Fiat was hit by 12.7mm rounds in the wing and nose, lost hydraulic pressure, and extinguished an associated fire. The pilot, Captain Nico, landed safely in Bissalanca.[32]

Because of the heavy ground resistance, the reserve was employed to alleviate the situation. The exchange of fire had ignited the dry vegetation of the bush, and the fire grew rapidly. At 1100 another flight of Fiats arrived over Quitafine and found two positions, BRAVO and ECHO, still active. The paras were now taking intense fire from the vicinity of target BRAVO, which was about 500 meters away. Another flight of two Fiats arrived, and one made a strafing run over BRAVO. It was hit by a 12.7mm round, lost power, and returned to Bissalanca.[33]

The situation was not developing as envisioned, even with the destruction of three antiaircraft emplacements and many enemy losses. The paras were stalemated with intense enemy fire, and the FAP was reduced to five Fiats. The T-6Gs were useless against the PAIGC guns and would be massacred, if they were thrown into the fray. Likewise the helicopters could not survive. Without the Fiats, recovery of the ground forces would be seriously compromised. Indeed, the entry of the para reserve brought new problems. The raging wildfire prevented the two forces, the primary and the reserve, from combining, and it required some hours for the two eventually to join. By then, around 1330 hours two Fiats arrived loaded with bombs, as rockets and machineguns were not effective against the emplacements. At 1415 a second flight of two Fiats arrived over Cassebeche with napalm and attempted to destroy BRAVO without success.[34]

Meanwhile the paras needed to take action to prevent a disaster. If they attacked the strong PAIGC positions, then they would certainly take casualties, and success was uncertain. It they remained overnight, then it was certain that they would endure mortar attacks with predictable consequences. It would be impossible to evacuate the wounded at night from the heavily wooded terrain. The last alternative was to withdraw to the north of the original landing zone and be evacuated to prepare for a later operation. So far the FAP had only one para lightly wounded, and two Fiats and a

31 Ibid., 17.
32 Ibid.
33 Ibid.
34 Ibid.

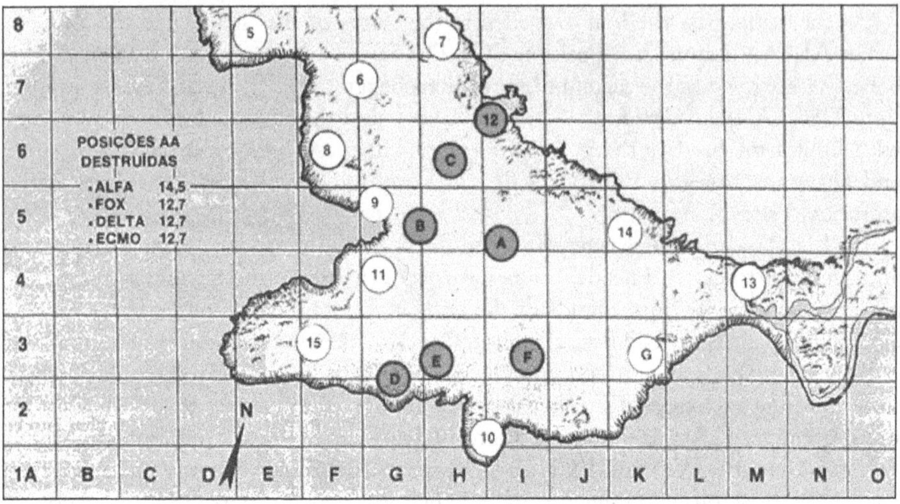

Map 13.2 Quitafine Peninsula Showing Location of Operation *Volcano* Targets. (Source: Força Aérea Portuguesa)

Dornier hit and out of action. The sensible decision was withdrawal and regroupment, yet Diogo Neto feared Spínola's criticism. Spínola proved understanding but cautioned Diogo Neto to be more careful in the future. Afterwards Diogo Neto prevailed on the FAP general staff in Lisbon to augment the Fiats with four additional aircraft and four qualified pilots. The paras also gained a company.[35]

Following the para withdrawal, a final flight of Fiats attacked position ECHO and destroyed it. Subsequent monitoring of PAIGC radio traffic revealed that it had suffered 19 dead and 32 wounded and that Cabral had ordered an inquiry into the "disaster at Cassebeche."[36] The systematic destruction of the PAIGC antiaircraft capability, particularly the ZPU-4 crewed by Cubans, obliged Cabral to seek a new weapon, and in 1973 the Strela appeared to challenge the free hand of the FAP. In the meantime, aviation continued to be the one means that created severe problems for the PAIGC.[37]

In reviewing Operation *Volcano*, it was clear that the ground force dropped into the midst of Quitafine was too small for the job at hand. In the face of an established enemy, a far larger force was required. Management and use of the reserve was ineffective, and the paras as light infantry were not adequately armed to go against the heavy weapons of the PAIGC. The terrain in which the paras were deployed was covered in dense vegetation and proved a major impediment to maneuver. The enemy, alerted

35 Ibid., 18.
36 Ibid., 19.
37 Ibid.

by the first attack on the ZPU-4, moved to consolidate his forces behind defensive earthworks and proved a difficult target for ground assault. He leveled his antiaircraft artillery and established nearly impregnable positions. The switch from bombs to rockets and machineguns on the Fiats ignored earlier experience and was a mistake, as the antiaircraft guns remained a threat, and the Fiats were unable to destroy them or to support the paras properly with the changed ordnance load. This experience proved an important lesson, and when the PAIGC air defense was reestablished in the months following *Volcano*, the FAP would craft a different, less risky, and more effective approach.[38]

Until the Strela arrived, the Fiats settled into a proven procedure during 1969 with which they systematically attacked and destroyed the PAIGC antiaircraft installations. Because the weapons were surrounded by two concentric earthwork parapets that protected the gun crew from a near miss, the Fiat pilots must achieve a direct hit within the defenses known as the "snail" because of its appearance from the air. A mission to destroy a gun emplacement began with intelligence that a gun was being moved into position, and this news was confirmed by a photoreconnaissance mission to determine the type and density of the weapons in the targeted area. The sites were very dangerous, as most of the gunners were Cubans who showed great courage in staying with the gun under air attack. A Fiat flight of four aircraft each armed with its maximum bomb load was then briefed and launched on the target. The Fiats on reaching the target formed a line astern and attacked in series from altitude at a dive angle of 45° with the sun at their backs to blind the gunners. The bombs were released in salvo at 450 knots in a single pass, and this procedure offered the highest percentage of direct hits inside the "snail." If the four Fiats did not reduce the gun emplacement completely, then a second flight was launched. The process was repeated until the gun was destroyed.[39]

Night attacks were attempted but were not particularly successful. Even if the weather was perfect, it was difficult for the pilots to judge distance and angle of attack at night. This was balanced with the fact that the PAIGC was only able to move heavy antiaircraft weapons into the Quitafine Peninsula and the area around Guileje in quantity, as these areas were immediately across the frontier, and the weapons could be introduced with relative ease. While there was clearly heightened danger in daylight attacks, it was these that were regularly successful in reducing the artillery. In some cases helicopter assaults with paras were used, but only against the lighter DShK guns around Cantanhez. In these cases, the helicopter pilots were able to reduce their exposure by approaching the target at treetop level and closing quickly across the open

38 Ibid., 20.

39 Diogo Neto, "Voo em três frentes" [Flight on three fronts] in *A Guerra de África 1961–1974, Volume 1* [The War in Africa 1961–1974, Volume 1] by José Freire Antunes (Lisbon: Temas & Debates, 1996), 321.

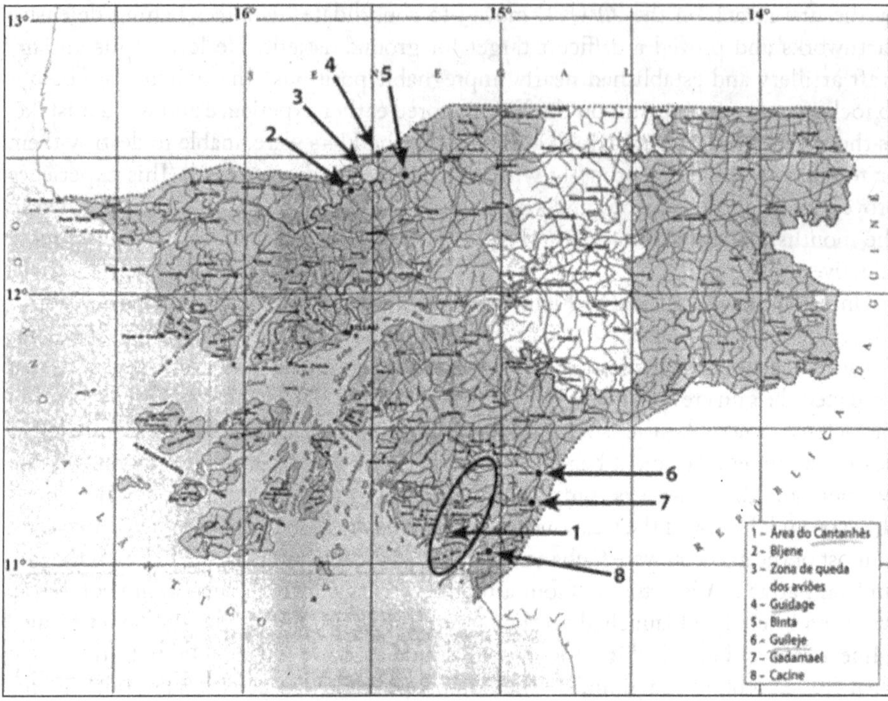

Map 13.3 PAIGC Assaults on Guiné, January 1970. (Source: Força Aérea Portuguesa)

area surrounding the gun to drop the assault troops before the gun crews could react effectively.[40]

In one mission in early 1969, Diogo Neto sent the Fiats to destroy an antiaircraft installation that the PAIGC had established on a dirt landing strip built for Dorniers near Sangonhá. Following the raid, ground forces entered the area and unearthed many caches of weapons, graves containing the dismembered remains of both natives and Cubans, and wrecked antiaircraft weapons.[41] The catalog of destruction wrought by air attack along the southern border throughout 1969 was enabled by the PAIGC and its regular introduction of antiaircraft artillery, munitions, and personnel as targets.

In January 1970 the PAIGC began a heightened offensive that would gather momentum over the next three years despite Spínola's best efforts to fight the organization politically with his "A Better Guiné" initiative and militarily with increased troop numbers. The assaults came from Senegal and Guinée-Conakry, as shown by

40 Ibid.
41 José Correia, correspondence with the author, 31 July 2012, Lisbon.

the nearby map, and in the north were supported by the PAIGC base of Cumbamory in Senegal. In the south, the PAIGC base of Kandiafara supported the corridor through Guileje that led to the interior. Saliguia supported the assault through the Gadamael area, and Sansalé sustained activity on the Quitafine between Cacoca and Sangonhá that was aimed at Cacine. The FAP would be under severe pressure over the next four years.

Arrival of the *Strela*

It is generally acknowledged that the introduction of the Soviet-manufactured SA-7 Strela into Guiné in the spring of 1973 represented a new and dangerous enemy anti-air threat. Certainly the missile had been around for a decade, as it was designed in 1959 as the first iteration of a Soviet hand-held anti-air missile. It entered service in 1968 and began series production in 1970. It made its first appearance in combat in April 1972 in South Vietnam and was an immediate threat to the slower flying aircraft, such as helicopters.[42] Western intelligence had known about the weapon for about three years prior, and after its introduction there had been reports from FAC aircraft that had observed "funny little black missiles following some of the fast-movers off the target."[43] Several A-1 Skyraiders and FAC aircraft were brought down before the evasion procedures and countermeasures were fully in place. Operating altitudes were raised, and crews kept alert for missile firings. When the missile was sighted, the pilot took evasive maneuvers, and these were generally effective, as a hard turn frequently allowed the fuselage and wings to shield the engine heat on which the missile homed.[44] Aircraft with flare dispensers also released flares to decoy the missiles.

Cabral was certainly aware of this development in Vietnam, and in December 1972, just eight months after its appearance there, he was in Moscow making his case for the missile. The Soviets had been generous with Cabral, as between 1967 and 1969 they had shipped the PAIGC numerous ZPU-4s and countless DShKs and from 1966 onward had arranged for Cuban advisors to assist in the antiaircraft defenses. There had even been the success of destroying a Fiat in June 1968, but this was not enough, as the FAP still had a relatively free hand. Cabral argued that Portugal depended on airpower to cling to the colony, and the Fiats located at Bissalanca were on ten-minute alert and able to be anywhere in Guiné within about the same time. The PAIGC need the capability to challenge this airpower decisively.

In mid-1973, the PAIGC sent a group of fighters to the Soviet Union for training in use of the Strela, and these returned to Conakry after several weeks. Meanwhile,

42 John A. Doglione, et al, *Airpower and the 1972 Spring Invasion* (Maxwell AFB: Air University Press, 1976), 43.
43 Ibid., 44.
44 Ibid.

the PIDE/DGS advised Spínola repeatedly on the missile, its imminent appearance in Guiné, and the possibility that the PAIGC was capable of operating a sophisticated weapon. Neither he nor Lieutenant Colonel José Almeida Brito, the commander of the air wing, were convinced.[45] Simultaneously in Lisbon defense officials discretely approached the United States in an attempt to obtain intelligence on the missile with little initial success.

There were also other intrigues. In Conakry when the 44 missiles arrived, Cabral hesitated to employ them immediately in the Quitafine, where their introduction would be simplest. He worried that at present the FAP enjoyed a relatively free hand in the air of Guiné delivering mail, evacuating wounded, and transporting supplies and troops. If he escalated the conflict by introducing the Strela, then Spínola could react by increasing his ground and air forces substantially and put the PAIGC in a dangerous position. Cabral also faced internal resistance from the Marxist faction led by Nino Vieira and Luís Cabral and struggled against rising difficulties with Sekou Touré and his military. It was he, however, who had the final say in this decision. This displeased the radical wing of the movement, and when Cabral died violently by the hands of these compatriots on 20 January 1973, there was no longer an obstacle to Strela deployment. Even in Bissalanca, FAP officers at the time were able to link the death of Cabral with the appearance of the missile in the third week of March, as there were only a few weeks between the two events.[46]

Indeed, its introduction came at the height of the rainy season when there were towering cloud formations that impeded the Strela in its proper functioning. This weather had much to do with the initial weak performance of the weapon, and a number of them fell from the sky and into the hands of Portuguese forces when their guidance systems were unable to acquire a target, and their rocket motors exhausted themselves.[47] This poor record can be traced to the missile team. The missile was entrusted to two shooters, and these and the missile itself were protected by an escort of 15 armed men. Altogether they formed an "SA-7 Group." The shooters were always searching for optimal missile launch conditions, and these were elusive, as it was extremely difficult to acquire a small and agile target, such as the Fiat, among the massive weather systems of the rainy season. With some five dozen Strelas delivered and used in the early months of 1973 and so few hits, the Soviet advisors in counsel with Moscow decided to refuse new deliveries of the missile. This explains why the pilots in Guiné saw so few missile launchings in the final months of 1973 and early 1974.[48]

45 José Manuel Correia, "Strela: A Ameaça ao Dominio dos Ceus no Ultramar, 1ª Parte" [Strela: The Threat to Domination of the Skies of the Ultramar, Part 1], *Mais Alto* 392 (July–August 2011): 32.
46 Ibid., 33.
47 Ibid.
48 Ibid.

The first Strela sighting occurred on 20 March when a flight of two Fiats piloted by Almeida Brito and Lieutenant Miguel Cardoso Pessoa were providing close air support in the area of Campada next to the frontier with Senegal. They were in loose formation, when suddenly a missile passed inexplicably between the two jets without exploding. Its shock wave rocked Almeida Brito's aircraft violently. The two pilots immediately attacked the likely launch site and in doing so remained unwittingly within the Strela acquisition envelope. Fortunately there were no more Strelas and the two returned safely to Bissalanca.[49]

Two days later a Dornier piloted by Sergeant Moreira was targeted over Bigene and, mistaking the Strela miss for an RPG shot, radioed for the alert aircraft to launch and bomb the source of the attack. Accordingly Lieutenants António Martins de Matos and Lourenço Marques launched for the area of the attack over which the Dornier continued to orbit. Moreira directed his comrades to the thicket from which the attack was launched, and as they searched, they saw on the ground the spent body of a Strela still smoking from its solid propellant. As the two Fiats began their attack, the pilots saw two new missiles launch from the bush towards them and leave long, conspicuous trails of smoke. Lieutenant Martins de Matos radioed for the second alert pair of aircraft armed with 50 and 200 kilogram bombs, but only Captain José Pinto Ferreira was able to launch, and when he arrived on station, he attacked from altitude with a steep dive angle. As he was exiting from his third pass, Pinto Ferreira saw a Strela launch at his Fiat. He was in a high-G turn and evaded the missile but not the aerodynamic shock of a near miss. Thinking that he had been hit, he turned toward Bissalanca escorted by Lieutenants Martins de Matos and Lourenço Marques and there landed without damage.[50]

Later on Sunday, 25 March, Lieutenants Miguel Pessoa and António Martins de Matos were assigned the alert duty when they received a call from the caserne at Guileje, which was under rocket attack. They launched into large cloud formations indicative of the rainy season. Visibility was poor and visual contact with the ground was intermittent. The ground troops directed the pair of Fiats to the old village of Gandembel, the likely source of the rocket launchings, and as they lost altitude approaching the border with Conakry, they leveled off at 100 feet looking for traces of the enemy. Suddenly the Fiat of Pessoa was hit by a Strela in the tail section causing an immediate flameout. Pessoa began to lose control of the aircraft, and as the engine failed, he descended rapidly. Faced with the certainty of a crash, he ejected at very low altitude and remembers that the aircraft did not immediately plummet like a rock but flew on for a short spell. Because of the low altitude, the chute did not fully deploy, and Pessoa fainted on landing because of the pain and beating his body absorbed. He awoke soon thereafter in the midst of thick vegetation that had likely broken his fall

49 Ibid.
50 Ibid., 34.

and absorbed a large part of his impact and thought that he had suffered a sprain in his leg.[51] He was unable to move.

Meanwhile, the alert aircraft launched and flew over both Guileje and Gandembel and the border area between them without seeing a trace of either Pessoa or his downed aircraft. A second flight was equally as fruitless, and only in the afternoon did the fourth alert aircraft piloted by Almeida Brito find a very light disturbance in an area of dense vegetation. This had been made by Pessoa falling and now he lay helpless on the ground with a broken leg and severe back pain caused by the brutal ejection. Martins de Matos, the alert pilot and relief for Almeida Brito, now appeared over the bush and orbited the site to mark the downed pilot's position. The next morning around 1100, after Pessoa had slept for about 20 hours, he was rescued by the African Commandos from Marcelino da Mata in an Alouette III and taken to Guileje. Later that day he arrived at Bissalanca to the joy of his comrades.[52]

Two days later on 28 March, after playing a key role in the rescue of Pessoa, Almeida Brito and Captain Pinto Ferreira were the alert pilots. While in the operations center, the pilots received a piece of intelligence, considered reliable, that described a meeting of certain PAIGC principals at Madina do Boé to proclaim that village the capital of the new "independent territory of Guiné." The two alert aircraft launched, flew south across the frontier, entered Conakry airspace, and overflew the PAIGC base of Kambera. Almeida Brito observed no activity at the base and turned north to Madina do Boé at an altitude of 1,000 feet. The two were flying in formation when the sky in front of Pinto Ferreira was illuminated with the brightness of Almeida Brito's Fiat exploding from a Strela hit. A second Strela was immediately launched at the remaining Fiat, which escaped and returned to Bissalanca.[53]

The death of Almeida Brito and the loss of two Fiats in the space of 72 hours caused great shock and consternation both among the airmen in Bissalanca and at the air staff in Lisbon. On 3 April, in retaliation the FAP attacked the PAIGC base of Cumbamory in Senegal and caused great loss. The Portuguese caserne at Guidage monitored the attack on the aircraft radios and actually heard the attack from its considerable distance. Three days later it too was attacked in return by the PAIGC.

However, 6 April was probably the blackest for the FAP of any in the fifteen years of air operations in Africa. In the morning a Dornier piloted by Sergeant Baltazar da Silva launched from Bissalanca at 0855 for the encampment at Guidage on a medical support and fresh food resupply mission. It passed Bigene but never arrived at Guidage and failed to respond to radio calls. A search and rescue mission was activated at the same time that a medical mission in a second Dornier with a para-nurse aboard was

51 Miguel Pessoa, "Um 'Fiat' abatido" [A "Fiat" hit] in *A Guerra de África 1961–1974, Volume 2* [The African War 1961–1974, Volume 2] by José Freire Antunes (Lisbon: Temas & Debates, 1996), 987–990.

52 Correia, "Strela," Part 1, 34.

53 Ibid., 35.

performing a medical evacuation. While en route, the second Dornier was hit by a Strela and damaged to such an extent that it was forced to land in Bigene. Still in an attempt to effect the medical evacuation, a third Dornier was sent twelve hours later piloted by Sergeant Carvalho Ferreira. It was able to land at Guidage and on its return flight to Bissalanca disappeared over Sambuiá. Two Fiats were sent to search for the missing Dornier and were attacked by Strelas. Both followed evasive procedures and eluded the missiles. The Dornier of Sergeant Baltazar da Silva was found on the same day with all occupants dead. The rescue operation with its helicopters was covered by a flight of two T-6Gs. One of the Harvards piloted by Major Mantovani Borges Filipe was hit by a Strela and crashed. His wingman, Sergeant Henriques Ferreira, escaped unharmed and landed at Bissalanca. With the loss of Major Mantovani, the total number of pilots killed was four in ten days, or about 10 percent of the pilots in Guiné.[54]

The shock of such loss reverberated throughout the FAP. The implications for future air operations in Guiné, Angola, and Mozambique were grave and troubling and potentially represented a strategic reversal, if the FAP no longer had a free hand in the air. Unfettered air operation in all three theaters held the insurgents at bay and were the trump card in the war. When the dust cleared, sound thinking prevailed, as after all this was a missile with severe limitations, and it could be beaten.

The FAP approached the US air attaché for advice, but the United States was slow to respond because of its annoyance over the B-26 acquisition. Even so it was known that it had lost several aircraft in Vietnam near the border with the North, and these had all been low-level, relatively slow, propeller-driven aircraft that had remained in the Strela acquisition envelope long enough for the operator to target them leisurely.[55] The answer for the FAP was obviously to avoid the Strela envelope and, when caught in it, execute maneuvers that would break its sensor lock on the aircraft. This would clearly be easy for jets and difficult for helicopters and the small propeller-driven aircraft, but solutions were to be found, as the Strela had limitations.

The solution

On 9 May 1973 an extraordinary meeting was held at Bissalanca to define a new series of anti-missile flight procedures. These were divided into two categories:

1. Procedures of a general character, and
2. Procedures of a specific character.

The first category would apply to all types of aircraft and all types of missions. Aircraft generally would fly above 6,000 feet or below 200 feet after taking off to

54 Ibid., 37.
55 Doglione, et al, 43–44.

avoid the acquisition envelope of the Strela. On approach and landing, aircraft would now spiral down from altitude over the field to reduce their exposure to the envelope danger. In their operations, they were expected to vary their routes and approaches to airfields and targets and no longer fly in flights of two—a practice that presented a larger, more vulnerable target. If a Strela was sighted, the code word "Strela" was to be broadcast on the frequency in use as a warning to take action and immediately execute evasion maneuvers.[56]

Second, in the class of specific procedures, the numerous small airfields in the northern and southern frontier areas were to be closed to the Dornier, and the T-6G would no longer be used as an attack aircraft. In the case of the Fiat, certain of its attack missions were redefined. Dive bombing was now to commence at 10,000 feet, and the bombs were to be released and the aircraft exiting the delivery by 6,000 feet, the "hard deck." In close air support the same parameters applied, and the ground troops were to indicate the targets with smoke rockets. Reconnaissance, armed or not, was also restricted.[57]

Preoccupation with these procedures was warranted, as on the night of 26 April a French commercial flight barely escaped a missile. The Fiats continued to be favored Strela targets, as they came under attack on 11, 19, and 20 May in the area of Binta, on 2, 6, and 10 June in the area of Gadamael, and again in the last two weeks of July also in Gadamael. Again after an incident on 10 August in Morés, two Fiats were taken under fire from Strelas as they completed their bomb delivery, the first on 7 September at Guileje and the second on 22 September at Tancroal. In this last case, the Fiat squadron retaliated with six sorties against the enemy group responsible.[58]

The antiaircraft anxiety rose to a new level now that the FAP had lost its free hand, and this resulted in several Fiat attacks on the anti-air defenses of Simbeli and Kandiafara after low-level reconnaissance of these PAIGC bases. On 27 November, Colonel Lemos Ferreira in a note to Spínola suggested that FAP bombardments of PAIGC infrastructure in Guinée-Conakry and in the PAIGC camps in Cantanhez, Sare-Moroso, Morés, and Tombali would send a psychological message as well as a physical one that the FAP was not neutralized. However, violations of Senegalese and Guinean airspace threatened to escalate the anti-air contest and take the struggle to a new level.[59]

On 31 January 1974, the air group received a request for close air support in the area of Canquelifá on the northern frontier with Senegal. The alert flight of two Fiats was launched with Lieutenant Colonel Jesus Vasquez and Lieutenant Castro Gil. The pilots overflew Copá and bombed the objective at 1730. In exiting the bomb run at

56 José Manuel Correia, "Strela: A Ameaça ao Dominio dos Ceus no Ultramar, 2ª Parte" [Strela: The Threat to Domination of the Skies of the Ultramar, Part 2], *Mais Alto* 393 (September–October 2011): 24.
57 Ibid.
58 Ibid.
59 Ibid., 26.

B-26 Invader with flat green anti-reflective paint scheme, Luanda 1975. (Source: Personal archive of Josê Nico)

6,000-foot floor, the Fiat of Castro Gil lost velocity because of a Strela hit, and he found himself on a heading into Senegal. After radioing Jesus Vasquez and informing him of his situation, he ejected in foreign territory. The following day a flight of two Fiats on alert in Nova Lamego helped by Alouette IIIs with paras searched for the pilot. The two Fiat pilots, Captain Pinto Ferreira and Lieutenant António Matos, found no trace whatever of either Castro Gil or his parachute. Meanwhile, Castro Gil crossed the frontier into Guiné at night, walked past a PAIGC patrol under the cover of darkness, and reached a village where he "rented" a ride on an ancient bicycle from the village chief for 500 escudos. The chief and Castro Gil made their way to Piche barracks, again passing several PAIGC patrols on the road. On arrival, Castro Gil was flown to Bissalanca, where on 1 February with great enthusiasm he celebrated his escape from the Strela and his successful flight across eastern Guiné.[60] He developed such enthusiasm in his merriment that he had to be taken to the military hospital, where he shouted that he should not be restrained because he was a jet pilot. The nurses, however, thought that he was a poor soldier hallucinating from heat stroke and put him in the psychiatric ward.[61]

Two days after Castro Gil's recovery, the navy intercepted a Soviet seagoing landing craft (LCI-435) on its way to Conakry with a cargo of BRM-2 armored vehicles

60 Ibid.
61 José Correia, correspondence with the author, 31 July 2012, Lisbon.

equipped with Strela missiles destined for the PAIGC and the Quitafine.[62] In Guiné, the Strela refused to disappear and on 6 April a civilian airliner was attacked. It managed to land at Farim and escaped unharmed.[63]

In addition to procedural measures, the FAP took physical steps to frustrate the heat-seeking Strela guidance mechanism. In March, the FAP introduced a new paint scheme for its aircraft in the three theaters that employed an anti-reflective or low-reflective paint in a dark green color. This dark-tinted paint diffused and dispersed the heat associated with the aircraft and presented an indistinct infrared picture that frustrated Strela guidance acquisition. The aircraft insignias were also reduced in size, moved to less exposed positions, eliminated, or some combination of the three. Additionally, any contrasting paint schemes on the aircraft were changed to deny an enemy gunner any aiming point. These adjustments in the physical appearance of the aircraft significantly reduced the infrared signature of the wings and fuselage and consequently the ability of the Strela to acquire and track its target. The priority given to the change in appearance had four stages:

1. Combat aircraft (G-91, B-26, PV-2),
2. Transport aircraft (Noratlas),
3. Aircraft critical to operations (Puma, Alouette III, *Dornier*), and
4. All remaining aircraft (T-6G, C-47, etc.).

Most FAP aircraft in the *ultramar* received the new paint in short order, as it was a relatively easy fix to the threat. This and the new procedures were able to contain the Strela problem; however, all insurgencies are political wars, and this aspect of the conflict needed to be addressed.

Spínola, Caetano, and Cabral

Beginning in late 1970, there were a number of developments that made the strategic resolution of Guiné vital to the success of any overall Portuguese solution. First, Spínola believed that Cabral was his own man and, despite a Marxist structure and philosophy in the PAIGC, wished to be free of Soviet influence in establishing a new

62 The Landing Craft Infantry (LCI) was a seagoing amphibious assault ship of World War II designed to transport and land large numbers of infantry directly onto beaches. Developed in response to a British request, it was a small steel ship of 234 tons (389 tons fully loaded) that could independently transport 200 troops some 4,000 NM at 12 knots directly from a rear area to a landing beach. Some 923 were built starting in 1943 and were used in all theaters. Of these, 30 were transferred to the Soviet Union in the spring and summer of 1945 in anticipation of its joining the war against Japan and invading southern Sakhalin and the Kuril islands.

63 Correia, "Strela," Part 2, 27.

regime in Guiné.[64] One solution to this desire was the offer of independence under some sort of Portuguese federation in which Guiné would be part of a Lusophone organization similar to the British Commonwealth of Nations. With this concept in mind, Spínola began a series of overtures to Cabral through Léopold Senghor, the respected president of Senegal, in an attempt to end the war honorably for Portugal and to obtain independence for Guiné in a prudent way.

Spínola's strategy at this time was to reduce substantially the fighting capability of the PAIGC and to tilt the military balance in Portuguese favor. As time passed, Spínola's plans, of which airpower was a vital part, proved successful in checking the PAIGC progress and gaining the respect of its leadership to such an extent that negotiations for some form of self-government with a Portuguese alignment began to make sense. By 1972 conditions were ripe for negotiations, and at the end of April, Spínola had his initial conversations with a Senegalese minister to arrange a secret meeting with Senghor. This took place on 18 May at Cap Skiring in Senegal, and Senghor expressed great sympathy for Portugal and praised its social programs in Guiné, as he was well aware that many of his people south of the Casamance River were regularly crossing the border to receive medical attention from Portuguese doctors. Spínola explained his concept of Guinean independence, and Senghor indicated that he thought Cabral would be receptive to the idea and that he would act as an intermediary. This prompted Spínola to suggest that Senghor meet with Caetano in Bissau or perhaps metropolitan Portugal. Later on 26 May in Lisbon in a meeting with Silva Cunha, the Portuguese Foreign Minister, and Caetano, the proposal was rejected out of hand under the rationale that such a move could not be controlled and that it would have grave consequences for the other colonies. He did not want Spínola sitting at the same table with Cabral, something that he saw as a humiliation. In Caetano's view, a military defeat was preferable. Spínola was shocked at such a warped position, and in his own words, "In reality, moreover, the last opportunity was lost to resolve the problem of Guiné with honor and dignity."[65] This experience prompted him to write his book *Portugal and the Future* in which he voices his frustrations.[66] Contact with Cabral continued through his brother Luís Cabral into early 1973.[67] Spínola and many others remained convinced that Sékou Touré had an active hand in fomenting and supporting much of the internal dissention in the PAIGC in order to weaken the moderate Cabral in favor of a hardline approach.[68] This activity led to

64 José Pedro Castanheira, "Ao Serviço de Spínola e Marcelo" [In the Service of Spínola and Marcelo], *Expresso* (20 September 1997), 28.

65 António de Spínola, *País sem Rumo: Contributo para a História de uma Revolução* [Country without Direction: Contribution to a History of a Revolution] (Lisbon: Editorial SCIRE, 1978), 42.

66 Castanheira, 30–31; Spínola, *País sem Rumo*, 25–28.

67 Guilherme Almor de Alpoim Calvão, *De Conakry ao M.D.L.P.* [From Conakry to the M.D.L.P.] (Lisbon: Editorial Intervenção, 1976), 88–96.

68 Castanheira, 31.

Cabral's assassination and with it any hope of negotiating an honorable conclusion for Portugal and a prudent one for Guiné.

After Caetano refused to authorize negotiations with Cabral, the campaign in Guiné entered a waiting period during which there seemed to be a general expectation that some significant event would end the conflict. Meanwhile, the PAIGC continued to apply pressure militarily. Aircraft readiness began to suffer because of weakening maintenance under the pressure of the many years of combat, and one of the more troubling outgrowths of this was a diminished ability to MEDEVAC the wounded by helicopter.[69] There were indeed cases that were placed on ships for a relatively lengthy voyage to Bissau for treatment.

At the point of Spínola's meeting with Caetano, he had gained enough credibility to bring Cabral to the negotiating table, a situation about as propitious as one could achieve in the besieged environment of Guiné. Any conflict resolution would now have to come from another direction, and the war labored on until 24 April 1974, when middle-ranking Portuguese officers effected a coup and toppled the government in Lisbon. In 1975 Portugal granted Guiné its independence, and the PAIGC assumed control of the country.

69 José Alberto de Moura Talheiros, "Guiné 1973: O 'Ano da Brasa'" [Guiné 1973: The "Year of the Grill"], *Boina Verde* 214 (January–March 2006): 22.

14

The Leverage of Airpower

In the *ultramar* Portugal had deployed by 1973 some 149,000 army troops, about 3,000 naval personnel, and about 6,000 airmen, including about 1,900 paratroops, for an approximate total of 158,000. When actual aviation personnel are considered, there were about 4,000 throughout the *ultramar* at the height of the conflict in 1973. Operating and maintaining aircraft and aviation facilities required aptitudes and skills well above the average military service entry in both the officer and enlisted ranks, and thus by the simple nature of its design, the FAP was an elite force of relatively small numbers. Yet its impact or leverage was anything but small and was felt by the enemy daily in every theater. Fighting insurgents is a complex business and requires a deep understanding of the use of air power in a supporting role. The FAP appreciated its mission and was able to execute it remarkably well time and time again with a variety of proven yet often obsolete aircraft assembled in a true counterinsurgency air force.

In the *ultramar*, the war might be termed intelligence intensive, as it was largely driven by this aspect. Insurgents crossing the borders generally formed columns or groups, and these had to travel relatively great distances to reach targets of any consequence. Even in the small territory of Guiné it took many days to penetrate inland. The terrain in all of the theaters was comparatively remote, and insurgent groups first needed to be located in this wilderness and then tracked to determine their direction and intentions. While PIDE/DGS agents in the sanctuary countries often provided critical information on the assault column, such as its size and departure details, once it had crossed the border, it could easily become "lost" in the dense vegetation that covered most of the battlefield in the *ultramar*. Here aviation reconnaissance in all of its forms played a vital role in developing the broader picture that reflected many other contributing sources, such as PIDE/DGS agents, paid informants, turned prisoners, captured documents, infantry patrols, and the like. However, it was Portuguese air reconnaissance that dramatically increased the intelligence capabilities of the ground forces. The UCTI in the north of Angola is but one example in which the heli-borne trackers enabled heli-born paras to swoop onto unsuspecting and vulnerable enemy columns isolated and far from help. Another was the "pirate" operations in the east of Angola in which heli-borne commandos supported by gunships acting on intelligence routinely destroyed insurgent penetrations. Airpower also brought devastation to these columns with deep strikes into the remote terrain where they were hiding. The Fiat armed reconnaissance missions into Téte represent the classic example of this potent capability.

While intelligence gathering was vital to locating and destroying insurgents, the war on the ground was one of light infantry. A soldier, whether he was a Portuguese or an insurgent, could carry only so much and be an effective fighter. Thus when Portuguese forces made contact with the enemy in the remoteness of the *ultramar*, heavy firepower needed to be brought to the fight quickly, and airpower proved indispensable in filling this important supporting role. Close air support made the difference in almost every case between the Portuguese prevailing or alternatively suffering casualties and allowing the insurgents to escape. This support function was ably performed in a variety of ways by the "low-tech" approach and illustrates the careful coordination of joint air-ground operations.

Portugal was not a wealthy country at the time and was forced to craft a counterinsurgency air force from obsolete but proven aircraft. To its enormous credit it was able to devise new and ingenious uses for this equipment. For instance, the use of the World War II-vintage T-6G Harvard trainer in a reconnaissance and attack role represents a clear example of this, and it proved its worth throughout the war. First, it was inexpensive and available. Second, it was ideal for the job, as it was slow and had a long loiter time, characteristics that lent it to spotting small terrorist bands from the air and marking targets for attack aircraft. Armed with bombs, rockets, and machineguns, the Harvard was also able to hit targets with deadly accuracy as a light attack aircraft.

Old transports, such as the C-47 Dakota, were also inexpensive and available and supported the population, supplied forward bases, dropped paras, performed reconnaissance and psychological support, and even acted as improvised bombers. The Noratlas, a recycled attempt at a hybrid civilian-military transport, proved indispensable with its short field capability and easy-access rear doors. The Dornier likewise with its STOL capability proved an inexpensive yet vital workhorse. The "front line" attack aircraft, the F-86, F-84, and G-91, were all obsolete and unable to go against the Soviet forces in Europe successfully. These castoffs carried major bomb loads on critical attack and close air support missions – all still very effective in the African environment.

Lastly and perhaps most importantly, airpower provided the Portuguese with the flexibility and initiative that normally belongs to the insurgent. Before airpower, insurgents used to move quickly, live off the land, and benefit from better intelligence than the government because of their contact with a local, supportive population. Insurgents could carefully choose where and when to strike, and the government was usually forced to react, often in futile chases scrambling after a long-vanished foe. Portuguese airpower changed this balance and shifted the advantage.

Airpower now allowed Portuguese troops to be deposited in remote areas and remain supplied and supported for long periods of time. Operations during the reoccupation of the north of Angola in the early years of the conflict illustrate this capability. Further, Portuguese airpower enabled ground forces to harry the enemy relentlessly and, when he was cornered, had the ability to deliver devastating and accurate firepower to destroy him. In summary, Portuguese airpower robbed the insurgent of his initiative and foreclosed the normal insurgent advantage of surprise.

This seemingly makeshift approach was anything but that and was part of a comprehensive overall strategy to defend the territorial borders and to destroy the enemy columns that violated them. Air and other military operations were conducted in coordination with government reforms, education and health advances, economic programs, and information campaigns to support these efforts. Spínola's "A Better Guiné" was the best known; however, those in Angola and Mozambique managed to generate economic growth of 11 percent and 9 percent respectively over many years. The people in these colonies were too busy with prosperity to subscribe to the insurgent messages.

Unfortunately, insurgencies are always long wars, and Portugal was not truly prepared either intellectually or materially to fight such a conflict. Long wars are frustrating to airmen because of the highly complex and technical nature of aviation and the expertise needed to manage even routine operations. It takes many years to develop an effective air force and long periods to adapt to fighting insurgents. The FAP in this case began with a sound foundation grounded in NATO and adapted remarkably well. The years of small strikes, heli-borne para raids, close air support missions, and troop transport and resupply missions were wearing on both men and machines with no end in sight. It was not only the FAP that was exhausted in the end, it was an entire nation. After Spínola's experience with Caetano, particularly after initially being his great admirer, it was apparent to the even lowest airman that change was needed. While FAP leadership had proved farsighted in both its preparation for and conduct of the war and had helped grind the enemy to a halt, the politicians had failed with complementary support. When the military ultimately intervened on 25 April 1974, it provided the missing political solution and made possible a transition to democracy.

Selected Bibliography

Government publications and documents

Almeida, Manuel de Campos, ed. *50 Anos de Administração Aeronáutica*. [50 Years of Aeronautical Administration]. Lisbon: Força Aérea Portuguesa, 2007.

Arriaga, Kaúlza de. "História do Batalhão de Caçadores Pára-Quedistas." Deposition of General Kaúlza de Arraiga, February 1986, Lisbon.

Cardoso, Edgar Pereira da Costa. *Presença da Força Aérea em Angola*. [Air Force Presence in Angola]. Lisbon: Secretaria de Estado de Aeronáutica, 1963.

Comissão Histórico-Cultural da Força Aérea Portuguesa. *Força Aérea 50 Anos*. Lisbon: [Portuguese Air Force, 1952–2002]. Lisbon: Comissão Histórico-Cultural da Força Aérea Portuguesa, 2002.

Estado-Maior do Exército. *Resenha Histórico-Militar das Campanhas de África, Vols. I–VI*. [Historical Military Report of the African Campaigns, Vol. I–VI]. Lisbon: Estado-Maior do Exército, 1989-1998.

Grão, Luís António Martinho. *História das Tropas Pára-Quedistas Portuguesas*. [History of the Portuguese Paratroops]. Tancos: Comando das Tropas Aerotransportadas, 1986.

Montalvão, José. Depoitmento do Cor Pilav José Montalvão que ilustra a importância do transporte aéreo durante a guerra do ultramar. [Deposition of Colonel Pilot-aviator José Montalvão, which illustrates the importance of air transport during the war of the *ultramar*]. Livro de Actos 1995, item 3, pages 135–138. Archivo Histórico de Força Aérea, Alfragide.

Monteiro, Rui Tavares. Depoimento do Gen. Tavares Monteiro em 05 Abril de 1995 Sobre a Ponte Aérea Metrópole – Ultramar. [Deposition of Gen. Tavares Monteiro on 05 April of 1995 on the Air Bridge *Metrópole – Ultramar*]. Livro de Actos 1997, item 6, page 35, annex III. Archivo Histórico de Força Aérea, Alfragide.

Neto, Manuel Diogo. Palavras Proferidas pelo General Diogo Neto no Jantar Comemorativo dos 35 Anos da Esquadra 92 – (BA9 Luanda), Equipada com Aviões Noratlas – Barriga de Ginguba), em 11 de Novembro de 1996 (Extraidas da cassette de video No 13). [Thoughts Offered by General Diogo Neto at the

Luncheon Commemorating the 35th Anniversary of Squadron 92 – (BA9 Luanda), Equipped with Noratlas aircraft – (Peanut Belly), on 11 November 1996 (Extracts from video cassette No 13)]. Livro de Actos 1997, item 3, page 2, Archivo Histórico de Força Aérea, Alfragide.

Proença, Casimiro Abreu. Os comentários do Gen. Proença, escritos à margem do texto da entrevista feita ao Gen. Tavares Monteiro en Abril de 1995. [The commentaries of General Proença, written in the margin of the text of the interview made with General Tavares Monteiro in April 1995]. Livro de Actos 1997, item 3, page 2. Archivo Histórico de Força Aérea, Alfragide.

Service Historique de l'Armée de l'Air. *Regards sur L'Aviation Militaire Française en Algérie, 1954–1962.* Château de Vincennes: Service Historique de l'Armée de l'Air, 2002.

Service Historique de l'Armée de l'Air. *Regards sur L'Aviation Militaire Française en Indochine, 1940–1954.* Château de Vincennes: Service Historique de l'Armée de l'Air, 1999.

Unpublished writings, interviews and correspondence

Arriaga, Kaúlza de. Interview by the author, 8 November 1994, Lisbon. Correspondence with the author, 9 November 1994, Lisbon.

——. *Celebrating the Past, Building the Future: Portuguese National Defense during the Last 40 Years and in the Future.* Lecture delivered at a conference by the same name on 20 October 1966.

Associação da Força Aérea Portuguesa (AFAP). *VI Encontro da AFAP: Infra-Estruturas Aeronáuticas em África.* [6th Meeting of the AFAP: Aviation Infrastructure in Africa]. Proceedings of a conference held at the Academia da Força Aérea, Granja do Marquês, on 18 October 1997.

Conceição Silva, General Tomás George. Interview by the author, 3 April 1995, Lisbon.

Cubas, Ricardo dos Santos, Geraldo Estevens, Guilherme Costa Santos, and José Augusto Barrigas Queiroga. "Ensinamentos da Guerra em Angola." [Teachings of the War in Angola]. Proceedings of a conference, *Almoços da AFAP*, Montsanto, January 2011.

Felgas, Hélio Augusto Esteves. *Relatório Pessoal Acerca do Congo Belga: Referente a Outubro 1959.* [Personal Report Concerning the Belgian Congo: Referring to October 1959]. Carmona, Governor of the District of the Congo, 4 November 1959, Resende personal papers, Caixa 17, Proc° 02, Arquivo Histórico de Força Aérea Portuguesa, Alfragide.

Ferreira, Ivo. "Transportes Estratégicos entre as Províncias de Angola e Moçambique." [Strategic Transport between the Provinces of Angola and Mozambique]. Pedrouços: Instituto de Altos Estudos Militares, 1964.

Krus Abecasis, José. Interview by the author, 18 July 1999 and 14 November 2001, Lisbon.

Morgado, António Fernandes. "Apoio Logístico na Guiné, 1973–1974." [Logistical Support in Guiné, 1973–1974]. Pedrouços: Instituto de Altos Estudos Militares, 1979.

Quintas. *Eclosão do Conflito em Angola*. [Emergence of the Conflict in Angola]. In the unpublished collection of the author. TMs [photocopy]. Lisbon: undated.

Qugiroga, José Augusto Barrigas. *Histórias vividas: O Meu Segredo*. [Living History: My Secret] In the unpublished collection of the author. TMs [photocopy]. Lisbon: March 2008.

——. *Histórias vividas: O Desquite*. [Living History: The Divorce] In the unpublished collection of the author. TMs [photocopy]. Lisbon: January 2006.

——. *Histórias vividas: Por Terras do Fim do Mundo, Uma viagem entre os 18 e os 9 graus de Latitude Sul*. [Living History: For the Lands of the End of the Earth, A trip between 18 and 9 degrees South Latitude]. In the unpublished collection of the author. TMs [photocopy]. Lisbon: December 2007.

——. *Histórias vividas: Uma Noite de Calema*. [Living History: Night of the Violent Tempest]. In the unpublished collection of the author. TMs [photocopy]. Lisbon: unknown date.

——. *Histórias vividas: A missão que não cumpri…!*. [Living History: The Mission that did not end…!]. In the unpublished collection of the author. TMs [photocopy]. Lisbon: December 2007.

——. *Histórias vividas: Un destacamento em Cabinda, 1970*. [Living History: Detachment in Cabinda, 1970]. In the unpublished collection of the author. TMs [photocopy]. Lisbon: December 2007.

——. *Histórias vividas: Há dias de Sorte!*. [Living History: There are Lucky Days]. In the unpublished collection of the author. TMs [photocopy]. Lisbon: November 2006.

——. *Histórias vividas: O 20 de Janeiro de 1974*. [Living History: 20 January 1974]. In the unpublished collection of the author. TMs [photocopy]. Lisbon: unknown date.

——. *Histórias vividas: A prenda de Natal!*. [Living History: The Christmas Present!]. In the unpublished collection of the author. TMs [photocopy]. Lisbon: unknown date.

——. *Histórias vividas: A mina de fósforo*. [Living History: The Phosphorus Mine]. In the unpublished collection of the author. TMs [photocopy]. Lisbon: date unknown.

——. *Histórias vividas: A necessidade aguça o engenho*. [Living History: Necessity is the Mother of Invention]. In the unpublished collection of the author. TMs [photocopy]. Lisbon: date unknown.

——. *Histórias vividas:Encontros Imediatos*. [Living History: Close Encounters]. In the unpublished collection of the author. TMs [photocopy]. Lisbon: date unknown.

——. *Histórias vividas: Uma Operação pouco ortodoxa*. [Living History: An Operation Slightly Orthodox]. In the unpublished collection of the author. TMs [photocopy]. Lisbon: 2005

Vizela Cardoso, José Armando. *A Guerra Ultramarina em Moçambique: Operações Militares*. [The overseas War in Mozambique: Military Operations]. Presentation given at the Instituto de Estudos Superiores Militares, Lisbon, 20 June 2012.

Portuguese literature

Afonso, Aniceto, and Carlos de Matos Gomes. *Guerra Colonial*. [Colonial War]. Lisbon: Notícias, 1998.

Alves, Carlos. *A Epopeia de Mucaba, Angola–1961*. [The Epic of Mucaba, Angola–1961]. Lisbon: Núcleo de Estudos Oliveira Salazar, 1998.

Antunes, José Freire. "Kennedy, Portugal and the Azores Base." In *John F. Kennedy and Europe*, edited by Douglas Brinkley and Richard T. Griffiths. Baton Rouge: Louisiana State University Press, 1999.

——. *Jorge Jardim Agente Secreto*. [Jorge Jardim Secret Agent]. Venda Nova: Bertrand Editora, 1996.

——. *A Guerra de África 1961–1974*. [The African War 1961–1974]. Lisbon: Temas e Debates, 1996.

Arnaut Monroy, José Pedro. "Alguns factos mais sobre o Fiat G-91 em Moçambique (1970–72)." [Some additional facts about the Fiat G-91 in Mozambique (1970–1972)]. *Mais Alto* 387 (September–October 2010): 32.

Arriaga, Kaúlza de. *Sinteses*. [Synthesis]. Lisbon: Privately printed, 1992.

Barros, Manuel Pinto Machado. *A Batalha pela Força Aérea*. [The Battle by the Air Force]. Lisbon: Privately printed, 1964.

Brandão Ferreira, João José. *Em Nome da Pátria: Portugal, O Ultramar e a Guerra Justa*.[In the Name of the Fatherland: Portugal, the *Ultramar* and the Just War]. Alfragide: Publicações Dom Quixote, 2009.

——. "A Heróica Resistência de Mucaba – Evocação." [The Heroic Resistance of Mucaba – Recollection]. *Combatente* 347 (March 2009): 39.

——. *À Frente do Tempo*. [On the Face of Time]. Lisbon: Prefácio, 2008.

——. "A Ocupação Efectiva do 'Império,' Panorâmica Geral." [Effective Occupation of the "Empire," General Perspective]. *Anais do Clube Militar Naval* 87 (April–June 2007): 359–385.

——. *As Reformas das Forças Armadas Portuguesas: Da Época Contemporânea até ao Fim das Campanhas de África do Seculo XX*. [The Reforms of the Portuguese Armed Forces: From the Contemporary Epoch until the End of the African Campaigns of the Twentieth Century]. Lisbon: Instituto Português da Conjuntura Estratégica, 2007.

——. "As Reformas da Força Aérea Portuguesa: Da Sua Fundação até à Descolonização." [The Reforms of the Portuguese Air Force: From its Founding Until Decolonization]. *Anais do Clube Militar Naval* 86 (April–June 2006): 321–350.

——. "As Principais Reorganizações da Armada no Portugal Contemporâneo (Do Princípio do Século XIX à Descolonização)." [The Reorganizing Principles of the

Navy in Contemporary Portugal (From the Beginning of the 19th Century to Decolonization)]. *Anais do Clube Militar Naval* 84 (January–March 2004): 23–78.

——. *A Evolução do Conceito Estratégico Ultramarino Português: Da Conferência de Berlim à Descolonização*. [Evolution of the Strategic Concept of the Portuguese Overseas Territories: From the Conference of Berlin to Decolonization]. Lisbon: Hugin Editores, 2002.

——. *A Evolução do Conceito Estratégico Ultramarino Português: Da Conquista de Ceuta à Conferência de Berlim*. [Evolution of the Strategic Concept of the Portuguese Overseas Territories: From the Conquest of Ceuta to the Conference of Berlin]. São Pedro do Estoril: Edições Atena, 2000.

Brochado de Miranda, Jorge M. "El Empleo del Arma Aérea en la Lucha Antiguerrilla." [Employment of the Air Arm in the Antiguerrilla Fight]. *Revista de Aeronáutica y Astronautica* 715 (July–August 2002): 594–599.

——. "A Utilização da Arma Aérea na Luta Antiguerrilha." [Utilization of the Air Arm in the Antiguerrilla Struggle]. *Mais Alto* 324 (March–April 2000): 16–23.

Cabral, Amílcar. *Guiné-Bissau–Nação Africana Forjada na Luta*. [Guiné-Bissau–African Nation Forged in Struggle]. Lisbon: Publicações Nova Auora, 1974.

——. *Textos Políticos*. [Political Texts]. Porto: Edições Afrontamento, 1974.

——. *Palavras de Ordem Gerais*. [Speeches on Overall Methods]. Bissau: PAIGC/Secretariado Geral, 1976.

Caetano, Marcello. *Depoimento*. [Deposition]. Rio de Jameiro: Distribuidora Record, 1974.

Caio, Horácio. *Angola, Os Dias do Desespero*. [Angola, The Days of Desperation]. Lisbon: 18ª Edição, 1966.

Cann, John P. *Contra-Subversão em África*. [Counterinsurgency in Africa]. Lisbon: Prefácio, 2005.

Cardoso, Adelino. "North-American XAT-6E 'Texan': Um Avião-Protótipo Muito Pouco Conhecido." [North American XAT-6E "Texan": A Little Understood Aircraft Prototype]. *Mais Alto* 386 (July–August 2010): 45–46.

Castanheira, José Pedro. "Ao Serviço de Spínola e Marcelo." [In the Service of Spínola e Marcelo]. *Expresso*, 20 September 1997, 56.

——. "Memórias da Guerra e da Paz: Spínola." [Memories of War and Peace: Spínola]. *Expresso Revista*, 30 April 1994, 26.

Castro, Nuno. *Heróis do ultramar*. [Heroes of the *Ultramar*] Afragide: Oficina do Livro, 2012.

Clostermann, Pierre. "A Aviação na Guerra Subversiva." [Aviation in Subversive War]. Reprint of a paper given at the Instituto de Altos Estudos Militares on 29 April 1960, *Boletim do Comando Militar de Moçambique* 5 (January–February 1961): 55–75.

Comacho, Francisco. "Baixa do Cassange: O Massacre que Veio do Céu." [Baixa do Cassange: The Massacre that Came from the Sky]. *Grande Reportagem* 137 (August 2002): 62–77.

Correia, Alexandra. "O Dia da Barbárie." [The Day of Barbary]. *Visão História* 12 (June 2011): 58–61.

Correia, José Manuel. "MDAP: A Génese da Força Aérea nos Anos de 1950–3ª Parte." [MDAP: The Genisis of the Air Force in the Years of 1950–3rd Part]. *Mais Alto* 403 (May–June 2013): 27–37.

——. "MDAP: A Génese da Força Aérea nos Anos de 1950–2ª Parte." [MDAP: The Genisis of the Air Force in the Years of 1950–2nd Part]. *Mais Alto* 402 (March–April 2013): 15–23.

——. "MDAP: A Génese da Força Aérea nos Anos de 1950–1ª Parte." [MDAP: The Genisis of the Air Force in the Years of 1950–1st Part]. *Mais Alto* 401 (January–February 2013): 31–41.

——. "Fiat G-91 na Guiné 1966-68, Os Dois Primeiros Anos de Combate, 2ª Parte." [Fiat G-91 in Guiné 1966-68, The First Two Years of Combat, Part II]. *Mais Alto* 399 (September-October 2012): 31-37.

——. "Fiat G-91 na Guiné 1966-68, Os Dois Primeiros Anos de Combate, 1ª Parte." [Fiat G-91 in Guiné 1966-68, The First Two Years of Combat, Part I]. *Mais Alto* 398 (July-August 2012): 31-36.

——. "Fiat G-91/R4 em Moçambique – 4ª Parte." [Fiat G-91/R4 in Mozambique – 4th Part]. *Mais Alto* 388 (November–December 2010): 34–40.

——. "Fiat G-91/R4 em Moçambique – 3ª Parte." [Fiat G-91/R4 in Mozambique – 3rd Part]. *Mais Alto* 387 (September–October 2010): 33–39.

——. "Fiat G-91/R4 em Moçambique – 2ª Parte." [Fiat G-91/R4 in Mozambique – 2nd Part]. *Mais Alto* 386 (July–August 2010): 32–39.

——. "Fiat G-91/R4 em Moçambique – 1ª Parte." [Fiat G-91/R4 in Mozambique – 1st Part]. *Mais Alto* 385 (May–June 2010): 26–30.

——. "B-26 Invader: Um Bombardeiro Misterioso na Força Aérea Portuguesa, 2° Parte" [B-26 Invader: A Mysterious Bomber in the Portuguese Air Force, 2nd Part] *Mais Alto* 37 (November–December 2008): 19–28.

——. "B-26 Invader: Um Bombardeiro Misterioso na Força Aérea Portuguesa, 1° Parte" [B-26 Invader: A Mysterious Bomber in the Portuguese Air Force, 1st Part] *Mais Alto* 375 (September–October 2008): 19–28.

——. "A Guerra do Algodão." [The Cotton War]. *Mais Alto* 359 (January–February 2006): 40–44.

——. "1953–55 F-84G, As Divergências Luso-Americanas na Constituição da Terceira Esquadra da Base da Ota." [1953–55 F-84G, The Luso-American Divergences in Constituting the Third Squadron at Ota Air Base]. *Mais Alto* 383 (January–February 2010): 39–41.

——. "Operação Barracuda: Os Fiat G-91 ao Ataque na Guiné" [Operation Barracuda: The Fiat G-91 on the Attack in Guiné], *Mais Alto* 367 (May–June 2007): 22–26.

——. "'Rotina' de Guerra: Os F-86F Sabre na Guiné." ["Route" to War: The F-86F Sabre in Guiné]. *Mais Alto* 358 (November–December 2005): 34–43.

——. "'Atlas' Nome de Código: Os F-86F Sabre na Guiné." [Code Name "Atlas": The F-86F Sabre in Guiné]. *Mais Alto* 357 (September–October 2005): 30–38.

——. "Ecos do Ataque ao Império: Invasão da Índia Portuguesa, A Componente Aérea da Operação Vijay." [Reflections on the Attack on the Empire: Invasion of Portuguese India, the Air Component of Operation Victorious]. *Mais Alto* 340 (November–December 2002): 30–38.

Cruz, Alfredo. "1969/2009, SA-330 Puma: Quarenta Anos e 70,000 Horas de Voo." [1969-2009, SA-330 Pima: Forty Years and 70,000 Flight Hours]. *Mais Alto* 382 (November–December 2009): 16–20.

Diniz, Mário, and Luís Proença. "AL III História breve de uma vida longa, 1963–2013." [AL III brief history of a long life, 1963–2013]. *Mais Alto* 404 (July–August 2013): 16–38.

——. "SA-330 Puma No 19512, A Chegada do Helicóptero a Moçombique e o Seu Percurso até ao Museu do Ar em Sintra, 2ª Parte." [SA-330 Puma No 19512, The Arrival of the Helicopter in Mozambique and Its Route to the Air Museum in Sintra, Part 2]. *Mais Alto* 402 (March–April 2013): 43–49.

——. "SA-330 Puma No 19512, Quatro Décadas na Vanguarda Operacional, 1° Parte." [SA-330 Puma No 19512, Four Decades in the Operational Vanguard, Part 1]. *Mais Alto* 401 (January–February 2013): 19–24.

——. "Douglas C-47A Skytrain, DAKOTA No 6157: Três Décadas de História nos Céus de Três Continentes, 2ª Parte." [Douglas C-47A Skytrain, DAKOTA No 6157: Three Decades of History in the Skies of Three Continents, 2nd Part]. *Mais Alto* 390 (March–April 2011): 39–43.

——. "Douglas C-47A Skytrain, DAKOTA No 6157: Três Décadas de História nos Céus de Três Continentes, 1ª Parte." [Douglas C-47A Skytrain, DAKOTA No 6157: Three Decades of History in the Skies of Three Continents, 1st Part]. *Mais Alto* 389 (January–February 2011): 41–43.

——. "Nord Aviation N 2502A, Noratlas No 6405: Um Avião, Duas Matrículas – A Epoepeia de um Elefante Africano – 2ª Parte." [Nord Aviation N 2502A, Noratlas No 6405: One Aircraft, Two Heritages – The Epic of an African Elephant – 2nd Part]. *Mais Alto* 388 (November–December 2010): 41–45.

——. " Nord Aviation N 2502A, Noratlas No 6405: Um Avião, Duas Matrículas – A Epoepeia de um Elefante Africano – 1ª Parte." [Nord Aviation N 2502A, Noratlas No 6405: One Aircraft, Two Heritages – The Epic of an African Elephant – 1st Part]. *Mais Alto* 387 (September–October 2010): 41–45.

——. "Junkers Ju-52 No 6304: Da Alemanha ao Museu do Ar um Percurso de 73 Anos." [Junkers Ju-52 No 6304: From Germany to the Air Museum a journey of 73 years]. *Mais Alto* 385 (May–June 2010): 39–44.

Diogo Neto, Manuel. "Recordando a Esquadra 92 da BA9 – Luanda." [Remembering Squadron 92 of BA9 – Luanda]. *Mais Alto* 264 (March–April 1990): 38–41.

——. "Operação da Serra da Canda." [Operation Serra da Canda]. *Mais Alto* 233 (January–February 1985): 10–13.

——. "Acontecimentos no Norte de Angola". [Events in the North of Angola]. *Mais Alto* 30 (June–July 1961): 8–9, 21.

Duarte Martins, Arménio. "Cotovia e Falcão." [Skylark and Hawk]. *Mais Alto* 311 (February–March 1998): 8–9.

Espírito Santo, Gabriel Augusto do. "A Força Aérea Portuguesa nas Últimas Campanhas do Império – Conclusões." [The Portuguese Air Force in the Last Campaigns of the Empire – Conclusions].

Felgas, Hélio. *Os Movimentos Terroristas*. [The Terrorist Movements]. Lisbon: Privately printed, 1966.

——. *Guerra em Angola*. Lisbon: Classica Editora, 1961.

——. *O Clima do Congo Português*. [The Climate of Portuguese Congo]. Carmona: Imprensa Angolana, 1959.

Ferreira, Ameadeu José. "Contribuição da Força Aérea para uma doutrina e uma solção militar." [Contribution of the Air Force towards a doctrine and a military solution]. *Revista do Ar* 270 (March 1960): 344–348.

Ferreira, Pedro Manuel. "PV-2 em Moçambique, 2ª Parte." [PV-2 in Mozambique, 2nd Part]. *Mais Alto* 364 (November–December 2006): 38–42.

——. "PV-2 em Moçambique, 1ª Parte." [PV-2 in Mozambique, 1st Part]. *Mais Alto* 363 (September–October 2006): 38–42.

——. "*Santa Maria*, A Força Aérea na Caça ao Paquete 'Rebelde'." [*Santa Maria*: The Air Force in the Hunt for the 'Rebel' Steamship]. *Mais Alto* 352 (November–December 2004):12–22.

——. "A missão que ficou por realizar 'Ibis' ou a pré-génese da instalacão da Força Aérea em Moçambique." [The Mission to Execute *Ibis* or the Groundwork for the Installation of the Air Force in Mozambique]. *Mais Alto* 350 (July–August 2004): 22–27.

——. "FAV Formações Aéreas Voluntárias, Uma Singularidade Portuguesa no Teatro de Guerra Africano, II Parte." [FAV Formations of Air Volunteers, A Portuguese Uniqueness in the African Theater of War, Part II]. *Mais Alto* 348 (March–April 2004): 30–41.

——. "FAV Formações Aéreas Voluntárias, Uma Singularidade Portuguesa no Teatro de Guerra Africano, I Parte." [FAV Formations of Air Volunteers, A Portuguese Uniqueness in the African Theater of War, Part I]. *Mais Alto* 348 (March–April 2004): 4–15.

——. "Ecos do Ataque ao Império: Invasão da Índia Portuguesa, O Confronto Diplomático." [Reflections on the Attack on the Empire: Invasion of Portuguese India, the Diplomatic Confrontation]. *Mais Alto* 340 (November–December 2002): 30–39.

Ferronha, António Luís. "Sincretismos afro-cristãos." [Syncretisms Afro-Christian]. *Grande Reportagem* 137 (August 2002): 72.

Fidalgo Ferreira, Rui Alberto. "Uma perspective da Logística da Força Aérea durante Operações em África entre 1961 e 1974." [A perspective on Air Force Logistics during Operations in Africa between 1961 and 1974]. *Revista Militar* 2507 (December 2010): 1359–1376.

Fraga, Luís Alves de. *A Força Aérea na Guerra em África: Angola, Guiné e Moçambique, 1961–1974*. [The Air Force in the War in Africa: Angola, Guinea and Mozambique, 1961–1974]. Lisbon: Prefácio, 2004.

Gomes, Adelino. "Exército e UNITA Colaboraram antes de 74." [Army and UNITA Collaborate before 74]. *Público* (19 December 1995): 2-4.

Grão, Luís António Martinho. *Enfermeiras Pára-quedistas, 1961–2002*. [Paranurses, 1961–2002]. Lisbon: Prefácio, 2006.

——. *História das Tropas Pára-Quiedistas Portuguesas*. [History of the Portuguese Paratroops]. Lisbon: Comando das Tropas Aerotransportadas, 1986.

Hipólito, Abel Barroso. *A Pacificação do Niassa: Um Caso Concreto de Contraguerrilha*. [The Pacification of Niassa: A Concrete Case of Counterguerrilla War]. Lisbon: privately published, 1970.

Jesus Bispo, António de. "A partocopação da Força Aérea na Guerra de África." [The participation of the Air Force in the War for Africa]. *Revista Militar* 2507 (December 2010): 1377–1406.

Krus Abecasis, José. "FAV 201." *Mais Alto* 350 (July–August 2004): 44–45.

——. "O 'Voo do Tempo', Escola de Aviação." [The "Flight of Time", School of Aviation] *Mais Alto* 307 (June–July 1997): 15–18.

——. "A Génese dos 'Transportes Aéreos Militares': As limitações, o ordem lógica e a competência técnica, 1947–1962." [The Genisis of the "Military Air Transports": The limitations, the logical order and the technical competence, 1947–1962]. *Mais Alto* 285 (September–October 1993): 13–19.

——. *Bordo de Ataque, Memórias de uma Caderneta de Voo e um Contributo para a História*. [Attack Heading, Memories from an Aviation Log Book and a Contribution to History]. Coimbra: Coimbra Editora, 1985.

Lemos Ferreira, José. "A Força Aérea em África – Missões Conjuntas e Próprias." [The Air Force in Africa – Joint and Service Missions]. *Revista Militar* 2507 (December 2010):1317–1320.

Lima Bacelar, Sérgio Augusto Margarido. *A Guerra em África 1961–1974: Estratégias Adoptadas pelas Forças Armadas*. [The War in Africa 1961–1974: Strategies Adopted by the Armed Forces]. Porto: Liga dos Amigos do Museu Militar do Porto and Universidade Portucalenese Infante D. Henrique, 2000.

Lobato, António. *Liberdade ou Evasão*. [Freedom or Evasion]. Amadora: Erasmos Editora, 1996.

Lopes, Mário Canongia. *Os Aviões da Cruiz de Cristo*. [The Aircraft of the Cross of Christ]. Lisbon: Dinalivro, 2001.

——. "Os Últimos Skymasters e os Searchmasters – Aviões Quadrimotores de Transporte da Aviação Militar Portuguesa (1952-1976) – III Parte." [The Last Skymasters and the Searchmasters – The Four-Engine Air Transports of Portuguese Military Aviation (1943–1951), Part III]. *Mais Alto* 326 (July–August 2000):19–23.

——. "Os Últimos Skymasters – Aviões Quadrimotores de Transporte da Aviação Militar Portuguesa (1952-1976) – II Parte." [The Last Skymasters – The

Four-Engine Air Transports of Portuguese Military Aviation (1943–1951), Part II]. *Mais Alto* 325 (May–June 2000): 20–23.

——. "Os Liberators e os Primeiros Skymasters, Aviões Quadrimotores de Transporte a Aviação Militar Portuguesa (1943–1951), I Parte." [The Liberators and the First Skymasters – The Four-Engine Air Transports of Portuguese Military Aviation (1943–1951), Part I]. *Mais Alto* 324 (March–April 2000): 34–40.

——. "Noratlas 1960–1977: Aviões bimotores de transported a Força Aérea Portuguesa, II Parte." [Noratlas 1960–1977: Twin engine transport aircraft of the Portuguese Air Force, Part II]. *Mais Alto* 313 (June–July 1998): 17–23.

——. "Noratlas 1960–1977: Aviões bimotores de transported a Força Aérea Portuguesa, I Parte." [Noratlas 1960–1977: Twin engine transport aircraft of the Portuguese Air Force, Part I]. Mais Alto 312 (April–May 1998): 3–7.

——. "Dakotas: Aviões bimotores de transported a Força Aérea Portuguesa–1944–1977, II Parte." [Dakotas: Twin engine transport aircraft of the Portuguese Air Force–1944–1977, Part II]. *Mais Alto* 311 (February–March 1998): 31–37.

——. "Dakotas: Aviões bimotores de transported a Força Aérea Portuguesa–1944–1977, I Parte." [Dakotas: Twin engine transport aircraft of the Portuguese Air Force–1944–1977, Part I]. *Mais Alto* 310 (December–January 1998): 9–14.

Machado, Miguel, and António Carmo. "Tropas Pára-Quedistas Portuguesas." [Portuguese Paratroops]. *Boina Verde* 158 (July–September 1991): 6–123.

Mateus Dalila Cabrita, and Álvaro Mateus. *Angola 61.* Alfragide: Texto Editores, 2011.

Mateus, Dalila Cabrita. *A PIDE/DGS na Guerra Colonial 1961–1974.* [The PIDE/DGS in the Colonial War 1961–1974]. Lisbon: Terramar, 2004.

Matos Gomes, Carlos de. *Moçambique 1970, Operação Nó Górdio.* [Mozambique 1970, Operation Gordian Knot]. Lisbon: Prefácio, 2002.

Matos, José. "Saga do Sabre Mk 6, 2ª Parte." [The Saga of the Sabre Mk 6, Part 2]. *Mais Alto* 395 (January–February 2012): 37–43

——. "A Saga do Sabre Mk 6, 1ª Parte." [The Saga of the Sabre Mk 6, Part 1]. *Mais Alto* 394 (November–December 2011): 37–41.

Mensurado, Joaquim M. *Os Páras na Guerra, 1961–63 • 1968–72.* [The Paras in War, 1961–63 • 1968–72]. Lisbon: Prefácio, 2002.

Mimoso e Carvalho, António. "F-86F Sabre: O Início na Força Aérea Portuguesa." [F-86F *Saber*: Introduction in the Portuguese Air Force]. *Mais Alto* 315 (October–November 1998): 24–32.

Monteiro, Duarte. "C-47 Dakota, II Parte." [C-47 Dakota, Part II]. *Mais Alto* 313 (June–July 1998): 48–49.

——. "C-47 Dakota, I Parte." [C-47 Dakota, Part I]. *Mais Alto* 311 (February–March 1998): 42–43.

Monteiro, Rui Tavares. "Transport Aéreo na FAP: Do zero aos 707." [Air Transport in FAP: from zero to 707]. *Mais Alto* 283 (May–June 1993): 3–7.

——. "Recordando os Noratlas." [Remembering the Noratlas] *Mais Alto* 279 (September–October 1992): 11–13.

——. "Skymaster: DC-4, C-54 ou SC-54." [Skymaster: DC-4, C-54 or SC-54]. *Mais Alto* 233 (January–February 1985): 24–26.

Munkelt Gonçalves, José. "Broussard, Aviões do Mato." [Broussard, Bush Aircraft]. *Mais Alto* 340 (November–December 2002): 4–14.

Negrão, Joaquim Vito Corte-Real. "Subsídios para o Reconhecimento Aéreo Visual." [Aids to Visual Aerial Reconnaissance]. *Boletim do Estado Maior da Força Aérea* 12 (1962), 2ª Repartição, Lisbon: 30–43.

Nico, José. "Pedaços das Nossas Vidas, III – O Rosto do Inimigo." [Pieces of Our Lives, III – The Face of the Enemy]. *Mais Alto* 363 (September-October 2006): 20-28.

——. "Pedaços das Nossas Vidas, II – O G-91 que não Trazia Correio." [Pieces of Our Lives, II – The G-91 That Did Not Bring the Mail]. *Mais Alto* 360 (March-April 2006): 30-33.

Pereira, Maria Arminda. "A Vida São Os Dias De Que Nos Lembramos." [Life Is The Days That We Remember]. *Boina Verde* 150 (September 1989): 8–10.

Pereira da Costa, José Emídio. "Informação e Contra- Informação Militar em Ambiente de Guerra Subversiva." [Military Intelligence and Counterintelligence in the Environment of Subversive War]. *Revista de Artilharia* 433–434 (September–October 1961): 147–152.

Pires Nunes, António Lopes. *Angola 1961: Da Baixa do Cassange a Nambuangongo.* [Angola 1961: From Baixa do Cassange to Nambuangongo]. Lisbon: Prefácio, 2005.

——. *Angola 1966–1974, Vitória Militar no Leste.* [Angola 1966–1974, Military Victory in the East]. Lisbon: Prefácio, 2002.

Reis, Maria Ivone Q. "As Nossas Aterragens." [Our Landings]. *Boina Verde* (September 1981):38.

Resende, Fernando Pinto de. "Lembrando O General Venâncio Deslandes." [I Remember General Venâncio Deslandes]. *Mais Alto* 239 (January–February 1986): 18–19.

——. "A Força Aérea em Angola (1960/1963)." [The Air Force in Angola (1960/1963)]. *Mais Alto* 228 (March–April 1984): 20–29.

Rodrigues, César. "Douglas C-47 'Dakota' e C-54 'Skymaster': Dois modelos de aeronave utilizados ne ponte aérea de Berlim que também serviram a Força Aérea Portuguesa." [Douglas C-47 Dakota and C-54 Skymaster: Two aircraft models utilized in the Berlin Airlift that also served the Portuguese Air Force]. *Mais Alto* 385 (May–June 2010): 35–36.

Rodrigues, Luís Nuno. *Salazar–Kennedy: A Crise de uma Aliança.* [Salazar–Kennedy: The Crisis of an Alliance]. Lisbon: Editorial Notícias, 2002.

Serras Pires, Adelino, and Fiona Claire Capstick. *The Winds of Havoc.* New York: St. Martin's Press, 2001.

Silva Cardoso, António. *Angola: Anatomia de uma Tragédia.* [Angola: Anatomy of a Tragedy]. Lisbon: Oficina de Livro, 2000.

Silva Cunha, Joaquim Moreira da. *O Ultramar, a Nação e o "25 de Abril."* [The Overseas Provinces, the Nation and the "25th of April"]. Coimbra: Atlântida Editora, 1977.

Soares de Moura, Augusto Cândido Pinto Coelha. "Mucaba, Uma Certidão Narrativa, II Parte." Mucaba, An Attested Narritive, Part II] *Mais Alto* 330 (January–February 2001): 30–39.

——. "Mucaba, Uma Certidão Narrativa, I Parte." Mucaba, An Attested Narritive, Part I] *Mais Alto* 329 (January–February 2001): 30–38.

——. "História do Aeródromo – Base No 3 (Negage)." [History of Aerodrome Base No 3 (Negage)]. *Combatente* 308 (March 1990): 1–3.

——. "Testemunho Norte de Angola (1961–1962)." [Testimony North of Angola (1961–1962)]. *Mais Alto* 228 (March–April 1984): 46–53.

Sousa, Vitor de. Dragões. "Um Voo sobre a História." [Dragons, A Flight through History]. *Mais Alto* 329 (January–February 2001): 3–13.

Spínola, António de. *País sem Rumo.* [Country without Direction]. Lisbon: Editorial SCIRE, 1978.

——. *Portugal e o Futuro.* [Portugal and the Future]. Lisbon: Editoria Arcádia, 1974.

——. *Por Uma Guiné Melhor.* [For a Better Guiné]. Lisbon: Agência-Geral do Ultramar, 1970.

Teixeira da Mota, Avelino. *Mar, Além Mar, Estudos e Ensaios de História e Geografia.* [The Sea, Beyond the Sea, Studies and Essays on History and Geography]. Lisbon: Junta de Investigações do Ultramar, 1972.

——. *Guiné Portuguesa.* [Portuguese Guiné]. Lisbon: Agência Geral do Ultramar, 1954.

——. *Guiné Portuguesa, II Volume.* [Portuguese Guiné, Volume II]. Lisbon: Agência Geral do Ultramar, 1954.

Vidal, João Machado. "As Guerras do Congo, 1960–1967." [Wars of the Congo, 1960–1967]. *Mais Alto* 314 (August–September 1998): 38–44.

——. "A Guerra do Biafra." [The War of Biafra]. *Mais Alto* 313 (June–July 1998): 32–41.

——. "T-6/SNJ Harvard, III Parte." *Mais Alto* 311 (February–March 1998): 11–14.

——. "T-6/SNJ Harvard, II Parte." *Mais Alto* 309 (October–November 1997): 31–35.

——. "T-6/SNJ Harvard, I Parte." *Mais Alto* 307 (June–July 1997): 3–6.

Viegas Ferreira de Almeida, Eduardo Alexandre. *Quarenta Anos de Aviação.* [Forty Years of Aviation]. Lisbon: Privately published, 1995.

Vilarinho Alves, Américo. "Aeródromo Base 3 – Negage, Súmula Histórica" [Airdrome Base 3 – Negage, Historical Summula]. Interview by Aniceto Ferreira de Carvalho, 31 May 2008, http://aerodino.no.sapo.pt/negage1.hmtl (accessed 24 August 2012).

Background literature

Abshire, David M., and Michael A. Samuels. *Portuguese Africa, A Handbook*. New York: Praeger Publishers, 1969.
Alata, Jean-Paul. *Prison D'Afrique*. Paris: Éditions du Seuil, 1976.
Armstrong, Sue. *In Search of Freedom*. Gibraltar: Ashanti Publishing, 1989.
Bruce, Neil. "Portugal's African Wars." *Conflict Studies* 34 (March 1973): 1–22.
Chabal, Patrick. *Amílcar Cabral*. Cambridge: Cambridge University Press, 1983.
Clostermann, Pierre-Henri. *Appui-Feu sur l'Oued Hallaïl*. Paris: Flammarion, 1960.
Collins, John M. *Military Geography*. Washington: National Defense University Press, 1998.
Corum, James S. "Airpower Roles and Missions in Counterinsurgency." In *No Clear Flight Plan, Counterinsurgency and Aerospace Power*, ed. James Fergusson and William March, 203–219. Winnipeg: Centre for Defense Studies, University of Manitoba, 2008.
Corum, James S., and Wray R. Johnson. *Airpower in Small Wars: Fighting Insurgents and Terrorists*. Lawrence: University Press of Kansas, 2003.
Cross, James Eliot. *Conflict in the Shadows: The Nature and Politics of Guerrilla War*. New York: Doubleday, 1963.
Davidson, Basil. *In the Eye of the Storm: Angola's People*. New York: Doubleday, 1972.
Davis, Jennifer. "Allies in Empire, Part I–U.S. Economic Involvement." *Africa Today* 17 (July–August 1970): 1–15.
Fall, Bernard B. *Street Without Joy*. New York: Schocken Books, 1972.
Fergusson, James, and William March, ed. *No Clear Flight Plan: Counterinsurgency and Aerospace Power*. Winnipeg: Center for Defense and Security Studies, University of Manitoba, 2008.
Forbath, Peter. *The River Congo*. New York: Harper & Row, 1977.
Francillon, René J. *Vietnam: The War in the Air*. New York: Arch Cape Press, 1987.
Gras, Philippe. *L'Armée de l'air en Indochine (1945–1954): L'impossible mission*. Paris: L'Harmattan, 2001.
Guest, Robert. *The Shackled Continent: Power, Corruption, and African Lives*. Washington: Smithsonian Books, 2004.
Hoar, Mike. *Congo Warriors*. London: Robert Hale, 1991.
Hochschild, Adam. *King Leopold's Ghost*. New York: Houghton Mifflin, 1998.
Horne, Alistair. *A Savage War of Peace: Algeria 1954–1962*. New York: Penguin Books, 1979.
Hutcheson, Keith. *Air Mobility: The Evolution of Global Reach*. Vienna, Virginia: Point One, 1999.
Johnson, David E. *Learning Large: The Evolving Roles of Ground Power and Air Power in the Post-Cold War Era*. Santa Monica: Rand, 2006.
Karan, Pradyumna P. "A Free Access to Colonial Enclaves." *Annals of the Association of American Geographers* 50, No 2 (June 1960): 188–190.
Kay, Hugh. *Salazar and Modern Portugal*. London: Eyre and Spottswoode, 1970.

Marcum, John. *The Angolan Revolution. Volume 1, The Anatomy of an Explosion (1950–1962).* Cambridge: M.I.T. Press, 1969.
Meredith, Martin. *The Fate of Africa.* New York: Public Affairs, 2005.
Minter, William. *Portuguese Africa and the West.* New York: Monthly Review Press, 1972.
——. "Allies in Empire, Part II–U.S. Military Involvement." *Africa Today,* Vol. 17, No. 4 (July–August 1970): 28–32.
——. "Allies in Empire, Part III–American Foreign Policy and Portuguese Colonialism." *Africa Today,* Vol 17, No. 4 (July–August 1970): 34–36.
Mondlane, Eduardo. *The Struggle for Mozambique.* London: Zed Press, 1969.
Newett, Malyn. *A History of Mozambique.* London: C. Hurst & Co., 1995.
——. *Portugal in Africa: The Last Hundred Years.* London: C. Hurst & Co., 1981.
Pélissier, René. *Naissance de la Guiné, Portugais et Africains en Sénégambie (1841–1936).* Orgeval: Pélissier, 1989.
——. *Naissance du Mozambique: Résistance et Révoltes Anticoloniales (1854–1918).* Orgeval: Pélissier, 1984.
——. *Le Naufrage des Caravelles, Etudes sur la Fin de l'Empire Portugais (1961–1975).* Orgeval: Editions Pélissier, 1979.
——. *La Colonie du Minotaure, Nationalismes et Révoltes en Angola (1926–1961).* Orgeval: Editions Pélissier, 1978.
Petter-Bowyer, Peter J. H. *Winds of Destruction.* Victoria: Trafford, 2003
Porch, Douglas. *The Portuguese Armed Forces and the Revolution.* Stanford: The Hoover Institution Press, 1977.
Robinson, Richard. *Contemporary Portugal.* London: George Allen & Unwin, 1979.
Shrader, Charles R. *The First Helicopter War: Logistics and Mobility in Algeria, 1954–1962.* Westport, Conn.: Praeger, 1999.
Thomas, Hugh. *The Slave Trade.* New York: Simon & Schuster, 1997.
van der Waals, W. S. *Portugal's War in Angola 1961–1974.* Rivonia: Ashanti, 1993.
Venter, Al J. *The Chopper Boys: Helicopter Warfare in Africa.* Gibraltar: Ashanti Publishing, 1994.
——. *The Zambesi Salient: Conflict in Southern Africa.* Cape Town: Howard Timmins, 1974.
——. *Portugal's Insurgent War, The Campaign for Africa.* Cape Town: John Malherbe, 1973.
Wheeler, Douglas L. "The Military and the Portuguese Dictatorship, 1926–1974: 'The Honor of the Army'." In *Contemporary Portugal* ed. Lawrence S. Graham and Harry M. Makler, 191–219. Austin: University of Texas Press, 1979.
——. "African Elements in Portugal's Armies in Africa (1961–1974)." *Armed Forces and Society* 2 (February 1976): 233–250.
——. "The Portuguese Army in Angola." *Modern African Studies* 7 (October 1969): 425–439.
Wheeler, Douglas L., and René Pélissier. *Angola.* London: Pall Mall Press, 1971.

Index